Chemical Thermodynamics for Earth Scientists

Chemical Thermodynamics
For Earth Scientists

Philip Fletcher

Series Advisory Editor:
Dr R C O Gill
Royal Holloway, University of London

Longman
Scientific &
Technical

Longman Scientific & Technical
Longman Group UK Limited,
Longman House, Burnt Mill, Harlow,
Essex CM20 2JE, England
and Associated Companies throughout the world.

Copublished in the United States with
John Wiley & Sons, Inc., 695 Third Avenue, New York, NY 10158

First published 1993

British Library Cataloguing in Publication Data
A catalogue record for this book is available from the British Library

ISBN 0582 06435 X

Library of Congress Cataloging-in-Publication Data
A catalogue record for this book is available from the Library of Congress

ISBN 0470 22072 4 (USA only)

Set by 6 in 10/12 Ehrhardt Ro.
Printed in Hong Kong
WP/01

Contents

Contents

Appendices

Preface

This book is concerned with the application of thermodynamics to chemical reactions occurring in the Earth's geological environment. The thermodynamic method, as developed by Josiah Willard Gibbs, shows that all substances can be characterised by a limited set of critical variables. Once determined these variables can be used to predict how any substance will respond to changes in its conditions. For example, we may predict whether or not a mineral will decompose if its temperature is raised, or we may predict the quantity of product formed by two reactive substances brought into contact.

Understanding issues such as these helps us interpret the behaviour of many natural systems; and for most of this century thermodynamics has been used to characterise chemically reacting substances ranging from pure inorganic salts to complicated mixtures such as soils or cements. In geochemical applications thermodynamics gives a basis from which to understand topics such as the weathering of rocks, the formation of new solid phases and factors controlling the composition of surface groundwaters. In addition to understanding natural systems or controlling industrial processes we have the responsibility to understand and regulate the impact of humans on natural systems. In view of its predictive capacity, thermodynamics can play a part in this activity. Consequently many undergraduate courses on geochemistry, soil science and environmental chemistry contain aspects of chemical thermodynamics.

This book, which is designed for such courses, covers three general topics. These are: the physical properties of relevant substances, the principles of thermodynamics, and the methods by which thermodynamic principles may be applied.

Chapters 1–6 constitute a foundation course on the composition, structural characteristics and reactivity of relevant substances. These include inorganic solids, gases, water, aqueous solutions and other mixtures. The basic features of chemical reactivity are introduced in chapter 1. Chapter 1–6 are essentially non-mathematical, with the emphasis placed on factors influencing chemical reactions.

Chapter 7 is concerned with energy and related properties and begins the introduction to chemical thermodynamics. Chapters 8 and 9 cover the basic principles of thermodynamics and are mathematical in nature. These principles are applied to the general properties of substances, including mixtures and chemically reacting systems, in chapters 10–12. In total, chapters 7–12 provide a typical course in chemical thermodynamics.

The final chapters, 13–19, are concerned with more specific properties of geochemical substances. In view of the ubiquity of water in natural systems, special attention is given to the thermodynamic properties of water and aqueous electrolytes. These are dealt with in chapters 13 and 14, which include a general description of the temperature and pressure dependence of the thermodynamic properties and describe the use of retrieval equations for their estimation. The properties of concentrated electrolyte solutions are dealt with in chapter 14. Similarly, the thermodynamic properties of solid materials, ion-exchange reactions and redox reactions are discussed in chapters 15–17. Up to this point in the text, the emphasis has been placed on methods for characterising the thermodynamic properties of substances. The final two chapters show how these properties can be used to predict the changing temperature and pressure. Consideration is given to conventional graphical techniques and also to computer-based techniques for simulating chemical equilibria in fluid/mineral assemblies.

The book is written so that a reader may 'dip' in at any selected point and follow the text with minimum reference to preceding chapters. This requires some repetition in the text which I hope is useful rather than confusing. The text is written at undergraduate level but at a sufficient depth to be useful in some aspects of geochemical or environmental research. The subject is mathematical and the reader will require a basic understanding of calculus although summaries of the necessary mathematical identities are provided.

My intention has been to outline the general principles of thermodynamics rather than to give comprehensive coverage. More comprehensive texts are named in the list of further reading matter. I have not included many thermodynamic data since such measurements are in a constant state of review. I have, however, included a list of data sources in Appendix 4, section A4.1.

My philosophy on units is to recommend use of the SI system. However, this text is designed to lead the reader into the literature which contains equations and data consistent with other units. I have therefore not adhered to SI entirely and have showed data in other units to maintain consistency with the literature.

This book could not have been completed without the critical reviews of the early drafts given by Tim Jones, Chris Hall, Peter Coveney and Bruce Yardley, to whom I am indebted. My thanks also go to Colin Atkinson, Paul Hammond, Richard Craster and John Sherwood for their help with some of the mathematical issues. I owe a large debt of gratitude to Paul and Helen Dastoor, who drew many of the diagrams in this book. I was particularly aware of their patience in the face of my constant demands for change. Special thanks and admiration must go to Diana Boatman for her skills in copy editing the final version. Any mistakes still remaining must be my own responsibility.

Finally, against my better judgement, I acknowledge Patrick Tomkins, Alex Wilson and Sarah Pelham, for their continual attempts to distract me with excessive quantities of the local beer.

Philip Fletcher
January 1992

Acknowledgements

We are indebted to the following for permission to reproduce copyright material:

American Journal of Science and the respective authors for Figs 13.2, 13.6, 13.7 and 13.9–13.31 (Helgeson and Kirkham, 1974a); Blackwell Scientific Publications Ltd for Figs 16.5 and 18.9 (Nordstrom and Munoz, 1986); Butterworth-Heinemann Ltd for Figs 12.10, 12.20, 12.23–12.28, 18.30, 18.32, 18.33 and 18.34 (Ferguson and Jones, 1973); Longman Group UK Ltd for Figs 10.5, 12.8 and 12.9 (Everett, 1959); Oxford University Press for Figs 8.4 and 9.1 (Atkins, 1973), Figs 13.4 and 13.8 (Eisenberg and Kauzmann, 1969); Prentice Hall Inc. for Figs 18.2, 18.3 and 18.6 (Drever, 1982) adapted by permission of Prentice Hall, Englewood Cliffs, New Jersey; John Wiley & Sons Ltd for Figs 3.2, 3.3 and 3.4 copyright 1974 John Wiley and Sons Ltd.

Whilst every effort has been made to trace the owners of copyright material, in a few cases this has proved impossible and we take this opportunity to offer our apologies to any copyright holders whose rights we may have unwittingly infringed.

Chapter 1

Introduction

Geochemistry is concerned with understanding the origins and transformations of the rocks and fluids of which the Earth is composed. These materials are complex mixtures of inorganic minerals, aqueous solutions and sometimes organic or colloidal substances. In the geological environment these already complicated systems are exposed to extremes of temperature and pressure compared with earth surface conditions

Ultimately, we are interested in fluid and mineral compositions, rates of change of composition and the transport of heat and matter. There is no single branch of physical chemistry which deals with all these issues. However, thermodynamics deals comprehensively with the factors controlling chemical transformations. It also plays an important supporting role in the description of kinetic and transport processes through the principles of statistical mechanics and irreversible thermodynamics.

The objective of chemical thermodynamics is to consolidate a set of well-defined experimental variables into a mathematical framework. This framework helps to predict whether or not a physical or chemical process can occur spontaneously and also the quantities of products formed by any reactions.

Central to the laws of thermodynamics are commonly understood terms and measurable properties. These are:

- **Energy**: The capacity to produce change. Energy is a physical quantity which can be assigned to any substance.
- **Force**: An agent which can impart motion to a body which is free to move. A force shows the presence of energy.
- **Mass**: The quantity of matter in a body.
- **Volume**: The amount of space occupied by a body.
- **Temperature**: The 'degree of hotness' as observed by the sense of touch.
- **Pressure**: Force per unit area.
- **Heat**: A form of energy transferred between bodies as a result of a temperature difference.
- **Heat capacity**: The amount of heat energy required to raise the temperature of a body by a specified amount.
- **Work**: A form of energy transferred between bodies as a result of a pressure difference. Work is associated with the change in the dimensions or position of a body.
- **Entropy**: A measure of the degree of disorder in a system. Entropy is a physical quantity which can be assigned to any substance.
- **Equilibrium**: A condition of balance between opposing physical effects. A state of equilibrium is independent of time or the history of the system.

A stable system is one at full equilibrium, whereas a metastable system is one that is not in the process of change but is not in the lowest possible energy state either

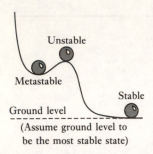

Unstable

Metastable

Stable

Ground level

(Assume ground level to
be the most stable state)

Figure 1.1
The states of equilibrium
(after Ehlers, 1972).

(figure 1.1). The natural direction of change is towards a stable state although metastable equilibria influence many geochemical processes.

These definitions are intended only to give a general idea of the meanings of the terms. More rigorous definitions will be developed later.

Thermodynamics may be seen as an exercise in the book-keeping of energy changes. The equations of thermodynamics are self-consistent and allow a system to be characterised by a limited set of critical measurements. Once determined, these critical properties can be used to predict other properties which may be more difficult to measure.

Thermodynamic measurements are **macroscopic**. That is, they are directly measurable bulk properties of materials, such as composition, temperature, pressure, mass, heat capacity or solubility. Thermodynamic measurements are made on systems of tangible size. The properties of the individual molecules or ions which make up the system are termed **molecular** or sometimes **microscopic** and include ionic radii, internuclear distances, crystallographic radii or quantum numbers, etc. The combined effects of the molecular properties control the macroscopic properties, but molecular properties themselves cannot be addressed thermodynamically. The relationships between microscopic and macroscopic properties are defined by the principles of statistical mechanics. This subject is not dealt with here.

In principle thermodynamics requires no assumptions about the molecular properties of matter. However, a progressively greater understanding of molecular behaviour is yielding equations which allow the thermodynamic properties of substances to be predicted by interpolation or extrapolation. This allows the use of thermodynamic variables under conditions where they have not been measured.

Chapter 2

Systems and properties

2.1 Systems

A thermodynamic **system** is any part of the universe we choose to study.

A system may be a pure substance or a mixture of substances. A system may be nothing more than a small quantity of liquid or it can be as extensive as a planet with an atmosphere undergoing continuous change. An example is a closed vessel containing a liquid and its vapour. The system could be (a) the liquid, (b) the vapour, (c) the container or (d) any combination of the three.

Any part of the universe which is not part of the system is termed the **surroundings**.

2.1.1 Properties of systems

Properties are defined as the measurable features of a system.

A **physical property** can be measured without changing the composition of the system, examples being temperature, pressure, density and mass. A **chemical property** may involve a change of state, for example the ability of a salt to dissolve in water is a chemical property. The techniques of thermodynamics use physical properties to characterise chemical properties.

There are two types of macroscopic physical properties. An **extensive** physical property is one for which the value of the property for the whole of the system is equal to the sum of the values for each component part. Examples of extensive properties are mass, volume and energy. Extensive properties are additive.

An **intensive** physical property is independent of the amount of material present. Examples are temperature, pressure and density. Intensive properties are not additive. **The quotient of any two extensive properties is an intensive property.** For example, density is equal to mass divided by volume.

Colligative properties constitute a set which have recognisable interrelationships. In general, they are a measure of the collective behaviour of the molecules in a system.

2.2 System boundaries

The boundaries of a system have properties of interest which will be used later to give a more rigorous basis to the concept of temperature and energy (chapter 7). The following definitions are a little abstract but help in the development of thermodynamic principles.

System boundaries may be thought of as the walls of a containing vessel and may be defined without explicitly using the concepts of energy or heat.

● If a system is in total internal equilibrium and enclosed so that **the only possible changes that can be made are mechanical changes**, the enclosure is termed **adiabatic** or **thermally insulating**.

Examples of mechanical changes are the changes in pressure or volume of the system.

● A boundary is **diathermal** or **thermally conducting** if it has the following property. If two systems, both in internal equilibrium but not in mutual equilibrium with each other, are separated by a diathermal wall, there will be a gradual change in their properties such that they achieve mutual equilibrium. The two systems are said to be in **thermal contact** and the final state achieved is said to be that of **thermal equilibrium**.

An **isolated system** is one which cannot be influenced by changes in the surroundings. Such a system would be enclosed by rigid adiabatic walls.

A **closed system** can exchange energy with its surroundings, but transfer of matter is prohibited. This does not preclude spontaneous composition changes within the system, such as those accompanying cooling lava. A closed system would be enclosed by diathermal walls.

An **open system** can exchange both energy and matter with its surroundings. An open system is not defined as a piece of matter; it is a region of space with geometrically defined boundaries across which both energy and matter can be transferred.

A **reservoir** is a system which is sufficiently large that its intensive properties do not change when matter or energy is transferred to or from the system. Oceans and the atmosphere are examples of large reservoirs.

2.3   The composition of systems

2.3.1   Phases

A **phase** is a physically distinct part of a system which is separated from other parts of the system by well defined boundaries.

A phase can be a pure substance or a homogeneous mixture. The mutual solubility of gases ensures that only one gas phase can exist in any one system. A mixture of miscible liquids is a single phase. However, some liquids are mutually insoluble, allowing the possibility that several liquid phases may coexist. Solid materials which differ from each other in chemical composition or crystal structure are different phases.

Using a more rigorous definition, a phase is a system or part of a system that has a homogeneous set of intensive properties within its boundaries. This means that properties of each part of a phase **must** be the same as for every other part. The discontinuity of macroscopic properties defines the phase boundaries.

A complete system can be composed of one or more phases. If a system contains more than one phase it is termed **a heterogeneous system**. It is common to define geochemical systems in terms of the relevant phases, treating all other substances as the surroundings.

Some examples of phases in systems are:

- **One-phase systems:**
  - A single substance; solid, liquid or gas.
  - A homogeneous mixture of any number of gases.
  - A fused melt of any number of miscible substances.
  - An aqueous electrolyte solution.
- **Two-phase systems:**
  - A single solid in contact with an electrolyte solution.
  - Two immiscible fluids such as a sulphide melt which has separated from basalt melt.
  - Ice + liquid water
- **Three-phase systems:**
  - Ice + liquid water + water vapour.
  - A saturated salt solution + its vapour in contact with a solid.
  - Two solids in contact with a gas such as:

$$MgCO_3(s) + MgO(s) + CO_2(g) \qquad (2.1)$$

  where (s) and (g) represent solid and gas phase respectively.
- **Four-phase systems:**
  - A mixed electrolyte solution + the vapour phase in contact with calcium carbonate and a clay mineral.

2.3.2   Chemical species and components

The term **elementary particle** is given to a single atom or molecule represented by an accepted formula or symbol.

A **species** is a chemical entity defined by a fixed chemical composition. A species may be an elementary particle or a molecule composed of several elementary particles. Examples of species are ions, gaseous or liquid molecules, solid phases, etc. In the system $CaSO_4 + H_2O$ there are many possible species, some of which are: liquid water $H_2O(l)$, gaseous water vapour $H_2O(g)$, the solids gypsum $CaSO_4 \cdot 2H_2O(s)$ and anhydrite $CaSO_4(s)$, plus a whole range of the aqueous ions such as Ca^{2+}, SO_4^{2-}, $CaSO_4^0$, H^+ and OH^-.

The term **component** is given to a chemical formula unit used to define the composition of a system or phase. In non-reactive mixtures the composition is defined by the relative proportions of the components. In chemically reactive systems, components are chosen such that **every species in the system can be written as the product of a reaction involving only the components.**

For example:

- Ice + liquid water + water vapour is a three-phase system with only **one** component since there are two reactions linking all three phases.

$$H_2O(l) \rightleftharpoons H_2O(g) \qquad (2.2)$$

$$H_2O(l) \rightleftharpoons H_2O(s) \qquad (2.3)$$

With reactions written this way, liquid water would be the one required component. However, the reactions could be rearranged to make any one of the pure H_2O

phases the required component. For example, $H_2O(l)$ could be eliminated by subtracting equation 2.2 from 2.3 to give:

$$H_2O(g) \rightleftharpoons H_2O(s) \tag{2.4}$$

Thus, equations 2.3 and 2.4 become the two independent equations with $H_2O(s)$ the one component common to both equations.

- The system $MgCO_3(s) + MgO(s) + CO_2(g)$ is a three-phase system. We require only two components to describe the composition of all three phases. For example, $MgO(s)$ and $CO_2(g)$ could be chosen since the carbonate can be described from the reaction:

$$MgCO_3(s) \rightleftharpoons MgO(s) + CO_2(g) \tag{2.5}$$

- An electrolyte solution containing dissolved NaCl and saturated with calcium carbonate in the presence of solid calcium carbonate is a two-phase system [liquid plus solid $CaCO_3(s)$]. There are three components: liquid water, dissolved NaCl, dissolved $CaCO_3$.

- An intermediate plagioclase can be described as a mixture of $NaAlSi_3O_8$ and $CaAl_2SiO_8$. This is a two-component system. It is tempting to describe this system as a five-component system in terms of the elements but we know that the two mineral phases are formed from reactive combination of the elements. The elements are in present in insignificant proportions.

The choice of components is arbitrary. However, the **number of components** in a system is the **minimum** number of independently variable constituents required to define completely the composition of all species and phases in the system.

THE GENERAL RULE IS AS FOLLOWS: IF A SYSTEM CONTAINS N SPECIES AND THERE ARE R INDEPENDENT REACTIONS BETWEEN THE SPECIES, THE NUMBER OF REQUIRED COMPONENTS C IS GIVEN BY:

$$C = N - R \tag{2.6}$$

In complex chemical and geochemical systems there may be many species and it becomes difficult to determine the number of independent reactions and therefore components. A simple scheme to help with this process is:

1. Identify the species as being present in significant amounts.
2. Write down one reaction for each species which describes the formation of the species from its constituent atoms and electrons.
3. Combine the formation reactions in such a way as to eliminate the atoms and electrons.

The resulting equations comprises the set of R independent reactions.

As an example, consider a system containing gypsum in contact with its saturated aqueous solution plus water vapour. The species present are: $CaSO_4 . 2H_2O(s)$, Ca^{2+}, SO_4^{2-}, $H_2O(l)$ and $H_2O(g)$. The reactions to form the species from the elements are:

$$Ca + S + 5O + 4H \rightleftharpoons CaSO_4 . 2H_2O(s) \tag{2.7}$$

$$2H + O \rightleftharpoons H_2O(l) \tag{2.8}$$

$$2H + O \rightleftharpoons H_2O(g) \tag{2.9}$$

$$Ca \rightleftharpoons Ca^{2+} + 2e^- \tag{2.10}$$

$$S + 4O + 2e^- \rightleftharpoons SO_4^{2-} \tag{2.11}$$

The atoms and electrons can be eliminated to give only two equations:

$$Ca^{2+} + SO_4^{2-} + 2H_2O(l) \rightleftharpoons CaSO_4 \cdot 2H_2O(s) \tag{2.12}$$

$$H_2O(l) \rightleftharpoons H_2O(g) \tag{2.13}$$

Therefore, $N = 5$, $R = 2$ and $C = 3$.

2.3.3 Chemical content of a phase of system

The amount n of a **pure** substance defines the quantity of a particular substance and is proportional to the number of chemical entities composing the substance. Chemical entities include elementary particles, species or components.

The amount of a substance is proportional to its mass in a phase or system. The most common unit for amount is the **mole**. This is the amount of substance which has the same number of chemical entities as there are atoms in 0.012 kg of carbon-12. The actual number of atoms of carbon in this amount of carbon-12 is called **Avogadro's constant** and has the symbol L. The currently accepted experimental value of L is 6.022045×10^{23} entities mol^{-1}. We may have a mole of any chemical entity that suits our purpose, such as electrons, molecules, ions or mineral components.

The mass of 1 mole of substance is equal to the formula weight expressed in grams. For example, nepheline may be defined in terms of the formula units $NaAlSiO_4(s)$ or $Na_2Al_2Si_2O_8(s)$. The formula weight of $NaAlSiO_4(s)$ is 142.06 g mol^{-1} whereas the formula weight of $Na_2Al_2Si_2O_8(s)$ is 284.12 g mol^{-1}; therefore, in a given quantity of nepheline there will be twice as many moles of $NaAlSiO_4(s)$ as of $Na_2Al_2Si_2O_8(s)$. This point emphasises the arbitrary nature of the definition of species and components which are chosen for convenience or by convention.

The **composition** defines the relative amounts of each substance in phase or system. Composition units for species and components can be chosen arbitrarily. However, there are three commonly used units.

- The **molal** unit is the number of moles of one substance per kilogram of a second substance. The second substance is termed the **solvent**, and is usually the one in greatest abundance; all other substances are termed **solutes**. Generally the molal unit is applied to solutions but the term may be equally well applied to intimate mixtures.
- The **molar** unit is the number of moles of a substance per litre (i.e. per dm^3) of solution. For dilute aqueous solutions at low temperatures and pressures molal and molar units are virtually the same numerically.
- The **mole fraction** is the number of moles of a substance divided by the total number of moles of all substances in the phase.

  The mole fraction is a dimensionless ratio, normally given the symbol x. In a system or phase containing N substances, the mole fraction of the jth substance is:

$$x_j = \frac{n_j}{\sum_i^N n_i} \tag{2.14}$$

where n represents the number of moles of the subscripted substance. For example, if a mixture contains n_1 moles of components A_1, n_2 moles of A_2 and n_3 moles of A_3, the mole fractions of A_1, A_2 and A_3 are:

$$x_1 = \frac{n_1}{n_1 + n_2 + n_3}, \; x_2 = \frac{n_2}{n_1 + n_2 + n_2}, \; x_3 = \frac{n_3}{n_1 + n_2 + n_3} \qquad (2.15)$$

The sum of the mole fractions in a system must be unity. Thus, for the three-component system, $x_1 + x_2 + x_3 = 1$.

In geochemical calculations the symbol X_j is used commonly to denote the mole fraction of components in solid solutions.

Molal and molar units are used mainly for aqueous solutions whereas mole fractions are used commonly for non-aqueous liquid solutions and solid solutions. The units can be converted from one to the other easily as follows.

If the molal and molar concentrations of the jth species in a multicomponent mixture are denoted by m_j and c_j respectively, the relationships between c_j and m_j are given by:

$$c_j = \frac{m_j \rho}{1 + 0.001 \sum_i m_i W_i} \qquad (2.16)$$

or

$$m_j = \frac{c_j}{\rho - 0.001 \sum_i c_i W_i} \qquad (2.17)$$

where the summation is over all the dissolved species excluding the solvent. The symbol ρ is the density of the mixture in $g\,cm^{-3}$ and W_i is the formula weight of the ith species in $g\,mol^{-1}$.

The relationship between the mole fraction and the molal concentration is:

$$x_j = \frac{m_j}{\sum_i m_i + 1000/W_s} \qquad (2.18)$$

where W_s is the formula weight of the solvent in $g\,mol^{-1}$. At very low concentrations the mole fraction of a solute can be approximated by:

$$x_j = \frac{n_j}{\sum_i n_i} \sim \frac{n_j}{n_s} \qquad (2.19)$$

where n_s is the number of moles of the solvent, from which:

$$x_j = \frac{m_j W_s}{1000} \qquad (2.20)$$

2.4   State functions

The **state** of a system of phase is completely defined by the numerical values of its properties.

The properties defining the state of a system are always macroscopic and are called **state variables** or **state functions**.

The change in any state function is **solely** dependent on the initial and final state of the system and is independent of how the change took place. This feature of state functions distinguishes them from **path functions** which depend on how a change took place.

By analogy, if a person travels from location A to location B, as defined by the coordinates on a map, the change in position is dependent only on the precise location of A and B. However, the actual distance travelled will be dependent on the route taken to get from A to B. In this sense, location is a state function whereas distance travelled is not. Similarly, if a gas is compressed from an initial pressure P_1 to a final pressure P_2, the change in pressure (ΔP) equals $P_2 - P_1$. Any intermediate pressures arising during the change do not enter into the calculation of ΔP. Pressure is therefore a state function. We will see later that energy is a state function, whereas work and heat are path functions.

Some systems cannot be characterised by state variables. The most common are those in the process of change. For example, if a gas contained within a cylinder is quickly compressed by the movement of a piston, the gas close to the piston may initially be more compressed than the gas at the far end of the cylinder. There is no single pressure which describes the state of the gas until all change has stopped and the gas is uniformly distributed within the cylinder. Systems where the state functions vary with time and space cannot be treated by classical thermodynamics. **Equilibrium thermodynamics deals only with systems where the state variables have uniform values throughout each phase in the system.**

2.5   A first look at equilibria and chemical reactions

2.5.1   Chemical reaction stoichiometry

A balanced chemical reaction equation defines the numbers of moles of each product species which can be obtained from a given number of moles of reactant species. The primary purpose of a reaction equation is to describe the condition of mass balance that must be imposed on any chemical change. By convention, species appearing on the right-hand side of the equation are termed products, those on the left reactants. For example, in a reaction between v_A moles of species A and v_B moles of B to produce v_C and v_D moles of products C and D, the equation is:

$$v_A A + v_B B \rightleftharpoons v_C C + v_D D \tag{2.21}$$

The symbol $\rightleftharpoons$ implies that such a process can proceed in both forward and reverse directions. A state of equilibrium represents a balance between the forward and reverse reactions.

Any number of balanced reaction equations can be combined, by addition or subtraction, to construct a further reaction. For example, gaseous carbon dioxide can be formed from either graphite or carbon monoxide:

$$C_{(graphite)} + O_2(g) \rightleftharpoons CO_2(g)$$

$$CO(g) + \tfrac{1}{2}O_2(g) \rightleftharpoons CO_2(g)$$

A third equation, for the formation of carbon monoxide from graphite, can be obtained by subtracting the second equation from the first. That is,

$$C_{(graphite)} + O_2(g) - CO(g) - \tfrac{1}{2}O_2(g) \rightleftharpoons CO_2(g) - CO_2(g)$$

which rearranges to give:

$$C_{(graphite)} + \tfrac{1}{2}O_2(g) \rightleftharpoons CO(g)$$

2.5.2   Le Chatelier's principle

When a reaction proceeds towards its equilibrium position the change in composition of the mixture may be accompanied by a change in the total volume of the system and the release or consumption of heat. If a reaction at equilibrium is perturbed by a change in a variable such as temperature, pressure or the amount of reactant or product, a new equilibrium will be established. A general rule which can help us predict the direction in which the equilibrium will move in response to changes in external stimuli is **Le Chatelier's principle**. This states that **when perturbed, a system changes to minimise the effect of the perturbation**. For example, if a reaction generates heat, then increasing the temperature of the system will retard the formation of products, whereas decreasing the temperature shifts the equilibrium position so that more products are formed. Similarly, if a reaction forms products which increase the pressure of the system, compressing the system will reverse the reaction and releasing the pressure will promote the reaction.

2.5.3   The law of mass action

The reaction defined by equation 2.21 can be driven in the reverse direction by the independent addition of species D to the system.

We introduce the **equilibrium constant** K_{eq}, which defines the relationship between the concentrations of reactive components **at equilibrium**. This quantity is equal to the product of the **activity** of the product species, raised to a power equal to the stoichiometric coefficient in the balanced equation, divided by a similar product for the reactant species, i.e.

$$K_{eq} = \frac{[C]^{\nu_C} \times [D]^{\nu_D}}{[A]^{\nu_A} \times [B]^{\nu_B}} \tag{2.22}$$

where [ ] denotes activity. The activity of a substance is defined rigorously in section 10.1.2. For components in the gas phase, activity is approximated by the partial pressure of the component. For species in the solution phase it is approximated by the molal or molar concentrations, or occasionally the mole fraction. For components in the solid phase it is approximated by the mole fraction of the component in the solid phase. When the above approximations are used for activity, the constant K_{eq} is called the mass action coefficient to distinguish it from the true equilibrium constant.

Generally, the mass action coefficient is dependent on temperature and pressure, but it may have a compositional dependence due to the approximations for activity. It embodies the concept of chemical equilibrium since any perturbation made to the equilibrium, by independent change in the activity of a component, will lead to re-equilibration to restore the quotient to the value of K_{eq}.

2.6   Thermodynamics in geochemistry

Geochemical and geological studies attempt to explain the distributions of elements and minerals in the earth, and chemical reactivity is a major factor controlling such distributions. The geochemist must therefore be able to characterise chemical reactions between geological substances. From section 2.5 we note that simple thermodynamics enable prediction of how the products from such reactions will change with variable conditions – such as temperature, pressure and the presence of other reactive substances. Later chapters on equilibrium thermodynamics will show more details on the methods for characterising chemical reactions and mixing processes. However, as an introduction to geochemical thermodynamics we will first consider some relevant reactions.

2.6.1   Some geochemical reactions and processes

Magmatic crystallisations

Magma is naturally occurring molten material generated within the earth. It is a fluid typically existing in the temperature range $500-1200\,^{\circ}C$. These temperatures may be found at lithostatic depths of around 30 km where pressures may be as high as 10 kbar. The reactivity of magma is varied and complex, although its primary role is as the parent material of igneous rocks. Magma is composed mainly of molten silicates. It is not pure, containing different chemical substances depending on how the magma was formed and on the history of its movements within the Earth's crust. As magma undergoes cooling, or changes in composition, igneous rocks may form by the process of crystallisation. Depending on the temperature, pressure and composition of the original magma, igneous rocks vary widely in composition, ranging from almost pure crystalline mineral phases through to complicated non-crystalline mineral assemblies. Despite the potential complexities, the crystallisation process is no more than a chemical reaction whereby a rock mineral is formed either by direct crystallisation from the magma or by a reaction between substances in the magma. For example, a magma containing molten silica SiO_2 may produce the mineral quartz $SiO_2(s)$ via the reaction:

$$SiO_2 \rightleftharpoons SiO_2(s)$$

Whereas in a more complex magma, containing the additional molten substance $KAlSiO_4$, potassium feldspar $[KAlSi_3O_8(s)]$ may also form via the reaction:

$$SiO_2 + KAlSiO_4 \rightleftharpoons KAlSi_3O_8(s)$$

The mineral assembly is therefore a mixture of two components. Many similar reactions accompany magmatic crystallisation.

Sedimentation processes

Igneous rocks are formed at high temperatures and are largely unstable under moderate Earth surface conditions and in the Earth's crust, especially when in contact with water. A notable exception is quartz which is stable at the Earth's surface. Sedimentation is the general term for the interaction of the hydrosphere and atmosphere on these unstable substances and includes phenomena such as diagenesis, weathering, erosion and deposition. At its simplest sedimentation concerns the alteration of minerals and the formation of new stable mineral or fluid phases by reaction with water, oxygen and dissolved gases, such as carbon dioxide.

Sedimentation in an ongoing process whereby minerals react to form new phases which themselves may be transported by the effect of wind, water flows or ice movement, or modified by mechanical errosion. Such perturbation from the equilibrium state leads to further reaction.

We must never neglect the fact that sedimentation is a dynamic process which may involve biological processes. To describe the purely chemical features of sedimentation it is necessary to address a variety of reactions. Of major interest are the dissolution/precipitation reactions of minerals whereby minerals may dissolve in water to yield a solution composed of water and ions. For example, the dissolution reactions of calcite $[CaCO_3(s)]$ and ferrosilite $[FeSiO_3(s)]$ are given by the following:

$$CaCO_3(s) \rightleftharpoons Ca^{2+} + CO_3^{2-}$$

$$FeSiO_3(s) + 2H^+ + H_2O \rightleftharpoons Fe^{2+} + H_4SiO_4$$

Subsequently the dissolved ions react with each other or with hydroxides or protons, or possibly with water molecules, to form one or more products. These products may be new mineral phases. For example the above reaction to dissolve calcite may be reversed and thought of as a secondary reaction to produce calcite if calcium and carbonate ions are released by some other reaction.

The complexity of multicomponent aqueous systems is increased further by additional reactions between dissolved ions to form species which remain in the liquid phase. This is called ion association and its products are called ion pairs or complexes. Examples include:

$$Ca^{2+} + CO_3^{2-} \rightleftharpoons CaCO_3^0$$

$$Ca^{2+} + CO_3^{2-} + H^+ \rightleftharpoons CaHCO_3^+$$

$$CO_3^{2-} + 2H^+ \rightleftharpoons H_2CO_3^0$$

$$Al^{3+} + OH^- \rightleftharpoons Al(OH)^{2+}$$

$$Al^{3+} + 2OH^- \rightleftharpoons Al(OH)_2^+$$

$$Al^{3+} + 3OH^- \rightleftharpoons Al(OH)_3^0$$

$$Al^{3+} + 4OH^- \rightleftharpoons Al(OH)_4^-$$

$$Ca^{2+} + SO_4^{2-} \rightleftharpoons CaSO_4^0$$

Ion association is important since it is thought to influence the solubilities of minerals in electrolytes. For example, gibbsite $[Al_2O_3(s)]$ dissolves to release aluminium into solution via the reaction:

$$Al_2O_3(s) + 3H_2O \rightleftharpoons 2Al^{3+} + 6OH^- \qquad (2.23)$$

If this reaction is performed in solutions containing the chloride ion a subsequent reaction is the formation of aqueous $AlCl^{2+}$ from the combination of Al^{3+} and Cl^-. This species is only one of several aluminium/chloride species that may exist. The formation of $AlCl^{2+}$ is equivalent to the consumption of aqueous aluminium and the solubility reaction (2.23) is driven over to the right-hand side, i.e. more gibbsite dissolves in accordance with Le Chatelier's principle.

In multicomponent electrolyte solutions, where there may be a variety of ion-pairs and complexes including those composed of a cation and the hydroxide ion. Consequently, **the solubility of a mineral may be subtly controlled by**

speciation. This phenomenon is apparent at all temperatures and pressures but is most noticeable at temperatures in excess of 400 °C and pressures below 1000 bars, the region in which inorganic ions associate most strongly.

The chemistry of dissolved aluminium is a particularly appropriate example since the ionic forms of dissolved aluminium (and silicon) are especially reactive and prone to form new mineral phases. In contrast, though they undergo ion association alkali metals and alkaline earth metals often remain in the solution phase.

Potentially, a large number of minerals can be formed from any one combination of ions. The number of reactions is increased further by the presence of dissolved gases which add more chemical species into the solution phase. Carbon dioxide, for example, dissolved to form carbonic acid by the reaction:

$$CO_2(g) + H_2O \rightleftharpoons H_2CO_3^0$$

The carbonic acid species itself will dissociate to some extent to release protons, bicarbonate ions and carbonate ions. These influence the solubilities of carbonate or bicarbonate minerals directly and have a more subtle effect on other mineral solubilities through ion association. In some situations gaseous carbon dioxide can react directly with a mineral, as in the carbonation of periclase (MgO):

$$MgO(s) + CO_s(g) \rightleftharpoons MgCO_3(s)$$

A final complexity arises from the presence of stable minerals which have a residual charge on their crystal surfaces. These charges are neutralised by loosely bound cations close to the mineral surfaces which can be replaced by other ions in the liquid phase. These **ion exchange** reactions have reaction equations of the form:

$$2[NaX] + Ca^{2+}(aq) \rightleftharpoons 2Na^+(aq) + [CaX_2]$$

This shows sodium ions on the surface of a charged solid 'X' being replaced by calcium ions from a nearby solution. Ion-exchanging minerals, such as clays, will modify solution concentrations, subsequently influencing other mineral solubilities in the same way as does ion association. For example, the presence of clay modifies the solubility of iron hydroxide compounds by the adsorption of Fe^{3+} and $Fe(OH)^{2+}$ species on the clay surface (Thompson and Tahir, 1991). Additionally, the presence of different ions bound to a clay mineral surface may even influence the solubility of the clay mineral itself by subtle modification of the crystal stability. Ion exchange reactions are dealt with in more detail in chapter 15.

Metamorphic processes   Metamorphism is the process by which rock material is transformed, in terms of texture and mineralogy, below the zone of weathering. During metamorphism one or more minerals will convert to others by a process of recrystallisation, with the rock remaining essentially solid. The process is simply the system tending towards a new equilibrium position as a result of changes in temperature, pressure and the introduction of new chemical substances. An example is the formation of wollastonite and anorthite from quartz and grossular:

$$SiO_2(s) + Ca_3Al_2Si_3O_{12}(s) \rightleftharpoons CaAl_2Si_2O_8(s) + 2CaSiO_3(s)$$

quartz + grossular = anorthite + wollastonite

Other metamorphic processes may produce or consume water or gases for example:

$$SiO_2(s) + Al_2Si_2O_5OH_4(s) \rightleftharpoons Al_2Si_4O_{10}(OH)_2(s) + H_2O$$

quartz   +   kaolinite   =   pyrophilite   +   water

or

$$CaCO_3(s) + SiO_2(s) \rightleftharpoons CaSiO_3(s) + CO_2(g)$$

calcite   +   quartz   =   wollastonite   +   carbon dioxide.

Typical metamorphic studies determine the conditions under which these processes can occur.

2.6.2   Some basic thermodynamic terms and their geochemical usage

There are many thermodynamic terms and concepts which will be defined and developed in later parts of the text. However, it is convenient to introduce some of these now in order to establish their geochemical applications. In particular, we will be concerned with a set of variables which help us characterise both the nature and quantity of substances produced by chemical reactions.

Standard Gibbs energy of reaction   In section 2.5.3 we saw that the equilibrium constant defines the overall equilibrium position. if this constant is large the reaction will produce large quantities of product. If the constant is very small the reactants will remain essentially unchanged. A more fundamental property is the **standard Gibbs energy** of the reaction

$$\Delta G^{\ominus} = -RT \ln K_{eq}, \tag{2.24}$$

where T is the absolute temperature (section 7.3.2) and R the gas constant (chapter 5). This shows that $\Delta G^{\ominus}$ characterises any reactive process in much the same way as K_{eq}. If $\Delta G^{\ominus}$ is large and negative the reaction will tend to form a large quantity of products. If $\Delta G^{\ominus}$ is large and positive only small quantities of product will ever be formed.

Effects due to changes in temperature and pressure   The value $\Delta G^{\ominus}$ gives a guide to the tendency of any reaction to proceed. For geochemical calculations it is especially important to know the temperature and pressure dependence of this variable for relevant reactions.

The temperature dependence of $\Delta G^{\ominus}$ is embodied in the **standard enthalpy change** for the reaction $\Delta H^{\ominus}$ through the equation:

$$\frac{d(\Delta G^{\ominus})}{dT} = -\frac{\Delta H^{\ominus}}{T^2} \tag{2.25}$$

Experimentally $\Delta H^{\ominus}$ is a measure of the total amount of heat **consumed** by a reaction as it progresses to completion at a specified temperature and pressure. If $\Delta H^{\ominus}$ is negative heat is released. Equation 2.25 shows that if $\Delta H^{\ominus}$ is positive heat is consumed and an increase in temperature makes $\Delta G^{\ominus}$ more negative. The reaction at higher temperature will therefore tend to form more products. Similarly, if $\Delta H^{\ominus}$ is negative an increase in temperature reduces the tendency to form products.

Equation 2.25 shows how $\Delta G^{\ominus}$ changes with temperature but gives no idea of the actual value of $\Delta G^{\ominus}$ at any temperature. This is given by the relationship:

$$\Delta G^{\ominus} = \Delta H^{\ominus} - T \Delta S^{\ominus}, \qquad (2.26)$$

where $\Delta S^{\ominus}$ is the **standard entropy change** for a reaction. The value of $\Delta S^{\ominus}$ is a measure of how the disorder or randomness of the system changes as the reaction proceeds to completion. Equation 2.26 shows that the value of $\Delta G^{\ominus}$ represents a balance between the changes in the standard entropy and the standard enthalpy of a system, both of which may take positive or negative values. More details on the factors affecting the sign of $\Delta G^{\ominus}$ in terms of $\Delta H^{\ominus}$ and $T \Delta S^{\ominus}$ are given in section 11.1.5 (table 11.2).

The pressure dependence of $\Delta G^{\ominus}$ is given by:

$$\frac{d(\Delta G^{\ominus})}{dP} = \Delta V^{\ominus}, \qquad (2.27)$$

where $\Delta V^{\ominus}$ is the volume change in the system as the reaction goes to completion at a specified temperature and pressure. If $\Delta V^{\ominus}$ is negative an increase in pressure makes the value of $\Delta G^{\ominus}$ more negative and the tendency to form products is increased.

The variables $\Delta G^{\ominus}$ and $\Delta H^{\ominus}$ and $\Delta V^{\ominus}$ constitute the quantitative basis to Le Chatelier's principle. In real chemical systems the value of $\Delta H^{\ominus}$ and $\Delta V^{\ominus}$ themselves have a dependence on temperature and pressure which must be properly accommodated. This issue is addressed in section 9.2.

The standard state properties of many minerals and fluid components have been determined at temperatures up to 1200 °C and pressures up to 10 kbar. These conditions are relevant to the lithosphere above the Mohorovičí discontinuity. Appropriate data for this domain are discussed in chapters 13 to 16. Thermodynamic data relevant to the Earth's mantle is more scarce. Recent measurements on the properties of $MgSiO_3$ and $MgSiO_3-FeSiO_3$ systems, in the temperature and pressure regime relevant to the transition zone and the upper part of the lower mantle (~ 200 kbar and ~ 1500 °C), demonstrate that high temperature/pressure measurements can be made (Akaogi et al., 1989; Fei et al., 1990). However, such encouraging developments are beyond the scope of this text.

Standard state properties
of species

It is clear that the value of $\Delta G^{\ominus}$ or the equilibrium constant are important quantities for any reaction. However, an alternative way of documenting thermodynamic data is by reference to the properties of the product and reactant species which constitute the reaction, rather than the reaction itself. For this purpose it is conventional to use the **standard Gibbs energy of formation** of the appropriate species ($\Delta G_f^{\ominus}$). This term, and an alternative commonly used in geochemistry, are defined in section 11.1. For the time being we will think of $\Delta G_f^{\ominus}$ as a measurable property of any species which can be determined at appropriate temperatures and pressures. If the values of $\Delta G_f^{\ominus}$ for **all** the reactant and product species are known the value of $\Delta G^{\ominus}$ for the reaction can be calculated for any reaction. For such a calculation we note that **the value $\Delta G^{\ominus}$ is the difference between the sum of the standard Gibbs energies of formation for the product minus the sum for the reactants.** For example, in equation 2.21 we see the formation of v_C moles of substance C and v_D moles of substance D from the reaction between v_A moles of substance A and v_B moles of B. The value of $\Delta G^{\ominus}$ for the reaction is given by:

$$\Delta G^{\ominus} = (v_C \Delta G_{f,C}^{\ominus} + v_D \Delta G_{f,D}^{\ominus}) - (v_A \Delta G_{f,A}^{\ominus} + v_B \Delta G_{f,B}^{\ominus}) \qquad (2.28)$$

where $\Delta G_{\mathrm{f,A}}^{\ominus}$ is the standard Gibbs energy of formation of substance A and so on. In section 11.1 we will see that an analogous expression exists for the standard enthalpy of reaction in terms of the standard enthalpies of formation for species A slightly different convention exists for entropy.

The main value in documenting properties of substances rather than reactions is that many reactions can be characterised using a combination of data for only a limited number of substances. Consequently, in this text emphasis is placed on methods for calculating and predicting standard thermodynamic data for selected substances. We will see that for water (Chapter 13) and many mineral phases (Chapter 16) the standard Gibbs energies of formation can be determined directly from the measurable properties of the pure substances. In contrast, the standard Gibbs energies of formation for dissolved ions must be determined from the properties of concentrated salt solutions extrapolated to the state of infinite dilution (sections 14.1 and 14.2).

Properties of mixtures   The equilibrium constant (equation 2.22) is expressed in terms of the activity of the reactive substances. As already stated, we may think of activity as the thermodynamic equivalent of concentration, but must recognise that in real systems activity and concentration are rarely equal. This is particularly true of ions dissolved in aqueous solutions and for components in liquid melts. It is also true of the minerals existing in complex mineral assemblies. These are formed by simultaneous crystallisation of several minerals from magma or metamorphic processes and may be thought of as solid solutions composed of many minerals. Consequently, sections 14.3 and 16.2.6 deal respectively with the thermodynamic properties of concentrated electrolytes and solid solutions and also consider activity calculations at elevated temperatures and pressures.

2.7   States of matter and chemical bonds

Chapters 3 to 5 give more details on the composition and nature of electrolytes, solids and gases. In particular, emphasis is placed on the microscopic properties that influence the thermodynamic properties. This current section provides a basic introduction to the bonding in such systems in support of chapters 3 to 5.

2.7.1   Solids, liquids and gases

Substances usually exist in one of three states of matter: solid, liquid or gas. A major difference between the states is due to the distance of separation between the molecules or atoms of the substance. In solids the particles are held together by forces originating from the particles themselves. The result is a regular arrangement of particles with each particle having little freedom of motion. In liquids the interparticle forces are less than in solids. The particles may move under thermal motion but are not totally unconstrained. In gases the interparticle forces are small and particles can move almost independently of each other.

2.7.2 Bonds between atoms

The following comments on the chemical bonds between atoms will help us understand the forces at work within chemical substances and how these forces influence chemical reactivity.

Ionic and covalent bonds

An ionic bond between atoms or molecules occurs as a result of the **total transfer** of one or more electrons from one atom to another. The atom losing the electrons becomes positively charged, while the atom gaining the electron becomes negatively charged, giving rise to electrostatic attraction. Ionic bonds are usually formed between metals and non-metals, the metal becoming a cation, the non-metal an anion.

A covalent bond is a chemical bond which is formed by the sharing of electrons. Covalent bonds are obtained when two atoms retain partial control of the two electrons forming the bond. Bonds between non-metals are usually predominantly covalent. A **dative bond** is a covalent bond where the bonding electrons are all dominated by one atom.

Polar molecules and electronegativity

Ionic and covalent bonds represent two extremes of bonding. In most cases chemical bonds are neither wholly covalent nor wholly ionic. An example is the bond in hydrogen fluoride where there is a partial transfer of electrons from the hydrogen to the fluorine atom. A way of viewing this is as a shift in the electron density away from the hydrogen atom. With this type of **unequal** sharing of electrons the electron distribution is **asymmetric** and the bond is termed **polar**. This asymmetry leads to a partial negative charge on one atom and a partial positive charge on the other. Such charge distributions are known as dipoles. Examples of dipolar bonds are:

The O–H bond $\quad$ $O^{\delta-}\!-\!H^{\delta+}$
The C–O double bond $\quad$ $C^{\delta+}\!=\!O^{\delta-}$
The H–F bond $\quad$ $H^{\delta+}\!-\!F^{\delta-}$
The C–Cl bond $\quad$ $C^{\delta+}\!-\!Cl^{\delta-}$

In this context the symbols $\delta+$ and $\delta-$ represent partial positive and negative charges respectively.

The term **dipole moment** is given to the product of the numerical value of one of the two equal and opposite charges and the distance between the charges. A dipole moment has units of coulomb metres ($C\,m$) although a commonly used unit is the **debye unit**. One debye unit equals $3.335\,64 \times 10^{-30}\,C\,m$.

A **quadrupole** is a distribution of charge expressed as an assembly of four charges, rather than two as in a dipole. For example, figure 2.1(b) depicts carbon monoxide gas as a quadrupole. here, each oxygen atom retains a partial negative charge and the carbon atom has the two partial positive charges.

The property we use to distinguish a covalent bond from an ionic bond is the **electronegativity** of the elements in the bond. The electronegativity of an atom is its ability to attract the electrons in a chemical bond towards itself.

Any atom which has a great tendency to attract electrons is highly electronegative.

Any atom which loses electrons easily has a low electronegativity.

The electronegativity of an element depends on its particular molecular environment and is not an invariant property of that element. Carbon, for example, will have slightly different electronegativities in alkanes of different chain lengths.

$\overset{\delta+\quad\delta-}{H\!-\!Cl}$

(a) A molecule with a dipole moment

$\overset{\delta-\quad 2\delta+\quad\delta-}{O\!=\!C\!=\!O}$

(b) A molecule with a quadrupole

Figure 2.1 Examples of polar molecules.

However, it is convenient to tabulate single values for elements in their common valence states.

Several empirical and theoretical scales of electronegativity have been devised. The original, devised by Pauling, states that the electronegativity of an element is proportional to the difference between the bond energy of an A—B$_n$ molecule, and the mean energies of the homopolar bonds A—A and B—B. The difference Δ is given by:

$$\Delta = U(A—B) - \left[\frac{U(A—A) + U(B—B)}{2} \right] \qquad (2.29)$$

Where $U(\ )$ denotes a specified bond energy and the electronegativities of elements A and B (χ_A and χ_B) are given by:

$$\chi_A - \chi_B = Q\Delta^{1/2} \qquad (2.30)$$

where Q is a constant of proportionality which converts units of energy to electron volts. Q has the value 0.208 if bond energies are in kilocalories per mole. If the electronegativity of one element is arbitrarily assigned, electronegativities of other elements can be calculated from suitable thermochemical data.

Table 2.1 shows the average relative electronegativities of some atoms. In the HF example, fluorine is more electronegative than the proton, and electrons are attracted to the fluorine in the hydrogen fluoride molecule.

Relative electronegativities are dimensionless. They are used to evaluate the degree of ionicity in any bond. There is no clear distinction between a covalent and an ionic bond. However, a rough guide is that a bond will be predominantly ionic if the difference in the electronegativities of atoms is 2.0 or more. For example, the molecule HCl has an electronegativity difference between H and Cl of 0.9. The bond therefore

Table 2.1   **Electronegativities of common atoms**

1A	2A	3B	4B	5B	6B	7B	.	8B	.	1B	2B	3A	4A	5A	6A	7A
H 2.1																
Li 1.0	Be 1.5											B 2.0	C 2.5	N 3.0	O 3.5	F 4.0
Na 0.9	Mg 1.2											Al 1.5	Si 1.8	P 2.1	S 2.5	Cl 3.0
K 0.8	Ca 1.0	Sc 1.3	Ti 1.5	V 1.6	Cr 1.6	Mn 1.5	Fe 1.8	Co 1.9	Ni 1.9	Cu 1.9	Zn 1.6	Ga 1.6	Ge 1.8	As 2.0	Se 2.5	Br 2.8
Rb 0.8	Sr 1.0	Y 1.2	Zr 1.4	Nb 1.6	Mo 1.8	Tc 1.9	Ru 2.2	Rh 2.2	Pd 2.2	Ag 1.9	Cd 1.7	In 1.7	Sn 1.8	Sb 1.9	Te 2.1	I 2.5
Cs 0.7	Ba 0.9	La–Lu 1.0–1.2	Hf 1.3	Ta 1.5	W 1.7	Re 1.9	Os 2.2	Ir 2.2	Pt 2.2	Au 2.4	Hg 1.9	Tl 1.8	Pb 1.9	Bi 1.9	Po 2.2	At 2.2
Fr 0.7	Ra 0.9															

has a substantial ionic contribution but is not completely ionic. The bond in sodium fluoride will be highly ionic since the electronegativity difference is 3.1. Bonds between like atoms are always covalent since the electronegativity difference is zero. Metals have low electronegativities and non-metals high electronegativities; therefore metal–non-metal bonds are predominantly ionic.

van der Waals forces

van der Waals forces are weak, short-range, attractive forces **between molecules** which play a significant role in the bonding of molecular solids and clay minerals, and in the structure of liquid water. These forces arise from interactions between dipoles and quadrupoles and are predominantly electrostatic. The potential energy of interaction is inversely related to the sixth power of the intermolecular distance and is clearly short-range.

There are three types of van der Waals forces:

- Forces which arise from interactions between permanent dipoles on different molecules.
- Permanent dipoles which can induce dipoles on adjacent non-polar molecules which can then experience attraction to the permanent dipole.
- The London dispersion for which is due to the mutual polarisation of molecules caused by fluctuations in the electronic charge distribution of neighbouring molecules. The fluctuations are on a timescale of less than 10^{-16} seconds and produce transient dipoles which are attracted to each other electrostatically.

The hydrogen bond

The term **hydrogen bond** is given to a specific type of dipole–dipole interaction. It is a bond between the hydrogen atom in a polar bond such as O—H and electronegative atoms such as oxygen, nitrogen or fluorine on neighbouring molecules.

The hydrogen bond has some degree of covalency in addition to van der Waals interactions. Consequently, a typical energy for the hydrogen bond is around 30 or $40 \, kJ \, mol^{-1}$, which is higher than for most dipole–dipole interactions. A bond of this strength will play a central role in determining the both the structure of compounds and also the properties which are dependent on the strength of intermolecular forces. Examples of such properties are boiling point, heat capacity and vapour pressure.

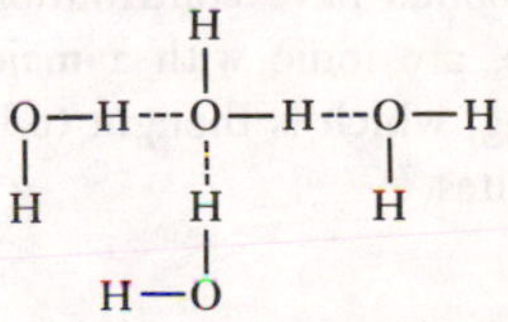

Hydrogen bonds are thought to be present in solids such as ice or solid HF, and also in liquids such as ammonia, liquid water and liquid HF.

It is common to depict a hydrogen bond with a dashed line between the hydrogen atoms and the electronegative atom on the neighbouring atom, as in figure 2.2.

A hydrogen bond is usually linear but may deviate by as much as $30°$ from linearity.

(a) Hydrogen bonds in liquid water

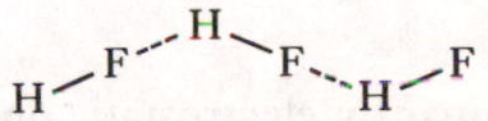

(b) Hydrogen bonds in solid HF

Figure 2.2 Hydrogen bonds.

Chapter 3

The properties of solid minerals

Crystals are solid materials in which molecules or ions occupy fixed positions in space. The ions or molecules are held in position by combinations of forces originating from the species themselves. The net result is an arrangement of ions, elements or molecules packed into a regular three-dimensional lattice which has structural rigidity.

3.1 Different types of solids

The structures and properties of crystals are determined by the nature of the forces holding the components together. Bond types are classified as ionic, covalent, molecular or metallic. In most substances the chemical bonds have contributions from more than one bond type. Sulphides, for example, are ionic with a major covalent contribution plus some degree of metallic bonding, which is thought to be responsible for the metallic lustre of minerals such as pyrites.

3.1.1 Ionically bonded solids

In an ionically bonded solid the cohesive forces are predominantly electrostatic ones between ions, with covalent and van der Waals forces playing a lesser role. A typical ionic solid is sodium chloride, which is shown in figure 3.1. It contains a uniform mixture of equal numbers of sodium and chloride ions in a cubic arrangement.

The structure is held together by electrostatic attractions between ions, because electrostatic attractions between opposite charges outweigh the repulsions between like charges.

Solid materials that are dominated by ionic bonds have high melting points – often higher than 1000 °C. Generally they are good conductors of heat and electricity, sometimes they have high compressive strength and hardness, but they may be quite brittle.

The solubilities of these materials in water is generally high, in the order of tens to hundreds of grams of solid per kilogram of water, the exceptions being carbonates, sulphides, phosphates, sulphites, and some sulphates of divalent cations. Solubilities of ionic materials often increase with temperature, although this is not a general rule.

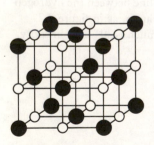

● Chloride ion

○ Sodium ion

Figure 3.1
The structure of sodium chloride.

3.1.2 Covalently bonded solids

Covalent bonds are found within many minerals of geological interest, such as quartz, aluminosilicate minerals and clays.

Covalent solids have similar physical properties to ionic solids, being good conductors of heat but poor conductors of electricity. They are generally harder than ionic solids and less brittle. The solubilities of covalent crystals in water are significantly lower than those of ionically bonded solids.

3.1.3 Molecular-bonded solids

In molecular crystals, the lattice points are occupied by molecules. The molecules themselves will contain components which are bonded together covalently. The attractive forces between the molecules are hydrogen bonding and/or van der Waals forces. Molecular bonds are weak compared with ionic or covalent bonds and molecular crystals are less stable, usually melting below 100 °C. Molecular-bonded solids are often soft but not brittle.

Such bonds are found in organic crystals such as sugars, although molecular bonds also play a role in the structure of clay minerals and ice.

3.1.4 Bonds in metals

The bonding in metals is quite different from that in other types of crystals. In metals, the atoms occupy lattice points in a regular crystal. However, the electrons from the atoms are delocalised over the entire crystal. It is common to think of a metallic crystal as a regular array of positive ions immersed in a 'sea' of mobile electrons. A cohesive force in metals arises from electrostatic attraction between the ions and electrons. The structure is held together because these attractive forces outweigh the repulsions between positive ions.

The great cohesive electrostatic force is responsible for the strength of metals, and the delocalisation or mobility of the electrons accounts for metals being soft and also good conductors of electricity and heat. Metals fall within a range of hardness.

3.1.5 Amorphous solids

Solids that are not crystalline in nature, lacking a regular three-dimensional arrangement of atoms, are said to be amorphous. Glass, a form of silicon dioxide, and opal ($SiO_2 . nH_2O$) are notable examples.

An amorphous solid is distinguishable from other amorphous materials on the basis of viscosity. An amorphous substance is considered to be solid if its shear viscosity exceeds $10^{13.6}$ P [1 P (poise) = 1 Pa s (pascal second)]. On this basis opal is an amorphous solid, whereas a form of hydrated amorphous silica precipitated from aqueous solutions is not a true solid.

The intermolecular forces in amorphous solids are van der Waals forces or hydrogen bonds. These are weaker than ionic or covalent bonds and are unable to create a regular crystal structure by fully overcoming thermal motion.

Amorphous solids are formed by vapour deposition, electrodeposition, anodisation, evaporation of a solvent, precipitation from solution and chemical oxidation of crystalline solids. A further class of amorphous solids can be prepared by rapidly cooling a molten material to below its freezing point. If crystallisation rates are fast enough, the material will form with a discontinuous change in volume, viscosity, heat capacity and other intensive and extensive properties with time. However, with slow crystallisation kinetics, the solid may form with a continuous change in these properties to form a glass. Since the transition occurs over a temperature range, the glass transition temperature is defined as equal to the temperature at which the viscosity equals $10^{13.6}$ P. This point is generally marked with a notable change in the heat capacity and thermal expansion coefficient.

3.1.6   Polymorphism

It is common to find that a given solid can exist in one of several forms each having a unique crystal structure. this is called **polymorphism**.

There are two type of polymorphism. **Enantiotropy** is polymorphism where a stable structural form can change to another form as a result of changing temperature. For example, quartz has two **enantiomers** which have a transition temperature at 575 °C. The term **allotrope** is used for an enantiomer of a pure element. **Monotropy** is the second type of polymorphism, in which no reversible change between crystalline forms can take place. One form, the metastable form, will always revert to the stable form at any temperature. The conversion may be slow, as in the case of yellow phosphorus (metastable) which converts to the red form (stable via the liquid phase. There are few examples of monotropes in naturally occurring systems. Different polymorphs of the same substance can often have notably different physical properties such as heat capacities or solubilities.

3.2   Imperfections in crystal structures

In reality, minerals are not perfectly regular solids with the ideal composition implied by their formulae. All crystals have some degree of imperfection.

3.2.1   Point defects

The simplest imperfection is the **point defect**, in which a deviation from the perfect regular lattice occurs at one lattice point. There are three possible types of point defects:

- **Vacancies**, which are empty sites on the lattice.
- **Interstitials**, which result from the insertion of an extra atom or ion between sites in the lattice.
- **Impurities**, which are foreign ions residing in the solid.

Schottky and Frenkel defects

One common point defect is a **Schottky defect**, which has an atom or ion transferred from a lattice site in the interior of a crystal to a lattice site on the surface of the crystal. A second common defect is the **Frenkel defect**, which is formed when an atom or ion on a lattice site is displaced to become an interstitial. Both of these defects, termed **intrinsic**, cause vacancies within the structure but leave the structure electrically neutral. The two defect types are depicted in figure 3.2.

In geochemical systems, Schottky and Frenkel defects are caused by:

- Increasing the temperature and therefore the amplitude of thermal vibrations.
- Quenching from high temperature, which causes defects present at high temperature to be retained at low temperature if the rate of cooling is high enough.

In other systems, defects may arise from additional sources, such as:

- Mechanical working by hammering or rolling which causes severe deformation of the crystal.
- Bombardment of crystals by high–energy particles, such as neutrons from a nuclear reactor, which can physically dislodge atoms from lattice sites.

Impurities

Impurity atoms or ions, foreign to the crystal structure, may reside within a crystal lattice. They can be accommodated in two different ways. First, they may reside in interstitial sites if small enough to fit into the cavities in the parent structure. Secondly, atoms or ions on lattice sites may be substituted for foreign species. This often occurs when crystals grow from mixed fluid systems. Both these defects can lead to structures becoming electrically charged as is the case in clays and zeolites, which show substitutions between cations of different charges. In these cases charge-neutralising ions and usually water molecules reside at particle surfaces. Such ions are loosely bound to the structure and can be replaced if particles are placed in contact with aqueous salt solutions. This phenomenon is known as **ion exchange**.

Most minerals contain impurities, although the amount of impurity is governed by the crystal structure, the radii of the ions and the location of the impurities in the structure. When one ion substitutes for another at a crystallographic site, the extent of replacement depends on how similar the ions are. If they are of similar charge and radius the substitution may occur extensively. For example, Ca^{2+} is known to replace Mn^{2+} in rhodachrosite [$MnCO_3(s)$]. The resulting solid can be thought of as a solid solution of $MnCO_3(s)$ and $CaCO_3(s)$. At elevated temperature the mutual solubility of solids becomes enhanced because thermal motion tends to minimise the difference between the effective radii of the ions. The effective ionic

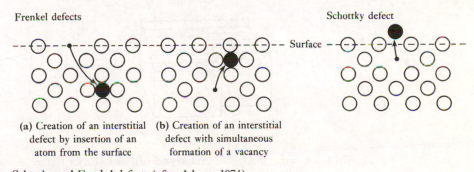

(a) Creation of an interstitial defect by insertion of an atom from the surface

(b) Creation of an interstitial defect with simultaneous formation of a vacancy

Figure 3.2   Schottky and Frenkel defects (after Adams, 1974).

radii of all ions increases with increasing temperature but lighter ions gain proportionally more thermal motion and their effective ionic radii increase more.

3.2.2   Linear defects

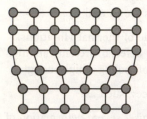

Figure 3.3
An edge dislocation in a crystal lattice (after Adams, 1974).

Dislocations in crystal structures occur due to 'mistakes' in crystal growth. They fall broadly into two categories, edge and screw dislocations, although combinations of these are known. An **edge dislocation** can be considered as the insertion of an extra plane of atoms into the crystal (figure 3.3). Distortion of the crystal structure is severe and the lattice is said to be **dilated**.

A similar type of imperfection is the **screw dislocation** associated with a sheared crystal lattice as indicated in figure 3.4(a). This imperfection may be thought of as a continuous spiral ramp of dislocated atoms extended through the crystal, as in figure 3.4(b).

3.2.3   The consequence of crystal imperfections

Point defects and dislocations modify the mechanical and electrical properties of materials. Some bulk chemical properties are also influenced by defects. For example, the presence of Schottky defects lowers the density of crystals because the volume is increased with no increase in mass. Similarly, isomorphic substitution between ions of different charge can influence density. For example, crystals of KCl grown in calcium chloride solutions show calcium ions occupying some potassium sites with no significant volume change. In order to retain electrical neutrality two potassium ions must be replaced by one calcium, leaving a vacancy in the crystal. Since one calcium ion is lighter than two potassium ions, the density decreases. When lattice defects are generated thermally, their energy of formation gives an extra contribution to the thermodynamic properties.

Properties of crystal surfaces, such as dissolution, precipitation or phase transformations, are also influenced by defects. An idealised concept is that crystal surfaces contain kinks and steps. The clusters of atoms at these features are of a higher potential energy than those surrounding them and become sites of enhanced reactivity. For example, in a dissolution process the atoms at kinks have two 'sides' exposed to the solution rather than one, and therefore detach first. Thus dissolution is preferentially from the kinks, causing the kinks to migrate along the dislocation

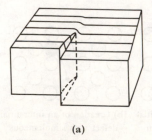
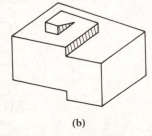

(a)   (b)

Figure 3.4   A screw dislocation in a crystal (a), and (b) its further development. The step height is usually one lattice spacing (from Adams, 1974).

leading to the formation of etched pits. Conversely, crystal growth occurs at steps which are regions of maximum local surface area. Through these mechanisms, crystal defects influence the kinetics of precipitation/dissolution reactions.

3.3   Phase transitions

As the temperature of any body is raised, transitions between phases are observed, familiar ones being the melting of ice to liquid water and the subsequent boiling of water on further increasing temperature. Transitions between polymorphs of a given solid are also common. Generally such transitions are accompanied by changes in volumes and heat capacities of the substances.

3.3.1   Latent heats of freezing/fusion

The transition from solid to liquid with increasing temperature is called **melting** or **fusion** and the reverse process called **freezing**.

When the temperature of a solid is gradually raised, the temperature at which fusion occurs is quite definite for most pure crystalline materials, although amorphous materials such as glass, wax or some metallic alloys gradually become plastic in character.

The definite temperature is the melting point. During the process of fusion a quantity of heat, known as the **latent heat**, is adsorbed. Consequently, energy may be provided to a system at its melting point and there will be no increase in temperature until the entire quantity of solid has undergone the phase transition. The reverse process is also observed, i.e. an identical quantity of heat is liberated at the freezing point of a liquid. The latent heat of fusion is a measure of the energy necessary to overcome the strong energies of interaction of between the molecules or ions in the solid. Latent heats are associated with first- and second-order phase transitions. The **molar heat of fusion** is the energy, usually in kilojoules, required to melt 1 mole of solid. The precise definition of a latent heat is given in section 12.2.3.

3.3.2   Types of transition

From measurements of heat capacities at different temperatures, two major types of phase transitions have been observed and are depicted in figure 3.5.

First-order transition   A **first-order transition** shows a continuous rise in heat capacity as the temperature approaches the transition temperature, at which point the heat capacity goes to infinity. A first-order transition is accompanied by a latent heat and discontinuity in volume. Most gas/liquid and solid/liquid phase transitions are first-order, as are many polymorphic phase transitions such as calcite/aragonite.

First-order phase polymorphic transitions are commonly associated with major

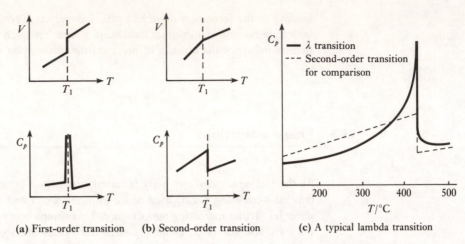

(a) First-order transition   **(b)** Second-order transition   **(c)** A typical lambda transition

Figure 3.5   Phase transitions.

structural rearrangements, which can take place slowly. This leaves the possibility of metastable persistence of some polymorphs across a phase transition.

Second-order transition

A **second-order transition** shows a smooth variation in heat capacity with temperature and a step change at the transition temperature. A second-order transition exhibits no latent heat and has a continuous volume change over the transition.

For polymorphic transitions in solids the precise nature of the transition may be difficult to determine experimentally. The conversion of α to β quartz is an example. The uncertainties arise because of the difficulties in making calorimetric measurements at high temperature and also because impurities and crystal imperfections alter the way temperature influences bond energies.

Order–disorder and λ transitions

A third type of phase transition is a **λ transition**. This is similar to a first-order transition since the heat capacity appears to go to infinity at the transition point. The difference is that the rate of increase in heat capacity starts to rise rapidly at temperatures well before the transition temperature. The shape of the plot of heat capacity versus temperature resembles the Greek letter λ – hence the name. Strictly, a λ transition is second-order because there is a continuous gradient of volume and no latent heat. Lambda transitions are rare compared with first-order transitions.

The rapid increase in heat capacity with temperature suggests that the λ phase transition is starting well before the transition temperature. This is an example of **order–disorder** phenomena. To understand the changes we consider a crystal composed of an equal number of atoms A and B. The crystal is said to be **ordered** if the arrangement of atoms throughout the lattice points is regular. In general this means that an atom of any one type has atoms of the second type as its nearest neighbours and vice versa. A crystal is **disordered** if atoms are randomly distributed throughout the lattice (figure 3.6). In order–disorder phenomena the crystal structure is thought not to change significantly.

Disorder results from increased thermal motion with increasing temperature and a perfect crystal will be completely ordered at the absolute zero of temperature. We may think of such an imperfect crystal as containing ordered and disordered domains throughout the crystal which change as a function of temperature. The phase

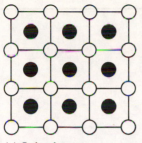

(a) Ordered state

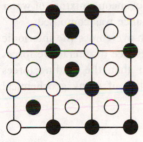

(b) Disordered state

Figure 3.6
Order–disorder in
crystals.

transition can be pictured as a formation of a new phase in an old phase, and the ordered domains favour the formation of the new phase. Since the order–disorder phenomenon is temperature-dependent, the formation of the new phase is 'smeared' out over a large temperature range and domains of the new phase grow in different regions of the old phase.

The precise type of phase transitions exhibited by minerals is related to the structure of the mineral, its impurities, the level of isomorphic substitution and the extent of crystal defects. Such detailed considerations are beyond the scope of this book. However, these factors must be borne in mind when the thermodynamic properties of a mineral are characterised at temperatures close to the transition temperature.

Aqueous solutions

Most solid, liquid and gaseous substances will dissolve to some extent when placed in contact with water. The resulting liquid phase is a homogeneous mixture of water plus the ionic or molecular entities from which the substance was composed.

Minerals are soluble materials and the action of water overcomes the interionic forces binding the mineral together. This implies the presence of interactive forces between water and the dissolved species that are greater than the binding forces within the minerals. These forces influence the properties of many geochemical systems and are discussed in this chapter.

4.1 The nature of liquid water

Liquid water is a condensed phase composed of molecules of H_2O. Water has a heat capacity of $4.184 \, J \, kg^{-1}$ at $25 \, °C$ and a boiling point at $100 \, °C$ at 1 atmosphere pressure. Both of these are higher than observed for similar molecular hydrides such as hydrogen fluoride, hydrogen sulphide, ammonia or methane. By comparison with these light molecules, water should be a gas at room temperature: this implies the existence of strong intermolecular attractions in water.

4.1.1 The water molecule

Figure 4.1
The water molecule.

The properties of liquid water may be understood from the structure of the water molecule, which is polar and non-linear. The bond angle is $108 \, °$ in the liquid phase, slightly larger than the value of $105 \, °$ found in the gas phase. It is usual to think of the water molecule as dipolar with a dipole moment of 1.87 debyes as depicted in figure 4.1.

4.1.2 Structure of liquid water

The **attractive** forces between water molecules comprise the van der Waals forces described in section 2.7.2. The hydrogen bond, in particular, plays a significant role in influencing properties of liquid water. The charge distribution in the hydrogen bond in water is given in figure 4.2. In water the most stable form of the hydrogen bond is believed to be linear.

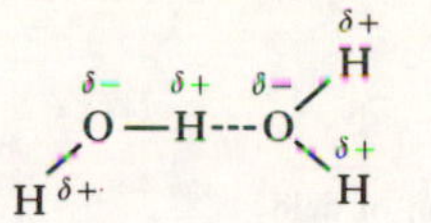

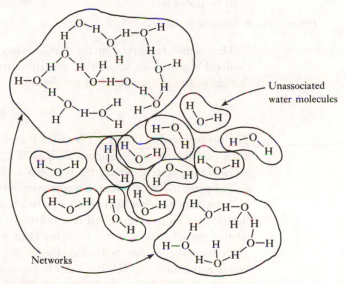

Figure 4.2
The charge distribution
in the hydrogen bond.

There are two major **repulsive** forces:

- Repulsion due to molecules of finite sizes moving together and undergoing internuclear repulsion at short internuclear distances.
- Dipole–dipole repulsion occurring when the geometrical orientation of molecules in space brings like-charge ends of different dipoles close together.

The consequence of the various attractions and repulsions is that liquid water has a 'partial structure'. The hydrogen bond imposes a relatively loose, open structure on the water molecules and X-ray diffraction experiments have shown that on average each water molecule is tetrahedrally coordinated by about four other water molecules at a distance of 0.28 mm with the next layer at 0.485 nm and a third layer at 0.58 nm.

The hydrogen bond extends through the structure although this is neither fixed nor static. Rather, it is a dynamic structure with hydrogen bonds constantly being broken and formed on the timescale of one every 3×10^{-12} seconds. Consequently, there is always a proportion of the water molecules 'free' or unassociated at any one time. These are believed to fill interstitial sites in an arrangement of networks of associated water molecules (figure 4.3).

Figure 4.3   Network arrangement in liquid water (from Bockris and Reddy, 1970).

4.1.3   Water as a dielectric

A dielectric medium is an **electrically insulating** material which can transmit an electric field to some extent. If a dielectric medium is placed in an electric field, as depicted in the parallel-plate capacitor in figure 4.4, the molecules of the material will be polarised by **induction** in the direction of the electric field. The effect of polarisation is to oppose the charges on the plates themselves and effectively reduce the transmission of the field through the medium. If the molecules themselves carry permanent dipoles, the external electric field will tend to align these dipoles; this imposes a further resistance to the transmission of the electric field. Random thermal motion opposes the orientation of permanent dipoles, however.

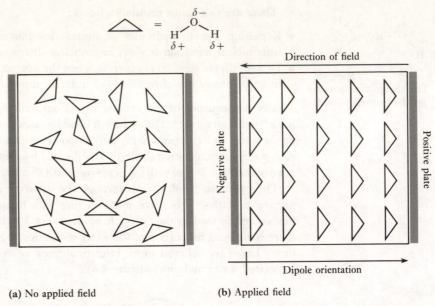

(a) No applied field   **(b)** Applied field

Figure 4.4   A dielectric in an electric field.

The overall resistance to the transmission of charge can be quantified using Coulomb's law. This states that the attractive or repulsive force between point charges q_1 and q_2, separated by the distance r, is given by:

$$F = \frac{q_1 q_2}{4\pi \varepsilon r^2} \tag{4.1}$$

The proportionality constant ε is called the **electric permittivity** of the medium. if the permittivity of a vacuum is designated ε_0, then for any medium the ratio $\varepsilon/\varepsilon_0$ is called the **relative permittivity** of the medium and is usually given the symbol ε_r. The permittivity of a vacuum is approximately equal to $(1/36\pi) \times 10^{-9}\,\mathrm{F\,m^{-1}}$.

A substance with a high relative permittivity can reduce the transmission of an electric field more effectively than a substance with a low relative permittivity. The relative permittivity of water is higher than would be expected if the value arose solely from the unrestricted orientation of the dipoles on the water molecule. For example, the molecule SO_2 has a dipole moment comparable with that for the water molecule but a relative permittivity at $298.15\,\mathrm{K}$ of 12.35 compared with 78.5 for water. The value is high because the dipoles on water molecules are already partially aligned within the loose water structure. The effect of increasing temperature leads to a dramatic decrease in the relative permittivity of water since thermal motion causes the breakdown of the water structure, therefore removing the enhancing effect of the structure.

Liquid water, with its high relative permittivity, is a good solvent since the electrostatic forces between dissolved ions are minimised, preventing the spatial arrangement of ions to form a crystal.

4.1.4   The autoprotolysis of water and pH

Acid/base behaviour   The Brönsted–Lowry definition of an **acid** is a substance which can transfer a proton to another substance. A **base** is a substance which can accept a proton from an acid. A generalised proton–transfer reaction is:

$$HA + B \rightleftharpoons A^- + BH^+ \qquad (4.2)$$

where HA is the acid and B the base. These may be dissolved substances or the molecules of the solvent itself. The product species BH^+ is also an acid since it has the capacity to transfer the proton it gained back to species A^-. These species are called the **conjugate acid** and the **conjugate base** respectively. In this notation the reaction of acetic acid with water is:

$$HC_2H_3O_2 + H_2O \rightleftharpoons C_2H_3O_2^- + H_3O^+ \qquad (4.3)$$

Similarly, the ionisation of the weak base, ammonia, is:

$$NH_3 + H_2O \rightleftharpoons OH^- + NH_4^+ \qquad (4.4)$$

The autoprotolysis of water   Liquid water is **amphiprotic**. It can act as both an acid and a base when one water molecule transfers a proton to a second water molecule via the reaction:

$$H_2O + H_2O \rightleftharpoons OH^- + H_3O^+ \qquad (4.5)$$

This reaction is called the **autoprotolysis** of water, and H_3O^+ is the hydronium ion.

Alternatively, we may consider the species H_3O^+ as a hydrated proton arising from the following reversible reaction:

$$H^+ + H_2O \rightleftharpoons H_3O^+ \qquad (4.6)$$

The second reaction may be subtracted from the first to give:

$$H_2O \rightleftharpoons H^+ + OH^- \qquad (4.7)$$

This equation depicts the reaction as a dissociation, and is the common way of writing the autoprotolysis reaction. We may think of liquid water as a mixture of undissociated water molecules plus protons and hydroxides.

At any specified temperature and pressure the following approximate relationship holds:

$$\frac{[H^+] \times [OH^-]}{[H_2O]} = K_c \qquad (4.8)$$

where the square brackets [ ] are used to denote the molar concentrations of proton, hydroxide and undissociated water and K_c is a constant. The dissociation of water is slight and $[H_2O]$ is almost constant. Therefore, the above equation becomes:

$$[H^+] \times [OH^-] = K_w$$

where

$$K_w = K_c[H_2O] \qquad (4.9)$$

The constant K_w is called the **ion-product constant** or the **ionisation product** of water; it has the value 1.0×10^{-14} mol dm^{-3} at 1 bar and 25 °C. The reason why this relationship holds is explained later; for the time being we will accept this as confirmed experimentally.

In practice we can adjust the concentrations of either proton or hydroxide by adding HCl or NaOH, for example. However, equation 4.9 restricts us from varying the concentrations independently. If an excess of hydroxide or proton is added to a solution, the autoprotolysis reaction will readjust the concentrations of proton and hydroxide until the condition in equation 4.9 is satisfied.

Whenever the molar concentrations of H^+ and OH^- are equal, the solution is said to be neutral. At $25\,°C$ neutrality occurs at a concentration of $1 \times 10^{-7}\,mol\,dm^{-3}$. The neutral concentration changes with temperature and pressure since the value of K_w is dramatically dependent on temperature and pressure.

pH and pOH The concentrations of proton and hydroxide are often small and inconvenient to work with. A more practical measure called **pH**, was devised by the Danish biochemist Sørensen. The pH of a solution is given by the **negative decadic logarithm of the molar hydrogen ion concentration**:

$$pH = -\log[H^+] \qquad (4.10)$$

The pOH of a solution is defined similarly as:

$$pOH = -\log[OH^-] \qquad (4.11)$$

The restriction imposed by the ionisation constant results in the relationship:

$$pH + pOH = -\log K_w = 14 \text{ at } 25\,°C \text{ and } 1 \text{ bar} \qquad (4.12)$$

Strictly, pH is defined as the negative decadic logarithm of the activity of the hydrogen ion. Activity is defined in section 10.1.2 and ion activities discussed in section 14.3.

4.2 The nature of water-soluble species

Water-soluble species can be derived from a range of sources including gases, inorganic salts, organic molecules or organic salts. We may categorise them as follows:

Ions These are atoms or molecules carrying integral charges per formula unit. They may be simple inorganic ions such as Na^+, Ca^{2+}, Cl^-, SO_4^{2-}, PO_4^{3-} or CO_3^{2-} which are derived from solids containing these ions in their structures. Ions may also be derived from soluble species which react with other dissolved species or water molecules. An example is dissolved acetic acid, some of which is converted to the acetate anion and the hydronium ion via the following reaction with water:

$$CH_3COOH + H_2O \rightleftharpoons CH_3COO^- + H_3O^+ \qquad (4.13)$$

Similarly, complexed ions such a $AlCl_6^{3-}$ can arise from a reaction such as:

$$Al^{3+} + 6Cl^- \rightleftharpoons AlCl_6^{3-} \qquad (4.14)$$

In order to simplify many theoretical treatments ions are often assumed to be spherical in shape. This is a good approximation for most inorganic ions and many complex ions such as $Cu(NH_3)_6^{2+}$, $Fe(CN)_6^{4-}$ or $AlCl_6^{3-}$. Spherical symmetry may be a poor approximation for some inorganic complexes and large organic ions with complicated molecular structures.

Similarly, the assumption that ionic charge is spherically distributed around the ion is reasonable for ions with spherical shapes. However, most ions are polarisable, i.e. their ionic charge distribution can be distorted electrostatically by the presence of an electric field. Small, highly charged cations, which are surrounded by an intense electric field, have a strong polarising power which can (a) distort the charge distribution of adjacent ions, forming induced dipoles, and (b) cause the alignment of permanent dipoles on other ions. Large anions with a low charge density are most susceptible to polarisation.

The sizes of ions are usually given by their crystallographic radii which, are evaluated from X-ray electron density maps of ionic solids. They are obtained by measuring interionic distances and assuming these to be the sum of ionic radii. Pauling evaluated such data for simple inorganic ions by assuming values for the ionic radii of fluoride and oxide ions and using these to estimate others from available X-ray data. Crystallographic radii do not represent the radii of ions in aqueous solutions but can be used to give a quantitative basis to electrolyte thermodynamics if certain modifications are made. These are discussed in sections 4.4.4 and 14.2.

Neutral molecules   These may be organic molecules such as acetic acid or complexed species such as $PbCl_2$, which arises from the reaction:

$$Pb^{2+} + 2Cl^- \rightleftharpoons PbCl_2^0 \qquad (4.15)$$

Neutral molecules may be non-polar, either from the complete absence of polar bonds or from the vector summation of polar bonds cancelling each other. Polar neutral molecules may exist having an overall dipole moment resulting from one or more dipolar bonds in the molecule. Neutral molecules should not generally be considered as spherically symmetrical or as having a symmetrical charge distribution.

4.3   Heats of solution and dilution

When substances are dissolved in water, or previously prepared solutions are diluted, there is often a liberation or consumption of heat accompanying the process.

- The **heat of solution** is the heat generated when a quantity of a solute is dissolved in a quantity of solvent. This phenomenon is observed for all solutes.
- The **heat of dilution** is the heat generated when a previously prepared solution is diluted with more solvent.

In the crystalline state, molecules or ions are bound together by strong forces. In the dissolved state, the same species are involved in complex interactions with each other and the solvent molecules. The heat of solution is a measure of the energy differences between the two states. The heat of dilution is a measure of how these solvent–solute and solute–solute interactions change with progressive dilution. Although complex, the range of interactions can be categorised as follows.

4.4 Ion–solvent interactions

The intense electric field close to dissolved ions gives rise to interactions between ions and water molecules. This in turn modifies the structure of liquid water in the near-vicinity of an ion.

4.4.1 Basic interactions

A major ion–water interaction is the electrostatic attraction between an ion and the permanent dipole on a water molecule. This causes the realignment of water molecules in the near-vicinity of an ion, as depicted for both cations and anions in figure 4.5. Therefore, ions exist within a sphere of water molecules closely bound to the ion. These water molecules have lost their independent translational motion and move around with the ion as a single entity. The bound molecules contribute to the size

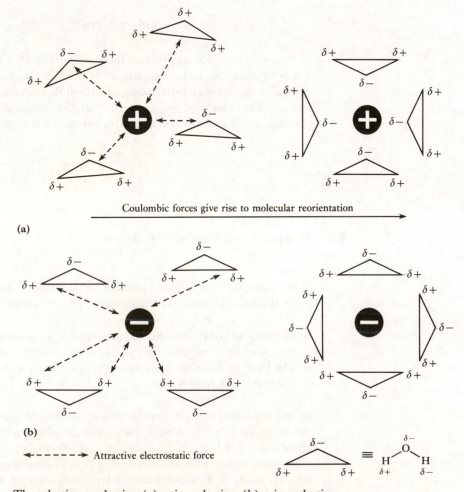

Figure 4.5   The solvation mechanism (a) cation solvation, (b) anion solvation.

of the mobile species, which thus has an effective radius greater than its bare crystallographic radius.

The total number of bound water molecules per ion is the **solvation number**. These bound molecules are usually distributed around an ion in successive shells or layers. Monatomic cations, for example, usually have either four or six water molecules in the first solvation shell, although little is usually known about the second shell.

In general, molecules outside the solvation sphere may still be weakly influenced by the central ion. There is no clear cut-off between bound water molecules and those simply influenced, although the distinction is vital in many theoretical treatments. To a first approximation the solvation number is independent of ion concentration.

4.4.2   The influence of ion solvation on the relative permittivity

We may think of solvation in terms of three regions of space (figure 4.6). These are: (a) the primary solvation shell, (b) the bulk unmodified solvent far from an

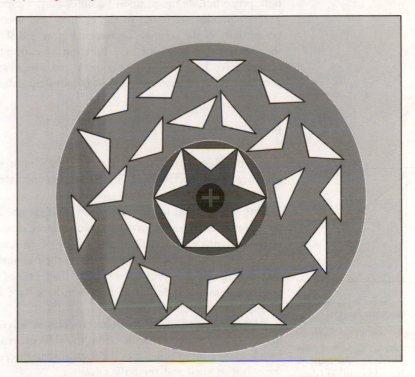

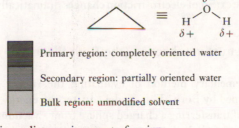

Primary region: completely oriented water

Secondary region: partially oriented water

Bulk region: unmodified solvent

Figure 4.6   The immediate environment of an ion.

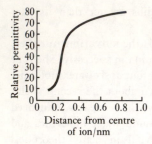

Figure 4.7
The variation of
permittivity around
an ion (data from
Bockris and Reddy, 1970).

ion and (c) a secondary region of space beyond the primary solvation shell where the orientation of the water molecules gradually changes from the highly oriented solvation shell to that of the structured solvent. This third region is expected to be highly disordered since the spherical symmetry of bound water is totally incompatible with the tetrahedral arrangement in the bulk solvent.

It is proposed that in the solvation shell the water molecules are so tightly bound that they are effectively unorientable and hardly contribute to the orientation polarisation of the solution. This is not completely true since water molecules may deform under the influence of an electric field, offering some degree of polarisation. However, with polarisation minimised, the relative permittivity of the hydration sheath is much lower than that of bulk water. As depicted in figure 4.7, the relative permittivity of water increases, as a function of distance from the ion, from a very low value in the solvation shell through a region of progressively greater ordering (the secondary region in figure 4.6) to the value of the bulk solvent. The relative permittivity of a solution is determined by the proportion of water molecules residing in the different regions. For very dilute solutions most water molecules are in the bulk phase but, as the solution becomes more concentrated, greater proportions of water molecules are found in the primary and secondary regions and the relative permittivity of the solution decreases. Figure 4.8 shows the relative permittivity of sodium chloride solutions as a function of concentration. Divalent and trivalent cations have high solvation numbers and are observed to decrease the relative permittivity of solutions more than monovalent cations at similar concentrations.

4.4.3   Electrostriction

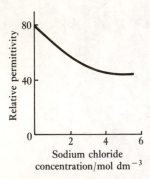

Figure 4.8 The relative permittivity, ε_r, versus concentration, for sodium chloride solution (data from Bockris and Reddy, 1970).

The term **electrostriction** is given to the compression of a material due to the presence of an electric field. This phenomenon is observed for water in aqueous solutions.

One contribution to electrostriction is solvation. In the solvation shell of an ion the water molecules are aligned and packed more closely. This leads to a local increase in the solvent density in the near-vicinity of an ion. This phenomenon influences the density of aqueous solutions and also their compressibilities since it is supposed that the already compressed solvation shell cannot be compressed further by an external pressure. Indeed, solvation numbers have been evaluated from compressibility measurements.

A second contribution to electrostriction is solvent collapse. This is caused by the removal of water molecules from the solvent structure by ion solvation. This in turn leads to a slight collapse of the solvent structure, especially in the region between the solvation shell and the bulk solvent.

The extent of electrostriction changes dramatically with temperature and pressure.

4.4.4   Non-structural theories (the Born theory)

An elementary method of evaluating the energy of ion–solvent interactions was developed by Born. This model equates the energy of ion–solvent interaction to the work of transferring a charged sphere from a vacuum, where there are no interactions, into an infinitely large continuum of known relative permittivity. The transfer is formalised as a three-part process. First is the removal of charge from a sphere of charge z and radius r. Second is the transfer of the uncharged sphere into the solvent, and third is the recharging of the sphere in the solvent. The first and third steps involve the work of discharging and charging the sphere and the energy of interaction is the sum of the two work functions. The work of charging a sphere, in a medium of relative permittivity ε_r, from zero to a final value ze is given by:

$$w = \frac{(ze)^2}{2\varepsilon_r r} \qquad (4.16)$$

where e is the charge on one electron and z is the integral number of electronic charges on one sphere. The work of discharging a sphere has the same form with a minus sign. Since the relative permittivity of a vacuum is unity, the energy of interaction, $U(\phi)$, for 1 mole of ions is:

$$U(\phi) = -L\frac{(ze)^2}{2r}\left[1 - \frac{1}{\varepsilon_{r(s)}}\right] \qquad (4.17)$$

where $\varepsilon_{r(s)}$ refers specifically to the relative permittivity of the solvent and L is Avogadro's constant.

Applications of this theory have been used to predict both the energy of interaction and the amount of heat liberated or consumed per mole of dissolved ions. Extensions of the theory predict that heats of ion–water interaction should vary with z^2/r, tending monotonically to zero as the ions' ionic radii approach infinity. If crystallographic radii are used, this linearity is not observed, and predicted energies of interaction are higher than those measured (figure 4.9). The deficiencies in the Born model lie in the assumption that the solvent is a continuous dielectric and also

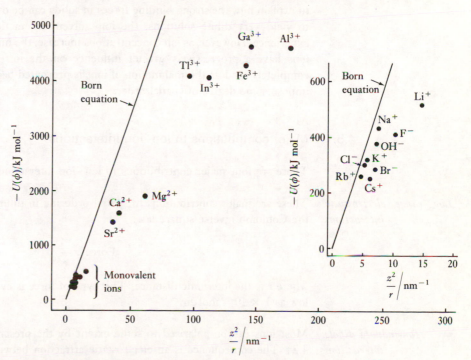

Figure 4.9   Energies of solvation for some ions (data from Friedman and Krishnan, 1973).

that ions are hard spheres of well-defined radii. This second assumption is especially poor when transition metals are being considered.

An interesting point, first noted by Latimer *et al.* (1939), was that if 0.089 nm was added to the crystallographic radii of cations and 0.01 nm was added to the crystallographic radii of anions, these hypothetical radii correctly predicted energies of solvation and removed the non-linearity from figure 4.9. More recently, Helgeson and Kirkham (1976) observed that similar corrections to crystallographic radii could be made to rationalise density data. The corrected radius, termed **the effective electrostatic radius**, is an idealised parameter which should not be thought of as a real physical radius. It is considered to represent the consequences of dielectric saturation, void space in the solvation shell and the finite size of solvent dipoles which are not explicitly included in the simple Born treatment. The Born treatment accommodates solvation but does not provide for solvent collapse. These contributions are addressed by modern theoretical treatments of ion–solvent interactions which are beyond the scope of this book. These important structural considerations are addressed by modern theoretical treatments of ion–solvent interactions which are beyond the scope of this book.

4.5   Ion–ion interactions

Ion–solvent interactions are a major factor in the dissolution of solids and are used to explain how the strong binding forces in solids can be overcome. If solids dissolve to yield very dilute solutions, the ion–solvent interactions are all that the ions experience. However, as ion concentrations increase, the mutual interactions between ions have a progressively greater influence on the ion properties. In particular, completely random thermal motion of ions is prevented because ion–ion interactions impose some degree of spatial order.

4.5.1   Major contributions to ion–ion interactions

There are four major contributions to ion–ion interactions.

Long-range electrostatic interactions

These are major contributions to spatial ordering in dilute electrolytes. They obey the Coulomb inverse square law:

$$\text{Force} \propto \frac{1}{r^2} \tag{4.18}$$

where r is the interionic distance. This type of force is evident at concentrations as low as $1 \times 10^{-5}\,\text{mol}\,\text{dm}^{-3}$.

Ion-induced dipole interactions

Most ions can be polarised to some extent by the presence of other ions (section 4.2). The consequence is an electrostatic attraction between an ion and the dipole it induces. The interaction is strongest over molecular distances.

Short-range attractions   These are electrostatic interactions working over interionic distances close to the dimensions of ionic radii themselves. They occur between ions of opposite sign.

Short-range repulsions   These occur when ions diffuse so close together that there is a degree of covalent interaction as the electron orbitals of interacting species overlap. Qualitatively, we may view this as a resistance to the penetration of the electron orbital from one species by the electron orbital from a second species.

4.5.2   Long-range electrostatic forces (Debye–Hückel theory)

These forces are the major ion–ion interactions in dilute electrolytes.

To understand the magnitude of these forces we consider the case of an aqueous solution containing an electrically neutral mixture of ions in thermal motion. If the ions were distributed randomly, the probability of finding a cation in the vicinity of a chosen 'reference' anion would be just the same as the probability of finding an anion. For such a random configuration there would be no net electrostatic potential produced at the reference ion since attractive forces would be exactly balanced by repulsive forces. This hypothetical situation only occurs in infinitely dilute solutions.

Ions of like charge repel each other and unlike charges attract each other. Therefore, in 'real' solutions, an ion of opposite charge to the reference ion is more likely to be found in the vicinity of the reference ion than an ion of like charge. In fact, the electrostatic interactions would give rise to an ordering of ions in solution, rather like a regular ionic crystal, were it not for thermal motion. The reality is a dynamic compromise between electrostatic forces producing ordering, and random thermal motion destroying the structure. The extent of ordering progressively increases as ion concentrations increase.

The essential step in evaluating electrostatic interactions is to **calculate the electrostatic potential at a reference ion due to its interaction with all the other ions in solution**. This is simple for a material such as a regular crystal where the spatial distribution of all ions is known. We simply use the law of superposition, which states that the potential at any point charge in an assembly of charges is the sum of the potentials arising from the reference ion and **each other ion in the system**. However, for solutions the actual ion distribution is not known and the problem is almost intractable. We must simplify the ideas to obtain a solution.

A useful simplifying concept is that of a reference ion surrounded by an **ionic atmosphere** composed of all the other ions in solution. Because the solution is electrically neutral overall, the charge on the reference ion is balanced by the net charge on the atmosphere. In view of the dynamic nature of the atmosphere, ions of opposite charge to the reference ion (called counter-ions) will be more likely to diffuse close to the reference ion than ions of like charge. The atmosphere is then thought of as a smeared-out cloud of charge with the charge density being greatest at points in space closest to the reference ion and tailing out to zero at infinite distance (figure 4.10).

The energy of interaction between a reference ion and the electrically charged ionic atmosphere is equal to the work of bringing the ion from infinity to the centre of the ionic atmosphere.

From this model, an approximate equation for the average energy of electrostatic

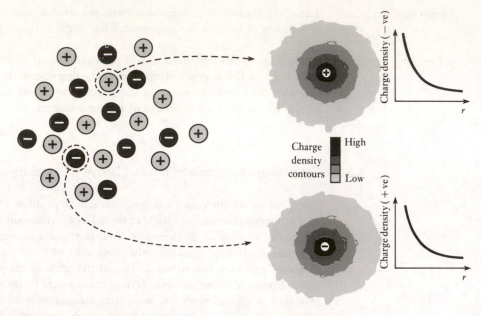

Figure 4.10   The ionic atmosphere.

interaction between a specified ion of charge z_j and the ionic atmosphere is:

$$U_{elect} = - \frac{z_j^2 e^2 \kappa}{8\pi\varepsilon_0 \varepsilon_r} \qquad (4.19)$$

where κ is defined as:

$$\kappa^2 = \frac{Le^2}{\varepsilon_0 \varepsilon_r kT} \sum_i c_i z_i^2 \qquad (4.20)$$

Here, L is Avogadro's constant, e the charge on one electron, ε_0 is the permittivity of a vacuum, ε_r is the relative permittivity of the medium (section 4.1.3), k is the Boltzmann constant and T is the absolute temperature; c_i is the molar concentration of all ions, κ has the dimensions of reciprocal length and is called the **Debye inverse length**. It is a measure of the distance over which the electrostatic field of the specified ion extends and is thought of as the reciprocal of the radius of the ionic atmosphere.

For more concentrated electrolytes we use:

$$U_{elect} = - \frac{z_j^2 e^2}{8\pi\varepsilon_0 \varepsilon_r} \left(\frac{\kappa}{1 + \kappa \mathring{a}} \right) \qquad (4.21)$$

where $\mathring{a}$ is a called **the distance of closest approach** for ions. It is sometimes interpreted as the diameter of the ion j, although a clear physical significance cannot be truly ascribed to this distance.

These equations play a central role in the quantitative prediction of chemical reactivity in aqueous media and therefore deserve special consideration. The interested reader may wish to consider their derivation as follows.

Box 4.1

Deriving the equations for long-range electrostatic forces

The work due to the transfer of electrostatic charge

A variable which is central to the following discussion is the **electrostatic potential**. It is defined as follows.

Electrostatic theory states that for two charges, q_1 and q_2, separated by a distance r, **the force along the line of separation is**:

$$F = \frac{q_1 \times q_2}{4\pi\varepsilon_0\varepsilon_r r^2},$$ (4.22)

The energy of electrostatic interaction between the two charge is equal to the work of bringing a charge q_1 from infinity to a distance r from a charge q_2. This is:

$$U(r) = w(r) = -\int_\infty^r F \, dr = -\frac{q_1 \times q_2}{4\pi\varepsilon_0\varepsilon_r} \int_\infty^r \frac{1}{r^2} \, dr$$ (4.23)

Therefore,

$$U(r) = \frac{q_1 \times q_2}{4\pi\varepsilon_0\varepsilon_r r}$$ (4.24)

This can be simplified by collecting some of the terms together to define the electrostatic potential at a distance r from q_2. This term is designated $\phi(r)$ and is equal to:

$$\phi(r) = \frac{q_2}{4\pi\varepsilon_0\varepsilon_r r}$$ (4.25)

Therefore, the work of transfer of electric charge equals:

$$U(r) = q_1 \times \phi(r)$$ (4.26)

A simple treatment of electrostatic forces between ions

The objective is to calculate the work of bringing the ion from infinity to the centre of the ionic atmosphere.

Equation 4.26 gives the energy of electrostatic interaction between point charges separated by a distance r. Therefore, for a reference ion of charge z_j,

$$U(r) = z_j e \times \phi_{ave}$$ (4.27)

where ϕ_{ave} is the **average electrostatic potential at the central ion due to all the other ions in the atmosphere**. Even with the simplified model based upon the ionic atmosphere, we seem to fall at the first hurdle. The problem is that the value of ϕ_{ave} is dependent on the actual charge distribution in the ionic atmosphere, and this charge distribution is controlled by the energy of electrostatic interaction which is the property we are trying to calculate.

This difficult problem was overcome by Debye and Hückel (1923). The detailed treatment is beyond the scope of this book, although a simplified overview is as follows.

The basic assumptions in the Debye–Hückel theory are:

1. All electrolytes are completely dissociated into ions.
2. All ions are spherically symmetrical unpolarisable charges.

continued

Box 4.1
continued

3. The solvent is a structureless continuum whose sole property is the bulk value of its permittivity.
4. Random thermal motion of ions is obtained.

The procedure involves the following steps:

- First we calculate the **total** electrical potential at any distance (r) from the centre of the reference ion [$\phi(r)_{total}$]. This is dependent on a contribution from the ionic atmosphere and a contribution from the central ion itself. The calculation of $\phi(r)_{total}$ was the central theme of Debye–Hückel in 1924.
- Secondly, the electrostatic potential at any distance from the central ion [$\phi(r)_{atmos}$], arising solely from the ionic atmosphere, is calculated. This function can be evaluated since all electric potentials are additive, i.e.

$$\phi(r)_{total} = \phi(r)_{ion} + \phi(r)_{atmos} \tag{4.28}$$

and the potential due to an ion itself (ϕ_{ion}) is given by:

$$\phi(r)_{ion} = \frac{z_j e}{4\pi\varepsilon_0\varepsilon_r r} \tag{4.29}$$

- Thirdly, from the distribution of electrostatic potential $\phi(r)_{atmos}$ we evaluate the mean potential ϕ_{ave}.

To perform the above three steps we start by using Poisson's equation from classical electrostatics. This relates the total electrical potential at any point to the charge density at that point for a continuous distribution of charge. When applied to a spherical shell of radius r and thickness dr with its centre coinciding with the centre of spherical symmetry the equation becomes:

$$-\frac{1}{r^2}\frac{d}{dr}\left[r^2\frac{d\phi(r)_{total}}{dr}\right] = \frac{\rho}{\varepsilon_0\varepsilon_r} \tag{4.30}$$

where ρ is the charge density.

To make use of equation 4.30 we need to express ρ as a function of ϕ_{total}. To do this we use the Maxwell–Boltzmann equation from classical statistical mechanics. Denoting $\bar{n}_i$ as the average number of ions of type i (anion or cation) at any point in the ionic atmosphere, and n_i as the bulk concentration of this ion, then

$$\bar{n}_i = n_i \exp\left[\frac{-ez_i\phi(r)_{total}}{kT}\right] \tag{4.31}$$

where k is the Boltzmann constant. The electric charge density ρ is equal to $\bar{n}_i$ multiplied by its electric charge ($z_i e$) and summed over all ions in the system, i.e. $\rho = \sum_i z_i\bar{n}_i e$. The Maxwell–Boltzmann equation becomes:

$$\rho = \sum_i z_i n_i e \exp\left[\frac{-ez_i\phi(r)_{total}}{kT}\right] \tag{4.32}$$

At this point the two equations 4.30 and 4.32 have two unknowns, ρ and $\phi(r)_{total}$ and are in principle soluble. Even so, an exact solution to the resulting second-order differential equation is not possible analytically, although it can be solved numerically with computers. Debye and Hückel gave an approximate solution to the equations treating the ions as rigid unpolarisable spheres separated by a minimum distance $\mathring{a}$ (the distance of closest approach).

The approximate solution, relating $\phi(r)_{total}$ to the distance from the central

continued

Box 4.1
continued

ion, is:

$$\phi(r)_{total} = \frac{z_j e}{4\pi\varepsilon_0\varepsilon_r}\left[\frac{\exp(\kappa\mathring{a})}{(1+\kappa\mathring{a})}\cdot\frac{\exp(-\kappa r)}{r}\right] \qquad (4.33)$$

The variable κ is defined by:

$$\kappa^2 = \frac{Le^2}{\varepsilon_0\varepsilon_r kT}\sum_i c_i z_i^2 \qquad (4.34)$$

where L is Avogadro's constant, and c_i is the molar concentration of ions which is related to the concentration n_i by $n_i = c_i L$. Taking a closer look at κ, we see the concentration terms can be converted to **molal** units since $c_i = m_i\rho_0$, where ρ_0 is the density of pure water. This gives:

$$\kappa^2 = \frac{Le^2\rho_0}{\varepsilon_0\varepsilon_r kT}\sum_i m_i z_i \qquad (4.35)$$

From equation 4.33, $\phi(r)_{atmos}$ can be calculated by subtracting the contribution from the central ion to yield:

$$\phi(r)_{atmos} = \frac{z_j e}{4\pi\varepsilon_0\varepsilon_r r}\left[\frac{e^{\kappa\mathring{a}}}{(1+\kappa\mathring{a})}\,e^{-\kappa r} - 1\right] \qquad (4.36)$$

The final step is to evaluate ϕ_{ave} from equation 4.36. Since no ions can be closer than the distance $\mathring{a}$ we assume that the value ϕ_{ave} is equal to the value of $\phi(r)_{atmos}$ with r set to $\mathring{a}$. That is,

$$\phi_{ave} = \phi(\mathring{a})_{atmos} = \frac{-z_j e}{4\pi\varepsilon_0\varepsilon_r}\left(\frac{\kappa}{1+\kappa\mathring{a}}\right). \qquad (4.37)$$

This equation is known as the approximate solution to the Poisson–Boltzmann equation.

In extremely dilute solutions the value of κ is less than $10^{-8}\,m^{-1}$ and with $\mathring{a}$ around $10^{-10}\,m$ the value of $\kappa\mathring{a}$ becomes negligible. This gives an approximate equation for ϕ_{ave}:

$$\phi_{ave} = \frac{-q\kappa}{4\pi\varepsilon_0\varepsilon_r} \qquad (4.38)$$

At this stage we have equations for the electrostatic potential at the reference ion. We can use these expressions to calculate the total energy of interaction U_{elect} for a single reference ion and its atmosphere. This is determined by summing the energies, over an infinite number of steps, as the charge on the ion, q, is increased from zero to $z_j e$. It is given by the integral:

$$U_{elect} = \int_0^{z_j e} \phi_{ave}\,dq \qquad (4.39)$$

To produce an equation valid at extreme dilution we integrate equation 4.38, which gives:

$$U_{elect} = -\frac{z_j^2 e^2\kappa}{8\pi\varepsilon_0\varepsilon_r} \qquad (4.40)$$

For more concentrated electrolytes we integrate equation 4.37, to produce:

$$U_{elect} = -\frac{z_j^2 e^2}{8\pi\varepsilon_0\varepsilon_r}\left(\frac{\kappa}{1+\kappa\mathring{a}}\right) \qquad (4.41)$$

General weaknesses in the
Debye–Hückel theory

1. Not all strong electrolytes are completely dissociated. Explicit provision must be made for the quantity of associated salt in solution. In this sense, random thermal motion of ions is not achieved.
2. Ions are not spherically symmetrical or unpolarisable. This undermines the use of the Poisson equation in its symmetrical form.
3. The solvent is not a structureless continuum with a uniform relative permittivity. The effects of ion–solvent interactions, electrostriction, solvent–solvent interactions and dielectric saturation should be included.
4. The charge density in the ionic atmosphere will not be symmetrically distributed round the central ion due to the lack of spherical symmetry in the ions. Additionally, the orientation of dipoles on water molecules and ions may shield counter-ions from the organising effect of the central ion and therefore affect the charge distribution in the ionic atmosphere.
5. The Poisson–Boltzmann equation in its full form is a contradiction of a basic law of electrostatics which is based on the linear superposition of fields. The Boltzmann equation is exponential and the Poisson equation is linear. Both should be linear to satisfy the basic law. To conform with the laws of electrostatics the Poisson–Boltzmann expression must be truncated to the first-order terms; this limits the application of the equation to very dilute solutions.
6. The distance of closest approach $\mathring{a}$ has no clear physical meaning. A consequence of the basic theory is that $\mathring{a}$ is the same for all ion combinations in a given solution: this is counter-intuitive.

4.5.3   A limitation in Debye–Hückel theory due to ion association

As we have seen, the major weakness of the Debye–Hückel theory is incomplete dissociation. We conclude this since the approximate solution to the Poisson–Boltzmann equation is valid for cases when **the electrostatic potential energy of interaction ϕ_{ave} is less than the thermal energy**, i.e.:

$$z_j e \phi_{\mathrm{ave}} \ll kT \tag{4.42}$$

The problem arises because the energy of ion–ion interaction is known to be comparable with thermal energy. By imposing this restriction to equation 4.36 we see that for a chosen value of $\mathring{a}$ there is a minimum value of r below which the theory is not valid. Ions closer together than this minimum value are said to **associated**.

Bjerrum (1926) suggested that the cut-off between associated and non-associated ions should be $2kT$, giving a critical distance q for two ions of charge z_i and z_j:

$$q = \frac{z_i z_j e^2}{8 \pi \varepsilon_0 \varepsilon_r kT} \tag{4.43}$$

Table 4.1 shows the variation of the Bjerrum critical with ion type for water at 25 °C.

Bjerrum (1926) suggested that ion-pairs closer than the critical distance can be treated as if they were single species having the combined charge of the ions involved. The effect is an apparent reduction in the concentration of the unassociated species. The concentration of these associated and unassociated ions can be included in the

Table 4.1   **The Bjerrum critical distance**

Ion Charges		q/nm
Cation	Anion	
1	1	0.357
2	1	7.14
3	1	10.81
2	2	14.08
3	2	21.42
3	3	32.04

term $\sum c_i z_i^2$ in equation 4.20 to give a modified value for κ. This allows the approximate solution to Debye–Hückel theory to be applied with greater confidence.

As an approximation, if $\mathring{a}$ (the distance of closest approach) is greater than the value of q then there is likely to be **no** ion association.

If $\mathring{a}$ is taken to represent the sum of the electrostatic radii of the chosen ions, which are of the order of $0.2-0.3$ nm, then table 4.1 suggests that for 1:1 electrolytes significant ion association is not expected. However, for other ion combinations, $\mathring{a}$ will always be less than q and ion association is expected at all concentrations.

Guggenheim and Stokes (1969) describe and refer to computer techniques used to generate **exact** numerical solutions the Poisson–Boltzmann equation for specified ion combinations. With such techniques it is possible to evaluate the proportion of cations and anions in the ionic atmosphere as a function of distance from the central ion. These confirm that, for **all but 1:1 electrolytes**, significant proportions of counter-ions exist closer to the central ion than the critical distance. This implies that ion association may arise from purely electrostatic interactions and does not necessarily involve the development of covalent bonds between ions or the existence of a specific molecule.

This type of ion-pair may be thought of as a measure of the inadequacy of the approximate Debye–Hückel treatment. The nature of ion-pairs and other complexed species is discussed in more depth in the next section.

4.6   Complexes, chelates and ion-pairs

An ion-pair is pair of ions closer together than the Bjerrum critical distance. It is a physical entity held together by electrostatic forces acting over short distances. These cohesive forces are sufficient to overcome the tendency of the participating ions to move around independently under thermal motion. The term 'ion-pair' does not imply the existence of permanent independent molecules. However, in many electrolyte solutions, particularly those containing transition metals, stable molecules are known to exist. In general, the formation of complexes and ion-pairs is called **speciation**.

The distinction between complexes and ion-pairs cannot be made unambiguously since many species may have characteristics of both to a greater or lesser extent. However, ion-pairs have some properties which differ from those of complexes and which are worth considering.

4.6.1   Ion-pairs

The electrostatic forces within the ion-pair are independent of the specific chemical nature of the ions and are dominated by the charge and size of the ions. Consequently, it is expected that ions of similar size and radius will have similar values for the mass action coefficient for the ion-pair formation. For example, this is observed for ion-pairs formed between sulphate and many divalent ions.

The term 'ion-pair' is somewhat imprecise since associated species need not be restricted to one cation plus one anion. In concentrated electrolyte solutions, species composed of a cation and several anions may occur. Indeed, in electrolytes containing a mixture of anions there may be associated species containing several anions. Associated species containing only ions of the same charge are possible but unlikely in anything other than very concentrated solutions, in view of the high repulsive forces in involved.

The structure and size of an ion-pair is strongly influenced by the solvation shell of the participating ions. Some ion-pairs are formed with little or no disruption of the solvation sheath of the ions, and water molecules exist between the ions. These are called **outer sphere** ion-pairs and are shown in figure 4.11(b).

In other ion-pairs the ions are so close together that water molecules are expelled from **between** the ions (figure 4.11(c)) and the ion-pair itself is completely solvated. This is called either a **contact** or an **inner sphere** ion-pair. The overall charge on an ion-pair is the algebraic sum of the charges on the participating ions; for example, the species NaCl has zero charge whereas $CaCl^+$ has a single positive charge. However, the charge distribution need not be symmetrical and even a neutral ion-pair may behave as a dipole or a charge-separated ion.

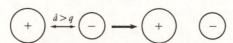

(a) No ion association $\mathring{a} > q$

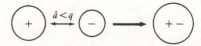

(c) Ion-pair formed with disruption of the solvation spheres of the individual ions. The bare ions are in contact.

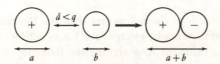

(b) Ion-pair formed with no disruption of the solvation spheres of the individual ions. Solvent is still present between the ions.

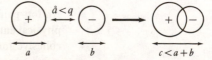

(d) Ion-pair formed with partial disruption of the solvation spheres of the ions. Some solvent remains between the ions, although some has been removed.

Figure 4.11   Ion association.

4.6.2   Complexes

A species formed from covalent bonding between ion–ion or ion–molecule combinations is called a **complex**. Many transition metals form complexes with I^-, F^- and Cl^- and to a lesser extent NO_3^-. Such reactions involve one or more anion molecules coordinated to a central cation. For example, Al^{3+} reacts with fluoride to form the species AlF^{2+}, AlF_2^+, AlF_3^0, and AlF_4^-. A striking example is the species $CoCl_4^{2-}$ formed when excess chloride ions are added to cobalt solutions. Solutions containing only aqueous Co^{2+} are pink; however, the presence of the complex imparts a blue colour to the solution.

The formation of such a species is normally associated with the attachment of an ion to a specific chemical site on a molecule. For example, formic acid can dissociate in water to form the formate ion $HCOO^-$ via the reaction

$$HCOOH + H_2O \rightleftharpoons HCOO^- + H_3O^+ \tag{4.44}$$

Cu^{2+} can react with the formate ion via the reaction:

$$Cu^{2+} + HCOO^- \rightleftharpoons HCOOCu^+ \tag{4.45}$$

The species $HCOOCu^+$ and the undissociated molecule $HCOOH$ can both be termed complexes since both the copper and the proton show chemical bonding to the oxygen on the formate ion.

The bonding in complexes will reflect the specific chemical nature of the reacting species, such as their electronic configuration, or the polarisability. In view of the variation of such properties between species it may be expected that mass action quotients will vary in magnitude from species to species and not show a simple dependence on charge and ion radius. This is observed with complexes formed between ions and organic molecules with reactive sites containing nitrogen or oxygen. For example, the ions Cu^{2+}, Co^{2+}, Ni^{2+}, Zn^{2+} and Mn^{2+} have similar charges and crystallographic radii but their measured mass action quotients for reactions with anions can vary over several orders of magnitude for the same anion. This implies complex formation rather than ion-pairing.

Complex formation is associated with significant disruption of the solvation sheath as a result of the strong covalent bonding. This is seen with the copper tetramine complex $Cu(NH_3)_4^{2+}$ formed from the addition of dilute ammonium solutions to solutions containing copper. If the ammonium is in excess most of the water molecules are expelled from the copper solvation sheath. Complexes should not generally be considered as spherically symmetrical or having a symmetrical charge distribution.

4.6.3   Chelates

If an anion has more than one charge, located in different places on the molecule, there is more than one point of attachment for a cation. An example is the reaction between divalent cations and EDTA (ethylenediaminetetra-acetic acid). The structure of the EDTA molecule is given in figure 4.12, where the four points of attachment are shown.

In solutions containing an excess concentration of OH^-, the EDTA molecule is fully dissociated and has four negatively charged carboxylic acid groups. The reaction

HOOC—CH$_2$ CH$_2$—COOH

　　　　　N—CH$_2$—CH$_2$—N

HOOC—CH$_2$ CH$_2$—COOH

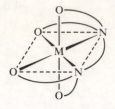

(a) A formal representation of the EDTA molecule   **(b)** An EDTA–metal chelate

Figure 4.12   The structure of EDTA (ethylenediaminetetra-acetic acid).

between a metal (M^{n+}) and dissociated EDTA, which is assigned the formula Y^{4-} for simplicity, is

$$M^{n+} + Y^{4-} \rightleftharpoons MY^{(n-4)+} \tag{4.46}$$

The species formed has a complicated structure with the EDTA molecule wrapped round the cation, which bonds covalently to the oxygen and nitrogen atoms on the EDTA. Formally, there are four independent reactions between Y^{4-} and either 1, 2, 3 or 4 moles of proton in accordance with the reactions:

$$H^+ + Y^{4-} \rightleftharpoons HY^{3+} \tag{4.47}$$

$$2H^+ + Y^{4-} \rightleftharpoons H_2Y^{2+} \tag{4.48}$$

$$3H^+ + Y^{4-} \rightleftharpoons H_3Y^+ \tag{4.49}$$

$$4H^+ + Y^{4-} \rightleftharpoons H_4Y^0 \tag{4.50}$$

A molecule capable of bonding to a cation via more than one reactive site is called a **complexone**. The complexed species so produced is called a **chelate**.

Naturally occurring organic chelates are thought to influence mineral solubilities under some conditions.

Chapter 5

Gases, liquid vapours and simple liquid mixtures

The fundamental properties of gases are few and simple but of great importance.

In geological systems at high temperatures and pressures, gases are less important than solids, electrolyte solutions or liquid melts. However, many properties of solids and liquids can be deduced from the properties of the vapour phase with which they are in equilibrium. This is convenient since the properties of gases are more easily measured than those of solids or liquids. Consequently, the study of gases has played a central role in the development and application of thermodynamics.

5.1 Pure gases and the ideal gas laws

The volumes of gases have a sensitive dependence on temperature and pressure. At low pressures and high temperatures all gases obey four laws which describe the relationships between temperature, pressure, volume and amount of gas. These are:

- **Boyle's law**: At constant temperature the volume of given mass of an ideal gas is inversely proportional to the pressure. Therefore, if a gas at pressure P_1 occupies a volume V_1 and at a pressure P_2 occupies a volume V_2, then:

$$P_1 \times V_1 = P_2 \times V_2 \tag{5.1}$$

- **Charles's law**: At constant pressure the volume of a given mass of gas is directly proportional to the absolute (Kelvin) temperature. This law is also ascribed to Gay-Lussac, and is expressed as:

$$\frac{V_1}{T_1} = \frac{V_2}{T_2} \tag{5.2}$$

- **The pressure–temperature relationship**: By combining Boyle's and Charles's laws, we see that at constant volume the pressure of a given mass of gas is directly proportional to the absolute temperature. Mathematically this is expressed as:

$$\frac{P_1}{T_1} = \frac{P_2}{T_2} \tag{5.3}$$

- **Avogadro's law**: At the same temperature and pressure, equal volumes of different gases contain the same number of elementary particles (molecules or atoms). That is,

$$V \propto n \tag{5.4}$$

where n is the number of moles of gas.

A relationship which links temperature, pressure and volume can be derived by combining any two of the first three gas laws. This is called the **combined gas law**:

$$\frac{P_1 V_1}{T_1} = \frac{P_2 V_2}{T_2} = \text{constant} \tag{5.5}$$

A more useful expression is obtained by combining the combined gas law with Avogadro's law. This is the **limiting law for ideal gases**:

$$PV = nRT \tag{5.6}$$

where r is an experimentally determined proportionality constant called the **gas constant**. The value of R is $8.314 \, \text{J} \, \text{K}^{-1} \, \text{mol}^{-1}$ and is independent of temperature, pressure and type of ideal gas.

When a gas obeys these laws it is said to be an **ideal** or **perfect** gas.

5.2   Mixtures of ideal gases – Dalton's law of partial pressures

We commonly encounter mixtures of several gases and it is useful to understand how the pressure of a mixture is related to the pressures of the individual substances in the mixture. This relationship was deduced by Dalton in 1801.

The **partial pressure** of a specified gas in a mixture is defined as the pressure that the gas would exert if it alone occupied the volume.

For a mixture of ideal gases, the total pressure is equal to the sum of the partial pressures of each gas.

Consider a mixture of n_A moles of A and n_B moles of B occupying a volume V, the partial pressures of the two gases, P_A and P_B, can be obtained from the ideal gas equation. That is,

$$P_A = n_A \frac{RT}{V} \tag{5.7}$$

and

$$P_B = n_B \frac{RT}{V} \tag{5.8}$$

The total pressure of the mixture P_T is equal to:

$$P_T = P_A + P_B = (n_A + n_B) \frac{RT}{V} \tag{5.9}$$

An alternative relationship between the total pressure and the partial pressure of any substance in the two–component mixture is derived by dividing P_T by P_A, i.e.

$$\frac{P_A}{P_T} = \left(\frac{n_A}{n_A + n_B} \right) \left(\frac{RT/V}{RT/V} \right) = \frac{n_A}{n_A + n_B} \tag{5.10}$$

The term $n_A/(n_A + n_B)$ is the mole fraction of substance A, as defined in section

2.3.3. Therefore,

$$P_A = x_A P_T \tag{5.11}$$

Similarly,

$$P_B = x_B P_T \tag{5.12}$$

The equations 5.9, 5.11 and 5.12 can be generalised to any number of substances in a mixture.

5.3 Equations of state for imperfect gases

The ideal gas law (equation 5.6) implies that a graph of pressure versus volume should be a hyperbola with mutually perpendicular asymptotes. For most gases this behaviour is observed at high temperatures and pressures below 10 bars, but as temperatures are lowered the gases depart from ideality. Figure 5.1 shows pressure/volume data for 1 mole of carbon dioxide at various temperatures and

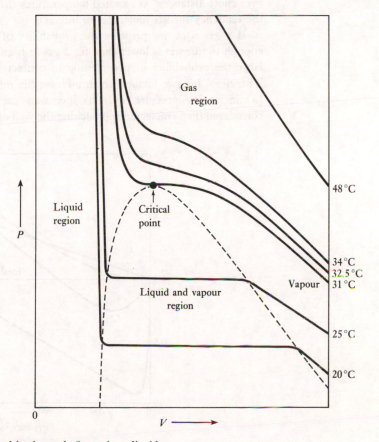

Figure 5.1   Andrews' isothermals for carbon dioxide.

demonstrates that at 48 °C the behaviour is almost ideal, whereas at 32.5 °C an anomalous point of inflexion is evident.

5.3.1   The compressibility factor

An alternative way of representing non-ideal behaviour is with the **compressibility factor** Z. For **1 mole** of any gas, Z is defined by:

$$Z = \frac{P\bar{V}}{RT} \tag{5.13}$$

where $\bar{V}$ is the molar volume. For an ideal gas Z is unity. Figure 5.2 shows the pressure dependence of the compressibility factors for several gases. The non-ideality is emphasised clearly and all gases tend towards ideal behaviour at low pressure.

To develop a qualitative understanding of non-ideal behaviour we may think of the pressure of a gas as a measure of the frequency with which gas particles collide with the boundaries of the containing vessel. This frequency increases with particle concentration and therefore **pressure is an idealised measure of particle concentration**.

The likely interactions between gas molecules are van der Waals forces which act over short distances. At elevated temperatures the increased thermal motion opposes the tendency for gas molecules to interact.

If a gas is at low pressure the probability of any two particles becoming close enough to interact is lower than for a gas at high pressure. Also, if the particles are large the probability of two coming in contact is higher than for small particles. Therefore, the two factors which increase the interparticle interactions in gases are (a) increasing pressure and (b) increased particle dimensions. Van der Waals considered that a molecule approaching the wall of a containing vessel will experience

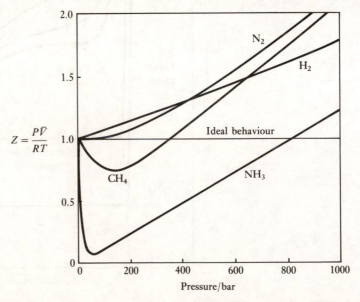

Figure 5.2   The compressibility factor for several gases at 298.15 K (from Moore, 1972).

unbalanced forces, there being more attractive forces pulling the molecule into the bulk than pulling the molecule to the wall. The mean velocity of a gas molecule near the wall, in the presence of attractive forces, will be less than in the absence of attractive forces. Consequently, the impact of any molecule on the containing vessel will be reduced and the pressure lower than if the gas were behaving ideally.

The limiting law is valid for gases where the particles have zero volume and no interparticle interaction. Although this situation is hypothetical, many gases do behave in accordance with this law at low pressures and high temperature.

5.3.2   Improved equations of state

A notable improvement over the ideal gas law is **van der Waals' equation** (equation 5.14). With this equation corrections are made for the reduced pressure due to interaction and the finite size of the molecules.

$$\left(P + \frac{an^2}{V^2}\right) \times (V - nb) = nRT \tag{5.14}$$

Corrected   Corrected
pressure   volume

where a and b are constants which are different for different gases. The volume correction is linearly proportional to the amount of gas present. The ideal gas law predicts that at $0\,K$ and any pressure, the volume of a gas is zero. The volume correction prevents this nonsensical extrapolation.

The pressure correction is a measure of how frequently any two molecules encounter each other. The frequency increases as the square of the number of moles per unit volume – hence the term $(n/V)^2$ multiplied by the proportionality constant a.

A similar theory to that of van der Waals was developed by Dieterici in 1898–9. His equation is:

$$P(V - n\beta) \exp(n\alpha/RTV) = nRT \tag{5.15}$$

where α and β are coefficients. This has the same form for the volume correction as the van der Waals equation. The pressure correction is based on a lower frequency of molecular collision with the containing walls, rather than the less energetic collisions. It is interesting to note that if the exponential in Dieterici's equation is expanded out in a power series in a/RTV and all terms after the second are neglected, a form of the van der Waals equation can be obtained.

The equations of van der Waals and Dieterici give good qualitative accounts of the behaviour of real gases. However, other equations, in particular the **virial equation**, are more important quantitatively. Initially, the virial equation was deduced empirically. The modern version has the form:

$$\frac{PV}{nRT} = 1 + \frac{nB}{V} + \frac{n^2C}{V^2} + \frac{n^3D}{V^3} \cdots \tag{5.16}$$

where B, C and D are temperature-dependent coefficients called the second, third and fourth virial coefficients. The expansion was later given a theoretical basis using statistical mechanics. It was shown that the second virial coefficient represents deviation from ideality due to two-molecule interactions, the third virial coefficient represents three-molecule interactions, etc.

A simple equation used by geochemists is the **Redlich–Kwong equation**.

$$P(V - nb) + \left[\frac{na(V - nb)}{T^{1/2}V(V + nb)} \right] = nRT \tag{5.17}$$

where a and b are fitting parameters specific to the gas.

5.4 Critical point phenomena

At a given temperature, a pressure may be found at which a gas undergoes liquefaction. This is a pressure-induced phase transition. As the temperature is elevated a **critical point** is reached at which the gas cannot be liquefied at any pressure. All gases have a critical point.

The **critical constants** which define the state of the system **at the critical point** are as follows.

- The **critical temperature** T_c is defined as the maximum temperature at which a gas can be liquefied by the application of pressure.
- The **critical pressure** P_c is the pressure which causes liquefaction at the critical temperature.
- The **critical volume** v_c is the volume of given mass of gas at its critical point.
- The **critical density** ρ_c is the density of a gas at its critical point. The densities of a gas and its liquid phase are identical at the critical point.

A good illustration of the critical point can be seen in $P-V-T$ diagram for an arbitrary substance (figure 5.3). The region above the critical point is called the **supercritical region**. In this region the gas and liquid phases of a substance are homogeneous and it is appropriate to use the term **fluid** rather than either gas or liquid.

The full complexity of the $P-V-T$ surface for real gases cannot be described with algebraic expressions. The best we can do is to use simple expressions, such as the van der Waals equation, which takes some account of the properties of a system near the regions of coexistence of two phases. The full three-dimensional representation of a system shows how simplified two-dimensional $P-T$ and $P-V$ plots fit into the overall picture. The $P-V$ projection shows us that the critical point is in fact a maximum in the $P-V$ plot. The critical point can be characterised by the two conditions:

$$\left(\frac{\partial P}{\partial V} \right)_T = 0 \tag{5.18}$$

$$\left(\frac{\partial^2 P}{\partial V^2} \right)_T = 0 \tag{5.19}$$

We can use the van der Waals equation to consider the properties of a fluid at its critical point. Rearranging equation 5.14 for 1 mole of a fluid using the critical constants gives:

$$P_c = \frac{RT_c}{(\bar{V}_c - b)} - \frac{a}{\bar{V}_c^2} \tag{5.20}$$

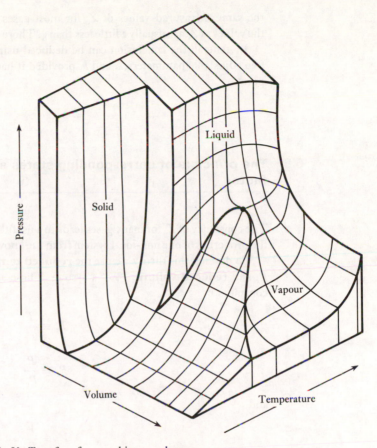

Figure 5.3 The $P-V-T$ surface for an arbitrary substance.

This yields:

$$\left(\frac{\partial P}{\partial V}\right)_T = \frac{-RT_c}{(\bar{V}_c - b)^2} - \frac{2a}{\bar{V}_c^3} = 0 \qquad (5.21)$$

and

$$\left(\frac{\partial^2 P}{\partial V^2}\right)_T = \frac{2RT_c}{(\bar{V}_c - b)^3} - \frac{6a}{\bar{V}_c^4} = 0 \qquad (5.22)$$

Solving these equations gives:

$$T_c = \frac{8a}{27bR}, \; \bar{V}_c = 3b, \; P_c = \frac{a}{27b^2} \qquad (5.23)$$

Combining the functions in equation 5.23 to eliminate a and b yields the relationship:

$$\frac{P_c \bar{V}_c}{RT_c} = \frac{3}{8} = 0.375 = Z_c \qquad (5.24)$$

This is saying that the compressibility factor for all fluids at their critical points is the same. Measured values of Z_c for most gases support this conclusion although the values of Z_c are usually a little less than $\frac{3}{8}$. There are only a few violations of the rule.

In general, the conclusion can be deduced using any equation of state with only two adjustable parameters, a and b, provided it has a point of inflexion at the critical point.

5.5 The principle of corresponding states and the compressibility factor

The constancy of Z_c for many gases leads to a useful approximate theory for predicting the properties of one non-ideal system from the known properties of a second system.

To develop this theory we use the **reduced temperature** T_r, **reduced pressure** P_r and **reduced volume** V_r of a system. These are defined in terms of the critical constants by:

$$T_r \equiv \frac{T}{T_c} \tag{5.25}$$

$$P_r \equiv \frac{P}{P_c} \tag{5.26}$$

and

$$\bar{V}_r \equiv \frac{\bar{V}}{\bar{V}_c} \tag{5.27}$$

The relationships in equation 5.24 can be combined with the reduced variables to give:

$$P = \frac{a}{27b^2} P_r, \quad \bar{V} = 3b\bar{V}_r, \quad \frac{8a}{27Rb} T_r \tag{5.28}$$

Using the expressions in equation 5.29, the van der Waals equation becomes:

$$\left(P_r + \frac{3}{\bar{V}_r^2}\right)\left(\bar{V}_r - \frac{1}{3}\right) = \frac{8}{3} T_r \tag{5.29}$$

That is, we have eliminated all the adjustable parameters to give a universal equation of state for all fluids and gases. This leads to the following conclusion called the **principle of corresponding states**.

FOR ANY FLUID OR GAS THE REDUCED PRESSURE IS RELATED TO THE REDUCED TEMPERATURE AND REDUCED VOLUME BY A UNIVERSAL FUNCTION.

The function can be represented by

$$P_r = F(T_r, \bar{V}_r) \tag{5.30}$$

Consequently, if $F(T_r, \bar{V}_r)$ can be established from measurements of the $P-V-T$

behaviour and the critical constants of one substance, the $P-V-T$ properties of any other substance can be predicted. Equation 5.29 is the function $F(T_r, \bar{V}_r)$ based on the van der Waals equation. In general, a function $F(T_r, \bar{V}_r)$ can be deduced using any equation of state with only two adjustable parameters, a and b, provided it shows a point of inflexion at the critical point.

If the compressibility factors for a range of gases are plotted against reduced pressure, the data at any single reduced temperature should fall on a single curve. This conclusion is supported by the data in figure 5.4.

Detailed statistical theories of the gas phase show that only the inert gases should conform with the principle of corresponding states. It is surprising to find that the principle is useful for substances having more complicated molecules. In general it

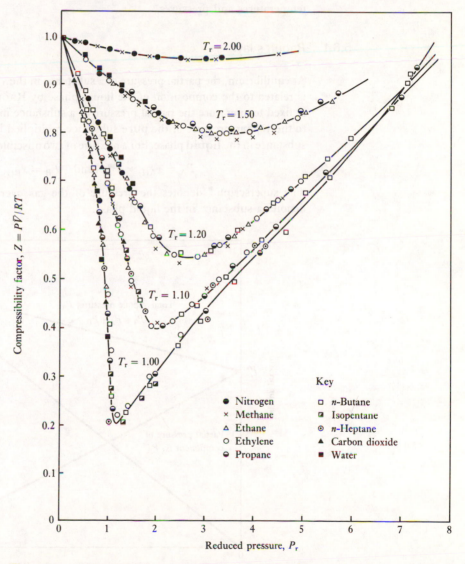

Figure 5.4   The compressibility factor for several gases versus reduced pressure (data from Gouq-Jen Su, 1946).

holds for spherically symmetrical molecules which have their centres of mass close to their centres of volume. Molecules which do not conform with the two restrictions or have large dipole moments generally deviate from the rule.

5.6   The properties of vapours above miscible liquids

The properties of vapours above liquid mixtures are discussed using simple chemical examples. It should be noted that all conclusions are valid for liquid melts at high temperatures and pressures.

5.6.1   Raoult's law

At equilibrium, the partial pressure of a substance in the vapour phase above a liquid is related to the composition of the liquid phase by **Raoult's law**. This states that at fixed temperature the partial pressure of a substance in the vapour phase is equal to the vapour pressure of the **pure** substance multiplied by the mole fraction of the substance in the **liquid** phase. For a mixture of two miscible liquids A and B we have:

$$P_A = x_{A(1)} \times P_A^\bullet \quad \text{and} \quad P_B = x_{B(1)} \times P_B^\bullet \tag{5.31}$$

The superscript $\bullet$ denotes the pressure of the gas over the **pure** liquid, and (1) denotes a substance in the liquid phase.

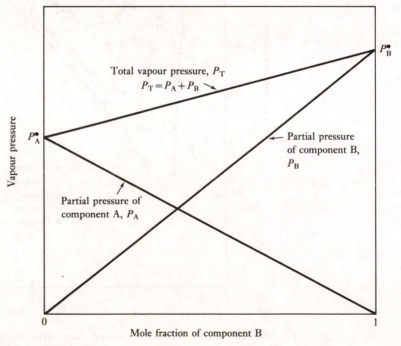

Figure 5.5   The vapour pressure above an ideal solution.

If the vapour phase is a mixture of ideal gases, $P_T = P_A + P_B$. Then,

$$P_T = x_{A(l)} \times P_A^{\bullet} + x_{B(l)} \times P_B^{\bullet} \tag{5.32}$$

Figure 5.5 shows the expected change in partial pressures as a function of liquid phase composition for a hypothetical ideal solution.

In a solution obeying Raoult's law the composition of the vapour phase need not be equal to the composition of the liquid phase. The mole fractions of substances in the vapour phase are given by:

$$x_{A(g)} = \frac{P_A}{P_A + P_B} \quad \text{and} \quad x_{B(g)} = \frac{P_B}{P_A + P_B} \tag{5.33}$$

Since $P_A = x_{A(l)} P_A^{\bullet}$ and $P_B = x_{B(l)} P_B^{\bullet}$, then:

$$x_{A(g)} = \frac{x_{A(l)} P_A^{\bullet}}{x_{A(l)} P_A^{\bullet} + x_{B(l)} P_B^{\bullet}} \quad \text{and} \quad x_{B(g)} = \frac{x_{B(l)} P_B^{\bullet}}{x_{A(l)} P_A^{\bullet} + x_{B(l)} P_B^{\bullet}} \tag{5.34}$$

This shows that only if $P_A^{\bullet} = P_B^{\bullet}$ will the liquid and vapour phases have the same composition. If $P_A^{\bullet}$ is higher than $P_B^{\bullet}$ then $x_{A(g)}$ will be greater than $x_{A(l)}$ and the gas will be richer in the more volatile component. This point is emphasised in figure 5.6.

5.6.2   Ideal mixing

Ideality or non-ideality is not an intrinsic property of any system. A mixture is ideal if its properties conform with some arbitrarily chosen reference behaviour. If Raoult's law is selected as the reference behaviour, then mixtures that obey Raoult's law over their **entire** composition range are ideal.

Ideality with respect to Raoult's law can be understood by recognising that the vapour pressure of a substance is a measure of the tendency of the molecules to escape from the liquid. The variation in the partial pressure of a substance is a measure of the extent to which the liquid hinders or helps the molecules to escape.

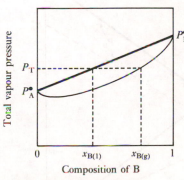

(a) Difference in vapour pressures of pure phases large.

Thus large difference in composition of phases.

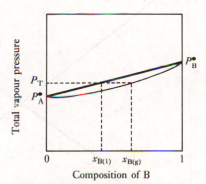

(b) Difference in vapour pressures small.

Thus small difference in composition of phases.

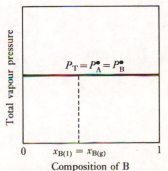

(c) No difference in vapour pressures. Thus no difference in composition of phases.

Figure 5.6   The vapour pressure over a mixture of two miscible liquids.

In mixtures, partial pressure is a measure of the change in the energy of interaction between molecules accompanying mixing. For example, in pure A there are A–A interactions, in pure B there are B–B interactions and in the mixture there will be additional A–B interactions. If the three cohesive forces are identical, the tendency of a molecule to escape will be independent of the molecules in its immediate environment. In this case the partial pressure of a substance will be dependent only on the mole fraction. This unlikely situation is observed for very few real mixtures.

5.6.3 Henry's law

In many single-phase systems, one substance is found in abundance compared with other substances and the system is best described in terms of a solvent plus solutes.

In this situation any solute molecule is completely surrounded by solvent, and the intermolecular interactions for the solute are dominated by solute–solvent interactions. If the solution is sufficiently dilute, these interactions will remain independent of composition. Therefore, the tendency of any molecule to escape becomes proportional to the mole fraction of the solute. This may be observed even though the solute–solvent interactions are nothing like the solute–solute or solvent–solvent interactions and the system is non-ideal with respect to Raoult's law.

Figure 5.7 shows this for a hypothetical system where the **vapour pressure of the solute varies linearly with the mole fraction of the solute but not in accordance with the Raoult's law line**. In such cases the partial pressure of the solute vapour is obeying Henry's law, which is given by the relationship:

$$P_B = x_{B(l)} \kappa_B \tag{5.35}$$

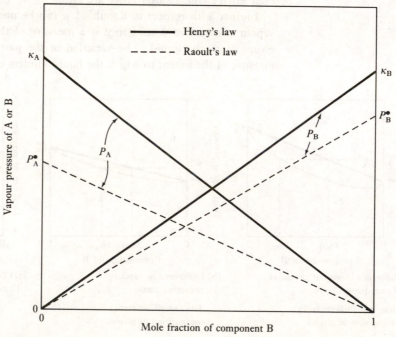

Figure 5.7 Henry's law.

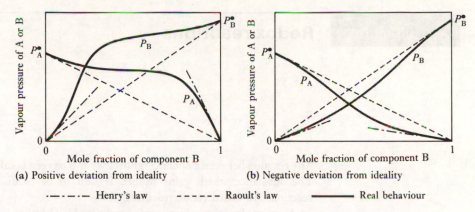

(a) Positive deviation from ideality (b) Negative deviation from ideality

—·—·— Henry's law   — — — — Raoult's law   ———— Real behaviour

Figure 5.8   Positive and negative deviations from ideality.

where κ_B is a constant with the dimensions of pressure. This law is a limiting law to which all solutes tend as their mole fraction tends to zero. Under these conditions the mole fraction of the solvent tends to unity and the behaviour of the solvent will approach Raoult's law.

Solutions where the solutes obey Henry's law are said to be **ideal dilute solutions**. Henry's law is an alternative to Raoult's law for ideality. Some systems, in particular aqueous electrolytes, conform with Henry's law at low solute concentrations.

5.6.4   Non-ideal solutions

For non-ideal mixtures the Henry's law line may be above or below the Raoult's law line. A positive deviation from Raoult's law occurs when the Henry's law line is above the Raoult's law line. Figure 5.8 shows both situations. In general, negative deviations are associated with strong solute–solvent interactions, where molecules are hindered from escaping from the liquid phase. Such mixtures often liberate heat during mixing and show a total volume which is less than the sum of the volumes of the pure substances. Positive deviations are associated with small solute–solvent interactions, leading to molecules easily transferring into the vapour phase. This type of system may even lead to immiscibility if the solute–solvent interactions are weak or even repulsive.

It should be stressed, however, that simple interpretations of positive and negative deviations may be inadequate for many systems. For example, in aqueous solutions the presence of inorganic electrolytes not only introduces strong solute–solvent interaction but also severely modifies the solvent–solvent interactions.

Chapter 6

Redox reactions

Any reaction between a chemical substance and oxygen is called an oxidation reaction. The substance which gains oxygen is said to be oxidised. Oxidation–reduction reactions are called redox reactions.

An example is the oxidation of magnetite ($Fe_2O_3(s)$) to hematite ($Fe_3O_4(s)$) via the reaction:

$$4Fe_3O_4(s) + O_2(g) \rightleftharpoons 6Fe_2O_3(s) \qquad (6.1)$$

However, we know that water dissociates to a small extent by the reaction:

$$2H_2O \rightleftharpoons 2H_2(g) + O_2(g) \qquad (6.2)$$

and may therefore write:

$$4Fe_3O_4(s) + 2H_2O \rightleftharpoons 6Fe_2O_3(s) + 2H_2(g) \qquad (6.3)$$

This implies that a species is oxidised by water with an accompanying loss of hydrogen. The reverse reaction between hematite and hydrogen to form magnetite is called a reduction and the hematite is said to have been reduced. Such reactions occur commonly in aqueous solutions and are known to influence the solubilities of minerals in rocks, soils and other natural systems. This occurs even though the amount of oxygen dissolved in fluids is low.

A redox reaction is characterised by a transfer of electrons. In order to discuss redox reactions from this viewpoint we adopt the following definitions.

A redox reaction is a chemical process in which one or more electrons are transferred completely from one species to another.

- **Oxidation** is a process of electron loss by the oxidised species.
- **Reduction** is a process of electron gain by the reduced species.

Reactions involving oxygen form one subset of the class of general redox reactions. Other common oxidising agents are ozone, hydrogen peroxide, carbon dioxide, chlorates, nitrates, permanganates, dichromates and chlorine. Common reducing agents are hydrogen, carbon, carbon monoxide, hydrogen sulphide and sulphur dioxide. For example, magnetite can be reduced in accordance with the reaction:

$$2Fe_3O_4(s) + CO_2(g) \rightleftharpoons 3Fe_2O_3(s) + CO(g) \qquad (6.4)$$

which is equivalent to reaction 6.1 in terms of the mineral oxidation but involves a different oxidising gas.

6.1 The ion-electron approach and half-cell reactions

In any redox reaction, electrons are transferred from a species or atom of higher electronegativity to one of lower electronegativity. One way of viewing any overall reaction is to split the reaction into two separate reactions which explicitly show the electrons involved and the ions in the liquid phase. These are called **half-cell reactions**.

6.1.1 Balancing half-cell reactions

The sum of the half-cell reactions gives the overall reaction. The general half-cell reduction reaction can be written:

$$mA_{(ox)} + nH^+ + e^- \rightleftharpoons pA_{(red)} + qH_2O \tag{6.5}$$

where (ox) and (red) denote the oxidised and reduced form of the species A respectively and m, n, p and q are stoichiometric coefficients. The species A can be an ion, element, molecule or mineral in any phase and the symbol e^- denotes 1 mole of electrons 'in the solution phase'. Half-cell reactions are always charge-balanced, i.e. the sum of the charges on all species, including electrons, on both sides of the equation is zero. Half-cell reactions do not occur in isolation because there is no such thing as a free electron existing in the solution phase. We use half-cell reactions to split any overall redox reaction use a reaction showing the source of electrons plus a reaction consuming an equivalent quantity of electrons. For example, the magnetite/hematite conversion can be obtained from combination of the following two half-cell reactions.

$$Fe_3O_4(s) + 8H^+ + 2e^- \rightleftharpoons 3Fe^{2+} + 4H_2O \tag{6.6}$$

$$Fe_2O_3(s) + 6H^+ + 2e^- \rightleftharpoons 2Fe^{2+} + 3H_2O \tag{6.7}$$

This gives:

$$2Fe_3O_4(s) + H_2O \rightleftharpoons 3Fe_2O_3(s) + 2H^+ + 2e^- \tag{6.8}$$

emphasising the fact that the oxidation of magnetite involves the loss of electrons.

Since any real chemical change must be written with sets of half-cell reactions that combine to eliminate the electrons, the complex redox reaction (equation 6.3) may be recovered from equation 6.8 from its combination with the reaction:

$$H_2(g) \rightleftharpoons 2H^+ + 2e^- \tag{6.9}$$

The above reaction equation should be noted, since it plays a central role in the thermodynamic characterisation of redox reactions. It can be combined with equation 6.2 to give the equation for the reduction of oxygen, i.e.

$$O_2(g) + 4H^+ + 4e^- \rightleftharpoons 2H_2O \tag{6.10}$$

Redox reactions are not limited to solids. Dissolved ions commonly undergo redox reactions. For example, Fe^{2+} can be oxidised to Fe^{3+} via the reaction:

$$Fe^{2+} \rightleftharpoons Fe^{3+} + e^- \tag{6.11}$$

The full redox reaction can be obtained by combining the above equation with equation 6.10, yielding:

$$O_2(g) + 4Fe^{2+} + 4H^+ \rightleftharpoons 4Fe^{3+} + 2H_2O \qquad (6.12)$$

In describing redox reactions by means of combinations of half-cell reactions we eliminate reference to molecular oxygen but retain aqueous electrons and protons as variables. The half-cell approach is sometimes convenient since many redox reactions do not involve molecular oxygen directly and, at low temperatures, reactions with oxygen may be kinetically slow. At elevated temperatures and pressures reactions occur faster, but reactions involving electrons are less well understood and oxygen may be the preferred variable.

Some common half-cell reactions which can be combined to give overall redox reactions as listed below:

$$\tfrac{1}{4}O_2(g) + H^+ + e^- \rightleftharpoons \tfrac{1}{2}H_2O$$

$$\tfrac{1}{2}Mn_3O_4(s) + 4H^+ + e^- \rightleftharpoons \tfrac{3}{2}Mn^{2+} + 2H_2O$$

$$\tfrac{1}{2}Mn_2O_3(s) + 3H^+ + e^- \rightleftharpoons Mn^{2+} + \tfrac{3}{2}H_2O$$

$$Fe(OH)_3(s) + 3H^+ + e^- \rightleftharpoons Fe^{2+} + 3H_2O$$

$$\tfrac{1}{2}Fe_3O_4(s) + 4H^+ + e^- \rightleftharpoons \tfrac{3}{2}Fe^{2+} + 2H_2O$$

$$\tfrac{1}{2}Fe_2O_3(s) + H^+ + CO_2 + e^- \rightleftharpoons FeCO_3(s) + \tfrac{1}{2}H_2O$$

$$\tfrac{1}{2}Fe_3O_4(s) + H^+ + \tfrac{3}{2}CO_2 + e^- \rightleftharpoons \tfrac{3}{2}FeCO_3(s) + \tfrac{1}{2}H_2O$$

$$\tfrac{1}{8}CO_2(g) + H^+ + e^- \rightleftharpoons \tfrac{1}{8}CH_4(g) + \tfrac{1}{4}H_2O$$

$$\tfrac{1}{8}SO_4^{2-} + H^+ + e^- \rightleftharpoons \tfrac{1}{8}S^{2-} + \tfrac{1}{2}H_2O$$

$$\tfrac{1}{8}SO_4^{2-} + \tfrac{9}{8}H^+ + e^- \rightleftharpoons \tfrac{1}{8}HS^- + \tfrac{1}{2}H_2O$$

$$\tfrac{1}{8}SO_4^{2-} + \tfrac{5}{4}H^+ + e^- \rightleftharpoons \tfrac{1}{8}H_2S(g) + \tfrac{1}{2}H_2O$$

$$\tfrac{1}{8}HSO_4^- + \tfrac{9}{8}H^+ + e^- \rightleftharpoons \tfrac{1}{8}H_2S + \tfrac{1}{2}H_2O$$

$$\tfrac{1}{2}S(s) + e^- \rightleftharpoons \tfrac{1}{2}S^{2-}$$

$$\tfrac{1}{2}S(s) + H^+ + e^- \rightleftharpoons \tfrac{1}{2}H_2S$$

$$\tfrac{1}{2}S(s) + \tfrac{1}{2}H^+ + e^- \rightleftharpoons \tfrac{1}{2}HS^-$$

$$\tfrac{1}{2}S^{2-} + e^- \rightleftharpoons \tfrac{1}{2}S(s)$$

$$\tfrac{1}{6}HSO_4^- + \tfrac{7}{6}H^+ + e^- \rightleftharpoons \tfrac{1}{6}S(s) + \tfrac{2}{3}H_2O$$

$$\tfrac{1}{6}SO_4^{2-} + \tfrac{4}{3}H^+ + e^- \rightleftharpoons \tfrac{1}{6}S(s) + \tfrac{2}{3}H_2O$$

$$\tfrac{1}{8}Fe_3O_4(s) + \tfrac{3}{16}S_2(g) + H^+ + e^- \rightleftharpoons \tfrac{3}{8}FeS(s) + \tfrac{1}{2}H_2O$$

$$\tfrac{1}{6}Fe_2O_3(s) + \tfrac{1}{6}S_2(g) + H^+ + e^- \rightleftharpoons \tfrac{1}{3}FeS(s) + \tfrac{1}{2}H_2O$$

$$\tfrac{1}{8}Fe_3O_4(s) + \tfrac{3}{8}S_2(g) + H^+ + e^- \rightleftharpoons \tfrac{3}{8}FeS_2(s) + \tfrac{1}{2}H_2O$$

$$\tfrac{1}{6}Fe_2O_3(s) + \tfrac{1}{3}S_2(g) + H^+ + e^- \rightleftharpoons \tfrac{1}{3}FeS_2(s) + \tfrac{1}{2}H_2O$$

6.1.2 Rules for constructing and using half-cell reactions

In some cases it is necessary to deduce the form of the half-cell reactions prior to determining the form of the overall redox reaction. The rules for deducing and combining half-cell reactions are as follows.

1. Construct an unbalanced skeleton equation containing all ionic and molecular reactants except water, H^+ and OH^-.
2. Identify one skeletal half-cell reaction for each atom or molecule undergoing change.
3. Balance all atoms in the skeletal half-cell reactions by adding the required number of water molecules, protons or hydroxides.
4. Balance the residual charge on the skeletal half-cell with the required number of electrons. This produces the full half-cell reactions.
5. Add the appropriate number of half-cell reactions together to eliminate the electrons. This gives the final fully charge- and mass-balanced redox reaction.

As an example we will consider the oxidation of sodium sulphide by potassium permanganate in a basic medium.

- **Step 1**: The unbalanced skeletal equation is:

$$2Na^+ + S^{2-} + K^+ + MnO_4^- \rightarrow S(s) + MnO_2(s) + 2Na^+ + K^+ \quad (6.13)$$

- **Step 2**: The half-cell reactions are:

$$S^{2-} \rightarrow S(s) \quad (6.14)$$

and

$$MnO_4^- \rightarrow MnO_2(s) \quad (6.15)$$

- **Step 3**: To mass-balance the half-cell reactions, first note that the sulphide reaction is already mass-balanced in this case. For the manganese reaction we add two water molecules to the right-hand side of the reaction to mass-balance the oxygens. This dictates that we must add four protons to the left-hand side of the resulting reaction:

$$S^{2-} \rightarrow S(s) \quad (6.16)$$

$$MnO_4^- + 4H^+ \rightarrow MnO_2(s) + 2H_2O \quad (6.17)$$

Noting that the reaction is in a basic solution we can further modify the manganese reaction by eliminating protons using the reaction $H_2O \rightarrow 2H^+ + OH^-$ to give:

$$MnO_4^- + 2H_2O \rightarrow MnO_2(s) + 4OH^- \quad (6.18)$$

- **Step 4**: Balancing the charge on the resulting half-cell reactions gives:

$$MnO_4^- + 2H_2O + 3e^- \rightarrow MnO_2(s) + 4OH^- \quad (6.19)$$

and

$$S^{2-} \rightarrow S(s) + 2e^- \quad (6.20)$$

● **Step 5**: Combining suitable multiples of the complete half-cell reactions to eliminate protons gives:

$$3S^{2-} + 2MnO_4^- + 4H_2O \rightarrow 3S(s) + 2MnO_2(s) + 8OH^- \qquad (6.21)$$

An alternative molecular reaction equation can be obtained by replacing the sodium and potassium, i.e.

$$3Na_2S + 2KMnO_4 + 4H_2O \rightarrow 3S(s) + 2MnO_2(s) + 6NaOH + 2KOH \qquad (6.22)$$

Chapter 7

Energy and related properties

Thermodynamics is concerned with the interconversion of heat, work and other forms of energy. In this chapter these terms are defined and their interrelationships explained.

7.1   Force, pressure and work

A **force** is an agent which imparts a constant acceleration to a body if the body is free to move. For a body of given mass:

$$\text{Force} = \text{Mass} \times \text{Acceleration} \tag{7.1}$$

This is Newton's first law of motion. An example is the force imposed on a body due to gravity, i.e.

$$\text{Force} = \text{Mass} \times g \tag{7.2}$$

where g is the acceleration in free fall.

Pressure P is defined as a force per unit area:

$$\text{Pressure} = \frac{\text{Force}}{\text{Area}} \tag{7.3}$$

Work has been done if the point of application of a force moves a finite distance in the direction of a force. The usual symbol for work done is w.

This is the conventional definition of work. However, it is convenient to determine work in terms of the displacement of a body **against** a force. An example is the lifting of a rock of mass M against gravitational force (see figure 7.1). If the body is displaced a distance dx against a force, the amount of work done **on** the body is given by:

$$dw \equiv -F\,dx \tag{7.4}$$

The term F denotes a force which, in this example, is independent of position. A governing equation is then:

$$w = -\int_{x_i}^{x_f} F\,dx \tag{7.5}$$

where x_f and x_i are the final and initial positions respectively. We evaluate the sign of the force by considering its direction relative to the direction of the displacement.

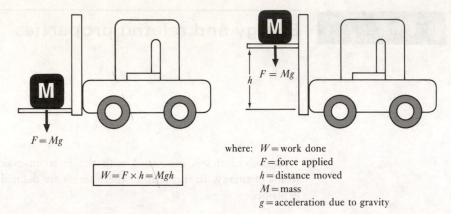

where: W = work done
F = force applied
h = distance moved
M = mass
g = acceleration due to gravity

$$W = F \times h = Mgh$$

Figure 7.1   The movement of a rock against gravity.

Therefore, F is a constant equal to $-M \times g$. With this simplification, equation 7.5 can now be integrated to give:

$$w = M \times g \times (x_f - x_i) \tag{7.6}$$

In this case the movement represents a positive displacement against the gravitational force; that is, $(x_f - x_i)$ has a positive value and work is therefore positive. This is consistent with the adopted sign convention for work, which is that:

- The sign of w is **positive** for work done **on** the system.
- The sign of w is **negative** for work done **by** the system.

The sign convention ensures that w always refers to work done **on** the system. It is essential to be clear about what we are defining as the system. In the gravitational case the system is the body; in figure 7.1 it is the rock.

In situations where the force varies with position x, we write $F(x)$ rather than just F. The general governing equations are then:

$$dw \equiv -F(x)\, dx \tag{7.7}$$

and

$$w = -\int_{x_i}^{x_f} F(x)\, dx \tag{7.8}$$

7.1.1   Work of compression

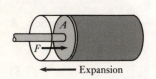

— Expansion

Figure 7.2
The expansion of a gas.

The sign convention becomes clear if equation 7.7 is applied to the expansion of a gas in a cylinder of area A against a constant force F (figure 7.2). We think of the gas as doing work on the piston, and the piston as the system. If we write equation 7.8 in the form:

$$w = -\int_{x_i}^{x_f} \frac{F \times A}{A}\, dx \tag{7.9}$$

this allows the replacement of force and area by pressure and volume since the pressure on the gas is $P_{ext} = -F/A$ and $dV = A \times dx$. Therefore,

$$w = \int_{V_i}^{V_f} P_{ext}\, dV \qquad (7.10)$$

Alternatively, for an infinitesimal change we may write:

$$dw = P_{ext}\, dV \qquad (7.11)$$

If the external pressure P_{ext} is constant, this integrates to give:

$$w = P_{ext}(V_f - V_i) = \text{Work on piston} \qquad (7.12)$$

This is the amount of work done on the piston by the gas. If we now state that the gas is the system, the work done on the gas is the negative of the work on the piston; therefore,

$$w = -P_{ext}(V_f - V_i) = \text{Work on gas} \qquad (7.13)$$

We conclude that **work is negative for the expansion of a gas** since the final volume V_f is greater than the initial volume V_i. In contrast, if the gas is compressed under a constant pressure, the surroundings do work on the system and this work is a positive quantity.

7.2   Temperature

Temperature is understood intuitively in terms of the degree of hotness as judged by the sense of touch. On the basis of touch, a set of bodies at different temperatures can be arranged in a unique series. This suggests that all bodies can be ascribed numbers which between them define a sequence. These numbers form a scale of temperature.

Experimental techniques exist which allow us to measure temperature or relative differences in temperature. This is our only requirement in order to use temperature within thermodynamics. However, temperature has subtle conceptual origins and the interested reader may wish to consider the concept in more detail in box 7.1.

Box 7.1

The zeroth law of thermodynamics

If two bodies, one of which feels hotter than the other, are placed in thermal contact the difference in hotness gradually becomes undetectable. At this point they have reached **thermal equilibrium**.

It is an experimental fact that

IF TWO BODIES ARE IN THERMAL EQUILIBRIUM WITH A THIRD BODY THEY WILL BE IN THERMAL EQUILIBRIUM WITH EACH OTHER:

This statement is called the **zeroth law of thermodynamics**, for which there is no *a priori* reason.

continued

Box 7.1
continued

- A consequence of the zeroth law is that two or more bodies in thermal equilibrium may be said to possess a property in common. This property is the function we call temperature. Temperature is an intensive property since it does not depend on the amount of substance being considered.
- Temperature is the property we use to determine whether or not a system is in thermal equilibrium with a second system. If two systems have the same temperature they are in thermal equilibrium. If they are not in thermal equilibrium their temperatures are different.

The two statements above are deduced from the following arguments. Consider two systems, each in internal equilibrium but not at equilibrium with each other. The systems could be two isolated gases or alternatively a gas-filled vessel and a column of mercury in glass. The thermodynamic state of each may be defined by any suitable properties. We may select pressure and volume as examples and use the symbols P and V for one system and **P** and **V** to represent the other. If the two systems are brought in contact via a diathermal wall, thermal equilibrium will be reached. We may denote the equilibrium pressures and volumes as (P', V') and $(\mathbf{P}', \mathbf{V}')$. Experimentally it will be found that the values of P and V can be adjusted to a second set of values (P'', V'') which are still at equilibrium with the system at $(\mathbf{P}', \mathbf{V}')$. In fact, a whole series of values of (P, V) can be found which retain equilibrium with the system at $(\mathbf{P}', \mathbf{V}')$. This series of values for (P, V) defines a curve called an **isothermal**. The zeroth law of thermodynamics dictates that all points on this curve are at equilibrium with each other since any point on the curve is at equilibrium with the system at $(\mathbf{P}', \mathbf{V}')$.

It appears that each point on an isothermal has a property in common. We deduce this by noting that any such curve, or straight line for that matter, can be represented by an equation of the form:

$$f(P, V) = \theta \tag{7.14}$$

where θ is a constant. Using the same logic it must be true that the system defined by **P**, **V** must have an isothermal which is in equilibrium with all points on the isothermal for the system defined by **P**, **V**. Thus, we may write $F(\mathbf{P}, \mathbf{V}) = $ constant and it must be true that:

$$f(P, V) = \theta = F(\mathbf{P}, \mathbf{V}) \tag{7.15}$$

Figure 7.3 shows these ideas schematically.

This is saying that there exists a function θ of pressure and volume which has a fixed value for all states of a system which are in thermal equilibrium. The function θ is also equal to a function of pressure and volume for a second system which is in equilibrium with the first. The function θ is temperature.

It is an experimental fact that there are states of the first system which are not in equilibrium with the second system when it is in the state (P', V'). An isothermal for such a system would be defined by:

$$f_2(P, V) = \theta_2 = F_2(\mathbf{P}, \mathbf{V}) \tag{7.16}$$

where θ_2 defines a different temperature. Indeed, there will be an infinite number of isothermals with all states on a corresponding curve in equilibrium with each other. Each isothermal defines a specific temperature.

From this point on we will assume that temperature is a fundamental property of all systems.

Box 7.1
continued

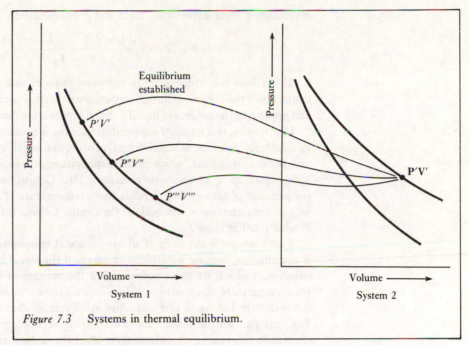

Figure 7.3   Systems in thermal equilibrium.

7.3   Quantitative temperature scales

7.3.1   Simple physical scales

Many physical properties of materials depend on whether the material is in contact with a lot or with a cold body. Examples are the volume of a gas or liquid, the length of a metal rod or the electrical resistance of a piece of wire. These observations allow us to devise convenient temperature scales. We select a property of a substance which changes with degree of hotness and devise a formula which relates the property to the numbers which we assign to the degree of hotness. Such a property is called a **thermometric** property. For example, if the length of a column of liquid is the thermometric property, we must select a relationship which relates the temperature to length. In the absence of any other information we may select a simple linear relationship given by:

$$T = \alpha L + \beta \tag{7.17}$$

where t is temperature, L length and α and β are constants. We must now select two fixed temperatures, assign arbitrary numbers to them, and measure the lengths in both instances. This allows us to set up two simultaneous equations to solve for α and β. That is,

$$T_1 = \alpha L_1 + \beta \quad \text{and} \quad T_2 = \alpha L_2 + \beta \tag{7.18}$$

which solve to give:

$$\alpha = \frac{T_1 - T_2}{L_1 - L_2} \quad \text{and} \quad \beta = \frac{T_2 L_1 - T_1 L_2}{L_1 - L_2} \tag{7.19}$$

Substituting these expressions for α and β into equation 7.17 gives:

$$T = \frac{T_1(L - L_2) - T_2(L - L_1)}{L_1 - L_2}$$

(7.20)

This allows the temperature of any body to be measured by placing the body in contact with the column of liquid, waiting until thermal equilibrium has been reached, and measuring the column of liquid L. Such a measuring device is a **thermometer**.

Historically, two standard temperatures are the **ice point**, which is the temperature at which ice remains in equilibrium with liquid water at 1 atmosphere pressure, and the **steam point**, which is the temperature at which steam is in equilibrium with liquid at 1 atmosphere pressure. The Celsius scale arbitrarily assigns a temperature of zero to the ice point and a temperature of 100 to the ice point. The unit of temperature on this scale is the degree Celsius ($^\circ$C), and is often given the symbol t rather than T.

This example is the basis of all experimental temperature scales. The technique is not absolute because we arbitrarily assumed the linear relationship in the general equation. Indeed, we must not state that the column of liquid expands uniformly with temperature. Such an observation appears to be true because we initially defined the system to behave in this way. Any other relationship could have been selected. For example, a logarithmic relationship such as $T = \alpha \ln L + \beta$ could have been chosen. In this case equal increments in length would not represent equal increments in temperature. This is not a problem; it is simply a feature of the selected temperature scale.

There are other properties of materials which can be used to construct thermometers.

- **The electrical resistance thermometer**: The electrical resistance, R, of a piece of metal wire held under tension is defined to vary linearly with temperature:

$$T = \alpha R + \beta$$

- **The constant-pressure gas thermometer**: The volume, V, of a quantity of gas kept at constant pressure is defined to increase linearly with temperature:

$$T = \alpha V + \beta$$

- **The constant-volume gas thermometer**: The pressure, P, of a quantity of gas kept at constant volume is defined to increase linearly with temperature:

$$T = \alpha P + \beta$$

The thermometers described above can be calibrated using the same reference temperatures. Even so, there is no reason why temperatures measured by any of the above scales should coincide at any points other than at the reference temperatures. In practice, measured temperatures correspond remarkably well, with discrepancies being never more than a few per cent over the range between the ice and steam points. However, differences do occur and can be detected, for example in liquid column thermometers when different liquids are used, and in resistance thermometers when different metals are used. Indeed, discrepancies become worse as the temperature ranges over which the scales are calibrated are extended. It is desirable to establish one scale which is absolute and valid over all conceivable ranges of temperature.

7.3.2   Gas thermometers and the absolute scale

The constant pressure and constant volume gas thermometers give the most consistent temperature scales, particularly when used with the so-called permanent gases such as hydrogen, oxygen, nitrogen or carbon dioxide.

Once the thermometer is suitably calibrated it is possible to plot the dependence of volume on temperature for a gas at fixed pressure, or the dependence of pressure on temperature for a gas at fixed volume. If $V-T$ or $P-T$ plots are constructed for **any gas**, the extrapolated lines all intersect at a point representing zero volume or pressure at a temperature of around $-273\,°C$ (figure 7.4).

This temperature at which gases exert no pressure is called the **absolute zero of temperature**. It is the temperature at which the molecules of a gas have no translational motion and therefore zero thermal energy. The term 'absolute' means simply that this temperature is not an arbitrary constant chosen for convenience, as is the ice point for example.

In 1848 Lord Kelvin proposed a temperature scale with absolute zero as the starting point, now called the **Kelvin temperature scale**. The unit of temperature on this scale is called the kelvin and has the symbol K. In principle, 1 degree Celsius is equal to 1 kelvin. Experimental variations of the temperature of absolute zero vary from -273.14 down to $-273.19\,°C$ with a mean value of $-273.15\,°C$. This error becomes significant when measured temperatures are close to absolute zero but it is less important when temperatures are large. To overcome this, Giauque proposed defining the ice point as $273.15\,K$ relative to absolute zero. This fixes the number of degrees Kelvin in the interval. Consequently, any experimental error is all transferred into the magnitude of the kelvin rather than the position of absolute

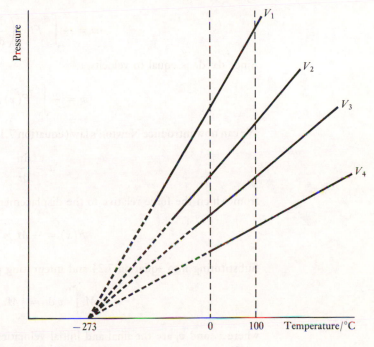

Figure 7.4   The absolute zero of temperature.

zero. Effectively, this may produce a unit of kelvin which is slightly different from the unit of Celsius. The cost is possible inaccuracies at high temperature where the effect is relatively insignificant.

The relationship between °C and K is:

$$T/K = t\,°C + 273.15 \tag{7.21}$$

7.4   Energy – some basic concepts

Energy is the capacity to produce change.

The idea of energy arose first in the field of mechanics. It is commonly understood but requires some elaboration.

7.4.1   Mechanical energy

To give the idea of energy a mathematical basis we can consider the work involved in moving a particle of mass M from position x_i to position x_f **against** a force $F(x)$. The governing equation (7.8) is:

$$w = -\int_{x_i}^{x_f} F(x)\,\mathrm{d}x \tag{7.22}$$

The integral over distance can be transformed to an integral over time t by multiplying by $\mathrm{d}t/\mathrm{d}t$ to give:

$$w = -\int_{x_i}^{x_f} F(x)\left(\frac{\mathrm{d}x}{\mathrm{d}t}\right)\mathrm{d}t \tag{7.23}$$

Since $\mathrm{d}x/\mathrm{d}t$ is equal to velocity v,

$$w = -\int_{x_i}^{x_f} F(x)v\,\mathrm{d}t \tag{7.24}$$

We can now introduce Newton's law (equation 7.1), including acceleration defined as

$$a = \frac{\mathrm{d}v}{\mathrm{d}t} \tag{7.25}$$

from which the force relative to the displacement is:

$$F(x) = -M \times \frac{\mathrm{d}v}{\mathrm{d}t} \tag{7.26}$$

Substituting into equation 7.24 and integrating gives:

$$w = M\int_{v_i}^{v_f} v\,\mathrm{d}t = \tfrac{1}{2}Mv_f^2 - \tfrac{1}{2}Mv_i^2 \tag{7.27}$$

where v_f and v_i are the final and initial velocities, respectively.

We see that two terms have arisen, $\tfrac{1}{2}Mv_f^2$ and $\tfrac{1}{2}Mv_i^2$, which are dependent on M

and v. They are properties of the system only and are therefore state functions. We can define the kinetic energy $E(k)$ as:

$$E(k) = \tfrac{1}{2}Mv^2 \tag{7.28}$$

The kinetic energy is therefore a state function.

7.4.2   Some general comments on energy

Energy is recognised by its effects and can only be observed when changes in energy occur. There are several forms of energy, all of which are state functions. These are summarised as follows.

- **Kinetic energy** is the energy of a body due to its motion. The amount of work necessary to bring the body to rest is the value of its kinetic energy.
- **Thermal energy** is associated with the random motion of atoms and molecules and is a force of kinetic energy. The greater the velocity of the elementary particles in a substance, the greater is the thermal energy. Thermal energy is independent of any translational movement of the substance as a whole.
- **Potential energy** is the energy possessed by a body by virtue of its position. In measuring potential energy it is necessary first to select a reference position in which the body is defined as having zero potential energy. The amount of work necessary to move the body to a second position is a measure of its potential energy. Potential energy is clearly a relative quantity. A good example is the movement of a mass against gravity, where the state of rest is conveniently defined as the state of zero potential energy.
- **Chemical energy** is a form of energy stored within the structural units of a body. Its quantity is dependent on the type of atoms and the arrangement of atoms in the substance. When substances react chemically, this form of energy can be released or stored as other forms of energy. In this sense chemical energy is a form of potential energy.
- **Mechanical energy** is associated with the change in physical dimensions of a body or its movement against a force. Mechanical energy is always transferred from a system of higher pressure to one of lower pressure.
- **Internal energy** is the sum of the total kinetic and potential energy of the atoms or molecules in a given substance.

In principle, all forms of energy can be converted from one form to another. The law of conservation of energy states that when energy transformations occur the total quantity of energy in the Universe remains constant.

7.5   Heat and heat capacity

7.5.1   Heat

Heat is the transfer of energy between two bodies which are at different temperatures. The usual symbol for quantity of heat transferred is q. Heat will flow from a hot body to a colder body, i.e. along a temperature gradient. This phenomenon is central to the second law of thermodynamics and is described in more depth later.

The sign convention we use for heat is:

- The sign of q is **positive** if the system absorbs heat from the surroundings. This is an **endothermic** process.
- The sign of q is **negative** if the surroundings absorb heat from the system. This is an **exothermic** process.

Heat is not a state function since we do not think of a body as possessing a unique quantity of heat. Therefore we cannot write $\Delta q = q_f - q_i$, where subscripts f and i denote a system in its final and initial state. The properties of heat and temperature are intimately related. Indeed, in the absence of heat transfer the concept of temperature breaks down.

7.5.2 Heat capacity

The fact that two or more substances have the same temperature does not mean they are thermally identical. It is an experimental fact that equal masses of **different** substances may require different amounts of heat to raise their respective temperatures by the same amount.

For an infinitesimal transfer of heat the rise in temperature is proportional to the heat supplied, i.e.

$$dq = [\text{Constant}] \times dT \tag{7.29}$$

This constant is called the heat capacity of the system. A substance with a high heat capacity will require the transfer of more heat to raise its temperature by a given amount than a substance with a small heat capacity.

By convention the **heat capacity** of a system is the number of joules needed to raise the temperature of a body by 1 K.

The **specific heat** of a substance is the number of joules required to raise the temperature of 1 gram of a substance by 1 K.

For pure substances the **molar heat capacity** is the heat capacity of 1 mole of substance.

Heat capacities are usually measured at either constant pressure, C_P, or constant volume, C_V. The heat capacity at constant volume is usually smaller than that at constant pressure. The difference arises because at constant pressure a system may expand upon heating, using some of the energy doing work on the surroundings, whereas at constant volume all the energy transferred produces a rise in temperature.

For many materials of chemical or geochemical interest the molar heat capacity rises steadily with temperature, although discontinuities are observed if phase transitions occur. Typically the molar heat capacity of a solid will double over the temperature range 0–1500 °C. For liquid water the molar heat capacity shows a complex dependence on both temperature and pressure. The molar heat capacities of substances dissolved in liquid water exhibit a complex dependence on temperature, pressure and the composition of the aqueous phase.

Chapter 8 — The principles of chemical thermodynamics

8.1 The mathematical characterisation of state functions

THE INTEGRATION OF A STATE FUNCTION BETWEEN ANY TWO BOUNDARY CONDITIONS IS INDEPENDENT OF THE INTEGRATION PATH.

This property distinguishes a state function from a path function. The value obtained by integrating a path function between its boundary conditions depends on the integration pathway.

Consider a system changing from an initial state (denoted by i), to a final state (denoted by f). For any state function X, the change in X in going from X_i to X_f (denoted ΔX) is given by:

$$\Delta X = X_f - X_i = \int_{X_i}^{X_f} \mathrm{d}X \tag{8.1}$$

For this type of relationship we say that the differential $\mathrm{d}X$ is **exact**. If the function X was energy and this relationship did not hold we could devise a sequence of changes which released more energy from the system than was put in. This desirable situation is unfortunately impossible.

To make the nature of path functions clear we can consider a gas taken from an initial state, characterised by temperature T_i, volume V_i and pressure P_i, to a state characterised by T_f, V_f and P_f. One way of performing such a change is adiabatically, where there is no heat flow since the system is thermally isolated from its surroundings. here, the work done w would be finite but the heat transformed q would be zero.

A second type of change is non-adiabatic, where heat may be lost or gained from the surroundings. For this type of change both w and q are finite. Both w and q are, of course, different from those for the adiabatic change even though the initial and final states of the system are the same. Thus, heat and work do not have exact differential. In contrast, any property X which is dependent only on the instantaneous state of the system and not on how that state was reached can be described by an exact differential.

Box 8.1   **Partial derivatives and exact differentials**

Exact differentials have some useful properties which are derived from the following basic theorem.

Suppose an **extensive** property u (as defined in section 2.1.1) is a function of the variables $x, y, z \ldots$ etc. For the time being we will consider only two variables x and y. If these are changed by the infinitesimal amounts dx and dy then the dependent variable u will change from $u(x, y)$ to $u(x + dx, y + dy)$. The modified value of u can be obtained from:

$$u(x + dx, y + dy) = u(x, y) + \left(\frac{\partial u}{\partial y}\right)_x dy + \left(\frac{\partial u}{\partial x}\right)_y dx \qquad (8.2)$$

Alternatively,

$$du = \left(\frac{\partial u}{\partial y}\right)_x dy + \left(\frac{\partial u}{\partial x}\right)_y dx \qquad (8.3)$$

where the symbol $(\partial u/\partial x)_y$ is the **partial derivative of u with respect to x at fixed** y. The partial derivatives play the same role as constants of proportionality, although in many cases they are not constant. The partial derivative is obtained by differentiating the function $u(x, y)$ with respect to x, treating y as a constant. The term **total differential of u** is given to du in equation 8.3. The equation may be extended if u is a function of n variables x_i where $i = 1-n$, in which case

$$du = \sum_{i=1}^{n} \left(\frac{\partial u}{\partial x_i}\right)_{x_j(j \neq i)} dx_i \qquad (8.4)$$

Each partial differentiation must be performed with **all other** variables held constant. This condition must be performed with **all other** variables held constant. This condition is denoted by the subscript $x_{j(j \neq i)}$ outside the brackets.

Some important rules of partial differentiation are as follows.

The chain rule

Provided the **same** variable is held constant in each derivative, partial derivatives may be multiplied together thus:

$$\left(\frac{\partial u}{\partial x}\right)_y = \left(\frac{\partial u}{\partial z}\right)_y \left(\frac{\partial z}{\partial x}\right)_y \qquad (8.5)$$

The commutative rule

Provided the function u is continuous the partial derivatives commute. This means that differentiation may be performed in any order, i.e.

$$\left[\frac{\partial}{\partial x}\left(\frac{\partial u}{\partial y}\right)_x\right]_y = \left[\frac{\partial}{\partial y}\left(\frac{\partial u}{\partial x}\right)_y\right]_x \qquad (8.6)$$

The inverter

Inverse relationships can be obtained from:

$$\left(\frac{\partial u}{\partial x}\right)_y = 1 \bigg/ \left(\frac{\partial x}{\partial u}\right)_y \qquad (8.7)$$

continued

Box 8.1
continued

The permuter

If the function $u(x, y)$ is continuous we assume that relationships exist for x as a function of y and u, and y as a function of x and u. We can define the full interrelationships by considering a line of constant u, where $du = 0$. If, in the basic theorem (equation 8.3), we set du to zero and divide by dx, then:

$$\left(\frac{\partial u}{\partial x}\right)_y = -\left(\frac{\partial u}{\partial y}\right)_x \left(\frac{\partial y}{\partial x}\right)_u \tag{8.8}$$

Euler's chain rule

By combining the permuter and the inverter we get the relationship:

$$\left(\frac{\partial u}{\partial x}\right)_y \left(\frac{\partial x}{\partial y}\right)_u \left(\frac{\partial y}{\partial u}\right)_x = -1 \tag{8.9}$$

This is sometimes called the -1 rule.

Changing the variable held constant

If we know that u is indeed dependent on a third variable z then, for the function to be continuous, z must be dependent on x and y. For example, many thermodynamic variables are dependent on volume and temperature, and it is known that volume is dependent on temperature and pressure. The rule to change the variables held constant from x and y to x and z is:

$$\left(\frac{\partial u}{\partial x}\right)_z = \left(\frac{\partial u}{\partial x}\right)_y + \left(\frac{\partial u}{\partial y}\right)_x \left(\frac{\partial y}{\partial x}\right)_z \tag{8.10}$$

Taylor's theorem

This is a useful general theorem for evaluating a complicated function in terms of a series of simple functions. It is used to ease computation or to make mathematical simplifications. If a function $f(x)$ is to be evaluated at a neighbouring point $f(x + p)$ where p is a small increment, then:

$$f(x + p) = f(x) + \frac{df(x)}{dx}p + \frac{d^2f(x)}{dx^2}\left(\frac{p^2}{2}\right) + \frac{d^3f(x)}{dx^3}\left(\frac{p^3}{3!}\right) \cdots \tag{8.11}$$

Taylor's theorem may be generalised to a function dependent on many variables, i.e. $f(x, y, \ldots)$ by considering terms of the same form as the right-hand side of equation 8.11 for each additional variable.

Maclaurin's expansion

This is a special case of Taylor's theorem which may be used to expand a function of a **single** variable into a series:

$$f(x) = f(0) + \frac{df(0)}{dx}x + \frac{d^2f(0)}{dx^2}\left(\frac{x^2}{2!}\right) + \frac{d^3f(0)}{dx^3}\left(\frac{x^3}{3!}\right) + \cdots + \frac{d^nf(0)}{dx^n}\left(\frac{x^n}{n!}\right) \tag{8.12}$$

The term $f(0)$ represents the function evaluated at $x = 0$. The expansion is valid when $f(x)$ and its derivatives are finite at $x = 0$.

continued

Box 8.1
continued

The exact differential

If any differential du is exact, its integration between defined states is independent of the integration path. Figure 8.1 shows a situation where two variables x and y can be changed from a position (x, y) to the position $(x + dx, y + dy)$. There are two possible paths by which this change can occur. Each path has two stages. The first path is via the point $(x, y + dy)$ whereas the second path is via the point $(x + dx, y)$. For the first path the change in u (Δu) is:

$$\Delta u_{(1)} = \int_{u(x, y)}^{u(x, y + dy)} du + \int_{u(x, y + dy)}^{u(x + dx, y + dy)} du \qquad (8.13)$$

For the second path:

$$\Delta u_{(2)} = \int_{u(x, y)}^{u(x + dx, y)} du + \int_{u(x + dx, y)}^{u(x + dx, y + dy)} du \qquad (8.14)$$

We are concerned with the condition where $\Delta u_{(1)} = \Delta u_{(2)}$. This required condition for exactness can be deduced by considering the basic equation 8.3. The details will not be presented here and we will simply explain the result. We start by simplifying equation 8.3 by substituting $P = (\partial u / \partial x)_y$ and $Q = (\partial u / \partial y)_x$ to give:

$$du = P(x, y) \, dx + Q(x, y) \, dy \qquad (8.15)$$

By integrating the above equation between $u(x, y)$ and $u(x + dx, y + dy)$ with the help of Taylor's theorem, and comparing the resulting equations, it can be shown that for both integration paths to be equivalent the following condition must hold.

$$\left[\frac{\partial P(x, y)}{\partial y} \right]_x = \left[\frac{\partial Q(x, y)}{\partial x} \right]_y \qquad (8.16)$$

That is, in addition to being a continuous function between the limits **the commutative rule must hold**. If, for any real function u, this can be proved, then the function $P(x, y) \, dx + Q(x, y) \, dy$ is exact and du may be integrated without specifying the path. However, if the function $P(x, y) \, dx + Q(x, y) \, dy$ is not exact, it is always possible to find a factor with which to multiply both $Q(x, y)$ and $P(x, y)$ in order to make the resulting function exact. That is, a function $\mu(x, y)$ will exist which

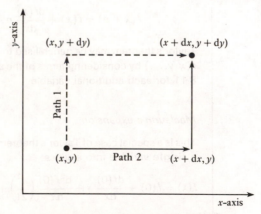

Figure 8.1 Alternative integration paths.

continued

Box 8.1
continued

ensures that the following relationship is true.

$$\left\{\frac{\partial[\mu(x,y)P(x,y)]}{\partial y}\right\}_x = \left\{\frac{\partial[\mu(x,y)Q(x,y)]}{\partial x}\right\}_y \tag{8.17}$$

Such a factor is called an **integrating factor**. For example, the equation

$$(y - x^2y)\frac{dy}{dx} + (x - xy^2) = 0 \tag{8.18}$$

is exact since $P(x, y) = x - xy^2$, $Q(x, y) = (y - x^2y)$ and

$$\left[\frac{\partial P(x,y)}{\partial y}\right]_x = \left[\frac{\partial Q(x,y)}{\partial x}\right]_y = -2xy.$$

However, the equation

$$2y \ln x \frac{dy}{dx} + y = 0 \tag{8.19}$$

is not exact since $P(x, y) = y$, $Q(x, y) = 2x \ln x$ and

$$\left[\frac{\partial P(x,y)}{\partial y}\right]_x \neq \left[\frac{\partial Q(x,y)}{\partial x}\right]_y \tag{8.20}$$

Yet, by multiplying through by y/x, the resulting equation

$$2y \ln x \frac{dy}{dx} + \frac{y^2}{x} = 0 \tag{8.21}$$

is exact since $P(x, y) = (y^2)/x$, $Q(x, y) = 2y \ln x$, and

$$\left[\frac{\partial P(x,y)}{\partial y}\right]_x = \left[\frac{\partial Q(x,y)}{\partial x}\right]_y = \frac{2y}{x} \tag{8.22}$$

For real functions, integrating factors may be difficult to find. Indeed, a major step in the development of thermodynamic theory was concerned with the determination of an integrating factor for work, which as we shall see cannot have an exact differential.

8.1.1 The mathematical test for exactness

We can show that work is not a state function by evaluating the properties of 1 mole of an ideal gas using $P\bar{V} = RT$, where $\bar{V}$ is the molar volume.

If $\bar{V}$ is a state function it has an exact differential with respect to temperature and pressure and the commutation rule must be satisfied, i.e.

$$\left[\frac{\partial}{\partial P}\left(\frac{\partial \bar{V}}{\partial T}\right)_P\right]_T = \left[\frac{\partial}{\partial T}\left(\frac{\partial \bar{V}}{\partial P}\right)_T\right]_P \tag{8.23}$$

The first derivative of the ideal gas law with respect to temperature and pressure gives:

$$\left(\frac{\partial \bar{V}}{\partial P}\right)_T = -\frac{RT}{P^2} \tag{8.24}$$

and

$$\left(\frac{\partial \bar{V}}{\partial T}\right)_P = \frac{R}{P} \tag{8.25}$$

Taking the second derivatives of the ideal gas law gives:

$$\left[\frac{\partial}{\partial T}\left(\frac{\partial \bar{V}}{\partial P}\right)_T\right]_P = \left[\frac{\partial}{\partial T}\left(\frac{-RT}{P^2}\right)\right]_P = -\frac{R}{P^2} \tag{8.26}$$

and

$$\left[\frac{\partial}{\partial P}\left(\frac{\partial \bar{V}}{\partial T}\right)_P\right]_T = \left[\frac{\partial}{\partial P}\left(\frac{R}{P}\right)\right]_T = -\frac{R}{P^2} \tag{8.27}$$

Both second derivatives are equal, and therefore $\bar{V}$ has an exact differential.

To evaluate the status of work we use the relationship:

$$dw = -P\,dV \tag{8.28}$$

The first step is to replace dV with terms in temperature and pressure using the total derivative for volume which is:

$$dV = \left(\frac{\partial V}{\partial P}\right)_T dP + \left(\frac{\partial V}{\partial T}\right)_P dT \tag{8.29}$$

Substituting equation 8.29 into equation 8.28 gives:

$$dw = -P\left[\left(\frac{\partial \bar{V}}{\partial P}\right)_T dP + \left(\frac{\partial \bar{V}}{\partial T}\right)_P dT\right] \tag{8.30}$$

Substituting terms from equations 8.24 and 8.25 gives the analogue of equation 8.29, i.e.

$$dw = -P\left[\left(\frac{RT}{P^2}\right)dP + \left(\frac{R}{P}\right)dT\right] \tag{8.31}$$

From this we conclude that:

$$\left(\frac{\partial w}{\partial T}\right)_P = -R \quad \text{and} \quad \left(\frac{\partial w}{\partial P}\right)_T = \frac{RT}{P} \tag{8.32}$$

These are the work analogues of equations 8.24 and 8.25. The test for exactness is that:

$$\left[\frac{\partial}{\partial P}\left(\frac{\partial w}{\partial T}\right)_P\right]_T = \left[\frac{\partial(-R)}{\partial P}\right]_T = 0 \tag{8.33}$$

must be equal to

$$\left[\frac{\partial}{\partial T}\left(\frac{\partial w}{\partial P}\right)_T\right]_P = \left[\frac{\partial(RT/P)}{\partial T}\right]_P = \frac{R}{P} \tag{8.34}$$

Clearly, the commutative rule does not hold; therefore dw is not an exact differential and is not a state function.

8.2   The first law of thermodynamics

We may now define the first law of thermodynamics. It is a statement of the conservation of energy which is written as:

ENERGY CAN BE CONVERTED FROM ONE FORM TO ANOTHER BUT CANNOT BE CREATED OR DESTROYED.

8.2.1   Internal energy

The first law of thermodynamics has several alternative mathematical forms. The most general definition for a closed system (constant mass) takes the form:

$$\oint dU = 0 \qquad (8.35)$$

where U is the total energy of a system. This states that for any **cyclic process** the energy change is zero. It is equivalent to saying that if a system is taken from state I to state II and back to state I there is no net change in energy, i.e. energy is conserved in the process. Equation 8.35 reaffirms that U is a state function and that dU is exact.

We can consider a system transferring a quantity of energy to its surroundings. In this case the first law becomes:

$$\Delta U_{sys} + \Delta U_{surr} = 0 \qquad (8.36)$$

where Δ denotes a change of measurable proportions and U_{sys} and U_{surr} are the **internal energies** of the system and surroundings respectively.

In chemical systems it is common to think of energy changes in terms of the system. Therefore, a useful alternative form of the first law is:

$$\Delta U_{sys} = q + w \qquad (8.37)$$

where q is the heat transferred to the system and w is the work done on the system.

Equation 8.37 says that any work done on the system or heat transferred to the system must appear as a change in the internal energy.

For convenience we drop the subscript sys with the understanding that U denotes the internal energy of the system unless otherwise specified.

A final form of the first law comes from equation 8.37 written for an infinitesimal change in U:

$$dU = đq + đw \qquad (8.38)$$

The symbols $đq$ and $đw$ are used to emphasise that dq **and** dw **are not exact differentials**. Some authors avoid using $đq$ and $đw$ by writing $dU = q + w$. Here, q and w need not even be small, provided that sum is infinitesimal. Other authors retain the use of dq and dw, stressing that they represent infinitesimal changes, not exact differentials. The alternative conventions are equivalent and the important feature is that we always avoid writing Δq and Δw.

In the case when the system is adiabatically enclosed there can be no heat exchange

(q or $\text{\dj}q = 0$). Therefore,

$$\Delta U = w \quad \text{and} \quad dU = \text{\dj}w \tag{8.39}$$

In the case when the system is diathermally bounded and mechanically isolated the system can do no work (w or $\text{\dj}w = 0$). Therefore,

$$\Delta U = q_V \quad \text{and} \quad dU = \text{\dj}q_V \tag{8.40}$$

where q_V represents heat transferred at constant volume.

8.2.2 Enthalpy

Two simple limiting conditions under which chemical reactions can occur are **constant pressure** and **constant volume**. Reactions at constant pressure are often encountered. Reactions at constant volume are rare in natural systems and difficult to construct experimentally.

In view of the importance of reactions at constant pressure, it is convenient to have a state function which is a measure of the heat transferred to a system at constant pressure. The state function created for this purpose is called **enthalpy** and is given the symbol H.

Defining enthalpy   For changes at constant pressure there may be an associated change in volume ΔV. We recall equation 7.11, which relates the work done on the system to the volume change in response to an external pressure P. From this relationship, the first law can be written:

$$dU = \text{\dj}q_P - P\,dV \tag{8.41}$$

where $\text{\dj}q_P$ denotes an infinitesimal transfer of heat a constant pressure. The form of equation 8.41 suggests that we may construct our required state function by simply adding PV to the energy U, i.e.

$$H = U + PV \tag{8.42}$$

This is a perfectly reasonable operation since U, P and V are all state properties of the system. In view of this, enthalpy is also a state function and has the same units as energy.

The differential of the enthalpy may be evaluated from:

$$H + dH = (U + dU) + (P + dP)(V + dV)$$
$$= U + dU + PV + P\,dV + V\,dP + dP\,dV \tag{8.43}$$

This can be simplified since the term $dP\,dV$ is negligible and the term $U + PV$ on the right-hand side can be replaced with the new function H. Therefore,

$$dH = dU + P\,dV + V\,dP \tag{8.44}$$

We are interested in the condition of constant pressure; therefore $V\,dP = 0$. By comparison of equations 8.44 and 8.41 it is clear that this condition leads to:

$$dH = \text{\dj}q_P \tag{8.45}$$

Alternatively, we may write for ΔH:

$$\Delta H = q_P \qquad (8.46)$$

This is derived by integrating equation 8.44 between the final and initial states (f and i) as follows:

$$\Delta H = \int_{H_i}^{H_f} dH = \int_{U_i}^{U_f} dU + \int_{V_i}^{V_f} P\,dV + \int_{P_i}^{P_f} V\,dP \qquad (8.47)$$

from which

$$\Delta H = \Delta U + \int_{V_i}^{V_f} P\,dV + \int_{P_i}^{P_f} V\,dP \qquad (8.48)$$

Equation 8.37 can be rewritten as:

$$\Delta U = q - \int_{V_i}^{V_f} P\,dV \qquad (8.49)$$

Equations 8.48 and 8.49 can now be combined to give the general expression for ΔH, i.e.

$$\Delta H = q + \int_{P_i}^{P_f} V\,dP \qquad (8.50)$$

Imposing our restriction of constant pressure gives the expected relationship between ΔH and q:

$$\Delta H = q_P \qquad (8.51)$$

The difference between **The change in the enthalpy of a system is equal to the heat transferred to**
ΔH and ΔU **the system at constant pressure.**

This statement is confirmed by equation 8.51. We can make a similar statement about internal energy.

The change in the internal energy of a system is equal to the heat transferred to the system at constant volume.

The value of ΔU differs from ΔH only by the PV products of the final and initial states. This can be deduced from equation 8.42 because:

$$\Delta H = \Delta U + \Delta(PV) \quad \text{or} \quad \Delta H = \Delta U + (P_f V_f) - (P_i V_i) \qquad (8.52)$$

For reactions between solids and liquids at constant pressure, $\Delta(PV)$ is close to zero; therefore $\Delta H \sim \Delta U$. However, for gaseous reactions $\Delta(PV)$ may not be negligible. This can be seen by applying the equation of state for ideal gases to a reaction of the form:

$$v_A A + v_B B \rightleftharpoons v_C C + v_D D \qquad (8.53)$$

Since $PV = v_i RT$ for any product or reactant,

$$\Delta(PV) = (v_C + v_D - v_A - v_B)RT \qquad (8.54)$$

from which:

$$\Delta H = \Delta U + (v_C + v_D - v_A - v_B)RT \qquad (8.55)$$

The value of the gas constant R is $8.314 \, \mathrm{J \, mol^{-1} K^{-1}}$. Therefore, even at room temperatures where $T \sim 300 \, \mathrm{K}$, the total PV term may be large compared with the value of ΔH.

8.2.3   Heat capacity

The two types of heat capacity C_V and C_P introduced in section 7.5.2 can be related to the state functions U and H respectively. The relationships are obtained since the heat capacity of a body may be defined as:

$$C = \frac{\mathrm{d}q}{\mathrm{d}T} \tag{8.56}$$

If heat capacity is measured at constant volume $\mathrm{d}q_V = \mathrm{d}U$ (see equation 8.40), so equation 8.56 becomes:

$$C_V = \frac{\mathrm{d}q_V}{\mathrm{d}T} = \left(\frac{\partial U}{\partial T}\right)_V \tag{8.57}$$

Similarly, at constant pressure $\mathrm{d}q_P = \mathrm{d}H$ (see equation 8.45), whereby

$$C_P = \frac{\mathrm{d}q_P}{\mathrm{d}T} = \left(\frac{\partial H}{\partial T}\right)_P \tag{8.58}$$

The difference between C_V and C_P for an ideal gas

The value of C_P is always larger than C_V since some of the transferred heat at constant pressure is converted to the work necessary to change the volume of the system. For relatively incompressible substances, such as minerals and liquids, C_P may be up to 30% larger than C_V. However, for gases the difference may be greater than this. For an ideal gas the relationship between C_V and C_P is derived simply by using equation 8.42 ($H = U + PV$), from which:

$$\frac{\mathrm{d}H}{\mathrm{d}T} = \frac{\mathrm{d}U}{\mathrm{d}T} + \frac{\mathrm{d}(PV)}{\mathrm{d}T} \tag{8.59}$$

We have used the total derivative because H, U and PV are functions of T only. However, from the definitions of C_P and C_V we note that:

$$\frac{\mathrm{d}H}{\mathrm{d}T} = \left(\frac{\partial H}{\partial T}\right)_P = C_P \quad \text{and} \quad \frac{\mathrm{d}U}{\mathrm{d}T} = \left(\frac{\partial U}{\partial T}\right)_V = C_V \tag{8.60}$$

For 1 mole of an ideal gas, where $P\bar{V} = RT$, equation 8.59 becomes:

$$C_P = C_V + \frac{\mathrm{d}(RT)}{\mathrm{d}T} \tag{8.61}$$

from which:

$$C_P = C_V + R \tag{8.62}$$

The ratio of heat capacities is often denoted by $\gamma = C_P / C_V$.

Since most gases have a C_V of between 10 and $40 \, \mathrm{J \, mol^{-1} K^{-1}}$, and R is $8.314 \, \mathrm{J \, mol^{-1} K^{-1}}$, the difference between C_P and C_V is significant. A rigorous and general relationship between C_V and C_P is given in section 9.2.7.

8.3 Mechanical equilibrium and reversibility

The goal of chemical thermodynamics is to generate a set of properties which help us predict the direction of change in any process. This is done via the second law of thermodynamics. However, to introduce the second law properly it is necessary to consider first the types of changes that can occur. Of particular importance is the **reversible** change.

8.3.1 Mechanical equilibrium

When a macroscopic mechanical system shows no tendency to change the spatial arrangement of its components, the system is said to have reached equilibrium. In this state all forces acting on the system are balanced. A commonly used example of mechanical equilibrium is that of two identical weights attached to each other via a **frictionless** pulley, as in figure 8.2. Here, equilibrium is established because of the balance between equal gravitational forces acting on the equal weights.

We will show later that the concept of mechanical equilibrium is not entirely adequate for a chemical equilibrium.

8.3.2 Reversibility

The pulley system in figure 8.2 can be harnessed to do work if the forces are no longer balanced. This is achieved by ensuring that the mass of one weight (M_1) is

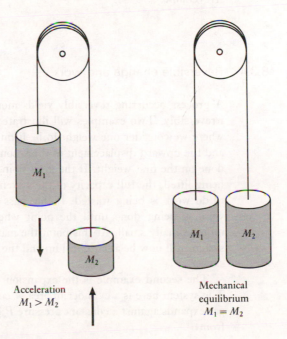

Acceleration
$M_1 > M_2$

Mechanical equilibrium
$M_1 = M_2$

Figure 8.2   A system of mechanical equilibrium.

greater than the other (M_2), in which case the pulley rotates. If the mass M_1 is only infinitesimally greater than M_2, i.e. $M_1 = M_2 + dM$, the pulley accelerates infinitely slowly. Also, the motion can be reversed by an infinitesimal change in the masses of the weights. A process which can be reversed in such a way is termed **reversible**.

The condition $M_1 = M_2 + dM$ is indistinguishable from the equilibrium position ($M_1 = M_2$). Therefore, a reversible process is a purely hypothetical process thought of as occurring through a series of equilibrium states.

8.3.3   Irreversibility

It is impossible for any real change to be reversible since a process cannot simultaneously be in a state of equilibrium and be changing. The reversible process may be thought of as a limiting case of real changes which are always **irreversible**.

If the pulley system in figure 8.2 were real, there would be friction at the pulley. Consequently, any motion due to a difference in the weights would give rise to the dissipation of heat. This dissipated energy is converted to thermal motion in the surroundings and is thought of as an amount of net work done by the system. Any changes cannot be completely reversed since dissipated heat cannot be transferred back to the system by simply changing the masses of the weights. Such a system cannot be reversed by infinitesimally small changes in the properties of the system.

A second example of irreversibility is a gas in a cylinder compressed very quickly. In this case eddies, shock waves or sound waves may be generated in the gas and transmitted as energy through the walls of the cylinder. This lost energy cannot be regained by simply moving the position of the piston. Therefore, the process is irreversible.

8.3.4   Reversible change and work

A process occurring reversibly yields more work than the same process occurring irreversibly. Two examples will illustrate this point. The first is the pulley system where we consider one weight to be lifting a second weight (figure 8.3). The mass and the upward displacement of the second weight are a measure of the work being done on the first weight. If the mass doing the lifting is much bigger than the mass being lifted, the full capacity of the system is not being used. Some of the potential to do work is being wasted. If the mass being lifted is gradually increased, more work is being done until the point when the difference between the masses is infinitesimally small. At this point the maximum work is being done. However, the system will now be at equilibrium and the acceleration of the policy occurs infinitely slowly.

The second example is the expansion of a gas against a piston as in figure 7.2. The system here is a cylinder of area A constrained by a **frictionless** piston. If the gas expands against a constant pressure P_{ext}, the work of expansion can be evaluated from:

$$dw = -P_{ext}\, dV \tag{8.63}$$

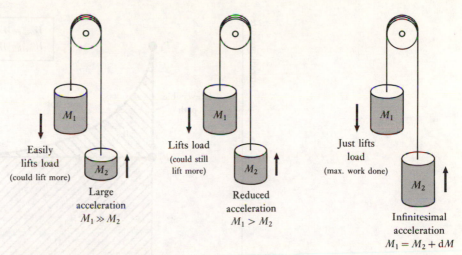

Figure 8.3   Maximum work from a pulley system.

whereby the work done on the system for a given volume change is:

$$w = -\int_{V_i}^{V_f} P_{ext}\, dV = -P_{ext} \times (V_f - V_i) \tag{8.64}$$

For this to occur, the value of the internal pressure must be greater than the external pressure. The criterion for mechanical equilibrium is that external and internal pressures are equal. Therefore this process is irreversible, i.e. it cannot be reversed by an infinitesimally small change in P_{ext}.

To make a volume change reversible we must set the external pressure to a value infinitesimally different from the internal pressure P_{in} at all times. This condition is indistinguishable from the condition $P_{ext} = P_{in}$. For this condition we drop the subscripts in and ext. The reversible work w_{rev} can be calculated from

$$dw_{rev} = -P\, dV \tag{8.65}$$

whereby

$$w_{rev} = -\int_{V_i}^{V_f} P\, dV \tag{8.66}$$

Equation 8.66 can be integrated if the equation of state for n moles of an ideal gas is used to relate the pressure of a gas to its volume. That is, $PV = nRT$, where R is the gas constant. From which,

$$w_{rev} = -\int_{V_i}^{V_f} \left(\frac{nRT}{V}\right) dV \tag{8.67}$$

The integration of equation 8.67 yields:

$$w_{rev} = -nRT \ln(V_f/V_i) \tag{8.68}$$

Figure 8.4 shows a schematic of pressure versus volume for both the irreversible and reversible expansions. From the integrations of these two plots it is apparent that reversible expansion yields more work than irreversible expansion. This is

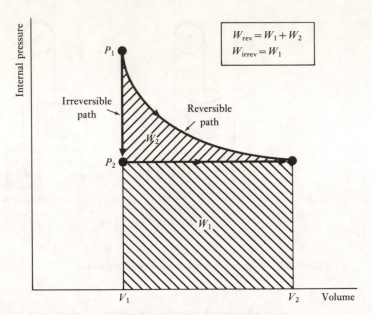

Figure 8.4   Reversible and irreversible expansion of piston (after Atkins, 1973).

because the irreversible expansion has an internal pressure greater than is necessary for the expansion and this additional capacity to expand is wasted. This is similar to the dissipated frictional heat loss in figure 8.2 which is also wasted energy.

The wasted energy is in fact a measure of the extent to which a system can undergo change. The gas in figure 7.2 will continue to expand spontaneously and irreversibly until its capacity for wasted energy is zero when $P_{in} = P_{ext}$. In section 8.4 this idea will be developed into the second law of thermodynamics, which gives the criterion for spontaneous change.

We may summarise by saying that **reversible changes occur infinitely slowly and yield the maximum amount of work**. All natural processes are irreversible and reversible processes are never realised. The reverse path is the limiting path for any irreversible process occurring under conditions progressively approaching equilibrium. A feature of the reversible change is that we can always calculate exactly the work done during the process.

The conditions that must be satisfied for the system to be in equilibrium are the same as those for the reversible change. Therefore, any conclusions we make about a reversible path are also conclusions about the state of equilibrium. This feature will be used later to define the thermodynamic properties of systems at equilibrium.

8.4   The second law of thermodynamics

The second law of thermodynamics is concerned with **spontaneous changes** which progress to a final state of equilibrium. The expansion of a gas into an evacuated chamber illustrates this point. Figure 8.5(a) represents two chambers connected by

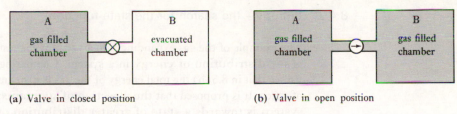

(a) Valve in closed position   (b) Valve in open position

Figure 8.5   The expansion of a gas into a vacuum.

a stopcock; one chamber is filled with a gas, the other evacuated. Upon opening the stopcock, the gas expands to fill both chambers (figure 8.5b). **There is no net transfer of energy to the system from its surroundings, yet the change inevitably occurs.** From total energy considerations the reverse process, in which all gas molecules diffuse into chamber A, is as likely as the equal distribution of gas through both chambers. This reverse process is never observed in practice and the spontaneous process is expansion. Other examples of spontaneous change are as follows.

- Heat energy flows spontaneously from a hot to a cold body with total conservation of energy.
- A mixture of hydrogen and oxygen, if sparked, will react spontaneously to form water with the liberation of heat. The reverse process, of water extracting heat from its surroundings to form oxygen and hydrogen, is not spontaneous.
- Water freezes spontaneously below 0 °C at atmospheric pressure.
- A salt in contact with a large mass of pure water will dissolve spontaneously, sometimes releasing heat and sometimes absorbing heat from the surroundings.

It is clear that the direction of spontaneous change is not simply towards a state of minimum internal energy for the system accompanied by the evolution of heat. It appears that there is some other driving force dictating the direction of spontaneous change. Historically, the development of the second law of thermodynamics was concerned with identifying this elusive driving force.

The second law of thermodynamics defines a new state function called **entropy**, which is given the symbol S. It will be shown later that entropy provides a general criterion for the direction of change in a system and defines the condition of equilibrium.

The second law can be written in many forms. However, a clear statement of the second law is best left until a definition of entropy is given.

8.4.1   Entropy

The existence of entropy was deduced experimentally in the 19th century by several workers. However, its existence can also be established a range of alternative methods including statistical considerations of the properties of matter. The alternative derivations of entropy and the logic required to show their equivalence are beyond the scope of this book. Consequently, the following section defines entropy, gives some idea of what it is and shows its place within the structure of the thermodynamics.

8.4.2   Entropy – the search for the state function

The example of the gas expanding into a vacuum suggests that entropy is a measure of the **distribution of energy** in a system. Comparing figures 8.5(a) and 8.5(b) we see that in 8.5(b) the total energy of the gas is distributed through a larger region of space. It is proposed that **the direction of spontaneous change in an isolated system is towards a state of greater distribution of energy**.

The entropy can be defined by considering the first law $dU = \mathrm{d}q + \mathrm{d}w$. Here, we have an exact differential (dU) equated to the sum of two inexact differentials, neither of which can be state functions. In section 8.3 we saw that an inexact differential $\mathrm{d}w_{\mathrm{rev}}$ could be related to the state function V by dividing $\mathrm{d}w_{\mathrm{rev}}$ by p. That is,

$$-\frac{\mathrm{d}w_{\mathrm{rev}}}{P} = \mathrm{d}V \tag{8.69}$$

In this context $1/P$ is called an **integrating factor** and has the property of transforming an inexact differential into an exact differential. If an integrating factor for $\mathrm{d}q_{\mathrm{rev}}$ can be found, then an exact differential can be constructed from this function. From a detailed mathematical treatment, Caratheodory deduced the required integrating factor and defined the conditions under which integrating factors can be applied. We will anticipate the result and define the required integrating factor as $1/T$. This means that the integral of $\mathrm{d}q_{\mathrm{rev}}/T$ is independent of the integration path, i.e. it is a state function. This new state function is defined as **the entropy of the system**:

$$\mathrm{d}S \equiv \frac{\mathrm{d}q_{\mathrm{rev}}}{T} \tag{8.70}$$

Entropy is an extensive quantity having the dimensions of energy per unit temperature. Therefore $T\,\mathrm{d}S$ has the dimensions of energy.

To put entropy in context we recall that:

● For a reversible change the work done **by** the system is greater than for the same change performed irreversibly. Since we have defined $\mathrm{d}w_{\mathrm{rev}}$ to be the reversible work done **on** the system, we can deduce the following inequality: $\mathrm{d}w_{\mathrm{rev}} < \mathrm{d}w$. Using the first law to define the relationship between heat and work we can conclude that:

$$\mathrm{d}q_{\mathrm{rev}} > \mathrm{d}q \tag{8.71}$$

Therefore, q_{rev}/T must be larger than q/T. This condition is called the **Clausius inequality**. Consequently, for any natural process occurring irreversibly,

$$\frac{\mathrm{d}q_{\mathrm{rev}}}{T} = \mathrm{d}S > \frac{\mathrm{d}q}{T} \tag{8.72}$$

Equation 8.72 is a mathematical statement of the second law. Alternatively, for any cyclic process, equation 8.72 can be written:

$$\oint \frac{\mathrm{d}q}{T} < 0 \tag{8.73}$$

We have shown that a reversible change can only occur quasi-statically. Therefore, the condition where an incremental change in entropy equals $\text{đ}q/T$ is a condition of reversibility and therefore a statement of equilibrium, i.e.

$$\text{d}S = \frac{\text{đ}q}{T} = \frac{\text{đ}q_{\text{rev}}}{T} \tag{8.74}$$

is a general definition of equilibrium.

Consider now an isolated system, i.e. one for which there is no exchange of heat with the surroundings. The equilibrium condition becomes

$$\text{d}S = 0 \tag{8.75}$$

This is consistent with $\text{d}S > 0$ for an observable change. Therefore, **in an isolated system, any spontaneous change will lead to an increase in the entropy of the system until the entropy reaches its maximum at the point of equilibrium.**

8.4.3 The degradation of energy

For a reversible change the first law can be written as $\text{đ}w_{\text{rev}} = \text{d}U - \text{đ}q_{\text{rev}}$. If $\text{đ}q_{\text{rev}}$ is replaced by $T\,\text{d}S$, then:

$$\text{đ}w_{\text{rev}} = \text{d}U - T\,\text{d}S \tag{8.76}$$

This equation shows that the maximum extractable work is equal to the total internal energy minus the energy term $T\,\text{d}S$. In this context $T\,\text{d}S$ represents a reduction in the work that can be extracted from the system. The greater $T\,\text{d}S$ is, the greater is the unavailability of some energy that could have been made to do work. This energy is not lost since the entropy change of the system plus its surroundings is zero. Rather it is **degraded** into a form which is not useful work. Its magnitude is a measure of the irreversibility of the change. An example is the reduction in the efficiency of a pulley due to fraction. Here the dissipated heat is a measure of the inefficiency of the system.

For any observable change, the accompanying change in entropy must be greater than zero. Therefore, the degraded energy cannot be recovered by any spontaneous process. This is a key point since it shows that a system cannot spontaneously regain the degraded energy which has been lost. These ideas are embodied in the experimental deductions of Clausius and Kelvin.

The Kelvin statement   No process can have the sole property of converting heat entirely to work. This means that a change in entropy accompanies the flow of heat.

The Clausius statement   No process can have the sole property of transferring heat from a colder body to a hotter body. This means that the direction of spontaneous change is to a state of higher system entropy.

8.4.4   A statement of the second law

A definition of the second law can now be given.

IN AN ISOLATED SYSTEM SPONTANEOUS CHANGES OCCUR IN THE DIRECTION OF INCREASING ENTROPY.

8.4.5   Consolidating the first and second laws

The first and second laws can be combined by considering the application of the first law to any change:

$$dU = {\rm d}q_{\rm rev} + {\rm d}w_{\rm rev} = {\rm d}q_{\rm irrev} + {\rm d}w_{\rm irrev} \qquad (8.77)$$

(Reversible)   (Irreversible)

Recalling from section 8.3 that for a reversible change ${\rm d}q_{\rm rev} = T\,{\rm d}S$ and ${\rm d}w_{\rm rev} = P\,{\rm d}V$, equation 8.77 becomes

$$dU = T\,{\rm d}S - P\,{\rm d}V \qquad (8.78)$$

Apparently this is an equation for reversible changes only. However, equation 8.78 is valid for any change since it contains only functions of state, which are independent of the way in which the transformations are carried out. This becomes clear if we note that, for an irreversible change, $T\,{\rm d}S$ is greater than ${\rm d}q_{\rm irrev}$ and $P\,{\rm d}V$ is greater than ${\rm d}w_{\rm irrev}$ by exactly compensating amounts. Equation 8.78 is called a **master equation** of thermodynamics.

8.4.6   Thermodynamic temperature and pressure

Up to this point we have viewed temperature and pressure as observed properties of a system. The master equation shows us an alternative way of considering both temperature and pressure.

Since the internal energy U is a state function, its total differential can be expressed in the general form given by equation 8.3 in box 8.1. To do this we must select a variable on which U is known to depend.

If we assume that entropy can be defined without reference to temperature, as in the statistical definition of entropy, then the variables which produce the simplest form of the differential are S and V. The total differential of U can be written as:

$$dU = \left(\frac{\partial U}{\partial S}\right)_V dS + \left(\frac{\partial U}{\partial V}\right)_S dV \qquad (8.79)$$

A comparison of equations 8.78 and 8.79 shows:

$$\left(\frac{\partial U}{\partial S}\right)_V = T \qquad (8.80)$$

and

$$\left(\frac{\partial U}{\partial V}\right)_S = -P \qquad (8.81)$$

These relationships show how the internal energy changes with volume and entropy respectively. They can be thought of as definitions of temperature and pressure. Alternatively we may write:

$$T = \left(\frac{\partial H}{\partial S}\right)_P \quad \text{and} \quad -P = \left(\frac{\partial A}{\partial V}\right)_T \tag{8.82}$$

8.4.7   Entropy – an exact differential

Up to this point we have accepted that $đq_{rev}/T$ and therefore dS are exact differentials. The general arguments to this end were given by Caratheodory and confirmed experimentally by measurements on the efficiency of heat engines. A simplified proof can be deduced by considering a reversible change in an ideal gas. For this situation the first law becomes $đq_{rev} = P\,dV + dU$. Dividing by T gives:

$$\frac{đq_{rev}}{T} = \frac{P\,dV}{T} + \frac{dU}{T} \tag{8.83}$$

Since the definition of heat capacity at constant volume yields $dU = C_V\,dT$ and the ideal gas law for 1 mole of gas is $P = RT/\bar{V}$, then:

$$dS = \frac{đq_{rev}}{T} = \left(\frac{C_V}{T}\right)dT + \left(\frac{R}{\bar{V}}\right)dV \tag{8.84}$$

It is clear that the integration of the right-hand side of equation 8.84 is path-independent. Consequently, the integration of the left-hand side must also be path-independent and $đq_{rev}/T$ is therefore an exact differential.

Since entropy is a state function we may define ΔS to be the difference in the entropy of a system between initial state i and final state f. A general relationship is

$$\Delta S = S_i - S_f = \int_i^f \frac{đq_{rev}}{T}$$

This shows that we compute ΔS by means of a reversible path where, for each infinitesimal step in the integration, we compute $đq/T$ and the sum of these quantities is ΔS. For a process at constant temperature the general expression is simply:

$$\Delta S = \int_i^f \frac{đq_{rev}}{T} = \frac{1}{T}\int_i^f đq_{rev} = \frac{q_{rev}}{T} \tag{8.85}$$

We emphasise that even though ΔS is computed via a reversible pathway, the value of ΔS will be the same even if an irreversible path is followed. It is worth noting that the path independence of ΔS is consistent with the fact that, for any infinitesimal change, $đq_{rev}/T$ does not equal $đq_{irrev}/T$. If, in contrast, the two functions were equal it would not be possible for two different pathways to lead to the same summations for $đq/T$. We stress that it is the function $đq_{irrev}/T$ which is path-dependent, not ΔS.

8.4.8　Entropy changes in systems and surroundings

The second law shows that spontaneous changes in an **isolated** system lead to an increase in the system entropy. For a system which can exchange heat and work with its surroundings we must extend the application of the second law to consider both. By defining a system plus its surroundings as the **universe**, an important general statement can be made. **The entropy of the universe is increasing**. This point makes it clear that the overall change in entropy dictates the direction of spontaneous change. Indeed, the entropy of a system may decrease as long as the entropy changes in the surroundings compensate. This explains endothermic reactions, such as the dissolution of a salt with the release of heat, where there is an increase in the entropy of the system but an overall net loss of entropy in the system plus its surroundings.

Consider a situation where the system transfers a quantity of heat to the surroundings. If the surroundings is an infinite reservoir it is always at equilibrium. Therefore, it is irrelevant whether the heat transfer is reversible or irreversible since both $đq_{irrev, surr}$ and $đq_{irrev, surr}$ are equal. If the heat transfer is actually reversible then the entropy change in the surroundings has the same magnitude as the entropy change in the system, i.e.

$$dS_{surr} = \frac{đq_{surr}}{T_{surr}} = -dS_{sys}, \quad \text{or} \quad dS_{surr} + dS_{sys} = 0 \qquad (8.86)$$

This confirms our previous statement of equilibrium (equation 8.75) and emphasises the need to define entropy in terms of the reversible heat change.

Since all spontaneous processes are irreversible, $dS_{sys} + dS_{surr} > 0$. Clearly, we need to consider both system and surroundings. It is convenient if we can devise a way of describing the entropy changes in the universe solely in terms of the entropy changes in the system. This would remove constant reference to the surroundings. To do this we consider a system and its surroundings to be in thermal equilibrium (i.e. $T_{surr} = T_{sys}$). The overall change in the entropy of the universe is:

$$dS_{univ} = dS_{sys} + dS_{surr} \qquad (8.87)$$

or, using equation 8.86,

$$dS_{univ} = dS_{sys} + \frac{đq_{surr}}{T_{surr}} \qquad (8.88)$$

If the temperature T_{surr} is replaced by T_{sys} and we note that the heat loss from the system equals the heat gained by the surroundings ($đq_{sys} = -đq_{surr}$), then equation 8.88 becomes:

$$dS_{unit} = dS_{sys} - \frac{đq_{sys}}{t_{sys}} \qquad (8.89)$$

In any spontaneous change the entropy in the universe must increase. Therefore:

$$dS_{sys} \geqslant \frac{đq_{sys}}{T_{sys}} \qquad \text{(spontaneous change)} \qquad (8.90)$$

This is the simplification we are looking for since it expresses spontaneous change solely in terms of the properties of the system. From this point onwards the subscript 'surr' will be dropped with the understanding that all reference is to the state properties of the system. From this point onwards, this convention applies to all state functions, not only entropy.

8.4.9 The entropy of expansion of an ideal gas

Consider the spontaneous isothermal expansion of an ideal gas from initial volume V_i to final volume V_f. The reversible work of expansion is given by equation 8.68, i.e.

$$w_{rev} = -nRT \ln(V_f/V_i) \tag{8.91}$$

There is **no net** energy change in the system ($\Delta U = 0$); therefore the heat gained from the surroundings is equal to the work done by the system ($q_{rev} = -w_{rev}$). Since $\Delta S = q_{rev}/T$, we may write:

$$\Delta S = nR \ln(V_f/V_i) \tag{8.92}$$

Since V_f is larger than V_i, the spontaneous change is associated with a positive entropy change, as expected.

Box 8.2

Entropy and disorder – some statistical concepts

The definition of entropy in section 8.4.2 is all we require to incorporate the second law of thermodynamics into the thermodynamic framework. However, the statistical definition of entropy provides insight into the physical significance of entropy.

The following argument shows that entropy is a measure of the **randomness** or **disorder** of the molecules in a system. The much-used example of a gas expanding into a vacuum (figure 8.5) provides the basis for this approach. The disorder in this system can be thought of as the number of positions in which a given molecule can be found. If there are many positions available, the probability that a specific molecule is in any one position is very low. The system is said to be disordered. A system is highly ordered if the number of possible positions is low.

The **probability** of finding a particular gas molecule in a given volume of space (V) is proportional to the magnitude of the volume. Defining the probability (or statistical weight) as ω, it can be written:

$$\omega = cV \tag{8.93}$$

where c is a constant of proportionality. If the system contained two molecules, the probability of finding both in the same volume element is the product of the individual probabilities. i.e. ω^2. For a system containing N molecules the probability of finding **all** the molecules in a given volume (Ω) is:

$$\Omega = \omega^N = (cV)^N \tag{8.94}$$

Considering figure 8.5 we have two options. The first is that all the molecules can remain in bulb A; this has a probability $\Omega_1 = c^N V_A^N$. The second option is that the molecules can be dispersed through both bulbs A and B. This has the probability $\Omega_2 = c^N (V_A + V_B)^N$. Clearly Ω_2 is greater than Ω_1. It is a fact that if

continued

Box 8.2
continued

all the molecules were initially placed in one bulb they would naturally diffuse to fill both bulbs. This suggests that **any spontaneous change will tend towards a state of highest probability or greatest disorder**. The final state is the equilibrium state.

We can now take the step of equating the **statistical entropy** of a system to some function of the probability Ω. The general relationship we choose is:

$$S = k \ln \Omega \tag{8.95}$$

The proportionality constant k is called Boltzmann's constant.

From equation 8.95, **spontaneous changes tend towards a state of greater statistical entropy**.

Entropy defined as the logarithm of probability has the property of additivity. We can see this because Ω is proportional to V^N. Thus, changing the number of molecules from N to $2N$, in the same volume, changes the value of Ω to Ω^2. The entropy therefore doubles from $S = k \ln \Omega$, to $S = 2k \ln \Omega$. This shows that statistical entropy is proportional to the amount of material present and is therefore extensive.

Liquids have lower entropies than their corresponding gases. In the gas phase the amount of available space per gas molecule is high. In the liquid phase the concentration of particles is high and the volume of the particles themselves reduces the availability of space for other molecules. Solids have lower entropies than their corresponding liquid phases. This is because the number of possible positions for a given molecule is greater in the liquid than in the solid, where particles are associated with specific crystallographic sites.

The equivalence of the statistical and thermodynamic entropies

The change in the thermodynamic entropy associated with the isothermal expansion of n moles of an ideal gas from volume V_i to volume V_f is given by equation 8.92, i.e.

$$\Delta S = nR \ln (V_f / V_i) \tag{8.96}$$

The change in the statistical entropy for the same process can be derived by equating ΔS to $S_f - S_i$. This gives:

$$\Delta S = k \ln (c V_f^N) - k \ln (c V_i^N) \tag{8.97}$$

$$= k \ln \left(\frac{c V_f^N}{c V_i^N} \right) \tag{8.98}$$

The c cancel and we can replace N with nL, where L is Avogadro's constant. Therefore:

$$\Delta S = nLk \ln \left(\frac{V_f}{V_i} \right) \tag{8.99}$$

Comparison of equations 8.92 and 8.99 shows that the statistical entropy and the thermodynamic entropy have the same form for the same process.

We will assume that **the statistical entropy and the thermodynamic entropy are identical**. This leads to the conclusion that $R = Lk$. It should be stressed that the similarity of equations 8.92 and 8.99 is instructive but does not constitute proof. The general proof of equivalence is beyond the scope of this book.

Chapter 9

Thermodynamic potentials and equilibria

In this chapter we will introduce an auxiliary set of thermodynamic potentials. These will be used to deduce a general condition for equilibrium. The interrelationships between these potentials, including the effects of changing temperature and pressure, will be discussed.

The contents of this chapter form the basis of the thermodynamics of chemically reacting systems and mixing phenomena discussed in chapters 10 and 11.

9.1   Thermodynamic potentials and their interrelationships

9.1.1   The Gibbs energy and the Helmholtz energy

The master equation (8.78) is sufficient for the application of thermodynamics to any real system. However, it is convenient to consider two other auxiliary state functions. These are the **Helmholtz energy** A and the **Gibbs energy** G, and are defined as:

$$A \equiv U - TS, \tag{9.1}$$

and

$$G \equiv U + PV - TS \tag{9.2}$$

Since the enthalpy is defined as $H = U + PV$, the Gibbs energy has the alternative definitions: $G \equiv H - TS$.

The new functions are useful because they simplify thermodynamic calculations. They help define the direction of change and the position of equilibrium for processes occurring spontaneously under conditions of constant temperature, pressure and volume. These are the conditions of greatest interest in chemical and geochemical systems.

Box 9.1

> **Massieu functions**
>
> We appear to have guessed at the definitions of G and A, with the only requirement being that of dimensional consistency. In fact new functions cannot be generated by random combination of variables. The form of the new functions can be obtained by applying Legendre transformations to all possible communications of the independent variables. This process allows a new set of functions to be identified which are consistent with a particular differential
>
> *continued*

Box 9.1
continued

expression; it also identifies the proper variables (defined below) which are unique to each new function.

There are several different auxiliary functions that could be generated; of these, A and G are the most useful. Others, which have the proper variables $1/T$ and $1/P$, include:

$$\Omega = S - \frac{P}{T}V \qquad (9.3)$$

$$\Psi = S - \frac{1}{T}U \qquad (9.4)$$

$$\Phi = S - \frac{P}{T}V - \frac{1}{T}U \qquad (9.5)$$

These are called Massieu functions, although Φ is sometimes called the Planck function. Functions with the proper variables $1/T$ and P/T arise naturally in statistical mechanics and irreversible thermodynamics. They are less valuable for chemical systems where the common proper variables are T and P.

A consolidated view of the potentials

The four primary thermodynamic state functions are called **thermodynamic potentials**. They are:

$$\left.\begin{array}{lll} \text{Internal energy:} & U = TS - PV \\ \text{Enthalpy:} & H = U + PV \\ \text{Helmholtz energy:} & A = U - TS \\ \text{Gibbs energy:} & G = U - TS + PV \end{array}\right\} \qquad (9.6)$$

The differential forms of the thermodynamic potentials give some insight into their usefulness. These are:

$$\left.\begin{array}{l} dU = T\,dS - P\,dV \\ dH = T\,dS + V\,dP \\ dA = -S\,dT - P\,dV \\ dG = -S\,dT + V\,dP \end{array}\right\} \qquad (9.7)$$

The terms on the right-hand side of the equations each have two degrees of freedom. The degrees of freedom are derived from the two pairs of fundamental variables related to each thermodynamic potential. For example, dU has the pairs (T, S) and (P, V). These variables are called **natural** or **proper variables**. The proper variables for U are S and V; therefore we write $U = U(S, V)$. Similarly, $H = H(S, P)$, $A = A(T, V)$ and $G = G(T, P)$. If **any one** of the potentials is known explicitly in terms of its proper variables then the thermodynamic information for the system is complete. Any equation which performs this function is called an **equation of state**.

A further set of relationships can be derived using the same logic as was used in section 8.4.6 to generate equations 8.80 (equation 9.8) and 8.81 (equation 9.9).

$$\left(\frac{\partial U}{\partial S}\right)_V = T \qquad (9.8)$$

$$\left(\frac{\partial U}{\partial V}\right)_S = -P \tag{9.9}$$

The other relationships are:

$$\left(\frac{\partial H}{\partial S}\right)_P = T \tag{9.10}$$

$$\left(\frac{\partial H}{\partial P}\right)_S = V \tag{9.11}$$

$$\left(\frac{\partial A}{\partial V}\right)_T = -P \tag{9.12}$$

$$\left(\frac{\partial A}{\partial T}\right)_V = -S \tag{9.13}$$

$$\left(\frac{\partial G}{\partial P}\right)_T = V \tag{9.14}$$

$$\left(\frac{\partial G}{\partial T}\right)_P = -S \tag{9.15}$$

A set of useful thermodynamic equalities can be derived from equations 9.8–9.15. For example, differentiating U with respect to S at constant V, and then V at constant S, gives:

$$\left[\frac{\partial}{\partial V}\left(\frac{\partial U}{\partial S}\right)_V\right]_S \tag{9.16}$$

Recalling from section 8.1.1 that a second derivative of any continuous function is independent of the order of differentiation, then:

$$\left[\frac{\partial}{\partial V}\left(\frac{\partial U}{\partial S}\right)_V\right]_S = \left[\frac{\partial}{\partial S}\left(\frac{\partial U}{\partial V}\right)_S\right]_V \tag{9.17}$$

From the definitions of temperature and pressure (9.8 and 9.9) we see that:

$$\left(\frac{\partial T}{\partial V}\right)_S = \left(\frac{\partial P}{\partial S}\right)_V \tag{9.18}$$

This is a **Maxwell equation**. There are three other Maxwell equations, i.e.

$$\left(\frac{\partial S}{\partial P}\right)_T = -\left(\frac{\partial V}{\partial T}\right)_P \tag{9.19}$$

$$\left(\frac{\partial T}{\partial P}\right)_S = \left(\frac{\partial V}{\partial S}\right)_P \tag{9.20}$$

and

$$\left(\frac{\partial S}{\partial V}\right)_T = \left(\frac{\partial P}{\partial T}\right)_V \tag{9.21}$$

Finally, heat capacities can be expressed in terms of entropy since at equilibrium

$dS = đq/T$. Therefore,

$$C_P = \left(\frac{\partial H}{\partial T}\right)_P = T\left(\frac{\partial S}{\partial T}\right)_P \qquad (9.22)$$

and

$$C_V = \left(\frac{\partial U}{\partial T}\right)_V = T\left(\frac{\partial S}{\partial T}\right)_V \qquad (9.23)$$

The above equalities in equations 9.8–9.15 and 9.18–9.21 may look a little obscure. However, these equations give us the means to generate expressions for the dependence of any chosen state function on any other extensive or intensive property. This allows us to develop working equations to predict the thermodynamic properties of substances under geological conditions. Such applications will be investigated in later sections.

Box 9.2

Comments on integrating the basic equations

The approach taken here has been first to define the thermodynamic potentials (equations 9.6) and then to differentiate these functions to obtain the useful equations 9.7. The opposite approach is possible, i.e. integrating equations 9.7 to obtain equations 9.6. The procedure is important since such integrations are performed commonly in thermodynamic calculations. The method is based on Euler's theory of homogeneous functions.

Consider the function:

$$dU = T\,dS - P\,dV \qquad (9.24)$$

To integrate this we note that U, S and V are **extensive functions** which increase as the size of the system increases. In contrast, the functions T and P are **intensive** (independent of amount).

If the system is increased in size by a factor λ then any extensive property is increased from X to λX. The change in such a property is $\lambda X - X$, or more simply $X(\lambda - 1)$. Therefore,

$$\Delta U = (\lambda - 1)U = T(\lambda - 1)S - P(\lambda - 1)V \qquad (9.25)$$

The term $(\lambda - 1)$ in equation 9.25 can be cancelled from both sides of the equation to yield:

$$U = TS - PV \qquad (9.26)$$

which is the desired relationship. All equations 9.7 can be derived in this way.

9.1.2 Conditions for spontaneous change and equilibrium

A major achievement of classical thermodynamics is the ability to define the direction of change and the final position of equilibrium in a spontaneous process. A set of general criteria for these purposes can be derived using the definitions of the Helmholtz energy, the Gibbs energy and the condition for spontaneous change in terms of entropy (section 8.4.8 and equation 8.90).

*Equilibrium in terms of
the Gibbs energy*

Consider the Gibbs energy defined as $G = H - TS$. For an infinitesimal change in G:

$$dG = dH - T\,dS - S\,dT \tag{9.27}$$

Imposing the restriction of constant temperature gives:

$$dG = dH - T\,dS \quad \text{(Constant } T) \tag{9.28}$$

In order to use the above equation to derive the conditions of spontaneous change we must perform a series of algebraic manipulations upon it using the previous definition of enthalpy and the first law. The definition of enthalpy yields:

$$dH = dU - P\,dV - V\,dP \quad \text{(Constant } T) \tag{9.29}$$

from which:

$$dG = dU - P\,dV - V\,dP - T\,dS \quad \text{(Constant } T) \tag{9.30}$$

The first law, expressed as $dU = đq + đw$, can be substituted into equation 9.30 to give:

$$dG = đq + đw - P\,dV - V\,dP - T\,dS \quad \text{(Constant } T) \tag{9.31}$$

Two further restrictions can be imposed at this point. One comes from the assumption that only work causing pressure–volume changes can be done on the system, which allows $đw$ to be replaced by $-P\,dV$. The second is the restriction of constant pressure. Therefore, equation 9.31 becomes:

$$dG = đq - T\,dS \quad \text{(Constant } T \text{ and } P) \tag{9.32}$$

This is the critical equation from which both the condition of equilibrium and the direction of spontaneous change can be derived.

The equilibrium position is equivalent to a reversible change, for which we know that from equation 8.74:

$$dq = đq_{rev} = T\,dS \tag{9.33}$$

Substituting equation 9.33 into 9.32 gives:

$$dG = 0 \quad \textbf{(Equilibrium at constant } \textbf{\textit{T}} \textbf{ and } \textbf{\textit{P}}). \tag{9.34}$$

The condition for spontaneous change can be derived simply by considering the condition for an irreversible change given in section 8.4.2, i.e.

$$đq_{irrev} < T\,dS \tag{9.35}$$

from which

$$dG < 0 \quad \textbf{(Spontaneous change at constant } \textbf{\textit{T}} \textbf{ and } \textbf{\textit{P}}) \tag{9.36}$$

These conditions are depicted graphically in figure 9.1 which shows a system, in this case a hypothetical reaction between species A and B to give products C and D. The position of equilibrium is the minimum in Gibbs energy for the whole system.

General conditions

The equilibrium position defined in terms of G is consistent with constant temperature and pressure. These are the most relevant conditions. However, the Clausius inequality may be applied to systems under other conditions. For completeness these are specified as shown in table 9.1.

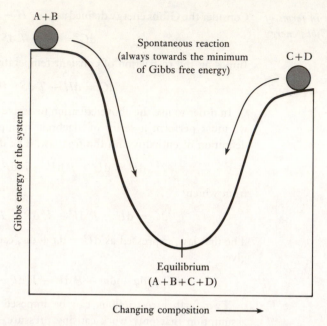

Figure 9.1   The direction of spontaneous change in a chemical reaction (after Atkins, 1973).

Table 9.1   **Conditions for equilibrium**

Variables Held Constant	*Conditions for Spontaneous Change*	*Equilibrium Conditions*
T and P	$(dG)_{T,P} < 0$	$(dG)_{T,P} = 0$
T and V	$(dA)_{T,V} < 0$	$(dA)_{T,V} = 0$
U and V	$(dS)_{U,V} > 0$	$(dS)_{U,V} = 0$
S and V	$(dU)_{S,V} < 0$	$(dU)_{S,V} = 0$
S and P	$(dH)_{S,P} < 0$	$(dH)_{S,P} = 0$
G and T	$(dP)_{G,T} < 0$	$(dP)_{G,T} = 0$
G and P	$(dT)_{G,P} < 0$	$(dT)_{G,P} = 0$
A and T	$(dV)_{A,T} < 0$	$(dV)_{A,T} = 0$
A and V	$(dT)_{A,V} < 0$	$(dT)_{A,V} = 0$
U and S	$(dV)_{U,S} < 0$	$(dV)_{U,S} = 0$
H and S	$(dS)_{H,S} < 0$	$(dS)_{H,S} = 0$
H and P	$(dP)_{H,P} < 0$	$(dP)_{H,P} = 0$

9.1.3   Some comments on the Gibbs energy

The Gibbs energy is the most useful thermodynamic potential since it is related to the proper variables T and P which are of major concern in chemical and geochemical systems. The Gibbs function also gives the conditions for equilibrium under constant temperature and pressure. The Gibbs energy is not directly measurable for any system. It is, therefore, useful to express the Gibbs energy in terms of variables which are measurable. For this purpose there are four much-used equations which can be derived from equations 9.6. These **master equations** are:

$$G = PV - TS \tag{9.37}$$

$$G = H - TS \tag{9.38}$$

$$dG = V\,dP - S\,dT \tag{9.39}$$

$$dG = dH = T\,dS \tag{9.40}$$

The enthalpy and entropy of a substance are not themselves directly measurable although both are related to properties which can easily be measured. For example, equations 9.22 and 9.23 show relationships with measurable heat capacities. The application of equations 9.38 and 9.40 and further links with measurable properties will be investigated in later sections.

9.1.4   Compression and expansion

The dimensions of systems change in response to changes in temperature and pressure. The characterisation of these changes is itself of technological importance. However, the parameters characterising these changes play a key role in classical thermodynamics by linking together some important state functions. An example is the exact relationship between C_V and C_P.

Isothermal compression   For any substance the **coefficient of isothermal compression** κ is defined as:

$$\kappa = -\frac{1}{V}\left(\frac{\partial V}{\partial P}\right)_T \tag{9.41}$$

An alternative form of the definition can be obtained since for any variable x, $d \ln x = (1/x)\,dx$. Therefore,

$$\kappa = -\left(\frac{\partial \ln V}{\partial P}\right)_T \tag{9.42}$$

The coefficient is a measure of the change in volume in response to a differential change in pressure. It has a negative sign in the definition so that κ is always positive when an increase in pressure causes a reduction in volume.

For the case when the pressure change is small, and $(\partial V/\partial P)_T$ is independent of pressure, we may write:

$$\Delta V = \left(\frac{\partial V}{\partial P}\right)_T \int dP \simeq -\kappa V \Delta P \tag{9.43}$$

Isobaric thermal expansion   The **coefficient of isobaric thermal expansion** α is defined as:

$$\alpha = \frac{1}{V}\left(\frac{\partial V}{\partial T}\right)_P = \left(\frac{\partial \ln V}{\partial T}\right)_P \tag{9.44}$$

It is the temperature-change analogue of the compressibility and is usually positive because an increase in volume accompanies an increase in temperature. If the temperature change is small and $(\partial V/\partial T)_P$ is independent of temperature, we may write:

$$\Delta V = \left(\frac{\partial V}{\partial T}\right)_P \int dT \simeq \alpha V \Delta T \qquad (9.45)$$

For solids both κ and α are relatively independent of pressure and temperature, but for liquid water both coefficients change significantly with variations in temperature and pressure.

The relationship between α and κ can be derived using the -1 rule (equation 8.9 in box 8.1), giving:

$$\frac{\alpha}{\kappa} = \left(\frac{\partial P}{\partial T}\right)_V \qquad (9.46)$$

Isenthalpic Joule–
Thomson expansion

The **Joule–Thomson coefficient** (μ_{JT}) is a measure of the change in temperature of a substance arising from a change in pressure. In the original experiment, Joule and Thomson measured the temperature change accompanying the expansion of the gas against a constant pressure. The work done on such a system is $(P_f V_f) - (P_i V_i)$. If the system is thermally insulated the work is equal to the change in energy of the system; therefore $(P_f V_f) - (P_i V_i) = U_f - U_i$. This is a condition of zero net change in enthalpy (equation 8.52). The expansion is therefore **isenthalpic**. Hence, the definition of μ_{JT} is:

$$\mu_{JT} = \left(\frac{\partial T}{\partial P}\right)_H \qquad (9.47)$$

The coefficient is a measure of the energy required to overcome molecular interactions in a system as it expands isothermally. For real gases the coefficient is non-zero, as can be observed when the rapid release of compressed gas from a cylinder gives a notable drop in temperature. For ideal gases there are no interactions and the coefficient is zero.

A positive value for μ_{JT} corresponds to cooling on expansion. For all gases μ_{JT} changes with temperature and pressure. Figure 9.2 shows relevant data for carbon dioxide. The temperature at which the value of μ_{JT} changes sign is called the **inversion temperature**.

9.1.5 Partial molar properties

Chemical systems are usually composed of many different species. For such systems it is convenient to express the total value of any extensive variable, such as U, S, V, H, A and G, in terms of a sum of the contributions from each different species. For this purpose we define **partial molar properties**.

Consider a multicomponent system which has an extensive variable X. If the system contains a specific substance, designated i, then the associated partial molar property of the substance i is defined as:

$$\bar{X}_i \equiv \left(\frac{\partial X}{\partial n_i}\right)_{T,P,n_j} \qquad (9.48)$$

where n_i is the amount of the substance i. The conditional subscript T,P,n_j indicates

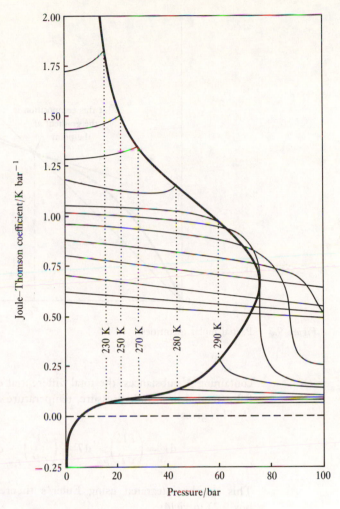

Figure 9.2   Schematic of the Joule–Thomson coefficient for carbon dioxide.

that the temperature, pressure and the amount of any **other** substance in the mixture ($n_{j(j \neq i)}$), are kept constant. **Partial molar properties are intensive**.

A partial molar property is thought of as the increase in the property of the system when 1 mole of the substance is added to an infinitely large quantity of the mixture so as not to change its overall composition. For example, the partial molar volume of a substance is defined as:

$$\bar{V}_i \equiv \left(\frac{\partial V}{\partial n_i} \right)_{T, P, n_j} \tag{9.49}$$

For a **multicomponent material any partial molar property may vary with composition as well as with temperature and pressure**. Figure 9.3 emphasises the point that the partial molar volume is the gradient of volume versus composition at a particular composition. It is **not** $\Delta V / \Delta n_i$, where ΔV is the finite change in volume accompanying a finite change in concentration.

Continuing with the example of volume changes, we note that the volume of a system is dependent on temperature, pressure and composition. For a system

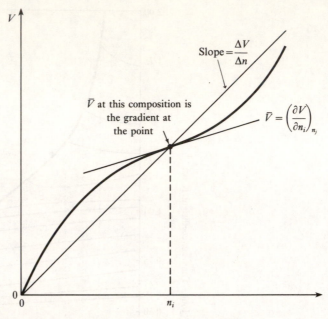

Figure 9.3 Partial molar quantities.

containing N substances, the total differential of V can be obtained from equation 8.3 extended to include pressure, temperature and all components.

This gives:

$$\mathrm{d}V = \left(\frac{\partial V}{\partial T}\right)_{P,n_j} \mathrm{d}T + \left(\frac{\partial V}{\partial P}\right)_{T,n_j} \mathrm{d}P + \sum_{i=1}^{N} \left(\frac{\partial V}{\partial n_i}\right)_{T,n_j} \mathrm{d}n_i \qquad (9.50)$$

This can be integrated using Euler's theorem of homogeneous functions (see box 9.2) to yield:

$$V = \sum_{i=1}^{N} n_j \left(\frac{\partial V}{\partial n_j}\right)_{T,n_j} \qquad (9.51)$$

Incorporating the definition of partial molar volume gives:

$$V = \sum_{i=1}^{N} n_i \bar{V}_i \qquad (9.52)$$

This states that **at any temperature and pressure the volume of a system can be expressed in terms of the sum of the partial molar volumes of its constituents.**

Since any **extensive** function can have an associated partial molar property, the conclusions drawn for volume, above, can be generalised. That is, any extensive property X can be related to the sum of the partial molar properties via:

$$X = \sum_{i=1}^{N} n_i \bar{X}_i \qquad (9.53)$$

9.1.6   Four fundamental equations for open systems

In order to extend the application of the four equations for the thermodynamic potentials to systems which may undergo a change in composition, Gibbs introduced the term **chemical potential**. This is the partial molar internal energy.

For a system comprising a single phase containing N substances, the chemical potential of the ith substance is:

$$\mu_i = \left(\frac{\partial U}{\partial n_i}\right)_{T,n_j} \tag{9.54}$$

It can be thought of as an increase in the internal energy of the system when 1 mole of the substance is added to an infinitely large quantity of the mixture so as not to change its overall composition.

Single-phase systems   The total differential of U with respect to entropy, volume and amount of each substance is then:

$$dU = \left(\frac{\partial U}{\partial S}\right)_{V,n_j} dS + \left(\frac{\partial U}{\partial V}\right)_{S,n_j} dP + \sum_i^N \mu_i\, dn_i \tag{9.55}$$

We know from equations 8.80 and 8.81 that $(\partial U/\partial S)_V = T$ and $(\partial U/\partial V)_S = -P$. Therefore,

$$dU = T\, dS - P\, dV + \sum_i^N \mu_i\, dn_i \tag{9.56}$$

This is known as a **fundamental equation of thermodynamics**. There are three others which can be derived using the auxiliary functions H, A and G. These are:

$$dH = T\, dS + V\, dP + \sum_i^N \mu_i\, dn_i \tag{9.57}$$

$$dA = -S\, dT - P\, dV + \sum_i^N \mu_i\, dn_i \tag{9.58}$$

$$dG = -S\, dT + V\, dP + \sum_i^N \mu_i\, dn_i \tag{9.59}$$

The fundamental equations show the characteristic state functions which must be used with a chosen set of independent intensive variables for open systems. In chemical applications the most useful set of independent variables are temperature and pressure and amount of substance. Consequently the Gibbs energy is an important quantity. This is essentially the same conclusion as for closed systems.

The above equations are valid for open or closed systems where the composition changes may occur through mass transfer or reactive combination of existing species, or both.

The four fundamental equations (9.56–9.59) can be integrated using the method described in box 9.2 to give:

$$U = TS - PV + \sum_i^N \mu_i n_i \tag{9.60}$$

$$H = TS + \sum_i^N \mu_i n_i \qquad (9.61)$$

$$A = -PV + \sum_i^N \mu_i n_i \qquad (9.62)$$

$$G = \sum_i^N \mu_i n_i \qquad (9.63)$$

Multiphase systems If a system has more than one phase, there are four separate fundamental equations for each phase. Consequently, the total derivative of U, H, G and A for the complete system is obtained by summing the right-hand sides of equations 9.56–9.59 over all phases. For example, if a system has K phases and N substances, each of which may exist in any phase, the total differential for U for the complete system is:

$$dU = \sum_l^K T_l \, dS_l - \sum_l^K P_l \, dV_l + \sum_l^K \sum_i^N \mu_{i,l} \, d_{n_{i,l}} \qquad (9.64)$$

where the subscript l denotes a property of the lth phase, and $\mu_{i,l}$ and $n_{i,l}$ are the chemical potential and amount of the ith substance in the lth phase. Similar equations may be generated for dH, dA and dG plus their integrated forms similar to equations 9.60–9.63.

9.1.7 Chemical potential

From the definition of partial molar properties in section 9.1.5 we can equate the Gibbs energy for a system to the sum of the partial molar Gibbs energies of constituents, i.e.

$$G = \sum_i^N n_i \left(\frac{\partial G}{\partial n_i} \right)_{T,P,n_j} \qquad (9.65)$$

From equations 9.63 and 9.65 we conclude that the chemical potential, as defined by Gibbs, is identical to the partial molar Gibbs energy.

$$\mu_i \equiv \bar{G}_i = \left(\frac{\partial G}{\partial n_i} \right)_{T,P,n_j} \qquad (9.66)$$

Under conditions of constant temperature and pressure the minimisation of the Gibbs function is the condition for equilibrium. Therefore, the chemical potential becomes a critical parameter in describing the position of equilibrium for a multicomponent system. With the exception of internal energy and entropy, the chemical potential is the most important thermodynamic parameter for most chemical systems.

9.1.8 A fifth fundamental equation for open systems: the Gibbs–Duhem equation

A fifth fundamental equation which supplements equations 9.56–9.59 can be derived from the differentiation of equation 9.63, i.e.

$$dG = \sum_i^N \mu_i \, dn_i + \sum_i^N n_i \, d\mu_i \qquad (9.67)$$

By substituting for dG in equation 9.67 using equation 9.59 we generate the relationship:

$$0 = S\,dT - V\,dP + \sum_i^N n_i\,d\mu_i \qquad (9.68)$$

This is called the **Gibbs–Duhem equation** for a single phase. In a multiphase system there is a Gibbs–Duhem equation for each phase.

The Gibbs–Duhem equation helps us simplify the application of thermodynamics to complex systems since it imposes a restriction on the system. For a single-phase system containing N substances there are $N + 2$ intensive properties (N chemical potentials plus T and P). The restriction is that only $N + 1$ intensive properties can be varied independently. This is of great value for systems at constant temperature and pressure where equation 9.68 becomes:

$$0 = \sum_i^N n_i\,d\mu_i \qquad (9.69)$$

For example, in a system containing a dissolved salt in water, the Gibbs–Duhem equation becomes:

$$n_s\,d\mu_S = -n_w\,d\mu_w \qquad (9.70)$$

where the subscripts s and w refer to dissolved salt and water respectively. Therefore, **the thermodynamic properties of one substance can be determined by measuring the properties of the other.** In a real application this requires careful choice of measurements and suitable integration of the Gibbs–Duhem equation. Such considerations will be discussed in later sections on specific materials.

9.1.9   A summary of the relationships between thermodynamic potentials

There are many relationships between the differential coefficients of thermodynamic variables. To derive them all is beyond the scope of this book. However, table 9.2 shows some of the most useful and includes many of those already given.

9.1.10   A summary of the relationships between partial molar quantities

The relationships between partial molar quantities mirror the relationships between thermodynamic potentials. A selected summary is displayed in table 9.3.

9.2   Changes in thermodynamic potentials over ranges of temperature and pressure

Temperature and pressure are the important independent variables for geochemical systems. Therefore, we will consider the effect of changes in these two variables on the thermodynamic potentials. To perform such calculations it is necessary to express

Table 9.2   **The relationships between thermodynamic potentials**

$$dU = T\,dS - P\,dV$$
$$dH = T\,dS + V\,dP$$
$$dG = V\,dP - S\,dT \tag{9.71}$$
$$dA = -P\,dV - S\,dT$$

$$\left(\frac{\partial T}{\partial V}\right)_S = -\left(\frac{\partial P}{\partial S}\right)_V$$

$$\left(\frac{\partial T}{\partial P}\right)_S = \left(\frac{\partial V}{\partial S}\right)_P$$

$$\left(\frac{\partial V}{\partial T}\right)_P = -\left(\frac{\partial S}{\partial P}\right)_T \tag{9.72}$$

$$\left(\frac{\partial P}{\partial T}\right)_V = \left(\frac{\partial S}{\partial V}\right)_T$$

$$\left(\frac{\partial H}{\partial S}\right)_P = T$$

$$\left(\frac{\partial U}{\partial S}\right)_V = T$$

$$\left(\frac{\partial H}{\partial P}\right)_S = V$$

$$\left(\frac{\partial G}{\partial P}\right)_T = V \tag{9.73}$$

$$\left(\frac{\partial A}{\partial V}\right)_T = -P$$

$$\left(\frac{\partial U}{\partial V}\right)_S = -P$$

$$\left(\frac{\partial A}{\partial T}\right)_V = -S$$

$$\left(\frac{\partial G}{\partial T}\right)_P = -S$$

$$dS = \frac{C_V}{T}\,dT + \left(\frac{\partial P}{\partial T}\right)_V dV$$

$$\left(\frac{\partial C_V}{\partial V}\right)_T = T\left[\frac{\partial}{\partial T}\left(\frac{\partial P}{\partial T}\right)\right]_V$$

$$dS = \frac{C_P}{T}\,dT - \left(\frac{\partial V}{\partial T}\right)_P dP$$

$$\left(\frac{\partial C_P}{\partial P}\right)_T = -T\left[\frac{\partial}{\partial T}\left(\frac{\partial V}{\partial T}\right)\right]_P$$

$$C_P - C_V = T\left(\frac{\partial P}{\partial T}\right)_V\left(\frac{\partial V}{\partial T}\right)_P$$

$$C_V = \left(\frac{\partial U}{\partial T}\right)_V$$

$$C_P = \left(\frac{\partial H}{\partial T}\right)_P$$

$$C_V = T\left(\frac{\partial S}{\partial T}\right)_V$$

Table 9.2   (*continued*)

$$C_P = T\left(\frac{\partial S}{\partial T}\right)_P$$

$$C_P = \left(\frac{\partial U}{\partial T}\right)_P + P\left(\frac{\partial V}{\partial T}\right)_P$$

$$\frac{C_V}{T}\left(\frac{\partial T}{\partial V}\right)_S = -\left(\frac{\partial P}{\partial T}\right)_V$$

$$\frac{C_P}{T}\left(\frac{\partial T}{\partial P}\right)_S = \left(\frac{\partial V}{\partial T}\right)_P$$

$$\gamma = \frac{C_P}{C_V}$$

$$\gamma = \left(\frac{\partial V}{\partial P}\right)_T\left(\frac{\partial P}{\partial V}\right)_S$$

$$\left(\frac{\partial U}{\partial V}\right)_T = T\left(\frac{\partial P}{\partial T}\right)_V - P$$

$$dU = C_V\, dT + \left[T\left(\frac{\partial P}{\partial T}\right)_V - P\right]dV$$

$$\left(\frac{\partial H}{\partial P}\right)_T = V - T\left(\frac{\partial V}{\partial T}\right)_P$$

$$dH = C_P\, dT + \left[V - T\left(\frac{\partial V}{\partial T}\right)_P\right]dP$$

$$\left(\frac{\partial T}{\partial P}\right)_H = \frac{1}{C_P}\left[T\left(\frac{\partial V}{\partial T}\right)_P - V\right]$$

$$\left(\frac{\partial T}{\partial P}\right)_H = -\frac{1}{C_P}\left(\frac{\partial H}{\partial P}\right)_T$$

$$\left(\frac{\partial T}{\partial P}\right)_H = -\frac{1}{C_P}\left[\left(\frac{\partial U}{\partial P}\right)_T + \left(\frac{\partial(PV)}{\partial P}\right)_T\right]$$

$$\left(\frac{\partial T}{\partial V}\right)_U = -\frac{1}{C_V}\left[T\left(\frac{\partial P}{\partial T}\right)_V - P\right]$$

$$\left(\frac{\partial T}{\partial V}\right)_U = -\frac{1}{C_V}\left(\frac{\partial U}{\partial V}\right)_T$$

$$\left(\frac{\partial H}{\partial S}\right)_T = T + V\left(\frac{\partial P}{\partial S}\right)_T$$

$$\left(\frac{\partial H}{\partial S}\right)_T = T - V\left(\frac{\partial T}{\partial V}\right)_P$$

$$\left(\frac{\partial H}{\partial S}\right)_V = T + V\left(\frac{\partial P}{\partial S}\right)_V$$

$$\left(\frac{\partial H}{\partial S}\right)_V = T - V\left(\frac{\partial T}{\partial V}\right)_S$$

$$\left(\frac{\partial T}{\partial S}\right)_H = -\frac{T}{V}\left(\frac{\partial T}{\partial P}\right) = \frac{T}{C_P}\left[1 - \frac{T}{V}\left(\frac{\partial V}{\partial T}\right)_P\right]$$

$$\left(\frac{\partial H}{\partial P}\right)_V = V + C_V\left(\frac{\partial T}{\partial P}\right)_V$$

(*continued*)

Table 9.2   (continued)

$$\left(\frac{\partial H}{\partial P}\right)_V = V - T\left(\frac{\partial V}{\partial T}\right)_S$$

$$\left(\frac{\partial U}{\partial P}\right)_T = -T\left(\frac{\partial V}{\partial T}\right)_P - P\left(\frac{\partial V}{\partial P}\right)_T$$

$$\left(\frac{\partial H}{\partial V}\right)_P = T\left(\frac{\partial P}{\partial T}\right)_S$$

$$\left(\frac{\partial U}{\partial T}\right)_P = \left(\frac{\partial U}{\partial T}\right)_V + \left(\frac{\partial U}{\partial V}\right)_T\left(\frac{\partial V}{\partial T}\right)_P$$

$$\left(\frac{\partial V}{\partial P}\right)_T = -V\kappa$$

$$\left(\frac{\partial V}{\partial T}\right)_P = V\alpha$$

$$\left(\frac{\partial P}{\partial T}\right)_V = \frac{\alpha}{\kappa}$$

Table 9.3   **The relationships between partial molar quantities**

$$\bar{G}_i = \left(\frac{\partial G}{\partial n_i}\right)_{T,P,n_j} = \mu_i \tag{9.74}$$

$$\bar{S}_i = \left(\frac{\partial S}{\partial n_i}\right)_{T,P,n_j} = \left(\frac{\partial \bar{G}_i}{\partial T}\right)_{P,n_j} \tag{9.75}$$

$$\bar{H}_i = \left(\frac{\partial H}{\partial n_i}\right)_{T,P,n_j} = \bar{G}_i + T\bar{S}_i \tag{9.76}$$

$$\bar{V}_i = \left(\frac{\partial V}{\partial n_i}\right)_{T,P,n_j} = \left(\frac{\partial \bar{G}_i}{\partial P}\right)_{T,n_j} \tag{9.77}$$

$$\bar{C}_{P,i} = \left(\frac{\partial C_P}{\partial n_i}\right)_{T,P,n_j} = \left(\frac{\partial \bar{H}_i}{\partial T}\right)_{P,n_j} = T\left(\frac{\partial \bar{S}_i}{\partial T}\right)_{P,n_j} \tag{9.78}$$

$$\bar{\alpha}_i = \left(\frac{\partial \alpha}{\partial n_i}\right)_{T,P,n_j} = \left(\frac{\partial \bar{V}_i}{\partial T}\right)_{P,n_j} \tag{9.79}$$

$$\bar{\kappa}_i = \left(\frac{\partial \kappa}{\partial n_i}\right)_{T,P,n_j} = \left(\frac{\partial \bar{V}_i}{\partial P}\right)_{T,n_j} \tag{9.80}$$

the appropriate thermodynamic potential in terms of other properties which can be easily measured over ranges of temperature and pressure. For geochemical substances these properties are the heat capacity at constant pressure C_P and the volume V. Temperature effects are usually related to C_P and pressure effects to V.

9.2.1   ### The effect of temperature and pressure on entropy

For any system, the temperature dependence of entropy at constant pressure is given by equation 9.22, rewritten as:

$$\left(\frac{\partial S}{\partial T}\right)_P = \frac{C_P}{T} \qquad \text{(constant pressure)} \tag{9.81}$$

Deriving the pressure dependence of entropy is a little more complicated since we need to determine $(\partial S/\partial P)_T$ in terms of volume. To do this we recall the Maxwell equation (9.19)

$$\left(\frac{\partial S}{\partial P}\right)_T = -\left(\frac{\partial V}{\partial T}\right)_P \tag{9.82}$$

Using the definition of the coefficient of isobaric expansion in equation 9.4.4, equation 9.82 can be written:

$$\left(\frac{\partial S}{\partial P}\right)_T = -V\alpha \qquad \text{(constant temperature)} \tag{9.83}$$

The temperature and pressure dependence of entropy can be combined since:

$$dS = \left(\frac{\partial S}{\partial T}\right)_P dT + \left(\frac{\partial S}{\partial P}\right)_T dP \tag{9.84}$$

Therefore,

$$dS = \frac{C_P}{T} dT - V\alpha \, dP \tag{9.85}$$

This can be integrated for a finite change $(T_1, P_1) \rightarrow (T_2, P_2)$ to give:

$$S_{(T_2, P_2)} - S_{(T_1, P_1)} = \int_{T_1}^{T_2} \frac{C_P}{T} dT - \int_{P_1}^{P_2} V\alpha \, dP \tag{9.86}$$

There are two integration paths which are practical (figure 9.4).

● First is the path:

$$(T_1, P_1) \rightarrow (T_2, P_1) \rightarrow (T_2, P_2)$$

This path requires a knowledge of C_P as a function of temperature at P_1 plus $V\alpha$ as a function of pressure at T_2.

● A second, equivalent, integration path is:

$$(T_1, P_1) \rightarrow (T_1, P_1) \rightarrow (T_2, P_2)$$

This path requires a knowledge of C_P as a function of temperature at P_2 plus $V\alpha$ as a function of pressure at T_1.

In practice it is difficult to carry out thermal measurements such as C_P at elevated pressure; therefore the first integration path is usually chosen. To be clear which integration path is chosen it is common to label all the variables. Therefore, equation 9.86 becomes:

$$S_{(T_2, P_2)} - S_{(T_1, P_2)} = \int_{T_1}^{T_1} \frac{C_{P(T,P_1)}}{T} dT - \int_{P_1}^{P_2} V_{(T_2, P)}\alpha_{(T_2, P)} \, dP \tag{9.87}$$

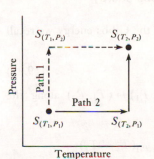

Figure 9.4
The path of entropy, a state function.

The importance of correctly defining the limits of integration and the integration path cannot be overstated.

The integrals can be evaluated directly if the temperature or pressure dependence of C_P or $V\alpha$ can be described with an algebraic expression such as a power series.

The integrations may be simplified if it is assumed that C_P and/or α are independent of temperature and pressure, in which case the C_P and α terms go outside the integrals. Further simplification can be introduced if the system is incompressible, in which case α is zero.

9.2.2 The effect of temperature and pressure on enthalpy

For a closed system:

$$dH = T\,dS + V\,dP \tag{9.88}$$

We already know that the temperature and pressure dependence of S is given by equation 9.85. Therefore,

$$dH = C_P\,dT + V(1 - \alpha T)\,dP \tag{9.89}$$

This can be integrated for a finite change $(T_1, P_1) \rightarrow (T_2, P_2)$ along the conventional integration path to give:

$$H_{(T_2,P_2)} - H_{(T_1,P_1)} = \int_{T_1}^{T_2} C_{P(T,P_1)}\,dT + \int_{P_1}^{P_2} V_{(T_2,P)}(1 - \alpha_{(T_2,P)})\,dP \tag{9.90}$$

9.2.3 The effect of temperature and pressure on the Gibbs energy

To determine the effect of temperature and pressure on the Gibbs energy we recall the relationship:

$$dG = V\,dP - S\,dT \tag{9.91}$$

Integrating equation 9.91 for the finite change $(T_1, P_1) \rightarrow (T_2, P_2)$ along the conventional integration path gives:

$$G_{(T_2,P_2)} - G_{(T_1,P_1)} = \int_{P_1}^{P_2} V_{(T_2,P)}\,dP - S_{(T_1,P_1)}(T_2 - T_1)$$
$$+ \int_{T_1}^{T_2} C_{P(T_1,P_1)}\left(1 - \frac{T_2}{T}\right)dT \tag{9.92}$$

Two other common variations of equation 9.92 are:

$$G_{(T_2,P_2)} - G_{(T_1,P_1)} = \int_{P_1}^{P_2} V_{(T_2,P)}\,dP - S_{(T_1,P_1)}(T_2 - T_1)$$
$$+ \int_{T_1}^{T_2} C_{P(T,P_1)}\,dT - T_2 \int_{T_1}^{T_2} \frac{C_{P(T,P_1)}}{T}\,dT \tag{9.93}$$

and

$$G_{(T_2,P_2)} - G_{(T_1,P_1)} = \int_{P_1}^{P_2} V_{(T_2,P)}\,\mathrm{d}P - S_{(T_1,P_1)}(T_2 - T_1)$$

$$+ \int_{T_1}^{T_2} C_{P(T,P_1)}\,\mathrm{d}T - T_2 \int_{T_1}^{T_2} C_{P(T,P_1)}\,\mathrm{d}\ln T \quad (9.94)$$

Clearly, to evaluate the change in the Gibbs energy as a function of temperature and pressure we need a value for the entropy at (T_1, P_1) plus information on the temperature dependence of C_P at the pressure P_1 and the pressure dependence of V at T_2. Symbols used for the limits of the integrals should be noted carefully since confusion may arise if they are not clearly distinguished from the symbols used for the differentials $\mathrm{d}P$ and $\mathrm{d}T$. For example, equation 9.92 contains the term

$$\int_{T_1}^{T_2} C_{P(T,P_1)}\left(1 - \frac{T_2}{T}\right)\mathrm{d}T$$

where T_2 and T_1 are fixed limits of the integration and T represents a variable which changes between the limits as the integration is performed. In some publications the symbol T is used for both the limit and the differential. If the alternative equation 9.94 is used, this potential source of confusion is less apparent since there are no terms in T_2 within the integral. It is always wise to make the notation unambiguous. However, in later sections we will use equations taken directly from the literature where the less rigorous notation is sometimes used.

The derivation of equations 9.92–9.94 is more difficult than that of the equivalent equations for either entropy or enthalpy. It is given below for completeness. The objective is to transform the basic relationships, without making approximations, to produce the simplest working equations.

Box 9.3 | **T and P dependence of the Gibbs energy**

The integration of equation 9.91 can be written:

$$G_{(T_2,P_2)} - G_{(T_1,P_2)} = \int_{P_1}^{P_2} V_{(T_2,P)}\,\mathrm{d}P - \int_{T_1}^{T_2} S_{(P_1,T)}\,\mathrm{d}T' \quad (9.95)$$

Note that we have used the symbol $\mathrm{d}T'$ rather than $\mathrm{d}T$. This is to distinguish between this integration with respect to temperature and a second integration step which will be introduced later.

The pressure term can be integrated simply if the dependence of volume on pressure is known at T_2.

Consider now the temperature dependence of G, which at the constant pressure P_1 becomes:

$$G_{(T_2,P_1)} - G_{(T_1,P_1)} = -\int_{T_1}^{T_2} S_{(P_1,T)}\,\mathrm{d}T' \quad (9.96)$$

This integration is complicated by the fact that the entropy has to be evaluated

continued

Box 9.3
continued

in terms of the measurable property C_P. Equation 9.91 gives the entropy in terms of the temperature dependence of C_P at the constant pressure P_1, i.e.

$$S_{(T_2,P_1)} = S_{(T_1,P_1)} + \int_{T'}^{T_2} \frac{C_{P(T,P_1)}}{T} \, dT \qquad (9.97)$$

This can be substituted into equation 9.86 to give an expression containing a double integral:

$$G_{(T_2,P_1)} - G_{(T_1,P_1)} = - \int_{T_1}^{T_2} S_{(T_1,P_1)} \, dT' + \int_{T_1}^{T_2} \left(\int_{T_1}^{T'} \frac{C_{P(P_1,T)}}{T} \, dT \right) dT' \qquad (9.98)$$

The first integral, containing the constant entropy term, can be integrated directly giving

$$G_{(T_2,P_1)} - G_{(T_1,P_1)} = -S_{(T_1,P_1)} (T_2 - T_1) - \int_{T_1}^{T_2} \left(\int_{T_1}^{T} \frac{C_{P(P_1,T)}}{T} \, dT \right) dT' \qquad (9.99)$$

In order to simplify the expression further it is necessary to consider the second double integral. In this form the double integral cannot be simplified because the inner integral cannot itself be simplified without including the full temperature dependence of $C_{P(P_1,T)}/T$ at this early stage. This difficulty arises because the inner integral is with respect to T and the function $C_{P(P_1,T)}/T$ is of course a function of T. The double integral represents a 'volume' bounded by the $C_{P(P_1,T)}/T$ surface over the triangular area bounded by the line $T = T'$ and the limits T_1, T_2. Figure 9.5(b) shows the integration path is equivalent to summing an infinite number of strips, of length $T - T_1$, at progressively increasing T'. Figure 9.5(c) shows that the same area can be covered by summing an infinite number of strips, of length $T_2 - T'$, at progressively increasing T.

This is the equivalent of changing the integration sequence by changing the integration limits to give:

$$G_{(T_2,P_1)} - G_{(T_1,P_1)} = -S_{(T_1,P_1)} (T_2 - T_1) - \int_{T_1}^{T_2} \left(\int_{T}^{T_2} \frac{C_{P(P_1,T)}}{T} \, dT' \right) dT \qquad (9.100)$$

In this form the inner integral has the term $(C_{P(P_1,T)}/T$ which is integrated with respect to dT'. However, $(C_{P(P_1,T)}/T)$ contains no terms in dT' and hence can be taken outside the inner integral to give:

$$G_{(T_2,P_1)} - G_{(T_1,P_1)} = -S_{(T_1,P_1)} (T_2 - T_1) - \int_{T_1}^{T_2} \left(\frac{C_{P(P_1,T)}}{T} \int_{T}^{T_2} dT' \right) dT \qquad (9.101)$$

The resulting inner integral can be evaluated simply, since

$$\int_{T}^{T_2} dT' = T_2 - T \qquad (9.102)$$

continued

Box 9.3
continued

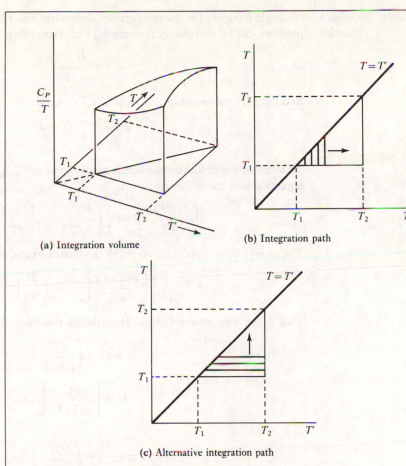

(a) Integration volume

(b) Integration path

(c) Alternative integration path

Figure 9.5   Transforming integration limits.

Therefore:

$$G_{(T_2, P_2)} - G_{(T_1, P_1)} = -S_{(T_1, P_1)}(T_2 - T_1) - \int_{T_1}^{T_2} \left[\frac{C_{P(P_1, T)}}{T}(T_2 - T) \right] dT$$

(9.103)

This simplification allows further rearrangement to give:

$$G_{(T_2, P_1)} - G_{(T_1, P_1)} = -S_{(T_1, P_1)}(T_2 - T_1) + \int_{T_1}^{T_2} C_{P(P_1, T)}\left(1 - \frac{T_2}{T}\right) dT$$

(9.104)

The final stage involves combining the temperature term (equation 9.104) with the pressure term in equation 9.85 to give:

$$G_{(T_2, P_2)} - G_{(T_1, P_1)} = \int_{P_1}^{P_2} V_{(T_2, P)} \, dP - S_{(T_1, P_1)}(T_2 - T_1) + \int_{T_1}^{T_2} C_{P(P_1, T)}\left(1 - \frac{T_2}{T}\right) dT$$

(9.105)

A simple equation for the temperature dependence of the Gibbs energy at constant pressure can be derived by rearranging the relationship $G = H - TS$ to give:

$$S = \frac{H - G}{T} \tag{9.106}$$

Recalling the relationship $-S = (\partial G / \partial T)_P$, then

$$\left(\frac{\partial G}{\partial T}\right)_P = \frac{G - H}{T} \tag{9.107}$$

This can be simplified by considering the expression $[\partial(G/T)/\partial T]_P$, which can be separated into the following terms:

$$\left[\frac{\partial(G/T)}{\partial T}\right]_P = \frac{1}{T}\left(\frac{\partial G}{\partial T}\right)_P + G\left[\frac{\partial}{\partial T}\left(\frac{1}{T}\right)\right]_P \tag{9.108}$$

The term $(\partial G / \partial T)_P$ in equation 9.108 can be substituted with equation 9.107 to yield:

$$\left[\frac{\partial(G/T)}{\partial T}\right]_P = -\left(\frac{H}{T^2}\right) \tag{9.109}$$

This is one form of the **Gibbs–Helmholtz** equation. Alternative forms are:

$$H = G - \left(\frac{\partial G}{\partial \ln T}\right)_P \tag{9.110}$$

$$H = \left[\frac{\partial G}{\partial(1/T)}\right]_P \tag{9.111}$$

and

$$H = -T^2\left(\frac{\partial G/T}{\partial T}\right)_P \tag{9.112}$$

These are all thermodynamic identities, valid for infinitesimal changes, but can be applied to finite changes over small temperature ranges where the heat capacity and therefore the enthalpy are constant. If applied over extensive temperature ranges it is necessary to account explicitly for the changes in C_P and H.

Figure 9.6 shows graphically the effect of temperature on the Gibbs energy.

9.2.4 The effect of temperature and pressure on chemical potential

The relationship between the chemical potential of the ith substance, temperature and pressure is given by:

$$d\mu_i = \bar{V}_i \, dP - \bar{S}_i \, dT \tag{9.113}$$

We may then deduce that the effect of temperature and pressure on the chemical potential of the ith substance is given by:

$$\mu_{i(T_2, P_2)} - \mu_{i(T_1, P_1)} = \int_{P_1}^{P_2} \bar{V}_{i(T_2, P)} \, dP - \int_{T_1}^{T_2} \bar{S}_{i(T_1, P_1)} \, dT$$

$$+ \int_{T_1}^{T_2} \bar{C}_{Pi(T, P_1)}\left(1 - \frac{T_2}{T}\right) dT \tag{9.114}$$

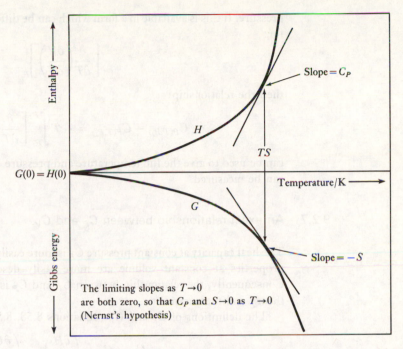

Figure 9.6   Variation of Gibbs energy G and enthalpy H of a pure substance with temperature at constant pressure (after Moore, 1972).

The effect of temperature and pressure on other partial molar properties can be evaluated with equations analogous to those in sections 9.2.1 and 9.2.2.

9.2.5   The effect of temperature and pressure on internal energy

It is sufficient to state that the effect of temperature and pressure on the internal energy can be derived from the following equations:

$$\left(\frac{\partial U}{\partial T}\right)_P = \left(\frac{\partial U}{\partial T}\right)_V + \left(\frac{\partial U}{\partial V}\right)_T\left(\frac{\partial V}{\partial T}\right)_P \tag{9.115}$$

$$\left(\frac{\partial U}{\partial P}\right)_T = -T\left(\frac{\partial V}{\partial T}\right)_P - P\left(\frac{\partial V}{\partial P}\right)_T \tag{9.116}$$

9.2.6   The effect of temperature and pressure on the heat capacity

If the temperature dependence of C_P at a known pressure P_1 can be determined, then C_P may be determined at any temperature and pressure using the identity:

$$\left(\frac{\partial C_P}{\partial P}\right)_T = -T\left[\frac{\partial}{\partial T}\left(\frac{\partial V}{\partial T}\right)\right]_P \tag{9.117}$$

The requirement is an expression for the dependence of volume on temperature and pressure. If this is available in a form which can be differentiated to give an expression for:

$$\left[\frac{\partial}{\partial T} \left(\frac{\partial V}{\partial T} \right) \right]_P \tag{9.118}$$

then the relationship:

$$C_{P(T,P_2)} - C_{P(T,P_1)} = -T \int_{P_1}^{P_2} \left[\frac{\partial}{\partial T} \left(\frac{\partial V}{\partial T} \right) \right]_P dP \tag{9.119}$$

can be used to give the full temperature and pressure dependence of C_P if $(\partial V/\partial T)_P$ can be measured.

9.2.7 An exact relationship between C_P and C_V

The heat capacity at constant pressure C_P is more easily measured than C_V. However, properties at constant volume are more easily described by theoretical physics. Consequently, the relationship between C_V and C_P is commonly a key link between theory and measurement.

The definitions of C_P and C_V (equations 8.58, 8.57) give:

$$C_P - C_V = \left(\frac{\partial H}{\partial T} \right)_P - \left(\frac{\partial U}{\partial T} \right)_V \tag{9.120}$$

Recalling that $H = U + PV$, we can replace the terms in H with terms in U and PV giving:

$$C_P - C_V = \left(\frac{\partial U}{\partial T} \right)_P - \left(\frac{\partial U}{\partial T} \right)_V + P \left(\frac{\partial V}{\partial T} \right)_P \tag{9.121}$$

Progress can be made if we evaluate an expression for $(\partial U/\partial T)_P$. We know that the internal energy is a function of temperature and volume, and volume is a function of pressure and temperature. Therefore:

$$dU = \left(\frac{\partial U}{\partial V} \right)_T dV + \left(\frac{\partial U}{\partial T} \right)_V dT \tag{9.122}$$

and

$$dV = \left(\frac{\partial V}{\partial T} \right)_P dT + \left(\frac{\partial V}{\partial P} \right)_T dP \tag{9.123}$$

Substituting the expression for dV into dU at constant P, the desired expression for $(\partial U/\partial T)_P$ is:

$$\left(\frac{\partial U}{\partial T} \right)_P = \left(\frac{\partial U}{\partial V} \right)_T \left(\frac{\partial V}{\partial T} \right)_P + \left(\frac{\partial U}{\partial T} \right)_V \tag{9.124}$$

The above expression can also be derived from equation 9.122 using the rule for changing the variable held constant (equation 8.10, box 8.1).

Equation 9.124 can be substituted into equation 9.121 to give:

$$C_P - C_V = \left[P + \left(\frac{\partial U}{\partial V} \right)_T \right] \left(\frac{\partial V}{\partial T} \right)_P \tag{9.125}$$

The term $(\partial U/\partial V)_T$ has the units of pressure and is called the **internal pressure** of the system. It is a measure of the cohesive forces between the molecules of a system, which for gases is small but for solids and liquids may be large. Consequently the term

$$\left(\frac{\partial U}{\partial V}\right)_T \left(\frac{\partial V}{\partial T}\right)_P$$

is a measure of the energy required to overcome the cohesive forces for a given volume change at constant temperature. For ideal gases the internal pressure is zero. In general the pressure terms in equation 9.125 support the concept that the difference in C_P is related to work of expansion at constant pressure.

There are several ways of expressing the exact relationship between C_P and C_V. The following derivation is a useful exercise in the manipulation of thermodynamic equations. It will generate an equation which relates the difference in the heat capacities to α and κ.

If we recall that

$$\alpha = \frac{1}{V}\left(\frac{\partial V}{\partial T}\right)_P$$

then equation 9.125 can be written:

$$C_P - C_V = \left[P + \left(\frac{\partial U}{\partial V}\right)_T \right]\alpha V \qquad (9.126)$$

Further development can be made if we find an alternative expression for $(\partial U/\partial V)_T$. This is achieved using the basic relationship:

$$dU = \left(\frac{\partial U}{\partial S}\right)_V dS + \left(\frac{\partial U}{\partial V}\right)_S dV \qquad (9.127)$$

The dV term can be eliminated using equation 9.123 with the condition of constant pressure to yield:

$$\left(\frac{\partial U}{\partial V}\right)_T = \left(\frac{\partial U}{\partial S}\right)_V\left(\frac{\partial S}{\partial V}\right)_T + \left(\frac{\partial U}{\partial V}\right)_S \qquad (9.128)$$

In equations 8.80 and 8.81 pressure and temperature were defined as $T = (\partial U/\partial S)_V$ and $P = -(\partial U/\partial V)_S$. These can be substituted to give:

$$\left(\frac{\partial U}{\partial V}\right)_T = T\left(\frac{\partial S}{\partial V}\right)_T - P \qquad (9.129)$$

A Maxwell relationship can be used for further simplification since:

$$\left(\frac{\partial S}{\partial V}\right)_T = \left(\frac{\partial P}{\partial T}\right)_V \qquad (9.130)$$

Therefore, the desired expression for $(\partial U/\partial V)_T$ is:

$$\left(\frac{\partial U}{\partial V}\right)_T = T\left(\frac{\partial P}{\partial T}\right)_V - P \qquad (9.131)$$

The next step is to eliminate $(\partial U/\partial V)_T$ from equation 9.126 to give:

$$C_P - C_V = \alpha V T \left(\frac{\partial P}{\partial T}\right)_V \qquad (9.132)$$

Finally we use the relationship defined in equation 9.46 to give:

$$C_P - C_V = \frac{\alpha^2 V T}{\kappa} \qquad (9.133)$$

We note that all the terms on the right-hand side of the above expression are always positive, which confirms that C_P is always bigger than C_V. Equation 9.133 gives a simple way of calculating C_V from C_P since α, κ and V are much easier to measure than is C_V itself.

9.2.8   The need for reference states

We have seen that the **changes** in the thermodynamic potential of a substance over a specified range of temperature and pressure can be evaluated if the heat capacity and volume of the substance can be measured over the range. However, to evaluate the **actual** value of a potential at (T_2, P_2) we require the value of the potential at a chosen state (T_1, P_1). Unfortunately, there is no way to measure the thermodynamic potentials directly. This is a similar problem to that faced by geographers expressing the height of mountains. Rather than trying to determine some type of absolute elevation, the problem is resolved by adopting an arbitrary reference point, sea level, with a defined elevation of zero feet or metres. Similarly, chemists have adopted arbitrary reference points from which to measure thermodynamic potentials. The reference states are different for different types of substances and will be discussed in more depth in later sections. However, a wealth of experimental measurements on heat capacities at low temperature have suggested that for entropy there may be an absolute standard. This gives rise to the third law of thermodynamics.

9.2.9   The third law of thermodynamics

The statistical entropy (box 8.2) is given by:

$$S = k \ln \Omega \qquad (9.134)$$

where Ω is the number of different microscopic states available to the system. If a state exists where $\Omega = 1$ then the entropy of the system equals zero. This condition exists in a perfect crystal at the absolute zero of temperature since all atoms are regularly distributed and have their lowest possible energies.

Nernst deduced theoretically that for any chemical reaction between several substances the value ΔS tends to zero at 0 K (Nernst's heat theorem). This does not prove that the entropy of each substance is zero since it is possible that non-zero entropies may sum to zero. However, this situation would be fortuitous and the most likely explanation is that entropies are in fact zero at the absolute zero of temperature.

A formal statement of the third law The above arguments, supported by a wealth of experimental data, support a statement of the third law of thermodynamics, i.e.

THE ENTROPY OF A PERFECT CRYSTAL OF A COMPOUND OR ELEMENT IS ZERO AT THE ABSOLUTE ZERO OF TEMPERATURE.

Examples of this are shown in figure 9.7, which depicts the heat capacities of copper and graphite as a function of absolute temperature. The tendency is to zero heat capacity at zero temperature. Recalling that, at constant pressure,

$$S_{(T_2)} = S_{(T_1)} + \int_{T_1}^{T_2} \frac{C_P}{T} \, \mathrm{d}T \qquad (9.135)$$

or alternatively

$$S_{(T_2)} = S_{(T_1)} + \int_{T_1}^{T_2} C_P \, \mathrm{d} \ln T \qquad (9.136)$$

then plots of C_P versus $\ln T$ or C_P/T versus T can be integrated graphically between the limits $T_1 = 0$, where $S(T_1) = 0$, and T_2 to give the absolute entropy of a substance. This is shown graphically in figure 9.8. Strictly the function C_P/T is indeterminate at $T = 0$ and an extrapolation of data to 0 K has to be made for this type of plot.

In practice the third law is no more than a convenient reference point and does not have same status as the first two laws. The third law defines only the **calorimetric** contributions to entropy. Other contributions to the third-law entropy are derived from various types of disorder that may exist in a real crystal at absolute zero. These are termed **residual entropies**. One such, the **configurational entropy**, arises from the possibility that a set of crystallographically equivalent sites may be populated by more than one type of atom. The number of possible configurations is related

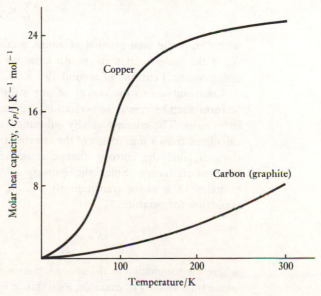

Figure 9.7 Molar heat capacity as a function of temperature (after Mahan, 1964).

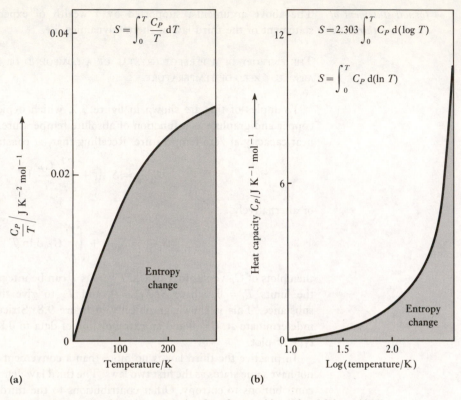

$$S = \int_0^T \frac{C_P}{T} \, dT$$

$$S = 2.303 \int_0^T C_P \, d(\log T)$$

$$S = \int_0^T C_P \, d(\ln T)$$

Figure 9.8   Evaluating entropy from heat capacity data for graphite (after Mahan, 1964).

to the mole fractions of the atoms occupying any one set of sites and the configurational entropy is given by:

$$S_{conf} = -R \sum_j m_j \sum X_{ij} \ln X_{ij} \tag{9.137}$$

where m_j is the total number of atoms occupying the jth crystallographic site and X_{ij} is the mole fraction of the ith atom on the jth site. The mineral albite has a configurational entropy of around $19 \, J \, K^{-1}$.

Contributions to the energy of any substance at absolute zero also arise from features such as crystal imperfection and magnetic spin ordering in paramagnetic substances. The entropy of any substance at a specific temperature, S_T, may be calculated from a summation of the calorimetric contributions, the residual contributions S_R and the entropy change associated with any phase transitions in the temperature range. Since the entropy of a phase transition is related to the enthalpy of a phase transition by $\Delta S_{trans} = \Delta H_{trans} / T_{trans}$, where T_{trans} is the transition temperature:

$$S_T = \int_0^T \frac{C_P}{T} + S_R + \sum \frac{\Delta H_{trans}}{T_{trans}} \tag{9.138}$$

where the summation in the above equation represents the inclusion of all observed phase transitions. For example, a substance undergoing both fusion and vaporisation during a temperature increase would require the addition of ΔH_{fus} and ΔH_{vap}.

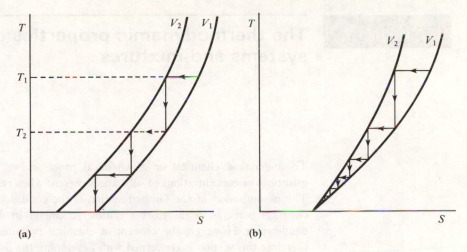

Figure 9.9   The unattainability of absolute zero.

An alternative, and more fundamental, statement of the third law is that:

IT IS IMPOSSIBLE, BY ANY PROCEDURE, TO REDUCE THE TEMPERATURE OF ANY SYSTEM TO ABSOLUTE ZERO BY A FINITE NUMBER OF STEPS.

This is an experimental observation consistent with the fact that the calorimetric entropies of all substances tend to the same value at absolute zero regardless of any other conditions. Suppose the state of a system is described by temperature T and any other quantitative parameter such as volume V. We may construct a single plot of T against S (entropy) for each value of V. Figure 9.9(a) shows this for two values of V. Consider a system defined by V_1, T_1 undergoing a reversible isothermal change to V_2, T_1 followed by a reversible adiabatic change. The second law requires that $\partial S/\partial T < 0$; therefore this process will lead to a spontaneous temperature drop, i.e. a change from V_2, T_1 to V_2, T_2. It appears that successive cycles such as this will give rise to a progressive lowering of temperature. This is the basis of the Hampson–Lindé process for liquefying gases where a gas is first compressed then passed through a narrow orifice where it expands adiabatically to undergo further cooling in accordance with the Joule–Thomson effect. The cooled gas is passed back into the system for further cycles. After repeated cycles the cooling effect is sufficient to liquefy a fraction of the gas.

A practical limit to the extent of cooling by such a system is imposed by the size of the required pumps. Similar techniques have shown that a further theoretical limit may also exist. Figure 9.9(a) suggests that the magnitude of any induced temperature drop is proportional to the entropy difference between the two curves. If the curves were closer together on the entropy axis the difference between T_1 and T_2 would be smaller. If the curves converge, then the progressive lowering of temperature leads to smaller entropy differences and progressively smaller temperature drops in subsequent cycles. It is therefore impossible to reach the point of intersection of the curves by a finite number of steps (figure. 9.9(b)). The asymptotic limit of such a cyclic process is the absolute zero of temperature. It has proved impossible to reach absolute zero by any such process and we conclude that the entropy of any substance is indeed fixed and positive at absolute zero. This deduction is consistent with Nernst's heat theorem.

Chapter 10

The thermodynamic properties of simple systems and mixtures

To understand chemical or geochemical processes we often need to know the quantities or concentrations of substances present when reactions are at equilibrium. The minimisation of the Gibbs function is the condition for equilibrium and the chemical potential is therefore a critical parameter in describing the position of equilibrium. However, the concept of chemical potential is formal and abstract; therefore this section is concerned with developing the links between concentration and chemical potential.

In this section the superscript $\ominus$ is generally used to denote a standard state. Occasionally other symbols are used when it is necessary to distinguish specific standard states that are different from each other.

We must make the distinction between a reference state and a standard state. In the reference state a numerical value may be assigned to a property by convention. In contrast, a standard state is simply an arbitrary set of conditions, possibly composition, temperature and pressure, which serves as a point from which we may make comparisons. We may never know the actual value of a property in a standard state since most measurements are concerned with the difference in properties between the standard state and the real conditions.

10.1 Non-reactive chemical mixtures

10.1.1 Equilibrium distribution of substances between phases

In some systems one or more components may exist in more than one phase. For example, liquid water may be at equilibrium with gaseous water vapour; or a soluble gas in contact with a solvent may exist in both the gas phase and in the dissolved state. A useful concept is that:

AT EQUILIBRIUM, THE CHEMICAL POTENTIAL OF ANY SUBSTANCE IS EQUAL IN ALL THE PHASES BETWEEN WHICH THE SUBSTANCE CAN PASS.

This can be deduced by considering a system containing two open phases α and β in thermal equilibrium. Each phase is composed of one or more substances which can move freely between phases.

If a small quantity, δn_A, of substance A is transferred from phase α to phase β, the overall change in the Gibbs energy is:

$$dG = dG^\alpha + dG^\beta = \delta n_A(\mu_A^\beta - \mu_A^\alpha) \tag{10.1}$$

The process will occur spontaneously if dG is less than zero, i.e.

$$\mu_A^\alpha > \mu_A^\beta \qquad (10.2)$$

The system will show no tendency to change if $dG = 0$. This condition occurs if:

$$\mu_A^\alpha = \mu_A^\beta \qquad (10.3)$$

This second condition is then the general statement of phase equilibrium and is of fundamental importance.

10.1.2 Absolute activity and activity

The chemical potentials of substances dictate the position of equilibrium in systems at constant temperature and pressure. However, it is often convenient to use a related parameter called the **absolute activity**, λ. For any substance, denoted A, λ_A is defined as:

$$\lambda_A \equiv \exp\left(\frac{\mu_A}{RT}\right) \qquad (10.4)$$

or

$$\mu_A = RT \ln \lambda_A \qquad (10.5)$$

If equation 10.5 is applied to a component in its particular state and to an arbitrarily chosen standard state, denoted by $\ominus$, the two resulting equations can be substracted to yield:

$$\mu_A - \mu_A^\ominus = RT \ln(\lambda_A / \lambda_A^\ominus) \qquad (10.6)$$

The ratio of absolute activities is defined as the **activity** a_A. That is:

$$a_A \equiv \frac{\lambda_A}{\lambda_A^\ominus} \qquad (10.7)$$

Consequently, equation 10.6 becomes:

$$\mu_A = \mu_A^\ominus + RT \ln a_A \qquad (10.8)$$

The activity is a measure of the difference in energy between a component in a system at known temperature and pressure and the same component in an arbitrary standard state. Therefore, **there is no unique value for the activity of a substance**.

Activity is dimensionless and has the value unity for any component in its standard state.

Activity is one of the most important variables in chemical thermodynamics. Its major role is to link the thermodynamic properties of a system to its composition. This application is possible because activity is usually measurable and can be related directly to the concentration of a substance in a mixture. Indeed, it is common to describe the activity of a substance as the 'thermodynamic equivalent of concentration'. This statement is merely illustrative and must not be thought of as the definition of activity.

The following sections describe the conventions adopted for activity, and the link with composition for some common chemical systems.

10.2   Gases

10.2.1   Pure ideal gases

The pressure dependence of the chemical potential is given by:

$$d\mu \equiv \bar{V} \, dP \tag{10.9}$$

The ideal gas law (equation 5.6) states that the molar volume of an ideal gas ($\bar{V}$) is given by:

$$\bar{V} = \frac{RT}{P} \tag{10.10}$$

For a system at fixed temperature the above equations can be combined to give:

$$d\mu = \left(\frac{RT}{P}\right) dP \tag{10.11}$$

An alternative expression which is often found useful is:

$$d\mu = RT \, d\ln P = \bar{V} \, dP \tag{10.12}$$

Equation 10.11 can be integrated between any pressure and a conveniently chosen standard state to give:

$$\mu - \mu^{\ominus} = RT \int_{P^{\ominus}}^{P} \left(\frac{1}{P}\right) dP' \tag{10.13}$$

where $\mu^{\ominus}$ is the **standard state chemical potential**. Note that we have replaced dP with dP' to avoid confusion with the symbol P which now represents a specified value of pressure at one limit on the integral. Then,

$$\mu - \mu^{\ominus} = RT \ln(P/P^{\ominus}) \tag{10.14}$$

The common standard state for an ideal gas is **1 bar pressure at any temperature**. Therefore,

$$\mu_{(T,P)} = \mu^{\ominus}_{(T,1\,\text{bar})} + RT \ln(P/1\,\text{bar}) \tag{10.15}$$

or

$$\mu_{(T,P)} = \mu^{\ominus}_{(T,1\,\text{bar})} + RT \ln P \tag{10.16}$$

The subscript (T, 1 bar) registers the fact that the standard chemical potential we have selected is independent of pressure but dependent on temperature. In contrast, the chemical potential which has the subscript (T, P) is dependent on both temperature and pressure. The use of these symbols makes the nomenclature clumsy; despite this we will retain them to avoid confusion.

A comparison of equations 10.8 and 10.14 shows that the activity of a **pure ideal gas** is equal to the ratio of the actual pressure to the standard pressure $P^{\ominus}$.

$$a = \frac{\lambda}{\lambda^{\ominus}} = \frac{P}{P^{\ominus}} \tag{10.17}$$

Using the convention $P^{\ominus} = 1$ bar, we see that $a = P$.

10.2.2   Pure non-ideal gases – fugacity

Equations 10.14 and 10.17 provides a good description of the chemical potential and activity of any gas at low pressure and high temperatures. In gases, at pressures above a few bars, the dependence of volume on pressure does not follow the ideal gas law and equation 10.15 is not applicable.

It is desirable to retain the simple algebraic form of equation 10.15 for non-ideal gases. This may be achieved by replacing pressure with a quantity called **fugacity**. Fugacity is an 'idealised' pressure usually given the symbol f. It is defined by an equation analogous to 10.12; i.e.

$$d\mu = RT\, d\ln f = \bar{V}\, dP \qquad (10.18)$$

We can take a further step and define a **fugacity coefficient** χ which is the ratio of the fugacity to the pressure, i.e.

$$\chi = \frac{f}{P} \qquad (10.19)$$

The value of χ approaches unity as pressure tends to zero and the gas becomes ideal. This condition is expressed mathematically as:

$$\frac{f}{P} \to 1 \quad \text{as} \quad P \to 0 \qquad (10.20)$$

Equation 10.18 can be integrated between any fugacity and a conveniently chosen reference state to give:

$$\mu_{(T,P)} = \mu^{\ominus}_{(T,f^{\ominus})} + RT\, \ln(f/f^{\ominus}) \qquad (10.21)$$

In this equation the symbol $f^{\ominus}$ represents the fugacity of a gas in a selected standard state.

The choice of a standard state adopted for a real gas is quite subtle. One option is to choose a standard state of unit fugacity. However, real gases may have states of unit fugacity in which the pressure is not equal to unity. This leads to different gases having different standard states.

A more convenient standard state is therefore **the state in which the gas has unit fugacity and behaves as an ideal gas**. In this state the fugacity, the pressure and therefore the fugacity coefficient are unity. Consequently,

$$\mu_{(T,P)} = \mu^{\ominus}_{(T,\,1\,\text{bar})} + RT\, \ln(f/1\,\text{bar})$$
$$= \mu^{\ominus}_{(T,\,1\,\text{bar})} + RT\, \ln f \qquad (10.22)$$

Alternatively:

$$\mu_{(T,P)} = \mu^{\ominus}_{(T,\,1\,\text{bar})} + RT\, \ln(P\chi) \qquad (10.23)$$

The selected standard state is a purely hypothetical reference state which is not a state of any real gas. Figure 10.1 shows the standard state graphically. The value of choosing such a standard state is that all deviation from ideality is embodied within the fugacity coefficient for the gas at the specified temperature and pressure. Since the standard state for all real gases is the same, the deviation from ideal behaviour for several gases may be composed by direct analysis of their fugacity coefficients.

Recalling the definition of activity in equations 10.7 and 10.8, we see that the

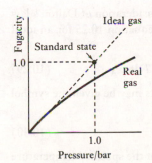

Figure 10.1
Standard state of unit fugacity.

activity is equal to the ratio of fugacities, i.e.

$$a = (\lambda/\lambda^{\ominus}) = (f/f^{\ominus}) \tag{10.24}$$

Using the convention $f^{\ominus} = 1$ bar we see that $a = f$.

10.2.3   Mixtures of gases

For any gas in an **ideal** mixture it follows from equation 10.15 that the chemical potential may be expressed in terms of its partial pressure P_A. For gas A in a mixture this becomes:

$$\mu_{A(T, P)} = \mu^{\ominus}_{A(T, 1\,\text{bar})} + RT \ln P_A/P^{\ominus}_A \tag{10.25}$$

where $\mu^{\ominus}_{A(T, 1\,\text{bar})}$ is the chemical potential of the pure ideal gas at 1 bar pressure and the specified temperature. It is the same standard state as for pure gases where $P^{\ominus}_A = 1$. By this convention, activity equals P_A, which is related to the total pressure P by Dalton's law of partial pressures (section 5.2), i.e. $P_A = x_A P$ where x_A is the mole fraction of A.

For a mixture of **non-ideal** gases we must consider equation 10.23 written in terms of the fugacity of any substance in the mixture. This is

$$\mu_{A(T, P)} = \mu^{\ominus}_{A(T, 1\,\text{bar})} + RT \ln f_A \tag{10.26}$$

The relationship between f_A and P_A is constrained by the conditions:

$$\frac{f_A}{p_A} \to 1 \quad \text{as} \quad P \to 0 \tag{10.27}$$

We may introduce the fugacity coefficient in the mixture as:

$$\chi_A = \frac{f_A}{P_A} = \frac{f_A}{x_A P} \quad \text{or} \quad f_A = \chi_A x_A P \tag{10.28}$$

By this convention activity equals f_A. We note that f_A is dependent on the pressure and composition primarily. However, χ_A will itself be dependent on composition and temperature and in real gases the compositional dependence will itself be temperature-dependent.

Alternative expressions can be generated by further consideration of Dalton's law of partial pressures. Since $P_A = x_A P$, substituting P_A in equation 10.25 for an ideal gas gives:

$$\mu_{A(T, P)} = [\mu^{\ominus}_{A(T, 1\,\text{bar})} + RT \ln P] + RT \ln x_A \tag{10.29}$$

The quantity $[\mu^{\ominus}_{A(T, 1\,\text{bar})} + RT \ln P]$ in the square brackets may be given the symbol $\mu^{\ominus}_{A(T, P)}$. This simplifies equation 10.29 to:

$$\mu_{A(T, P)} = \mu^{\ominus}_{A(T, P)} + RT \ln x_A \tag{10.30}$$

The term $\mu^{\ominus}_{A(T, P)}$ is the chemical potential of the pure gas at the specified temperature and pressure. This can be thought of as an alternative to the conventional standard state $\mu^{\ominus}_{A(T, 1\,\text{bar})}$. With this alternative standard state, the activity of an ideal gas in a mixture is equal to the mole fraction.

For non-ideal gaseous mixtures we use the expression:

$$\mu_{A(T,P)} = \mu_{A(T,P)}^{\ominus} + RT \ln(x_A \chi_A) \tag{10.31}$$

By this convention activity equals $(x_A \chi_A)$. The variable x_A is of course independent of temperature and pressure; however, it must be remembered that χ_A may have a notable dependence on temperature, pressure and composition.

The above is a simple view of the thermodynamic properties of gases and the reader is referred to more comprehensive treatment listed in Appendix 4.

10.2.4 The relationship between pressure and fugacity

For many pure gases the molar volume is known over large ranges of temperature and pressure. In order to calculate fugacity from such data it is necessary to know the precise relationship between P and f. To establish this relationship we define a coefficient α, which is the difference between the molar volume of a real gas and the volume that gas would have if it were ideal.

$$\alpha = (\bar{V}_{\text{ideal}} - \bar{V}_{\text{real}}) = \frac{RT}{P} - \bar{V} \tag{10.32}$$

Equation 10.32 can be rearranged and substituted into equation 10.18 to give:

$$d\mu = RT\,d\ln f = RT\,d\ln P + \alpha\,dP \tag{10.33}$$

This equation can be integrated from $P' = P$ to $P' = 0$ to give:

$$RT \int_{f(P=0)}^{f(P)} d\ln f' = RT \int_{P=0}^{P} d\ln P' - \int_{P=0}^{P} \alpha\,dP' \tag{10.34}$$

As the pressure approaches zero, $f \to P$ and the lower limits of the two integrals,

$$RT \int_{f(P=0)}^{f(P)} d\ln f' \quad \text{and} \quad RT \int_{P=0}^{P} d\ln P' \tag{10.35}$$

are the same. Therefore, equation 10.34 becomes:

$$RT \ln f = RT \ln P - \int_{P=0}^{P} \alpha\,dP' \tag{10.36}$$

or alternatively:

$$RT \ln \chi = RT \ln \frac{f}{P} = - \int_{P=0}^{P} \alpha P'\,d\ln P' \tag{10.37}$$

A common form of equation 10.37, in terms of the compressibility factor $Z = P\bar{V}/RT$ (section 5.3.1), can be obtained by eliminating $\alpha P'$. This is:

$$\ln \chi = \int_{P=0}^{P} (Z - 1)\,\frac{dP'}{P'} \tag{10.38}$$

A form of the above equation for the ith component in a gaseous mixture is:

$$\ln \chi_i = \int_{P=0}^{P} (Z_i - 1)\,\frac{dP'}{P'} \tag{10.39}$$

If an analytical expression for α or Z can be found, the integral in equation 10.37 can be evaluated and an exact relationship between f and P produced. Selecting the Redlich–Kwong equation (section 5.3) as a suitable equation of state for a pure gas gives:

$$\ln \chi = \ln \frac{f}{P} = Z - 1 - \ln\left(Z - \frac{bP}{RT}\right) - \frac{a}{RT^{3/2}b} \ln\left(1 + \frac{bP}{ZRT}\right) \quad (10.40)$$

10.3   Mixtures of condensed phases

Section 10.2 refers only to the properties of gas phases. However, at equilibrium the chemical potential of any substance in a condensed phase is equal to the chemical potential of the substance in its vapour phase with which it coexists (section 10.1.1). This allows us to address the properties of solid and liquid mixtures via the properties of their vapour phases.

In this section we will show that the chemical potential of a substance in a mixture can be related to the composition of the mixture.

The deductive process involves the following three steps.

- Select some **reference behaviour** which relates the fugacity of a substance in the vapour phase to the composition of the condensed phase.
- Insert this relationship into equation 10.14 for the chemical potential of any substance in a mixture of real gases.

  For thus purpose, equation 10.14 can be written:

$$\mu_{(g, T, P)} = \mu^{\ominus}_{(g, T, f^\ominus)} + RT \ln(f/f^\ominus) \quad (10.41)$$

Note that we have included the subscript g to specify the gas phase.

- Equate the chemical potential for the substance in the gas phase to that in the condensed phase (denoted by the subscript c).

These steps allow us to define the 'ideal behaviour' of the system. Such behaviour is 'baseline' behaviour with which the properties of the real system may be compared.

The selection of reference behaviour and techniques to characterise deviations from ideal behaviour are dependent on the type of system in question. The following sections describe those selected for common condensed phases.

The relationships to be developed are valid even when the partial pressure of a substance over the condensed phase is immeasurably small, such as over a mixture of miscible solids or solids dissolved in liquids.

10.3.1   Systems which conform with Raoult's law

Consider a two-component mixture of A and B existing in both gas and condensed phase. From Dalton's law of partial pressures (section 5.2) the total pressure in the gas phase, P, is given by:

$$P = P_A + P_B \quad (10.42)$$

If the vapour phase is behaving ideally, the partial pressures in the gas phase are related to the mole fractions of the substances in the condensed phase by Raoult's law (section 5.6.1). That is,

$$P_A = x_{A(c)} P_A^{\bullet} = f_A \tag{10.43}$$

and

$$P_B = x_{B(c)} P_B^{\bullet} = f_B \tag{10.44}$$

where $P_A^{\bullet}$ and $P_B^{\bullet}$ are the vapour pressures of the pure component. Equations 10.43 and 10.44 are suitable for reference behaviour.

For component A we can substitute equation 10.43 into equation 10.41 to give:

$$\mu_{A(g, T, P)} = \mu_{A(g, T, P^{\ominus})}^{\ominus} + RT \ln(x_{A(c)} P_A^{\bullet}) - RT \ln P_A^{\ominus} \tag{10.45}$$

Since, at equilibrium, the chemical potential of any substance must be equal in all phases, we recognise that:

$$\mu_{A(c, T, P)} = \mu_{A(g, T, P)} \tag{10.46}$$

Therefore,

$$\mu_{A(c, T, P)} = [\mu_{A(g, T, P^{\ominus})}^{\ominus} + RT \ln P_A^{\bullet} - RT \ln P_A^{\ominus}] + RT \ln x_{A(c)} \tag{10.47}$$

The term in the square bracket is equal to the **chemical potential of the pure substance in the condensed phase when $x_{A(c)}$ equals unity**. It is also equal to the chemical potential of pure A in the gas phase under the same conditions. If we give this term the symbol $\mu_{A(c, T, P)}^{\bullet}$, then:

$$\mu_{A(c, T, P)} = \mu_{A(c, T, P)}^{\bullet} + RT \ln x_{A(c)} \tag{10.48}$$

The pure substance at temperature T and pressure P is a convenient choice for a standard state. Thus $\mu_{A(c, T, P)}^{\bullet}$ becomes the standard-state chemical potential for condensed phases.

Equation 10.48 is the fundamental equation for the chemical potential of an ideal mixture. Some authors choose to define ideality by equation 10.48, from which Raoult's law may then be derived. This is an acceptable alternative to deriving equation 10.48 from Raoult's law.

10.3.2 The Margules equation for non-ideal systems

An empirical modification of Raoult's law was produced by Margules. Here, the partial pressure of component A in a binary mixture is given by:

$$P_A = P_A^{\bullet} x_A \exp[\alpha(1 - x_A)^2] \tag{10.49}$$

where $\alpha = W/RT$ and W is the energy of mixing per mole. The term α compensates for non-zero volume and enthalpy changes during mixing. For ideal solutions W and α are zero, the term $\exp[\alpha(1 - x_A)^2]$ is unity and equation 10.49 becomes Raoult's law. Regardless of the value of W, when x_A approaches unity the partial pressure conforms to Raoult's law. At low values of x_A the partial pressure conforms to Henry's law, since:

$$P_A \sim P_A^{\bullet} x_A e^{\alpha} = k_A x_A \tag{10.50}$$

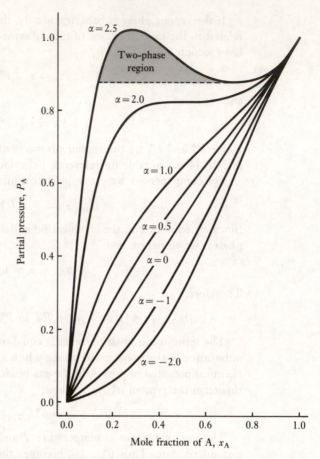

Figure 10.2   Properties of the Margules equation.

The changes in partial pressure due to changes in α are shown in figure 10.2. As α is decreased the system shows positive deviation from Raoult's law. As α is increased the system shows negative deviation from Raoult's law, eventually exhibiting a maximum and a minimum in the plot of P_A versus mole fraction. This last situation is not physically meaningful since it implies that pressures exist at which the vapour is in equilibrium with more than one composition. In reality such a mixture becomes immiscible and separates into two liquid phases. Consequently, we replace the S-bend with a horizontal line at the minimum which defines the region of immiscibility.

10.3.3   A general treatment for systems which deviate from Raoult's law

To describe properly any deviation from ideality, we define the **activity coefficient** γ, which for any substance A is defined as:

$$\gamma_A = \frac{a_A}{x_{A(c)}} \tag{10.51}$$

Including the activity coefficient, equation 10.48 becomes:

$$\mu_{A(c, T, P)} = \mu^{\bullet}_{A(c, T, P)} + RT \ln(x_{A(c)} \gamma_A) \tag{10.52}$$

The value of γ_A obeys the following rules.

- In the selected standard state both the mole fraction and the activity equal unity, by definition. Therefore, γ_A approaches unity as the mole fraction approaches unity. That is:

$$\gamma_A \to 1 \quad \text{as} \quad x_{A(c)} \to 1 \tag{10.53}$$

- γ_A is unity in all states for ideal solutions.

An activity coefficient defined in terms of mole fractions is called a **rational activity coefficient**.

This type of activity coefficient is appropriate for any substance in any mixture. However, it is best suited to systems that approximate to Raoult's law or solvents which, due to their high concentration, are approaching the limit ($x_{A(c)} \to 1$).

10.3.4 An alternative standard state for Raoult's law

In equation 10.48 the subscript (c, T, p) registers the fact that the standard state is both pressure- and temperature-dependent. However, if the standard state were chosen as the pure substance at 1 bar pressure (designated by $\circ$), then a pressure correction would be required to obtain $\mu^{\bullet}_{A(c, T, P)}$ at pressure P, i.e.

$$\mu^{\bullet}_{A(c, T, P)} = \mu^{\circ}_{A(c, T, 1\,bar)} + \left(\frac{\partial \mu^{\circ}_{A(c, T, 1\,bar)}}{\partial P}\right) \Delta P \tag{10.54}$$

We recall from section 9.1.11 and equation 10.12 that the pressure dependence of the chemical potential is related to the partial molar volume, i.e.

$$\left(\frac{\partial \mu^{\circ}_{A(c, T, 1\,bar)}}{\partial P}\right) = \bar{V}^{\circ}_{A(c, T, 1\,bar)} \tag{10.55}$$

Therefore,

$$\mu_{A(c, T, P)} = \mu^{\circ}_{A(c, T, 1\,bar)} + \bar{V}^{\circ}_{A(c, T, 1\,bar)} \Delta P + RT \ln x_{A(c)} \tag{10.56}$$

With this alternative selection of standard state, the activity for the ideal mixture becomes equal to:

$$RT \ln a_A = \bar{V}^{\circ}_{A(c, T, 1\,bar)} \Delta P + RT \ln x_{A(c)} \tag{10.57}$$

Therefore, the pressure dependence of the chemical potential is included in the activity term. We can correct for non-ideality by introducing an activity coefficient defined as:

$$\gamma_A = \frac{a_A}{x_{A(c)}} \tag{10.58}$$

In this case the activity coefficient not only takes into account deviations from Raoult's law but also contains the pressure term $\bar{V}^{\circ}_{A(c, T, 1\,bar)} \Delta P$. Thus, the activity coefficient will tend to unity as the mole fraction tends to unity **and the pressure** tends to 1 bar. At any other external pressure the activity coefficient will be different from unity, even at unit mole fraction. This situation is different from that represented by equation 10.52, where the pressure term is contained within the standard-state chemical potential. This reinforces the point that **the magnitudes of the activity and the activity coefficient are dependent on the choice of standard states**.

10.3.5   The activity–fugacity relationship

Recalling the definition of activity in equation 10.7 we see that, for a mixture obeying Raoult's law, the activity of a substance equals its mole fraction. Therefore:

$$\mu_{A(c, T, P)} = \mu_{A(c, T, P)}^{\bullet} + RT \ln a_A \tag{10.59}$$

Also, for an ideal gas phase, $P^{\bullet} = P_A x_{A(c)}$. Therefore,

$$a_A = \frac{P_A}{P^{\bullet}} \tag{10.60}$$

We conclude that:

THE ACTIVITY OF A SUBSTANCE IN THE CONDENSED PHASE IS EQUAL TO THE PARTIAL PRESSURE OF THE SUBSTANCE IN THE VAPOUR PHASE DIVIDED BY THE VAPOUR PRESSURE OF THE PURE SUBSTANCE AT THE SAME TEMPERATURE AND TOTAL PRESSURE.

If the vapour phase is non-ideal we use the exact relationship,

$$a_A = \frac{f_A}{f_A^{\bullet}} \tag{10.61}$$

This equation is valid regardless of the composition or ideality of the condensed phase.

This relationship between activity and fugacity activity applies to any choice of standard state, not only to the pure component, as in equation 10.61.*

We may write the general equation for the composition dependence of the chemical potential as:

$$\mu_{A(c, T, P)} = \mu_{A(c)}^{\ominus} + RT \ln\left(\frac{f_A}{f_A^{\ominus}}\right) \tag{10.62}$$

where $\mu_{A(c)}^{\ominus}$ represents an **arbitrarily chosen standard state** under any selected condition of composition, temperature or pressure.

10.3.6   Systems which conform with Henry's law

In homogeneous mixtures where one substance is in great excess (the solvent), the minor substances obey Henry's law rather than Raoult's law (section 5.7). Henry's law is therefore a better choice to describe the reference behaviour of solutes.

---

*The general relationship may be proved by following the method in developing equations 10.44–10.48 for any choice of standard state or reference behaviour. There are many options that can be chosen and it is an unnecessary exercise to go through all of them.

Henry's law states that the partial pressure of a substance in the vapour phase above a mixture is given by:

$$P_B = x_{B(c)} \times \kappa_{B(T, P)} \tag{10.63}$$

where $\kappa_{B(T, P)}$ is a temperature- and pressure-dependent constant which has dimensions of pressure. A convenient variant of Henry's law is obtained by replacing pressure with fugacity, whereby

$$f_B = x_{B(c)} \times \kappa_{B(T, P)} \tag{10.64}$$

This law is a limiting law to which all solutes tend as their mole fraction tends to zero. In the limiting case where all **solutes** in a mixture tend to zero concentration, the **solvent** obeys Raoult's law.

Solutions where the solutes obey Henry's law are said to be **ideal dilute solutions**. This term distinguishes such solutions from truly ideal solutions, which conform to Raoult's law.

Using Henry's law, the composition dependence of the chemical potential can be defined by following a similar procedure to that for Raoult's law (see the development of equation 10.44–10.48). However, we may short-cut this procedure by using the general expression for chemical potential given by equation 10.62. Substitution of equation 10.64 into equation 10.62 written for component B gives:

$$\mu_{B(c, T, P)} = \mu_B^{\ominus} + RT \ln\left(\frac{x_{B(c)} \kappa_{B(T, P)}}{f_B^{\ominus}}\right) \tag{10.65}$$

We select the standard state as **the pure substance, at the same temperature and pressure as the mixture, in which the fugacity of the solute conforms to Henry's law**. We will denote this standard state with the symbol †, and note that $x_B^{\dagger} = 1$. Then:

$$f_B^{\dagger} = x_B^{\dagger} \kappa_B \tag{10.66}$$

from which

$$\mu_{B(c, T, P)} = \mu_{B(c, T, P)}^{\dagger} + RT \ln\left(\frac{x_{B(c)} \kappa_{B(T, P)}}{x_{B(c)}^{\dagger} \kappa_{B(T, P)}}\right) \tag{10.67}$$

or

$$\mu_{B(c, T, P)} = \mu_{B(c, T, P)}^{\dagger} + RT \ln x_{B(c)} \tag{10.68}$$

This term $\mu_{B(c, T, P)}^{\dagger}$ corresponds to the chemical potential of substance B at unit mole fraction in the condensed phase. It is obtained by an extrapolation of Henry's law from the infinitely dilute solution. The state therefore, represents a solution for which the concentration term $x_{B(c)}$ is unity but in which the environment of the solute molecules is the same as in the infinitely dilute solution. This is a purely hypothetical standard state and is shown graphically in figure 10.3.

Superficially, equation 10.68 looks like equation 10.48 based on Raoult's law. However, the Raoult's law standard-state chemical potential, $\mu_{B(c, T, P)}$, will only be the same as the Henry's law standard-state chemical potential if the Henry's law constant is unity.

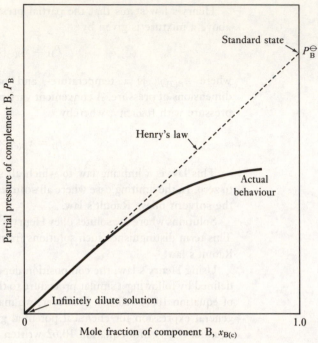

Figure 10.3 Henry's law standard state.

10.3.7   Systems which deviate from Henry's law

To compensate for any deviation from Henry's law we introduce an appropriate activity coefficient for the solute, γ_B. That is, for a component B in a mixture:

$$\gamma_B = \frac{a_B}{x_{B(c)}} \tag{10.69}$$

The value of γ_B obeys the following rules.

- Henry's law is an extrapolation of behaviour from infinitely dilute solution. Therefore, as a substance approaches zero concentration, its activity coefficient tends to unity.

$$\gamma_B \to 1 \quad \text{as} \quad x_{B(c)} \to 0 \tag{10.70}$$

- γ_B is unity for ideal dilute solutions.

In the Henry's law standard state, both the mole fraction and the activity equal unity, by definition. Therefore, γ_B approaches unity as the mole fraction approaches unity. However, we must remember that the Henry's law standard state is purely hypothetical. If the mole fraction of the solute was approaching unity, the substance would be conforming to Raoult's law rather than Henry's law. In this case Henry's law would **not** be appropriate reference behaviour.

An alternative standard state can be selected for solutes conforming to Henry's law. If the standard state were chosen as the pure substance at 1 bar pressure, designated by ‡, then a pressure correction would be required to obtain $\mu_{B(c, T, P)}^{\ddagger}$ at pressure P. The pressure correction takes the form:

$$\mu^{\ddagger}_{B(c, T, P)} = \mu^{\ddagger}_{B(c, T, 1\,bar)} + \left(\frac{\partial \mu^{\ddagger}_{B(c, T, 1\,bar)}}{\partial P}\right) \Delta P \qquad (10.71)$$

Therefore,

$$\mu_{B(c, T, P)} = \mu^{\ddagger}_{B(c, T, 1\,bar)} + \bar{V}^{\ddagger}_{B(c, T, 1\,bar)} \Delta P + RT \ln x_{B(c)} \qquad (10.72)$$

The activity coefficient, defined as:

$$\gamma_B = \frac{a_B}{x_{B(c)}} \qquad (10.73)$$

contains the pressure term $\bar{V}^{\ddagger}_{B(c, T, 1\,bar)} \Delta P$. Thus, the activity coefficient will tend to unity as the mole fraction of the solute tends to unity **and** the pressure tends to 1 bar.

10.3.8   Variants of Henry's law used as reference behaviour

All the previous conventions for non-ideal systems use mole fractions as the composition units. In many cases the units of molality or molarity are more convenient to use than mole fractions.

In these cases alternative variants of Henry's law are used as reference behaviour. One variant can be written as:

$$f_B = m_B \times \kappa^m_{B(T, P)} \qquad (10.74)$$

where m_B is the molal concentration of B and $\kappa^m_{B(T, P)}$ is a Henry's law constant. The superscript m is used to discriminate this Henry's law constant from κ used in the mole fraction convention.

A convenient standard state is a 1 molal solution of solute B, at the same temperature and pressure as the mixture, which behaves in accordance with equation 10.74. In this state, designated by the symbol *, $m_B^* = 1 \text{ mol kg}^{-1}$ and:

$$f_B^* = m_B^* \kappa^m_{B(T, P)} \qquad (10.75)$$

The resulting equation for ideal behaviour is:

$$\mu_{B(c, T, P)} = \mu^*_{B(c, T, P)} + RT \ln\left(\frac{m_B}{m_B^*}\right) \qquad (10.76)$$

The standard state is called the hypothetical ideal molal solution and is depicted in figure 10.4. It is common to neglect the term m_B^* for this standard state, since it is unity. However, in general, the term should be retained.

If the condensed phase behaves non-ideally then:

$$\mu_{B(c, T, P)} = \mu^*_{B(c, T, P)} + RT \ln a_B \qquad (10.77)$$

where

$$a_B = \left(\frac{m_B \gamma_B}{m_B^*}\right) \quad \text{or} \quad \gamma_B = \frac{a_B m_B^*}{m_B} \qquad (10.78)$$

This yields one of the most important equations of thermodynamics:

$$\mu_{B(c, T, P)} = \mu^*_{B(c, T, P)} + RT \ln(m_B \gamma_B) \qquad (10.79)$$

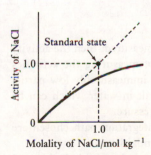

Figure 10.4
The hypothetical ideal
molal solution.

The activity coefficient γ_B is dimensionless and tends to unity as the **molality** of the solute tends to zero. An activity coefficient defined in terms of molal units is called a **practical activity coefficient**.

Molal units are often chosen for composition. If the concentrations of B is $c_B \text{ mol dm}^{-3}$, then:

$$\mu_{B(c, T, P)} = \mu_{B(c, T, P)}^{\diamondsuit} + RT \ln a_B \tag{10.80}$$

where the symbol $\diamondsuit$ designates a standard state, and

$$a_B = \left(\frac{c_B \gamma_B}{c_B^{\diamondsuit}} \right) \quad \text{or} \quad \gamma_B = \frac{a_B \times c_B^{\diamondsuit}}{c_B} \tag{10.81}$$

A convenient standard state is the hypothetical ideal molar solution in which $c_B^{\diamondsuit} = 1$. The activity coefficient γ_B is dimensionless and tends to unity as the **molarity** of the solute tends to zero.

10.3.9   The link between the activities of solutes and the solvent

The activities of the solvent and the solutes in mixtures are **not** free to vary independently of each other. The Gibbs–Duhem equation (section 9.1.8) imposes a constraint on the system which allows us to reduce the number of independently variable species by one in each phase.

Consider a condensed phase composed of A and B. The Gibbs–Duhem equation for the single phase is:

$$n_A \, d\mu_A + n_B \, d\mu_B = 0 \tag{10.82}$$

where n_A and n_B are the numbers of moles of A and B in the system. The general equation for the chemical potential, $\mu_{(c, T, P)} = \mu_{(c, T, P)}^{\ominus} + RT \ln a$, can be substituted into equation 10.82 for both A and B. Differentiation of the resulting equation yields:

$$n_A \, d \ln a_A + n_B \, d \ln a_B = 0 \tag{10.83}$$

Dividing by $n_A + n_B$, and recalling the definition of a mole fraction, we obtain:

$$x_A \, d \ln a_A + x_B \, d \ln a_B = 0 \tag{10.84}$$

This can be rearranged in terms of the activity of the solute to give:

$$\int d \ln a_B = - \int \left(\frac{x_A}{x_B} \right) d \ln a_A \tag{10.85}$$

This equation has two useful applications.

- If the activity of one component in a binary mixture is measured, then the activity of the second is known. This provides a useful way of measuring the activities of substances in mixtures where one substance has an immeasurably low partial pressure. An example is the activity of an inorganic salt in water which can be calculated from a measurement of the water vapour pressure.

  The exact limits on the integrals and the precise integration path chosen are dependent on the type of system and the choice of standard states.

- Equation 10.85 is an equation of constraint which can be used to check the internal consistency of thermodynamic data. If measurements or theoretical predictions of thermodynamic data do not conform to the constraint, something is wrong.

10.3.10   Osmotic coefficients

In many mixtures the mole fraction of solvent may be close to unity and its partial pressure closely approximated by Raoult's law. However, the solute may still be quite concentrated and deviating from Henry's law. This is the case for aqueous electrolyte solution where solutes deviate from Henry's law at concentrations as low as 10^{-5} mol kg^{-1}. For example, in an aqueous solution of 1 molal sodium chloride, the sodium chloride will be far from its reference state and behaving non-ideally. The water, by contrast, has a mole fraction of 0.982 and is approaching Raoult's law behaviour. At 25 °C the rational activity coefficient of the water in this system is 0.984. This number does not emphasise the degree of non-ideality in the **whole** system. It is desirable to have a parameter which is a more sensitive indicator of the non-ideality of the solvent.

The basic equation for the ideal behaviour of a solvent A is:

$$\mu_{A(c, T, P)} = \mu^{\bullet}_{A(c, T, P)} + RT \ln x_A \qquad (10.86)$$

Thus, a plot of chemical potential versus $\ln x$ should give a straight line of slope RT and intercept $\mu^{\bullet}_{A(c, T, P)}$ at $\ln x_A = 0$ (where $x = 1$). This is depicted in figure 10.5, which also shows non-ideal behaviour.

In the non-ideal case, a line between the intercept and any point $(\mu_{A(c, T, P)}, \ln x)$ will have a slope which is different from RT at each point along the non-ideal curve. Non-ideality is therefore associated with departure from the ideal slope. We can define the non-ideal slope at any point as $g_A RT$; then,

$$\mu_{A(c, T, P)} = \mu^{\bullet}_{A(c, T, P)} + g_A RT \ln x_A \qquad (10.87)$$

where g_A is the **rational osmotic coefficient**.

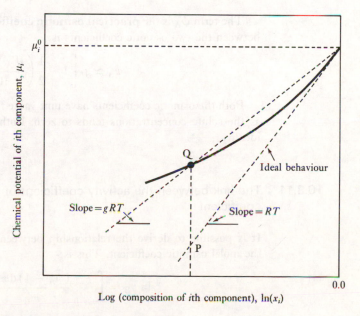

Figure 10.5   Non-ideality of a solvent (after Everett, 1959).

A second type of osmotic coefficient can be defined, since equation 10.87 yields the relationship:

$$\ln a_A = g_A \ln x_A \tag{10.88}$$

For a solution containing several solutes, the mole fraction can be expressed as:

$$x_A = \left(\frac{1000/W_A}{1000/W_A + \sum_B m_B} \right)$$

$$= \left(\frac{1}{1 + \dfrac{W_A \sum_B m_B}{1000}} \right) \tag{10.89}$$

where W_A is the formula weight of the solvent and $\sum_B m_B$ is the sum of the molalities of all the solutes in solution. By substitution, the natural log of the solute activity becomes:

$$\ln a_A = -a_A \ln\left(1 + \frac{W_A \sum_B m_B}{1000} \right) \tag{10.90}$$

The expression within the logarithm can be expanded as a power series using Maclaurin's theorem (box 8.1) to give:

$$\ln a_A = -g_A \ln\left[\frac{W_A \sum_B m_B}{1000} + \left(-\frac{1}{2} \right)\left(\frac{W_A \sum_B m_B^2}{1000} \right) + \cdots \right] \tag{10.91}$$

If the higher-order terms in equation 10.91 are neglected, the resulting approximation suggests an alternative type of osmotic coefficient defined by:

$$\ln a_A \equiv -\phi_A\left(\frac{W_A \sum_B m_B}{1000} \right) \tag{10.92}$$

The term ϕ_A is the **practical osmotic coefficient**. An approximate relationship between the two osmotic coefficients is:

$$\phi_A \simeq g_A\left[1 - \frac{1}{2}\left(\frac{W_A \sum_B m_B}{1000} \right) \right] \tag{10.93}$$

Both the osmotic coefficients have unit value for ideal mixtures and tend to unity as the solute concentrations tends to zero. Both are numerically sensitive to non-ideality.

10.3.11 The link between the activity coefficient of a solute and the osmotic coefficient

It is possible to derive the relationship between the molal activity coefficient and the molal osmotic coefficient. This is:

$$\ln \gamma_B = \int_{m_B=0}^{m_B} \frac{(\phi_A - 1)\,dm_B'}{m_B'} + (\phi_A - 1) \tag{10.94}$$

We can see that evaluating the activity coefficient of a solute at a known concentration

m_B requires the measurement of the osmotic coefficient over the **entire** range m_B to $m_B = 0$. This allows the integral in equation 10.94 to be evaluated. Integration can be performed numerically or graphically. An analytical solution may be obtained if a suitable expression can be fitted to the concentration dependence of ϕ_B. The above equation is useful since osmotic coefficients for solvents can be measured relatively easily from vapour pressure measurements.

Alternatively, we may wish to express the osmotic coefficient as a function of the activity coefficient of the solute. This relationship is obtained easily since $m_B\phi_A = -(1000/W_A)\ln a_A$, from which

$$m_B\phi_A = \int_0^{m_B} m_B \, d \ln(\gamma_B m_B) \tag{10.95}$$

whence

$$\phi_A = 1 + \frac{1}{m_B} \int_0^{m_B} m_B \, d \ln \gamma_B \tag{10.96}$$

In addition to predicting osmotic coefficients from activity coefficient measurements, the above equation can be used as an equation of constraint. This allows us to check on the accuracy of theoretical equations for activity coefficients. The theoretical expression for γ_B as a function of m_B would be substituted into equation 10.96 and the integral evaluated analytically. The predicted osmotic coefficients may then be compared with measured values.

The above expressions lead to a general relationship between the osmotic coefficient and the activity coefficients of the solute species in a multicomponent solution. This is:

$$d\left\{(\phi - 1) \sum_i m_i\right\} = \sum_i m_i \, d \ln \gamma_i \tag{10.97}$$

which gives:

$$\phi = 1 + \frac{1}{\sum_i m_i} \sum_i \int_0^{m_i} m_i \, d \ln \gamma_i \tag{10.98}$$

10.3.12 Changes in the Gibbs energy due to mixing

When substances mix, their thermodynamic properties differ from the thermodynamic properties of the same substances in an unmixed state. This is true for ideal as well as non–ideal mixtures.

The **total Gibbs energy** for a mixture, as defined in section 9.1.5, is the sum of the chemical potentials of each substance multiplied by the amount of substance, n_i, i.e.

$$G = \sum_i \mu_i n_i \tag{10.99}$$

The Gibbs energy **per mole** of solution, $\bar{G}$, can be obtained by dividing the right-hand side of equation 10.99 by $\sum_i n_i$, i.e.

$$\bar{G} = \sum_i \mu_i \left(n_i / \sum_i n_i\right) \tag{10.100}$$

The term $(n_i/\sum_i n_i)$ is equal to the mole fraction of the ith substance x_i. Therefore, the molal Gibbs energy becomes:

$$\bar{G} = \sum_i \mu_i x_i \tag{10.101}$$

An important property is the **change** in the molar Gibbs energy due to mixing, ΔG_{mix}. This is the difference between the molar Gibbs energy of the mixture and the molar Gibbs energy for same quantities of the pure unmixed substances (denoted by $G^\bullet$), and is given by:

$$\Delta G_{mix} = \bar{G} - \bar{G}^\bullet \tag{10.102}$$

where

$$\bar{G}^\bullet = \sum_i \mu_i^\bullet x_i \tag{10.103}$$

This gives:

$$\Delta G_{mix} = \sum_i \mu_i x_i - \sum_i \mu_i^\bullet x_i \tag{10.104}$$

The above equations are valid for ideal or non-ideal mixtures.

10.4   The thermodynamics of ideal mixing

Ideal mixing corresponds to random mixing of substances without changes in the interparticle forces. Such mixing is rarely encountered and represents a limiting case for mixtures of liquids and gases.

10.4.1   The Gibbs energy of ideal mixing

We recall that the chemical potential of a substance in an **ideal** mixture is related to the chemical potential of the pure substance by:*

$$\mu_i = \mu_i^\bullet + RT \ln x_i \tag{10.105}$$

Then,

$$\Delta G_{mix}^{id} = RT \sum_i x_i \ln x_i \tag{10.106}$$

The superscript id identifies the variable as a property of ideal mixtures. Equation 10.106 shows that, for a binary mixture, **the molar Gibbs energy of ideal mixing is a symmetrical function of mole fraction** (figure 10.6).

---

* We should use the designator (T, P) to show that equation 10.105 is strictly valid at constant temperature and pressure only, but this has been dropped to simplify the equations.

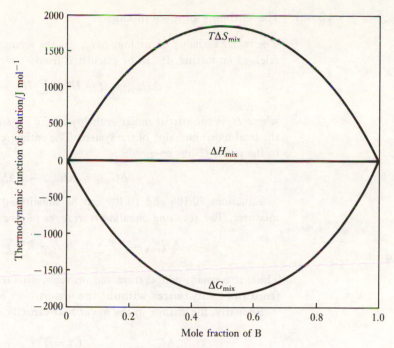

Figure 10.6   Thermodynamic functions for ideal mixing.

In any mixture the mole fraction of any substance will be less than unity. Therefore, the change in the Gibbs energy on mixing is negative. This confirms the intuitive view that **ideal mixing is a spontaneous process**, whereas de-mixing is not spontaneous.

10.4.2   The entropy of ideal mixing

A molar entropy change due to mixing can be calculated from the relationship:

$$\Delta S_{\text{mix}} = \bar{S} - \bar{S}^\bullet = \sum_i \bar{S}_i x_i - \sum_i \bar{S}_i^\bullet x_i \qquad (10.107)$$

where $\bar{S}_i$ is the partial molar entropy of the ith substance in a mixture and $\bar{S}$ is the total molar entropy of the system. The molar entropy change for an ideal mixture can be related to the change in the molar Gibbs energy by:

$$\Delta S_{\text{mix}}^{\text{id}} = -\left(\frac{\partial \Delta G_{\text{mix}}^{\text{id}}}{\partial T}\right)_P \qquad (10.108)$$

This gives:

$$\Delta S_{\text{mix}}^{\text{id}} = -R \sum_i x_i \ln x_i \qquad (10.109)$$

At any composition the **entropy of ideal mixing is positive**. This again confirms the intuitive view that ideal mixing is spontaneous.

10.4.3   The enthalpy of ideal mixing

The molar enthalpy of mixing, ΔH_{mix}, is the amount of heat per mole absorbed or released on mixing. It can be calculated from:

$$\Delta H_{\mathrm{mix}} = H - H^{\bullet} = \sum_i H_i x_i - \sum_i H_i^{\bullet} x_i \tag{10.110}$$

where H_i is the partial molar enthalpy of the ith substance in a mixture and H is the total molar enthalpy of the system. The enthalpy change is related to the change in the molar Gibbs energy by:

$$\Delta G_{\mathrm{mix}} = \Delta H_{\mathrm{mix}} - T\Delta S_{\mathrm{mix}} \tag{10.111}$$

Equations 10.106 and 10.109 can be substituted into equation 10.111 for ideal mixtures. The resulting equation rearranges to give:

$$\Delta H_{\mathrm{mix}}^{\mathrm{id}} = RT \sum_i x_i \ln x_i - RT \sum_i x_i \ln x_i = 0 \tag{10.112}$$

Thus, at constant temperature and pressure, **an ideal mixture can be prepared from pure substances without the absorption or release of heat**.

Similarly, the change in the molar heat capacity due to mixing is given by:

$$\Delta C_{P,\mathrm{mix}} = \bar{C}_P - \bar{C}_P^{\bullet} = \sum_i \bar{C}_{P,i} x_i - \sum_i \bar{C}_{P,i}^{\bullet} x_i \tag{10.113}$$

The relationship between $\Delta C_{P,\mathrm{mix}}$ and ΔH_{mix} is:

$$\Delta C_{P,\mathrm{mix}} = \left(\frac{\partial \Delta H_{\mathrm{mix}}}{\partial T} \right)_P \tag{10.114}$$

This shows that the value of $\Delta C_{P,\mathrm{mix}}$ for an ideal solution is zero:

$$\Delta C_{P,\mathrm{mix}}^{\mathrm{id}} = 0 \tag{10.115}$$

10.4.4   Volume changes due to ideal mixing

The molar volume change on mixing, ΔV_{mix}, can be calculated from:

$$\Delta V_{\mathrm{mix}} = \bar{V} - \bar{V}^{\bullet} = \sum_i \bar{V}_i x_i - \sum_i \bar{V}_i^{\bullet} x_i \tag{10.116}$$

where $\bar{V}_i$ is the partial molar volume of the ith substance in a mixture and $\bar{V}$ is the total molar volume of the substance. The total volume change for an ideal mixture is related to the change in the molar Gibbs energy by:

$$\Delta V_{\mathrm{mix}}^{\mathrm{id}} = \left(\frac{\partial \Delta G_{\mathrm{mix}}^{\mathrm{id}}}{\partial P} \right)_T \tag{10.117}$$

This gives:

$$\Delta V_{\mathrm{mix}}^{\mathrm{id}} = \left(\frac{\partial RT \sum x_i \ln x_i}{\partial P} \right)_T = 0 \tag{10.118}$$

Thus, at constant temperature and pressure, **an ideal mixture can be prepared from pure substances with no change in total volume**.

10.5 Non-ideal mixtures and excess functions

10.5.1 Excess functions

Figure 10.7(a–f) shows the mixing functions versus composition for some typical binary mixtures which show moderate deviation from ideal behaviour. We note that ΔG_{mix} is negative and ΔS_{mix} positive as expected, but the compositional dependencies of both functions are non-symmetrical. In contrast ΔH_{mix} may be positive or negative, or even change sign, as the composition changes.

To characterise such behaviour, recall that the chemical potential of a substance in a **non-ideal** mixture is related to the chemical potential of the pure substance by

$$\mu_i = \mu_i^{\bullet} + RT \ln x_i + RT \ln \gamma_i \tag{10.119}$$

where γ_i is the activity coefficient. Therefore, the composition dependence of ΔG_{mix} (equation 10.104) becomes:

$$\Delta G_{mix} = RT \sum_i x_i \ln x_i + RT \sum_i x_i \ln \gamma_i \tag{10.120}$$

This equation can be compared with the analogous equation for ideal mixing:

$$\Delta G_{mix}^{id} = RT \sum_i x_i \ln x_i \tag{10.121}$$

The difference between ΔG_{mix}^{id} and ΔG_{mix} is defined as **excess molar Gibbs energy**. It is denoted by the symbol G^E and is sometimes called the excess energy of mixing.

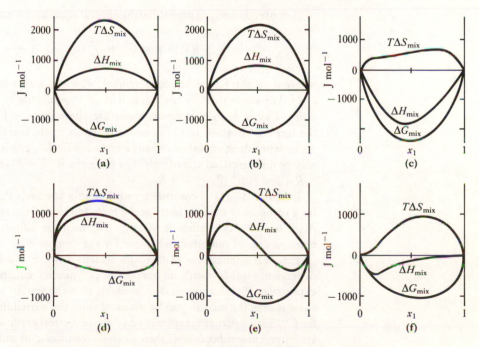

Figure 10.7 Energies of mixing for real solutions (after Abbott and Van Ness, 1989).

The term G^E is defined by:

$$G^E \equiv \Delta G_{mix} - \Delta G_{mix}^{id} \tag{10.122}$$

and is equal to the sum of the activity coefficient terms in equation 10.134, i.e.

$$G^E = RT \sum_i x_i \ln \gamma_i \tag{10.123}$$

The deviation from ideality due to **all** the components in a mixture is embodied in the compositional dependence of the single parameter G^E. Thus, **the excess molar Gibbs energy is a convenient and compact way of describing the non-ideality of multicomponent systems.**

Equation 10.123 gives G^E in terms of activity coefficients. However, it is possible to express the activity coefficient of any substance in a mixture in terms of the value of G^E as follows. If equation 10.123 is multiplied throughout by the sum of the number of moles of all substances in the system, denoted n_T, we have:

$$n_T G^E = RT \sum_i n_T x_i \ln \gamma_i \tag{10.124}$$

Noting that $x_i = n_i/n_T$, then:

$$n_T G^E = RT \sum_i n_i \ln \gamma_i \tag{10.125}$$

The partial differentiation of $n_T G^E$ with respect to n_i gives:

$$RT \ln \gamma_i = \left[\frac{\partial}{\partial n_i} (n_T G^E) \right]_{T, P, n_{j \neq i}} \tag{10.126}$$

A convenient way of presenting the general equation for any activity coefficient is:

$$RT \ln \gamma_i = G^E + \sum_{j=1}^{N} (\delta_{ij} - x_j) \left(\frac{\partial G^E}{\partial x_j} \right)_{x_k} \tag{10.127}$$

where δ_{ij} is the Kronecker delta symbol, which is zero when $i \neq j$ and unity when $i = j$. We know that the mole fractions of components in a mixture must sum to unity. That is, in an N-component mixture, there are $N - 1$ true variables. However, the form of equation 10.127 is such that **each mole fraction may be treated as an independent variable** and any expression for G^E, given as a function of $x_1 \ldots x_N$, may be differentiated accordingly. For example, if $g^E = R_{12} z_1 z_2$, then $(\partial G^E/\partial x_1)_{x_2} = x_2 R_{12}$ and $(\partial G^E/\partial x_2)_{x_1} = x_1 R_{12}$.

Equation 10.141 is consistent with Raoult's law such that the activity coefficient for a component tends to unity as the mole fraction of the component tends to unity. Also, the activity coefficient of any component at infinite dilution represents the extreme case of non-ideal behaviour for the component.

Practical applications of this idea usually involve expressing the compositional dependence of G^E with an empirical polynomial equation or other theoretical expression. If such an equation is available then the activity coefficient for any specific substance in the mixture can be obtained from the partial differential of the equation for G^E. Such equations may be selected on theoretical grounds but empirical equations are often chosen because of their flexibility and ease of differentiation.

In a similar way to the development of G^E we may also derive expressions for the excess molar entropy, enthalpy, heat capacity and volume of mixing. These are:

$$S^{E} = -\left(\frac{\partial G^{E}}{\partial T}\right)_{P} = -\left(\frac{\partial RT\sum_{i} x_{i}\ln\gamma_{i}}{\partial T}\right)_{P}$$

$$H^{E} = G^{E} - TS^{E} \quad \text{or} \quad H^{E} = -RT^{2}\left[\frac{\partial(G^{E}/RT)}{\partial T}\right]_{P}$$

$$C_{P}^{E} = -T\left(\frac{\partial^{2}G^{E}}{\partial T^{2}}\right)_{P}$$

$$V^{E} = \left(\frac{\partial G^{E}}{\partial P}\right)_{T} = \left(\frac{\partial RT\sum_{i} x_{i}\ln\gamma_{i}}{\partial P}\right)_{T}. \tag{10.128}$$

In differentiating equations 10.128 and 10.129 to obtain S^{E}, H^{E}, C_{P}^{E} and V^{E} we must remember that the temperature and pressure dependence of G^{E} will have a contribution from the temperature and pressure dependence of the activity coefficients. In this sense, **the values of S^{E}, H^{E}, C_{P}^{E} and V^{E} are a measure of the temperature and pressure dependence of the total deviation from ideality.**

Figure 10.8(a–f) shows some common features of excess functions for real binary organic mixtures. These are consistent with the energies of mixing in figure 10.7. All excess properties tend to zero as the systems approach purity. We see that G^{E} often has a near-parabolic dependence on composition but H^{E} and S^{E} show individualistic composition dependencies. Inorganic mixtures, in particular solid solutions, generally show more complex behaviour than organic mixtures.

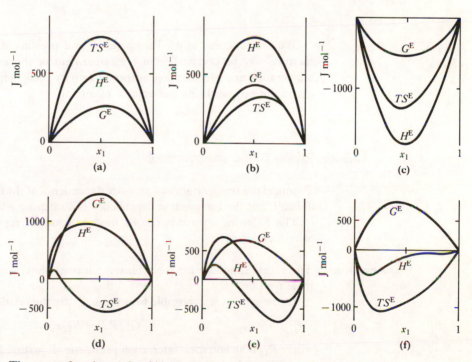

Figure 10.8   The excess functions (after Abbott and Van Ness, 1989).

Other variables which are useful in geochemical calculations or in tabulations of data are the excess functions for the individual substances in a mixture. For the ith component these are:

$$G_i^E = RT \ln \gamma_i \tag{10.129}$$

$$S_i^E = -R\left[\frac{\partial(T \ln \gamma_i)}{\partial T}\right]_P \tag{10.130}$$

$$H_i^E = R\left[\frac{\partial \ln \gamma_i}{\partial(1/T)}\right]_P \tag{10.131}$$

$$V_i^E = RT\left[\frac{\partial(\ln \gamma_i)}{\partial P}\right]_T \tag{10.132}$$

Consequently,

$$\ln \gamma_i = \frac{H_i^E}{RT} - \frac{S_i^E}{R} \tag{10.133}$$

The excess functions for the individual substances are related to the excess functions for the total mixture by:

$$G^E = \sum_i x_i G_i^E, \quad S^E = \sum_i x_i S_i^E \tag{10.134}$$

$$H^E = \sum_i x_i H_i^E, \quad C_P^E = \sum_i x_i C_{P,i}^E \tag{10.135}$$

$$V^E = \sum_i x_i V_i^E \tag{10.136}$$

We conclude that at fixed temperature and pressure the activity coefficient in a mixture may be obtained from a measurement of the excess Gibbs energy. The temperature and pressure dependence of activity coefficients is embodied within the excess entropy, enthalpy and volume of mixing.

10.5.2   Excess Gibbs-energy models

To model the temperature and pressure dependence of the excess functions is usually difficult, but the composition dependence can often be addressed.

The following equations can be used as compact representations of non-ideal behaviour.

Simple equations   The excess functions for some binary solutions show a symmetric dependence on mole fraction.

In these cases it is possible to express G^E by the relationship:

$$G^E/RT = R_{12}x_1x_2 \tag{10.137}$$

where R_{12} is an empirical interaction parameter. Equation 10.137 is sometimes called a **regular solution** model. The use of a single parameter enforces a symmetrical dependence of the excess functions on composition at fixed temperature and pressure. This behaviour is like that depicted in figure 10.8(a).

In the original definition of a regular solution, by Hildebrand, the restriction was that a single parameter characterised each pairwise interaction **and** the excess entropy of mixing was zero. This led to the conditions that $\Delta S_{mix} = \Delta S_{mix}^{id}$ and $H^E = G^E$.

Many other meanings have been given to the term 'regular', not always conforming to the restriction of zero S^E.

Once a model for G^E has been determined, the individual exchanger-phase activity coefficients may then be calculated using equation 10.127, which gives:

$$\ln \gamma_1 = x_2^2 R_{12} \quad \text{and} \quad \ln \gamma_2 = x_1^2 R_{12} \tag{10.138}$$

If we remove the restriction of zero excess entropy it is possible to calculate other thermodynamic properties from equation 10.128. For example,

$$H^E = -RT^2 \left(\frac{\partial R_{12}}{\partial T} \right)_P x_1 x_2 \tag{10.139}$$

$$S^E = -R \left[R_{12} + T \left(\frac{\partial R_{12}}{\partial T} \right)_P \right] x_1 x_2 \tag{10.140}$$

and

$$V^E = RT \left(\frac{\partial R_{12}}{\partial P} \right)_T x_1 x_2 \tag{10.141}$$

Because only one parameter R_{12} is used, G^E, H^E, S^E and V^E all show a parabolic dependence on composition, symmetrical about the point $x_1 = x_2 = 0.5$. However, we can see that the compositional dependence of behaviour of H^E, S^E and V^E is controlled by the partial derivative of R_{12}. Therefore G^E, H^E, S^E and V^E may be positive or negative, depending on the sign and magnitude of R_{12} and its derivatives, but they cannot change sign as composition changes.

The simple model for two components gives:

$$\ln a_1 = \ln x_1 + R_{12} x_2^2 \tag{10.142}$$

Writing $x_2 = 1 - x_1$ we see that:

$$\ln a_1 = \ln x_1 + R_{12} (1 - x_1)^2 \tag{10.143}$$

The above equation has a notable similarity to the simple Margules equation for the partial pressure of a vapour (equation 10.49). Since the activity of a substance 1 is equal to $P_1/P_1^{\bullet}$ we may write the Margules equation as:

$$a_1 = \frac{P_1}{P_1^{\bullet}} = x_1 \exp[\alpha(1 - x_1)^2] \quad \text{or} \quad \ln a_1 = \ln x_1 + \alpha(1 - x_1)^2 \tag{10.144}$$

We see that $\alpha = r_{12}$. For systems which exhibit positive deviation from Raoult's law, values of α fall between 0 and 2. Values of α above 2 are consistent with immiscibility, as indicated in figure 10.2 (section 10.3.2).

For a solution with N components, the generalised regular solution model has the form:

$$\frac{G^E}{RT} = \sum_{i=1}^{N-1} \sum_{j>i}^{N} R_{ij} x_i x_j \tag{10.145}$$

It is assumed that the coefficients R_{ij} are symmetric, i.e. $R_{12} = R_{21}$, etc.

The simple models serve to rationalize trends rather than being precise expressions for any one system. Such models have had some successful applications to simple mixtures of organic liquids but are limited in their use for solid solutions. An example of the use of equation 10.145 and its complementary activity coefficients with ion-exchange reactions is given in section 15.5.

Polynomial expressions   If a suitable model for the excess Gibbs energy for **multicomponent** systems can be found that contains only terms for **binary** systems, it is possible to evaluate the coefficients from binary-system data and use them to predict the activity coefficients in multicomponent systems. For this purpose any model must be precise, and it is often necessary to use flexible polynomial expressions to account for significant deviation from ideality. These are termed sub-regular models. Neglecting temperature or pressure derivatives, a simple sub-regular model for a two-component system has two parameters, S_{12} and S_{21}, e.g.

$$\frac{G^{E}}{RT} = x_1 x_2 (S_{12} x_1 + S_{21} x_2) \tag{10.146}$$

The activity coefficients are:

$$\ln \gamma_1 = S_{21} x_2^2 \left[1 + 2x_1 \left(\frac{S_{12}}{S_{21}} - 1 \right) \right] \tag{10.147}$$

and

$$\ln \gamma_2 = S_{12} x_1^2 \left[1 + 2x_2 \left(\frac{S_{21}}{S_{12}} - 1 \right) \right] \tag{10.148}$$

When written this way the similarity to the activity coefficients from the one-parameter model can be seen clearly.

The features and limitations of the two-parameter model can be seen if equation 10.146 is rewritten as:

$$\frac{G^{E}}{RTx_1 x_2} = S_{21} + (S_{12} - S_{21})x_1 \tag{10.149}$$

Clearly, a plot of $G^{E}/RTx_1 x_2$ versus x_1 is a straight line of intercept S_{21} regardless of the values selected for S_{12} or S_{21}. Inspection of equations 10.147 and 10.148 show that S_{21} is equal to $\ln \gamma_1$ as x_1 tends to zero and S_{12} is equal to $\ln \gamma_2$ as x_2 tends to zero. This behaviour offers a convenient way of determining the parameters from plots of measured activity coefficients extrapolated to infinite dilution. Examples are given in figure 10.9, which shows data for mixtures of benzene and isopropanol at 45 °C. In this case activity coefficients were calculated from experimental vapour pressure measurements from which the values of $G^{E}/RTx_1 x_2$ were computed. We see that changes in $G^{E}/RTx_1 x_2$ versus x_1 are inconsistent with the simple two-parameter polynomial model. This type of failure is common for many mixtures.

For a multicomponent system, the two-parameter model can be extended to:

$$\frac{\bar{G}^{E}}{RT} = \sum_{i=1}^{N-1} \sum_{j>i}^{N} x_i x_j (S_{ij} x_i + S_{ij} x_j) \tag{10.150}$$

This model still have only two parameters characterising each pairwise interaction between components and is subject to the same limitations as the binary model. In

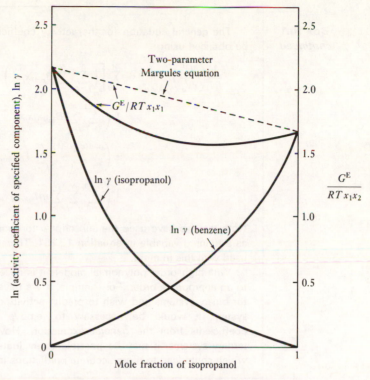

Figure 10.9   Excess Gibbs energy for an isopropanol/benzene mixture (after Abbott and Van Ness, 1989).

the generalised equation 10.150 we do not make the assumption of symmetry in the coefficients S_{ij}. Consequently, the generalised equation for the activity coefficient of the kth component is:

$$\ln \gamma_k = G^E + \sum_{p=1}^{N} \left[(\delta_{kp} - x_p) \sum_{j=1}^{N} (2S_{pj}x_p x_j + S_{jp}x_j^2) \right]$$

Box 10.1   **The Margules equation and activity coefficient**

To overcome the limitations due to pairwise interactions these polynomial models may be expanded indefinitely yielding the general Margules equation:

$$\frac{G^E}{RT} = \sum_{i=1}^{N-1} \sum_{j>i}^{N} \left\{ x_i x_j \left[\sum_{k=1}^{N} \left(\sum_{l=0}^{M} B_{ijkl} x_k^l \right) \right] \right\}$$

$$+ \sum_{i=1}^{N-1} \sum_{j>i}^{N} \sum_{k>j}^{N} \left\{ x_i x_j x_k \left[\sum_{l=1}^{N} \left(\sum_{m=0}^{M} T_{ijklm} x_l^m \right) \right] \right\} + \ldots \quad (10.151)$$

B_{ijkl} is a binary coefficient for interactions between the ith and jth components where the subscript l defines the power to which the mole fraction of the kth component is raised. Similarly, t_{ijklm} is a ternary coefficient for interactions between the ith, jth and kth component.

continued

Box 10.1
continued

The general equation for the activity coefficient for the pth component can be obtained using:

$$\frac{1}{RT}\left(\frac{\partial G^E}{\partial x_p}\right)_{x_k, k \neq p, i} = \sum_{l=0}^{M}\left[\left(\sum_{j>p}^{N}\sum_{k=1}^{N}B_{pjkl}x_jx_k^l\right) + \left(\sum_{i=1}^{\min(p, N-1)}\sum_{k=1}^{N}B_{ipkl}x_ix_k^l\right)\right.$$

$$+ \left(\sum_{i=1}^{N-1}\sum_{j>i}^{N}lB_{ijpl}x_ix_jx_p^{l-1}\right)\right] + \sum_{m=0}^{M}\left[\left(\sum_{j>p}^{N}\sum_{k>j}^{N}T_{pjklm}x_jx_kx_l^m\right)\right.$$

$$+ \left(\sum_{i=1}^{\min(p, N-1)}\sum_{k>p}^{N}T_{ipklm}x_ix_kx_l^m\right) + \left(\sum_{i=1}^{N-1}\sum_{j>i}^{p}T_{ijplm}x_ix_jx_l^m\right)$$

$$+ \left(\sum_{i=1}^{N-1}\sum_{j>i}^{N}\sum_{k>j}^{N}mT_{ijkpm}x_ix_jx_kx_p^{m-1}\right)\right] \tag{10.152}$$

Note that we have used the subscript p to denote a component since i is used as a running variable in equation 10.151. Therefore, equation 10.127 should be used with this in mind.

With high-order polynomial models it is necessary to truncate the polynomial to an appropriate order. For example, if we have reliable activity coefficient data for binary systems and wish to predict activity coefficients in a multicomponent system, it would be necessary to remove all but the binary interaction coefficients from the general expression. However, if data are available on ternary systems, it may be possible to evaluate ternary interaction coefficients which could lead to more accurate predictions in multicomponent systems.

Fractional models

Fractional equations have been used to describe phase equilibria of non-electrolyte solutions. An illustrative example is the two-parameter van Laar equation, which can be written:

$$\frac{G^E}{RTx_1x_2} = \frac{1}{\dfrac{1}{Q_{21}} + \left(\dfrac{1}{Q_{12}} - \dfrac{1}{Q_{21}}\right)x_1} \tag{10.153}$$

where Q_{12} and Q_{21} are fitting parameters. By comparing the above equation with the two-parameter Margules equation (equation 10.147) it is clear that the van Laar equation does not impose linearity on the plot of G^E/RTx_1x_2 versus x_1. Only in the unique case when $S_{12}/S_{21} = Q_{12}/Q_{21} = 1$, where the value of G^E/RTx_1x_2 is constant, does the van Laar equation become identical to the Margules equation.

In general the activity coefficients from the van Laar equation are:

$$\ln \gamma_1 = \frac{Q_{21}x_2^2}{\left[1 + \left(\dfrac{Q_{21}}{Q_{12}} - 1\right)x_1\right]^2} \tag{10.154}$$

and

$$\ln \gamma_2 = \frac{Q_{12}x_1^2}{\left[1 + \left(\dfrac{Q_{12}}{Q_{21}} - 1\right)x_2\right]^2} \tag{10.155}$$

Equations 10.154 and 10.155 show that Q_{21} is equal to $\ln \gamma_1$ as x_1 tends to zero and Q_{12} is equal to $\ln \gamma_2$ as x_2 tends to zero. The activity coefficients given by the van

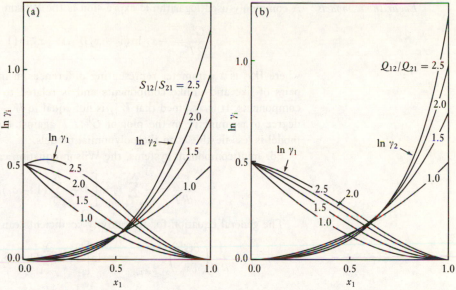

Figure 10.10   Activity coefficients from alternative models. (a) Two parameter margules equation (S_{12}/S_{21} values on the curve). (b) Van Laar equation (Q_{12}/Q_{21} values on the curve) (after Abbott and Van Ness, 1989).

Laar equation approach their limiting values at infinite dilution with greater steepness than do those given by the Margules equation. A further difference is that the Margules activity coefficients exhibit a greater dependence on the ratios of the parameters than the van Laar activity coefficients. These sensitivities are displayed in figure 10.10. Between them, the van Laar and the two-parameter Margules equations provide adequate descriptions of a considerable range of behaviour.

Box 10.2

A generalised fractional equation for G^E and activity coefficients

An even more flexible expression for G^E/RT is:

$$\frac{G^E}{RT} = \frac{x_1 x_2 q_1 q_2 2a_{12}}{x_1 q_1 + x_2 q_2},\qquad(10.156)$$

where q_1 and a_{12} are the model parameters. This can be generalised for multicomponent systems by

$$\frac{G^E}{RT} = \frac{\sum_{i=1}^{N=1}\sum_{j>i}^{N} x_i x_j q_i q_j 2a_{ij}}{\sum_{i=1}^{N} x_i q_i}\qquad(10.157)$$

The general equation for the activity coefficients of the pth component can be obtained from

$$\frac{1}{RT}\left(\frac{\partial G^E}{\partial x_p}\right)_{x_{k,k\neq p,i}} = \left(\frac{1}{\sum_{i=1}^{N} x_i q_i}\right)^2 \Bigg[\left(\frac{1}{\sum_{i=1}^{N} x_i q_i}\right)$$

$$\times\left(\sum_{j>p}^{N} x_j q_p q_j 2a_{pj} + \sum_{i=1}^{\min(p,N-1)} x_i q_i q_p 2a_{ip}\right)$$

$$- q_p\left(\sum_{i=1}^{N-1}\sum_{j>i}^{N} x_j x_i q_i q_j 2a_{ij}\right)\Bigg]\qquad(10.158)$$

Logarithmic models   A commonly used logarithmic expression is the Wilson equation:

$$\frac{G^{E}}{RT} = -x_1 \ln(1 - x_2 W_{21}) - x_2 \ln(1 - x_1 W_{12}) \qquad (10.159)$$

where W_{12} is a parameter representing differences in interaction energies between pairs of like and unlike components and is related to the molar volumes of the components. It is assumed that W_{12} is not equal to W_{21} which accommodates some degree of asymmetry in the plot of G^{E}/RT against mole fraction. However, this model is less flexible than the polynomial models.

For multicomponent systems, the Wilson equation can be generalised to:

$$\frac{G^{E}}{RT} = -\sum_{i=1}^{N} x_i \ln\left[1 - \sum_{j=1}^{N} (1 - \delta_{ij}) x_j W_{ji} \right] \qquad (10.160)$$

The general equation for the activity coefficients can be obtained using:

$$\frac{1}{RT}\left(\frac{\partial G^{E}}{\partial x_p}\right)_{x_k,\, k \neq p, i} = -\ln\left[1 - \sum_{j=1}^{N} (1 - \delta_{pj}) x_j W_{jp} \right]$$

$$+ \sum_{i=1}^{N}\left[\frac{x_i(1 - \delta_{ip})W_{pi}}{1 - \sum_{j=1}^{N}(1 - \delta_{ij})x_j W_{ji}} \right] \qquad (10.161)$$

where δ_{ip} is the Kronecker delta function.

Chapter 11

The thermodynamics of chemical reactions

An understanding of the thermodynamics of chemical reactions is important in interpreting and predicting geochemical phenomena. This chapter applies the ideas of chapters 8–10 to reactive equilibria.

11.1   The energetics of chemical reactions

11.1.1   Enthalpy changes for reactions

If the thermodynamic potentials of substances are known, then it is a simple matter to calculate the change in any thermodynamic potential during a chemical reaction.

Heats of reaction
The enthalpy of 1 mole of a pure substance at any specified temperature and pressure is called the **molar enthalpy** of the substance. If the state of specified temperature and pressure is chosen as a reference state the term **standard molar enthalpy** is used and is denoted by the superscript $\ominus$.

Consider a reaction between v_A moles of A and v_B moles of B to give v_C moles of C plus v_D moles of D which can be described by:

$$v_A A + v_B B \rightleftharpoons v_C C + v_D D \qquad (11.1)$$

The **enthalpy change** for the reaction (ΔH) can be calculated by subtracting the sum of the molar enthalpies of the products from the sum over the reactants, i.e.

$$\Delta H = (v_D H_D + v_C H_C) - (v_A H_A + v_B H_B) \qquad (11.2)$$

The **standard enthalpy change** for the reaction is:

$$\Delta H^{\ominus} = (v_A H_D^{\ominus} + v_A H_C^{\ominus}) - (v_A H_A^{\ominus} + v_A H_B^{\ominus}) \qquad (11.3)$$

The standard enthalpy change for a reaction is the total heat absorbed during the **complete conversion** of the reactants in their standard states to the products in their standard states. Note that the convention above is used to emphasise that the overall function is a summation of the properties of the products minus those of the reactants.

Hess's law of constant heat summation
For a reaction which goes to completion it is possible to measure the enthalpy of reaction using calorimetry. An example is the reaction:

$$C_{graphite} + O_2(g) \rightleftharpoons CO_2(g) \qquad (11.4)$$

which goes to completion at 298 K with essentially no complicating side reactions. The value of ΔH for this reaction as measured by calorimetry at 298 K is -94.05 kcal (-393.5 kJ). In contrast there are many reactions which do not go to completion and cannot be measured directly. An example is the formation of carbon monoxide via the reaction:

$$C_{graphite} + \tfrac{1}{2}O_2(g) \rightleftharpoons CO(g) \tag{11.5}$$

which is always accompanied by the formation of carbon dioxide. However, the enthalpy change for reaction 11.5 can be determined if we think of the formation of carbon monoxide (reaction 11.5) as one step in the formation of carbon dioxide from graphite. The sequence of reaction is shown graphically in figure 11.1. Since enthalpy is a function of state, and therefore independent of path, the enthalpy of combustion of carbon to carbon dioxide must equal the sum of the ΔH values for the combustion of carbon to carbon monoxide and of carbon monoxide to carbon dioxide. Therefore, for the following reactions:

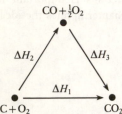
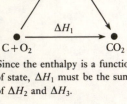

Since the enthalpy is a function of state, ΔH_1 must be the sum of ΔH_2 and ΔH_3.

Figure 11.1
Different paths for the conversion of carbon and oxygen to carbon dioxide (after Mahan, 1964).

$$\left. \begin{array}{l} C_{graphite} + O_2(g) \overset{\Delta H_1}{\rightleftharpoons} CO_2(g) \\[2ex] C_{graphite} + \tfrac{1}{2}O_2(g) \overset{\Delta H_2}{\rightleftharpoons} CO(g) \\[2ex] \tfrac{1}{2}O_2(g) + CO(g) \overset{\Delta H_3}{\rightleftharpoons} CO_2(g) \end{array} \right\} \tag{11.6}$$

the sum of the enthalpies of reaction is:

$$\Delta H_1 = \Delta H_2 + \Delta H_3 \tag{11.7}$$

The measurable quantities are ΔH_1 and ΔH_3, from which ΔH_2 can be calculated by difference.

This is the basis of **Hess's law of constant heat summation**, which can be generalised into the form:

THE ENTHALPY CHANGE OF ANY REACTION CAN BE EXPRESSED AS THE SUM OF THE ENTHALPY CHANGES FOR A SERIES OF REACTIONS INTO WHICH THE OVERALL REACTION MAY BE SUBDIVIDED.

Some conventions for standard states

The conventions used to define the standard states for pure substances and solutes are summarised in table 11.1. These form the basis of most practical conventions but may be modified for specific systems, as will be discussed later.

Enthalpies of formation

Molar enthalpies of reaction are measurable quantities representing the difference between two states. They can be measured without specific reference to the absolute enthalpies from which they are derived. For the purpose of tabulation and reference it is convenient to assign to each compound in a reaction some form of absolute value. In order to do this we must choose some state or condition of a substance and define this as having zero enthalpy.

By convention,

● The **reference state** of each element and compound is the most stable physical form at a pressure of 1 bar (20^5 Pa) and temperature 298.15 K.
● Each **element** in its reference state is assigned an enthalpy of zero.

Table 11.1   **Standard states**

Substance	Standard State
Gas	1 bar (10^5 Pa)
Liquid	Pure liquid at any temperature and pressure
Solid	Pure solid at any temperature and pressure
Solute	1 mol kg^{-1} solution at any temperature and pressure

Only in the reference state will the standard enthalpy of an element be zero. At other temperatures and pressures the enthalpy, relative to the reference state, can be calculated by equation 9.90.

In addition to the reference state there is an important definition:

THE STANDARD ENTHALPY OF FORMATION IS THE ENTHALPY CHANGE THAT ACCOMPANIES THE FORMATION OF 1 MOLE OF A COMPOUND, IN ITS STANDARD STATE, FROM ELEMENTS IN THEIR STANDARD STATES.

The standard states for any reactant or product, including those in formation reactions, are chosen in accordance with table 11.1. For example, in the reaction:

$$C_{graphite} + O_2(g) \rightleftharpoons CO_2(g) \qquad (11.8)$$

at **298.15 K and 1 bar**, the substances $C_{graphite}$ and $O_2(g)$ are in their reference states and also in their standard states at the specified temperature and pressure. The gas $CO_2(g)$ is in its standard state; therefore the standard enthalpy of formation is equal to the enthalpy change for the complete reaction, i.e.

$$\Delta H_f^\ominus = \Delta H \text{ (reaction)} \qquad (11.9)$$

The subscript f refers to the formation of the compound. A point to note is that if the reaction was measured under conditions other than 298.15 K and 1 bar, the enthalpies of the elements would not be zero and the standard enthalpy of formation would not be equal to the enthalpy of reaction. To calculate the enthalpy of formation would require a simple correction for the sum of the enthalpies of the elements (see section 11.1.4).

Some enthalpies of formation can be measured directly; others have to be calculated, using Hess's law, from those that can be measured. An important application of enthalpies of formation is in carrying out thermochemical calculations since:

THE STANDARD ENTHALPY CHANGE FOR ANY REACTION IS EQUAL TO THE DIFFERENCE BETWEEN THE SUMS OF THE STANDARD HEATS OF FORMATION OF THE PRODUCTS AND OF THE REACTANTS.

That is,

$$\Delta H^\ominus = \sum v \Delta H_f^\ominus \text{(products)} - \sum v \Delta H_f^\ominus \text{(reactants)} \qquad (11.10)$$

where v denotes the number of moles of substance in the reaction.

The concept of an enthalpy of formation can be applied to any substance including gas-phase or liquid molecules, atoms, ions or minerals.

11.1.2   Entropy changes for reactions

The third law of thermodynamics allows the assignment of **absolute** values of entropy to substances. The third-law entropy of 1 mole of a substance in its standard state (see table 11.1) is called the **standard molar entropy**.

Standard molar entropies of substances can be used directly to calculate the standard entropy change for any reaction, since

THE STANDARD ENTROPY CHANGE FOR ANY REACTION IS EQUAL TO THE DIFFERENCE BETWEEN THE STANDARD MOLAR ENTROPIES OF THE PRODUCTS AND OF THE REACTANTS.

That is,

$$\Delta S^{\ominus}(\text{reaction}) = \sum v \overline{S}^{\ominus}(\text{products}) - \sum v \overline{S}^{\ominus}(\text{reactants}) \qquad 11.11$$

This is the equivalent of Hess's law of heat summation for entropy changes.

In tabulations of thermodynamic data for geochemical substances the third-law entropy is invariably quoted. However, a **standard entropy of formation** may be defined as the entropy change that accompanies the formation of 1 mole of a compound, in its standard state, from elements in their standard state. It should be noted that the standard molar entropy of a substance is **not** equal to the standard molar entropy of formation from the elements, since both elements and compounds have non-zero entropies at 298.15 K.

11.1.3   Changes in the Gibbs energy for reactions

In direct parallel with standard enthalpy changes for reactions, we may define the following:

- Each **element** in its reference state is assigned a Gibbs energy of zero.
- The **standard Gibbs energy of formation** is the change in the Gibbs energy that accompanies the formation of 1 mole of a compound in its standard state from elements in their standard states.
- The **standard Gibbs energy of reaction** is equal to the difference between the standard energies of formation of the products and of the reactants. That is,

$$\Delta G^{\ominus}(\text{reaction}) = \sum v \Delta G_{f}^{\ominus}(\text{products}) - \sum v \Delta G_{f}^{\ominus}(\text{reactants}) \quad (11.12)$$

In accordance with the relationship $G = H - TS$, we may write for any chemical change at constant temperature:

$$\Delta G = \Delta H - T\Delta S \qquad (11.13)$$

or for a standard-state change:

$$\Delta G^{\ominus} = \Delta H^{\ominus} - T\Delta S^{\ominus} \qquad (11.14)$$

The value of $\Delta G^{\ominus}$ can be calculated from measured values of $\Delta H^{\ominus}$ and $\Delta S^{\ominus}$.

The value of $\Delta G^{\ominus}$ can also be measured directly using a relationship to be described in section 11.2

The change in **any** state function of a reaction under standard–state conditions can be defined. These including the following:

$$\Delta V^{\ominus}(\text{reaction}) = \sum v\bar{V}^{\ominus}(\text{products}) - \sum v\bar{V}^{\ominus}(\text{reactants}) \qquad (11.15)$$

$$\Delta C_{P}^{\ominus}(\text{reaction}) = \sum v\bar{C}_{P}^{\ominus}(\text{products}) - \sum v\bar{C}_{P}^{\ominus}(\text{reactants}) \qquad (11.16)$$

11.1.4 Apparent energies of formation

The enthalpies of all elements vary with temperature and pressure, as do the enthalpies of all substances. Consequently, the standard enthalpies of formation will also vary with temperature and pressure. In order to perform calculations with standard enthalpies of formation for a substance over a range of geochemical conditions, it would be necessary to measure and document the enthalpies of the elements which comprise that substance over the same ranges of temperature and pressure.

This is inconvenient and unnecessary because the contributions from the enthalpies of the elements to the standard enthalpies for a reaction cancel out. For example, the reaction between anhydrite and water to form gypsum is given by:

$$CaSO_4 + 2H_2O \rightleftharpoons CaSO_4 . 2H_2O(s). \qquad (11.17)$$

The standard enthalpy change in terms of the standard enthalpies of formation is given by:

$$\Delta H^{\ominus}(\text{reaction}) = \Delta H^{\ominus}_{f,CaSO_4} - \Delta H^{\ominus}_{f,CaSO_4.2H_2O(s)} - 2\Delta H^{\ominus}_{f,H_2O} \qquad (11.18)$$

where the standard enthalpies of formation of anhydrite, gypsum and water respectively are given by:

$$\Delta H^{\ominus}_{f,CaSO_4} = H^{\ominus}_{CaSO_4} - H^{\ominus}_{Ca} - H^{\ominus}_{S} - 4H^{\ominus}_{O} \qquad (11.19)$$

$$\Delta H^{\ominus}_{f,CaSO_4.2H_2O} = H^{\ominus}_{f,CaSO_4.2H_2O(s)} - H^{\ominus}_{Ca} - H^{\ominus}_{S} - 6H^{\ominus}_{O} - 4H^{\ominus}_{H} \qquad (11.20)$$

and

$$\Delta H^{\ominus}_{f,H_2O} = H^{\ominus}_{H_2O} - 2H^{\ominus}_{H} - H^{\ominus}_{O} \qquad (11.21)$$

The values of $\Delta H^{\ominus}_{f,CaSO_4}$, $\Delta H^{\ominus}_{f,CaSO_4.2H_2O(s)}$ and $\Delta H^{\ominus}_{f,H_2O}$ can be eliminated from equation 11.18 using equations 11.19, 11.20 and 11.21. The contributions for the elements cancel out, giving:

$$\Delta H^{\ominus}(\text{reaction}) = H^{\ominus}_{CaSO_4} - H^{\ominus}_{CaSO_4.2H_2O(s)} - 2H^{\ominus}_{H_2O} \qquad (11.22)$$

as expected. At different temperatures and pressures the values of $\Delta H^{\ominus}$ (reaction) will be different, solely because of changes in the enthalpies of the reactive substances, since the enthalpies of the elements cancel out. We cannot of course evaluate the enthalpies of the individual substances, but we may remove all reference to the properties of the elements by defining the **apparent enthalpy of formation** $\Delta H_{(T,P)}$. The apparent enthalpy of formation at a temperature and pressure T, P is equal to the standard enthalpy of formation at a reference temperature and pressure T_r, P_r, **plus** a term equal to the difference in the enthalpy of the **substance** between the two states. That is, there are no terms included for the enthalpies of the elements.

Expressed mathematically,

$$\Delta H_{(T,P)} \equiv \Delta H^{\ominus}_{f(T_r,P_r)} + [\bar{H}_{(T,P)} - \bar{H}_{(T_r,P_r)}] \tag{11.23}$$

where $[\bar{H}_{(T,P)} - \bar{H}_{(T_r,P_r)}]$ is the difference in the enthalpy of a substance between the two states. For many chemical and geochemical applications it is common to define the standard states for minerals, water and dissolved ions to be at any temperature and pressure. In these cases the apparent enthalpy of formation term will carry the superscript, i.e. $\Delta H^{\ominus}_{(T,P)}$. In later sections we shall see the use of other subscripts in order to be even more specific about the state of the substance.

Apparent enthalpies of formation become equal to true enthalpies of formation at the reference temperature and pressure. They may be used in exactly the same way as the true functions. A further simplification arises because the function $[\bar{H}_{(T,P)} - \bar{H}_{(T_r,P_r)}]$ is a property of substances that can be measured or computed quite easily, even though the enthalpies themselves cannot be measured. Such calculations are described in section 9.2.

When standard states of reacting substances are unrestricted in temperature and pressure, the standard enthalpy changes for reactions are themselves temperature- and pressure-dependent. We may evaluate $\Delta H^{\ominus}_{(T,P)}$ (reaction) at any temperature and pressure from the apparent standard enthalpies of substances, using:

$$\Delta H^{\ominus}_{(T,P)} = \sum v \Delta H_{(T,P)}(\text{products}) - \sum v \Delta H_{(T,P)}(\text{reactants}) \tag{11.24}$$

This formalism ensures that the standard enthalpy change for a reaction is recognized to be temperature- and pressure-dependent.

Similarly, we may define **apparent standard Gibbs energy of formation** as:

$$\Delta G^{\ominus}_{(T,P)} \equiv \Delta G^{\ominus}_{f,(T_r,P_r)} + [\bar{G}_{(T,P)} - \bar{G}_{(T_r,P_r)}] \tag{11.25}$$

The **apparent standard Helmholtz energy of formation** is:

$$\Delta A^{\ominus}_{(T,P)} \equiv \Delta A^{\ominus}_{f,(T_r,P_r)} + [\bar{A}_{(T,P)} - \bar{A}_{(T_r,P_r)}] \tag{11.26}$$

The **apparent standard internal energy of formation** is:

$$\Delta U^{\ominus}_{(T,P)} \equiv \Delta U^{\ominus}_{f,(T_r,P_r)} + [\bar{U}_{(T,P)} - \bar{U}_{(T_r,P_r)}] \tag{11.27}$$

We may use the above apparent standard properties in equations analogous to equations 11.12, 11.15 and 11.16 to calculate $\Delta G^{\ominus}$(reaction), $\Delta V^{\ominus}$(reaction) and $\Delta C^{\ominus}_P$(reaction). Since the standard third-law molar entropy for any substance is measurable, there is no need to use a standard entropy of formation. The value of $\Delta S^{\ominus}$(reaction) can be calculated directly from the standard third-law entropies of the reactants and products. Throughout the literature different conventions are used and the reader should take care to ensure consistency at all times.

Sections 9.2.1–9.2.6 give equations for the determination of S, C_P and α as a function of temperature and pressure. These can be used to calculate the temperature and pressure dependence of the apparent Gibbs energy of formation and the apparent enthalpy of formation since the following relationships hold:

$$\left[\frac{d(\Delta G^{\ominus})}{dT} \right]_P = -\bar{S}^{\ominus} \tag{11.28}$$

$$\left[\frac{d(\Delta G^{\ominus})}{dP} \right]_T = \bar{V}^{\ominus} \tag{11.29}$$

$$\left[\frac{d(\Delta H^{\ominus})}{dT}\right]_{P} = \bar{C}_{P}^{\ominus} \tag{11.30}$$

$$\left[\frac{d(\Delta H^{\ominus})}{dP}\right]_{T} = \bar{V}^{\ominus}(1 - \alpha T) \tag{11.31}$$

11.1.5 Qualitative interpretation of changes in the thermodynamic potentials for reactions

The sign and magnitude of the changes in ΔH, ΔS and ΔG for a reaction can be used to show whether the reaction is spontaneous or not.

The standard enthalpy change for a reaction is the total heat absorbed during the process. A large positive value for ΔH (i.e. exothermicity) usually favours the spontaneity of a reaction but does not guarantee it. Many endothermic reactions are spontaneous. Care must also be taken in interpreting the sign and magnitude of standard entropy changes since ΔS for a reaction refers to changes in the system, not the system plus the surroundings. The entropy change for a system may be negative if a change in the entropy of the surroundings compensates. Consequently many spontaneous processes have negative values for ΔS. The function that gives the direction of spontaneous change at constant temperature is ΔG. The interpretation of ΔG is as follows.

- $\Delta G < 0$: A spontaneous reaction.
- $\Delta G > 0$: Not a spontaneous reaction – the reverse reaction is spontaneous.

The relationship

$$\Delta G = \Delta H - T\Delta S \tag{11.32}$$

shows that the value of ΔG represents a balance between the changes in the entropy and the enthalpy of a system. We may recall from equation 8.86 in section 8.4.8 that the entropy change in the surroundings multiplied by the absolute temperature equals the enthalpy change in the system, i.e.

$$T\Delta S_{surr} = \Delta H \tag{11.33}$$

Then the value of $-\Delta G^{\ominus}/T$ may be thought of as the summation of the entropy

Table 11.2   **Factors affecting the sign and magnitude of ΔG**

ΔH	$T\Delta S$	Comments
−	−	ΔG will be negative when $T\Delta S$ is smaller than ΔH
		This condition will be favoured at low temperatures
+	−	ΔG will be positive regardless of temperature
−	+	ΔG will be negative regardless of temperature
+	+	ΔG will be negative when $T\Delta S$ is greater than ΔH
		This condition will be favoured at high temperatures

changes in both system and surroundings accompanying the specified change. This sum must be positive for a spontaneous change.

Table 11.2 lists the factors affecting the sign of ΔG in terms of ΔH and $T\Delta S$.

For reactions carried out under standard state conditions, i.e. conversion of reactants in their standard states to products in their standard states, the previous interpretations are valid. However, if a value of $\Delta G^\ominus$ is positive it does not mean that the reaction will not progress at all. The interpretation is that the reaction will not spontaneously produce the full quantity of products in their standard state. The primary use of $\Delta G^\ominus$ is to help predict equilibrium compositions for chemical reactions or mixtures of interacting substances. This specific application is discussed in section 11.2.

11.2 Equilibria in chemically reacting systems

The **equilibrium constant** describes the relationship between the concentration of mutually reacting species at equilibrium. The **affinity** of a reaction is a term used to describe the tendency of a reaction to proceed. This section explains the quantitative basis to these terms.

11.2.1 A generalised notation for chemical reactions

Any chemical reaction can be written using the following generalised equation:

$$\sum v_i A_i = 0 \tag{11.34}$$

where A_i designates the formula of the ith substance (either reactant or product) and v_i the number of moles of the ith substance in the reaction. The convention is that stoichiometric mole numbers v_i are **positive for products** and **negative for reactants**.

For example, a reaction between 1 mole of A_1 and 2 moles of A_2 to give 1 mole of A_3 plus 2 moles of A_4 would normally be written:

$$A_1 + 2A_2 \rightleftharpoons A_3 + 2A_4 \tag{11.35}$$

Using the general notation the reaction is written:

$$v_1 A_1 + v_2 A_2 + v_3 A_3 + v_4 A_4 = 0 \tag{11.36}$$

where $v_1 = -1$, $v_2 = -2$, $v_3 = 1$ and $v_4 = 2$. That is,

$$A_3 + 2A_4 - A_1 - 2A_2 = 0 \tag{11.37}$$

The general notation applies to the use of any thermodynamic property Ξ, i.e.

$$\Delta\Xi = \sum v_i \Xi_i \tag{11.38}$$

For example, the standard Gibbs energy change for a reaction is given by:

$$\Delta G^\ominus = \sum v_i \Delta G^\ominus_{f,i} \tag{11.39}$$

where $\Delta G_{f,i}^{\ominus}$ is the true standard Gibbs energy of formation. Similar expressions apply to values of $\Delta H^{\ominus}$, $\Delta U^{\ominus}$, $\Delta A^{\ominus}$, $\Delta C_P^{\ominus}$ or $\Delta S^{\ominus}$. The rules for using standard (or apparent) energies of formation are given in section 11.1.4.

11.2.2 The extent of reaction

Consider the reaction in equation 11.35. If quantities of A_1 and A_2 are mixed, the reaction produces quantities of A_3 and A_4 as it proceeds. We designate the concentrations of the reactants or products, at some **arbitrary** reference time in the process, with the symbols n_1^0, n_2^0, n_3^0 and n_4^0. As the reaction proceeds by an arbitrary time increment, the changes in the concentrations are given by Δn_1, Δn_2, Δn_3 and Δn_4. The concentrations at the end of the time increment are:

$$n_1 = n_1^0 + \Delta n_1 \tag{11.40}$$

$$n_2 = n_2^0 + \Delta n_2 \tag{11.41}$$

$$n_3 = n_3^0 + \Delta n_3 \tag{11.42}$$

$$n_4 = n_4^0 + \Delta n_4 \tag{11.43}$$

Note that Δn_1 and Δn_2 will have negative values since the reactants are consumed in the reaction.

This is the equivalent of saying that $|\Delta n_1|$ moles of A_1 have reacted with $|\Delta n_2|$ moles of A_2 to produce $|\Delta n_3|$ moles of A_3 and $|\Delta n_4|$ moles of A_4.

The algebraic relationships between the quantities of material consumed and produced is:

$$\frac{\Delta n_1}{\nu_1} = \frac{\Delta n_2}{\nu_2} = \frac{\Delta n_3}{\nu_3} = \frac{\Delta n_4}{\nu_4} = \text{constant} \tag{11.44}$$

That is,

$$\frac{\Delta n_1}{-1} = \frac{\Delta n_2}{-2} = \frac{\Delta n_3}{3} = \frac{\Delta n_4}{2} = \text{constant} \tag{11.45}$$

The constant is a measure of the change in the extent of reaction. We give the **extent of reaction** the symbol ξ and define the constant in equation 11.44 as $\Delta \xi$. The general relationship becomes:

$$\Delta \xi = \frac{\Delta n_i}{\nu_i} \tag{11.46}$$

The units of ξ are moles.

The extent of reaction is a useful parameter since it is a single variable with which to define reaction progress. Using it we avoid having to refer to the concentrations of the individual reacting substances. This is convenient for reactions involving many species.

An alternative way of defining the extent of reaction is

$$\xi = (n_i - n_i^0)/\nu_i \tag{11.47}$$

where the reference time is the very beginning the reaction. At this time value of ξ

and the concentrations of all product species equal zero. Using this definition of ξ we have $\Delta\xi \equiv \xi$ and we may use the two conventions interchangeably.

A change in the extent of reaction from ξ^0 to $\xi^0 + \Delta\xi$ means that $\Delta n_i / v_i$ moles of the subscripted substance have been produced or consumed.

11.2.3   The position of equilibrium

The objective of this section is to show that as a chemical reaction proceeds the Gibbs energy for the system decreases to a minimum. At this point the reaction proceeds no further.

To develop these ideas we must consider a chemical reaction to be the production of a mixture of reactants and products. A simple example, of one reactant forming one product, is adequate to illustrate this point.

The reaction:

$$A_1 \rightleftharpoons A_2 \tag{11.48}$$

can be thought of as a mixture of A_1 and A_2 which changes composition as the reaction proceeds.

If there are no other substances in the system, the Gibbs energy is given by:

$$G = n_1\mu_1 + n_2\mu_2 \tag{11.49}$$

If the system is ideal, the chemical potentials may be replaced by:

$$\mu_1 = \mu_1^\ominus + RT\ln x_1 \quad \text{and} \quad \mu_2 = \mu_2^\ominus + RT\ln x_2 \tag{11.50}$$

Combining equations 11.50 and 11.51 gives:

$$G = n_1\mu_1^\ominus + n_2\mu_2^\ominus + [n_1 RT\ln x_1 + n_2 RT\ln x_2] \tag{11.51}$$

The mole fraction terms may be expanded, since $x_1 = n_1/(n_1 + n_2)$ and $x_2 = n_2/(n_1 + n_2)$, to give:

$$G = n_1\mu_1^\ominus + n_2\mu_2^\ominus + \left[n_1 RT\ln\left(\frac{n_1}{n_1 + n_2}\right) + n_2 RT\ln\left(\frac{n_2}{n_1 + n_2}\right) \right] \tag{11.52}$$

We may now introduce the extent of reaction as defined by:

$$\xi = (n_1 - n_1^0)/v_1 = (n_2 - n_2^0)/v_2 \tag{11.53}$$

where n_1^0 and n_2^0 are the amounts of reactant and product at the commencement of the reaction. Consequently,

$$n_1 = n_1^0 - \xi \quad \text{and} \quad n_2 = n_2^0 + \xi \tag{11.54}$$

This allows us to replace all the n_1 and n_2 terms in equation 11.52 with values of n_i^0 and ξ; i.e.

$$G = (n_1^0 - \xi)\mu_1^\ominus + (n_2^0 + \xi)\mu_2^\ominus$$
$$+ \left[(n_1^0 - \xi)RT\ln\left(\frac{n_1^0 - \xi}{n_1^0 + n_2^0}\right) + (n_2^0 + \xi)RT\ln\left(\frac{n_2^0 + \xi}{n_1^0 + n_2^0}\right) \right] \tag{11.55}$$

In order to calculate the Gibbs energy as it varies with the extent of reaction, we require the values of n_1^0 and n_2^0 in addition to $\mu_1^\ominus$ and $\mu_2^\ominus$. Figure 11.2(a) shows

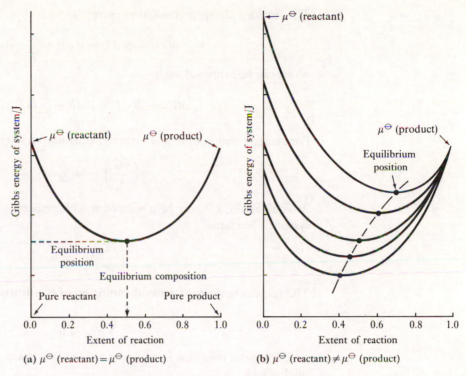

Figure 11.2   Gibbs energy versus extent of reaction: ●, equilibrium position.

such a plot for the hypothetical situation where the reaction begins with $n_1^0 = 1$, $n_2^0 = 0$ and the standard chemical potentials are equal. In this case the extent of reaction can vary between 0 and 1 mole. At the onset of reaction, where $\xi = 0$, the spontaneous process is the production of product, since this leads to a decrease in the Gibbs energy. As the extent of reaction increases, a minimum in the Gibbs function is reached. At this point, any further increase in the extent of reaction would give rise to an increase in the Gibbs energy which is not a spontaneous process. The minimum in the Gibbs energy is therefore a position of equilibrium and the composition of the mixture the equilibrium composition.

Figure 11.2(b) shows the plot of G versus extent of reaction for situations where $\mu_1^\ominus$ is not equal to $\mu_2^\ominus$. We see that the position of equilibrium is dependent on the difference between $\mu_1^\ominus$ and $\mu_2^\ominus$. This point is developed further in section 11.2.5.

11.2.4   The condition of equilibrium

The condition of equilibrium is obtained by minimising the Gibbs energy with respect to ξ. We may describe this situation for the general reaction equation 11.34.

Consider an infinitesimal change in ξ for any reaction constrained by:

$$d\xi = \frac{dn_i}{v_i} \tag{11.56}$$

The resulting change in the Gibbs energy is:

$$dG = -S\,dT + V\,dP + \sum_i \mu_i\,dn_i \tag{11.57}$$

which can be expressed as:

$$dG = -S\,dT + V\,dP + \sum_i (v_i\mu_i)\,d\xi_i \tag{11.58}$$

For constant temperature and pressure, equation 11.59 rearranges to:

$$\left(\frac{\partial G}{\partial \xi}\right)_{T,P} = \sum v_i\mu_i \tag{11.59}$$

At equilibrium, G must be a minimum with respect to any displacement of the system. Therefore,

$$\left(\frac{\partial G}{\partial \xi}\right)_{T,P} = 0 \tag{11.60}$$

This leads us to the most **general condition of equilibrium for a reaction**, which is:

$$\sum v_i\mu_i = 0 \tag{11.61}$$

The general condition is valid whether the substances exist in gaseous, liquid or solid phases.

The Belgian physicist Theophilé de Donder defined the **affinity** of a reaction as:

$$A = -\left(\frac{\partial G}{\partial \xi}\right)_{T,P} = -\sum v_i\mu_i \tag{11.62}$$

Consequently, there are three important points to be remembered:

- **A chemical reaction will proceed spontaneously if the affinity is greater than zero.** In this situation the sum of the chemical potentials of the reactants is greater than the sum of the chemical potentials of the products.
- **The reverse reaction will occur spontaneously if the affinity is less than zero.**
- **At equilibrium the affinity is zero.**

11.2.5   The equilibrium constant

The form of the equilibrium constant

A final step is to define the equilibrium constant using the general condition of equilibrium.

For each substance in a real mixture, the chemical potential is given by:

$$\mu_i = \mu_i^{\ominus} + RT \ln a_i \tag{11.63}$$

where, we recall, $\mu_i^{\ominus}$ is the chemical potential of the ith substance in a chosen standard state. This equation can be substituted directly into equation 11.61, which rearranges to give:

$$\sum_i v_i\mu_i^{\ominus} = -RT \sum_i v_i \ln a_i \tag{11.64}$$

The sum of the standard-state chemical potentials is called the standard energy change for the reaction and is equal to $\Delta G^{\ominus}$ as defined in section 11.1.3. The value of $\Delta G^{\ominus}$ is dependent only on the choice of standard states for the substances. If these standard states are temperature- and pressure-dependent, then so is $\Delta G^{\ominus}$.

The sum of the **equilibrium values** for the $\ln a_i$ terms defines the **equilibrium constant**. It is given the symbol K_{eq}.

$$\ln K_{eq} = \sum_i v_i \ln a_i \tag{11.65}$$

For the example reaction in equation 11.35, the value of K_{eq} is given by:

$$K_{eq} = \frac{a_3^{|v_3|} \times a_4^{|v_4|}}{a_1^{|v_1|} \times a_2^{|v_2|}} \tag{11.66}$$

where $|v_1|$ is the modulus of the stoichiometry number for the subscripted species. The equilibrium constant is related to the standard Gibbs energy change by:

$$\Delta G^{\ominus} = -RT \ln K_{eq} \tag{11.67}$$

Like $\Delta G^{\ominus}$, the equilibrium constant is dependent only on the choice of standard states for the substances involved.

The equilibrium constant and the affinity

By substituting $\mu_i = \mu_i^{\ominus} + RT \ln a_i$ into equation 11.62, and noting that $\Sigma_i \mu_i^{\ominus} = -RT \ln K_{eq}$, the chemical affinity and the equilibrium constant can be related by:

$$A = RT \left(\ln K_{eq} - \sum_i \ln a_i^{v_i} \right) \tag{11.68}$$

or

$$A = -RT \ln \left(\frac{\Pi a_i^{v_i}}{K_{eq}} \right) \tag{11.69}$$

Summarising comments on the equilibrium constant

To assist calculations it is common to separate the equilibrium constant into two terms, one containing only concentrations, the other the product of the activity coefficients, i.e.

$$\ln K_{eq} = \ln K_m + \sum_i \ln \gamma_i^{v_i} \tag{11.70}$$

where K_m is the mass action quotient defined as:

$$\ln K_m = \sum_i \ln C_i^{v_i} \tag{11.71}$$

where C_i is the selected concentration of compound i. The value of the mass action quotient controls the relative concentrations of species present at equilibrium.

It is normal to calculate the value of $\Delta G^{\ominus}$ or $\ln K_{eq}$, at specified temperatures and pressures, from the sum of the values of the standard Gibbs energies of formation for each species in the reaction. The fundamental relationship is:

$$\Delta G^{\ominus} = \sum_i v_i \Delta G_{f,i}^{\ominus} \tag{11.72}$$

where $\Delta G_{f,i}^{\ominus}$ is the standard Gibbs energy of formation of the ith species at the specified temperature and pressure. Alternatively, we may replace the standard Gibbs energy of formation with the apparent standard Gibbs energy of formation ΔG_i and write:

$$\Delta G^{\ominus} = \sum_i \nu_i \Delta G_i^{\ominus} \tag{11.73}$$

Thus, in order to characterise the distribution of species at equilibrium we must know at least the values of the apparent Gibbs energies of formation of all species in the reaction, in addition to the values of the activity coefficients at the equilibrium composition.

Knowing values of the activities for species in systems which are **not at equilibrium** is useful for determining the direction of change. We know that the value of $\Delta G^{\ominus}$ defines the extent to which a reaction can occur under standard-state conditions. Also, for a reaction to proceed spontaneously the value of ΔG for the reaction should be negative. The relationship between these two variables is:

$$\Delta G = \Delta G^{\ominus} + RT \ln(\Pi a_i^{\nu_i}) \tag{11.74}$$

This means that a mixture of reactive species will spontaneously readjust their concentrations (activities), by reaction, proceeding towards a state where the value of ΔG is zero. If the activities of species are known at any selected time the value of ΔG can be calculated. If this value is negative more products will be formed; if it is positive the reaction will proceed in the reverse direction.

The value of $\Delta G^{\ominus}$ is of course related to $\Delta H^{\ominus}$ and $\Delta S^{\ominus}$ using:

$$\Delta G^{\ominus} = \Delta H^{\ominus} - T\Delta S^{\ominus} \tag{11.75}$$

where $\Delta H^{\ominus}$ can be calculated from either the true or the apparent enthalpies of formation, and $\Delta S^{\ominus}$ from the standard third-law entropies of substances (section 11.1.4).

11.2.6   The effect of temperature and pressure on the equilibrium constant

The simplest way to evaluate the isobaric temperature dependence of an equilibrium constant is from the relationships: $\Delta G^{\ominus} = \Delta H^{\ominus} - T\Delta S^{\ominus}$ and $\Delta G^{\ominus} = -RT \ln K$. Assuming both $\Delta H^{\ominus}$ and $\Delta S^{\ominus}$ are independent of temperature, then:

$$\ln K = -\frac{\Delta H^{\ominus}}{RT} + \frac{\Delta S^{\ominus}}{R} \tag{11.76}$$

This is the equation of a straight line between $\ln K$ and the reciprocal absolute temperature, with slope $-\Delta H^{\ominus}/R$ and intercept $\Delta S^{\ominus}/R$. Typically, the approximation is useful over a temperature range about $20\,^{\circ}\text{C}$ for solubility reactions. When applied over a range of $100\,^{\circ}\text{C}$, errors of up to one natural logarithmic unit may be encountered for some reactions.

Assuming the standard states for a reaction are pressure-dependent, the simplest way to evaluate the isothermal pressure dependence of an equilibrium constant is from the relationship:

$$\left[\frac{\partial(\ln K)}{\partial P}\right]_T = \Delta V^{\ominus} \tag{11.77}$$

Assuming that $\Delta V^\ominus$ is independent of pressure, the above equation integrates to give a simple linear relationship:

$$\ln K_{P_2} = \ln K_{P_1} - \frac{\Delta V^\ominus}{RT}(P_2 - P_1) \tag{11.78}$$

The assumption of constant $\Delta V^\ominus$ is a reasonable first approximation for many geochemical reactions at most pressures when the temperature is above 100 °C. Below 100 °C the assumption is good at pressures less than 1000 bar. Indeed, the linear pressure dependence of equilibrium constants is often used to calculate volumes of reaction.

If the standard state is at a fixed pressure, then the equilibrium constant has no pressure dependence and all the pressure dependence is incorporated into the activity coefficient of the components.

The most comprehensive way of evaluating the effect of temperature and pressure on an equilibrium constant is to determine their effect on the Gibbs energies of formation of the components in the reaction. For this purpose, equation 9.92 would be used with suitable expressions for the temperature and pressure dependence of molar volume and heat capacity of the components.

Alternatively, equation 9.91 can be reformulated in terms of standard–state changes for the reaction to give:

$$\Delta G^\ominus_{(T_2, P_2)} - \Delta G^\ominus_{(T_1, P_1)} = \int_{P_1}^{P_2} \Delta V^\ominus_{(T_2, P)} \, \mathrm{d}P - \int_{T_1}^{T_2} \Delta S^\ominus_{(P_1, T)} \, \mathrm{d}T' \tag{11.79}$$

The integrals can be rearranged to give:

$$\Delta G^\ominus_{(T_2, P_2)} - \Delta G^\ominus_{(T_1, P_1)} = \int_{P_1}^{P_2} \Delta V^\ominus_{(T_2, P)} \, \mathrm{d}P - \Delta S^\ominus_{(T_1, P_1)}(T_2 - T_1)$$
$$+ \int_{T_1}^{T_2} \Delta C^\ominus_{P(T, P_1)}\left(1 - \frac{T_2}{T}\right) \mathrm{d}T \tag{11.80}$$

Since $\Delta G^\ominus = -RT \ln K$ we can write:

$$\ln K_{(T_2, P_2)} = \frac{T_1}{T_2} \ln K_{(T_1, P_1)} + \frac{1}{RT_2} \int_{P_1}^{P_2} \Delta V^\ominus_{(T_2, P)} \, \mathrm{d}P + \frac{\Delta S^\ominus_{(T_1, P_1)}}{RT_2}(T_2 - T_1)$$
$$- \frac{1}{RT_2} \int_{T_1}^{T_2} \Delta C^\ominus_{P(T, P_1)}\left(1 - \frac{T_2}{T}\right) \mathrm{d}T \tag{11.81}$$

A more common form of the above equation is obtained by substituting $\Delta S_{(T_1, P_1)}/R$ from the relationship

$$\ln K_{(T_1, P_1)} = -\frac{\Delta H_{(T_1, P_1)}}{RT_1} + \frac{\Delta S_{(T_1, P_1)}}{R}$$

This gives:

$$\ln K_{(T_2, P_2)} = \ln K_{(T_1, P_1)} - \frac{1}{RT_2} \int_{P_1}^{P_2} \Delta V^\ominus_{(T_2, P)} \, \mathrm{d}P + \frac{\Delta H^\ominus_{(T_1, P_1)}}{R}\left(\frac{1}{T_1} - \frac{1}{T_2}\right)$$
$$- \frac{1}{RT_2} \int_{T_1}^{T_2} \Delta C^\ominus_{P(T, P_1)}\left(1 - \frac{T_2}{T}\right) \mathrm{d}T \tag{11.82}$$

The above equations are comprehensive. Unfortunately, they require a knowledge of $\ln K$, $\Delta S^{\ominus}$ or $\Delta H^{\ominus}$ at one temperature and pressure **plus** expressions for the pressure dependence of $\Delta V^{\ominus}_{(T_2, P)}$ and the temperature dependence of $\Delta C^{\ominus}_{P(T, P_1)}$. Quite often data on the temperature and pressure dependencies are not available, as is the case for many reactions involving aqueous ions. Consequently, we often need to make simplifying assumptions to perform calculations. As expected, equation 11.82 reduces to equations 11.76 and 11.77 on the assumption that ΔC_P is always zero and $\Delta V^{\ominus}_{(T_2, P)}$ is constant.

In reality, $\Delta C^{\ominus}_{P(T, P_1)}$ for any reaction is rarely zero and usually has a temperature dependence. Consequently, a plot of $\ln K$ versus T/K is usually non-linear. Empirical expressions for the temperature dependence of $\Delta C^{\ominus}_{P(T, P_1)}$ may be deduced from such plots.

Chapter 12

Phase transitions and phase equilibria

Changing the temperature and pressure of mineral assemblies may cause solid-to-solid-phase transactions in some minerals. When heated to very high temperatures rocks may melt to form magmas. However, when cooled, magmas may crystallize to form complex mineral assemblies. In other contexts, sedimentary processes concern the dissolution and reprecipitation of minerals by the action of water. These are all examples of geologically relevant phase transitions and phase equilibria.

The laws of chemical thermodynamics impose restrictions on the number and type of phases that can exist under specified conditions of temperature and overall composition. This chapter is concerned with these conditions, which often play a vital role in geological investigation. The rules governing these conditions are quite general and equally applicable to other chemical systems.

12.1 The Gibbs phase rule

In complex systems there may be many phases with many species distributed between the phases. At equilibrium, the chemical potential of any one substance is the same in any phase. This condition leads to the **Gibbs phase rule**, which enables us to evaluate the types of systems that can or cannot exist.

12.1.1 Degrees of freedom

The complete set of intensive properties for any one phase may be large in number. For example, a gas may be categorised by its temperature, pressure, refractive index, density and molar heat capacity or other properties.

Experience tells us that there is **a minimum number of intensive properties that must be specified in order to fix the values of all other intensive variables**. This is the number of **degrees of freedom**, or variance, and usually has the symbol F. The value of F for a pure single phase is 2. The density of pure water, for example, is fixed if the temperature and pressure are specified.

Usually, the intensive variables we choose are temperature and pressure plus others which are easily measured. If phases are composed of several components, the concentrations of the components in each phase are additional intensive variables. The pressure on the systems arises from the vapour pressure of the components or by mechanical means.

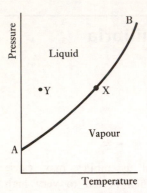

Figure 12.1
The vapour pressure
curve for water.

In multiphase or multicomponent systems the number of degrees of freedom is less easily evaluated than for single-component, single-phase systems. An example is the two-phase system of pure liquid water plus water vapour. This is a single-component system. To discover the minimum number of intensive variables required to define the system completely we may analyse the experimental $P–T$ diagram of water (figure 12.1). We observe that as temperature and pressure are varied, the phase composition of the system varies. At some values of P and T the system is completely liquid, at others completely solid or gaseous. The line AB defines the temperatures and pressures at which the two phases can coexist. If two phases exist (point X) and either the temperature or pressure is specified, then the other is fixed by the position on the line. If only one phase exists, at point Y for example, then we require a knowledge of both temperature and pressure to fix the position. The arguments above suggest there may be a rule relating the number of degrees of freedom to the number of phases plus the number of components. The rule is derived as follows.

12.1.2   The derivation of the phase rule

If a single phase has C **components** there are $C + 2$ intensive properties for the system; these are the chemical potentials of each component plus temperature and pressure. However, the Gibbs–Duhem equation (section 9.18) reduces by 1 the number of intensive properties that can be varied. If P phases are present, there are P Gibbs–Duhem equations and the number of independent variables is reduced by P. Thus, if F is the number of independent variables that can be varied we may write:

$$F = C + 2 - P \tag{12.1}$$

This is the **Gibbs phase rule**.

An alternative form of the phase rule, in terms of the number of species in existence N and the number of independent reactions between the species R, is:

$$F = N + 2 - R - P \tag{12.2}$$

Table 12.1 shows some examples of the phase rule applied to selected systems.

Table 12.1   **The phase rule applied**

	$F = C + 2 - P$			
System	*P*	*F*	*C*	*Possible Variables*
Pure gas in a cylinder	1	1	2	P and T
Liquid + gas in a cylinder	2	1	1	P or T
Ice/water/water vapour	3	1	0	None
$CaCO_3(s) \rightleftharpoons CaO(s) + CO_2(g)$	3	2	1	P or T
$NH_4Cl(s) \rightleftharpoons NH_3(g) + HCl(g)$	2	1	1	P or T

12.1.3 Phase diagrams

A phase diagram for a system is a graphical representation of the assembly of phases that exist as a function of imposed constraints. Common constraints are temperature, pressure and some kind of composition variable, usually the mole fractions of relevant species. Phase diagrams usually represent equilibrium systems, where the number of phases and their individual compositions are determined by the Gibbs energy for the system being at a minimum. Under these conditions the number of phases is defined by the Gibbs phase rule. Some phase diagrams can be used to show non-equilibrium states.

A phase diagram may be determined experimentally or predicted from the known thermodynamic properties of the components.

12.2 Properties of single-component systems

We will consider single-component systems with reference to solids, liquids and vapours.

12.2.1 General comments

In single-component systems the only equilibrium reactions involve boiling, melting, polymorphic phase transitions and sublimation. Two-dimensional phase diagrams for unary systems denote phase stability fields in terms of temperature and pressure or temperature and volume. Occasionally, two-dimensional representations involving three variables (temperature, pressure and volume) are used. These are difficult to construct but are quite revealing in terms of phase stability.

A complete pressure–temperature diagram for a one-component system can be constructed from the following measurements.

- The vapour pressure curve for the solid.
- The vapour pressure curve for the liquid.
- The variation in melting point with pressure.
- The values of P and T for the solid/liquid/vapour equilibrium point.
- Data for solid phase transitions if they occur.

In a one-component system there may be one, two or three phases. If $P = 1$ then $F = 2$ and the system is **bivariant**. If $P = 2$ then $F = 1$ and the system is **univariant**. If $P = 3$ then $F = 0$ and the system is **invariant**.

An example is the system composed of the single component water which may exist as any combination of pure liquid water, ice and water vapour (figure 12.2). We observe that the composition of the system varies as temperature and pressure are changed.

The bivariant regions of the phase diagram are where the component is entirely solid, liquid or gaseous. Temperature and pressure can be varied independently, without a change of phase, but both must be specified to define the state of the system.

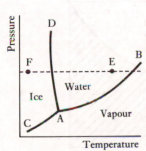

Figure 12.2
The phase diagram for water.

The lines AB, AC and AD define the temperatures and pressures at which two phases can coexist; these are the univariant lines on the phase diagram where we need specify only temperature **or** pressure.

There is only one condition of temperature and pressure at which the three phases can coexist. This is a univariant three-phase system, where once we have specified the number of components and the number of phases we need **not** specify the temperature or pressure of our system, since these values are fixed by the system itself. The values of temperature and pressure for this condition define the **triple point** for the substance.

12.2.2   Examples of phase transitions in single-component systems

The ice system

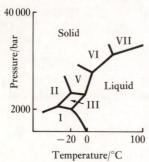

Figure 12.3
The $P-T$ projection for ice at high pressures.

Many substances, including ice, sulphur and phosphorus, show **high-pressure polymorphism** in the absence of a vapour phase.

There are six polymorphs of ice that can exist as a single phase or in equilibrium with liquid water. Figure 12.3 shows the $P-T$ diagram of the ice system, which is a small part, looked at in detail, of the complete water system.

High pressure results in the formation of structures which are more closely packed than ordinary ice. The densities of all six polymorphs are greater than that of liquid water, which results in the melting points of these polymorphs increasing with increasing pressure.

The molar volumes of the polymorphs are all less than that of water; consequently their melting points increase with increasing pressure. A striking example is ice VII, which melts at $200\,°C$ at $40\,000$ atm.

The carbon dioxide system

The $P-T$ diagram of carbon dioxide (figure 12.4) is similar to that for water. There are three areas in which only one phase exists. There is an invariant triple point at $-57.6\,°C$ and 5.11 bar plus a critical point at $31.1\,°C$ and 76 bar. The major difference with water is that dP/dT for the liquid/solid coexistence line is positive. This means

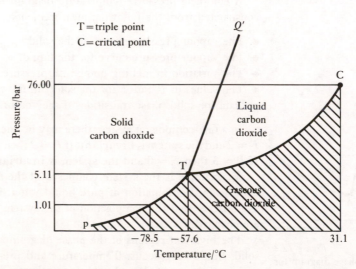

Figure 12.4   The $P-T$ projection for carbon dioxide.

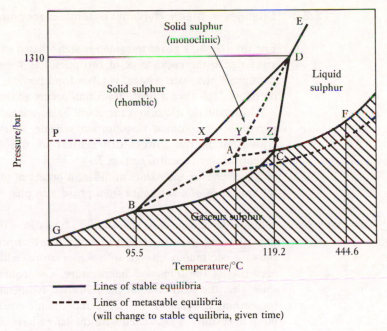

Figure 12.5   The $P-T$ projection for sulphur.

that increasing the pressure causes an increase in the melting point. This reflects the fact that the liquid contracts on freezing, which is most common but different from water. Below 5.18 bar, solid carbon dioxide sublimes: this property is used in refrigeration.

The sulphur system   Solid sulphur is an example of a substance that has several allotropes (section 3.1.6). The $P-T$ diagram (figure 12.5) shows that there are six two-phase equilibria, which are:

- rhombic/monoclinic (curve BD);
- rhombic/vapour (curve GB);
- rhombic/liquid (curve DE);
- monoclinic/liquid (curve CD);
- monoclinic/vapour (curve BC);
- liquid/vapour (curve CF).

There is a metastable triple point A which is only seen when rapid heating prevents the transition from rhombic to monoclinic to occur normally. These metastable equilibria are shown with broken lines. An example of the effects due to metastable equilibria is in the rapid heating of rhombic sulphur at pressure P. If heated slowly the rhombic form converts to monoclinic sulphur at point X which then melts at point Z. Rapid heating causes rhombic sulphur to be stable to point Y, at which point it melts.

The chemistry of sulphur is very complex: there are even more solid phases than are depicted on this diagram, plus two forms of the element existing in the liquid phase.

12.2.3 Changes in thermodynamic potentials accompanying phase transitions

For any system, a phase transition is seen when a change in temperature or pressure gives rise to the coexistence of two phases at some point during the change. For example, if the water system is taken from point E to point F via the defined path on figure 12.2, then a phase transition occurs as the system crosses the line AD.

During phase transitions there will be associated changes in the thermodynamic potentials. A common classification scheme states that **the order of a phase transition is the lowest differential of the Gibbs energy that shows a discontinuity** (see also section 3.3).

To emphasise the nature of different orders of phase transitions we will consider the transition of a substance from phase α to phase β.

First-order transitions   A first-order transition is characterised by a discontinuity in the **first derivative** of the Gibbs energy with respect to temperature or pressure. The Gibbs energy itself shows a continuous change with temperature, with the discontinuity in gradient occurring at the transition temperature. Consequently, entropy and volume both show clear discontinuities at the transition temperature. These reflect the fact that the entropy is equal to $(\partial G/\partial T)_P$ and volume equal to $(\partial G/\partial P)_T$. Physically, the jump in entropy is associated with the latent heat of the phase transition. Any heat added to a system at the transition temperature (T_c) is used in driving the phase change rather than raising the temperature. Therefore, a first-order transition is consistent with an infinite molar heat capacity at the transition temperature, as shown in figure 12.6.

At the transition point the thermodynamic potentials and their first derivatives are as follows:

$$G^\beta - G^\alpha = 0 \tag{12.3}$$

$$S^\beta - S^\alpha = \left(\frac{\partial G^\beta}{\partial T}\right)_P - \left(\frac{\partial G^\alpha}{\partial T}\right)_P = \frac{\Delta H}{T_c} \tag{12.4}$$

$$V^\beta - V^\alpha = \left(\frac{\partial G^\beta}{\partial P}\right)_T - \left(\frac{\partial G^\alpha}{\partial P}\right)_T \tag{12.5}$$

where ΔH is the latent heat of the phase transition for a given quantity of substance. If the specified quantity is 1 mole, ΔH is called the **molar enthalpy** of the phase transition.

Second-order transitions   A second-order transition is characterised by a discontinuity in the **second derivative** of the Gibbs energy with respect to temperature or pressure. The Gibbs energy and its first derivatives will be continuous.

At the transition point the thermodynamic potentials and their first derivatives are as follows:

$$G^\beta - G^\alpha = 0 \tag{12.6}$$

$$S^\beta - S^\alpha = \left(\frac{\partial G^\beta}{\partial T}\right)_P - \left(\frac{\partial G^\alpha}{\partial T}\right)_P = \frac{\Delta H}{T_c} = 0 \tag{12.7}$$

$$V^\beta - V^\alpha = \left(\frac{\partial G^\beta}{\partial P}\right)_T - \left(\frac{\partial G^\alpha}{\partial P}\right)_T = 0 \tag{12.8}$$

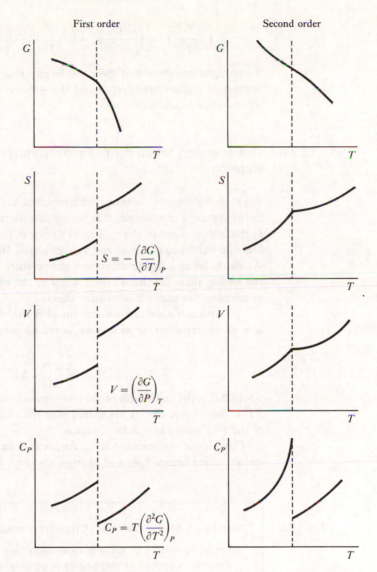

Figure 12.6   First- and second-order transitions.

Clearly, there will be no discontinuities in the volume or entropy and **no observable latent heat**; any heat added to the system at the transition temperature will produce an increase in temperature. Consequently the heat capacity will be finite at the transition temperature.

At the transition point the second derivatives of the thermodynamic potentials are as follows:

$$\left(\frac{\partial^2 G^{\beta}}{\partial T^2}\right)_P - \left(\frac{\partial^2 G^{\alpha}}{\partial T^2}\right)_P = \frac{C_P^{\beta}}{T_c} - \frac{C_P^{\alpha}}{T_c} \tag{12.9}$$

$$\left(\frac{\partial^2 G^{\beta}}{\partial P^2}\right)_T - \left(\frac{\partial^2 G^{\alpha}}{\partial P^2}\right)_T = \left(\frac{\partial V^{\beta}}{\partial P}\right)_T - \left(\frac{\partial V^{\alpha}}{\partial P}\right)_T = V(\kappa^{\beta} - \kappa^{\alpha}) \tag{12.10}$$

$$\left(\frac{\partial^2 G^\beta}{\partial T^2}\right)_P - \left(\frac{\partial^2 G^\alpha}{\partial T^2}\right)_P = \left(\frac{\partial V^\beta}{\partial T}\right)_P - \left(\frac{\partial V^\alpha}{\partial T}\right)_P = V(\alpha^\beta - \alpha^\alpha) \qquad (12.11)$$

These equations show that there will be step changes in the heat capacity (C_P). The isothermal compressibility (κ) and the isobaric compressibility (α) accompanying second-order transitions.

12.2.4   Characterising first-order phase transitions – the Clausius–Clapeyron equation

For many substances the transition temperature is dependent on the pressure applied to the system. For example, it is known that increasing pressure leads to an increase in the boiling point of water. This is shown in figure 12.2, where the line AB is the line of coexistence of liquid and vapour phase. Increasing the pressure requires the system to be at a higher minimum temperature for two phases to exist; therefore the boiling point has been raised. Clearly, the slope of the $P-T$ curve is required to calculate the magnitude of such effects.

The equation which allows us to quantify the effect of pressure on the temperature of a phase transition, or vice versa, is the Clausius–Clapeyron equation, which is:

$$\frac{\mathrm{d}P}{\mathrm{d}T} = \frac{\Delta H}{T\Delta V} \qquad (12.12)$$

where ΔH is the latent heat of the phase transition for a given quantity of substance, ΔV is the volume change associated with the phase change and $\mathrm{d}P/\mathrm{d}T$ is the slope of the $P-T$ curve in a defined region.

Thus, if we can measure the volumes of a known mass of ice and liquid water and the latent heat of fusion of ice, then the slope of the $P-T$ curve can be evaluated.

Box 12.1

Derivation of the Clausius–Clapeyron equation

The objective is to find a useful expression for $\mathrm{d}P/\mathrm{d}T$.

Consider a system of two phases in equilibrium, where temperature and pressure are varied infinitesimally so that the two phases remain in equilibrium. The resulting changes in the chemical potential of the two phases must be given by:

$$\mathrm{d}\mu^\beta_{(T,P)} = \mathrm{d}\mu^\alpha_{(T,P)} \qquad (12.13)$$

where the superscripts α and β distinguish the two phases.

We may relate $\mathrm{d}\mu$ to the molar entropy ($\bar{S}$) and the molar volume ($\bar{V}$) of a phase with the Gibbs–Duhem equation (equation 9.59) written for 1 mol of a substance, i.e. $\mathrm{d}\mu = \bar{V}\,\mathrm{d}P - \bar{S}\,\mathrm{d}T$. Then:

$$-\bar{S}^\beta\,\mathrm{d}T + \bar{V}^\beta\,\mathrm{d}P = -\bar{S}^\alpha\,\mathrm{d}T + \bar{V}^\alpha\,\mathrm{d}P \qquad (12.14)$$

Defining the molar entropy change and the molar volume change for the phase transition as:

$$\Delta\bar{S} = \bar{S}^\beta - \bar{S}^\alpha \quad \text{and} \quad \Delta V = \bar{V}^\beta - \bar{V}^\alpha \qquad (12.15)$$

continued

Box 12.1
continued

equation 12.14 becomes:

$$\frac{dP}{dT} = \frac{\Delta S}{\Delta V} \tag{12.16}$$

This is the **Clapeyron equation.**

This equation can be extended since we may write:

$$\mu^{\beta}_{(T,P)} = \bar{H}^{\beta} - T\bar{S}^{\beta} \quad \text{and} \quad \mu^{\alpha}_{(T,P)} = \bar{H}^{\alpha} - T\bar{S}^{\alpha} \tag{12.17}$$

It follows that at equilibrium $\bar{H}^{\beta} - T\bar{S}^{\beta} = \bar{H}^{\alpha} - T\bar{S}^{\alpha}$; therefore,

$$T\Delta S = \Delta H \tag{12.18}$$

where ΔH is the molar enthalpy of the phase change.

Combining equations 12.16 and 12.18 yields the Clausius–Clapeyron equation:

$$\frac{dP}{dT} = \frac{\Delta H}{T\Delta V} \tag{12.19}$$

This equation is sometimes called **Clapeyron's relation.**

12.2.5 The temperature and pressure dependence of enthalpies of phase transitions

The general relationship describing the effect of temperature and pressure on the molar enthalpy of a phase transition (ΔH) is:

$$d(\Delta H) = \left(\frac{\partial \Delta H}{\partial T}\right)_P dT + \left(\frac{\partial \Delta H}{\partial P}\right)_T dP \tag{12.20}$$

This was developed into a more useful formula by Planck, i.e.:

$$\frac{d(\Delta H)}{dT} = \Delta C_P + \frac{\Delta H}{T} - \Delta H\left(\frac{d \ln \Delta V}{dT}\right)_P \tag{12.21}$$

where ΔC_P and ΔV are both molar quantities. The Planck equations helps to produce working equations for the relationship between temperature and pressure at phase transitions.

A further simplification can be made for solid–vapour and liquid–vapour transitions by assuming that the molar volume of the liquid or solid is small compared with the molar volume of the vapour. Therefore, the volume change is almost equal to the volume of gas formed and ΔV can be approximated by RT/P per mole. In this case the term $\Delta H(d \ln \Delta V/dT)_P$ in equation 12.21 approximates to $\Delta H/T$. Equation 12.21 then approximates to:

$$\frac{d(\Delta H)}{dT} = \Delta C_P \tag{12.22}$$

which can be integrated between the limits of T_f and T_i, where f and i stand for final and initial temperatures, to give:

$$\Delta H_{T_f} = \Delta H_{T_i} + \int_{T_i}^{T_f} \Delta C_P \, dT \tag{12.23}$$

This equation is **inapplicable to solid–liquid or solid–solid transitions** because of the assumption about the magnitude of molar volumes.

12.2.6   Some integrated forms of the Clausius–Clapeyron equation

The Clausius–Clapeyron equation (12.19) is thermodynamically exact and applicable to any phase transition. However, in order to use this equation to predict the effect of pressure on the temperature of a phase transition, or vice versa, it must be integrated between selected limits. To do this it is necessary to introduce approximations for the temperature and pressure dependence of the molar enthalpy and molar volume.

There are several equations which can be generated using approximations valid for different systems.

The simplest equation for liquid–vapour or solid–vapour transitions

The following assumptions are made:

- Both ΔH and ΔV are independent of temperature and pressure.
- The vapour behaves ideally and ΔV can be approximated by RT/P, per mole.

Substituting RT/P into equation 12.19 gives:

$$\frac{1}{P}\left(\frac{dP}{dT}\right) = \frac{\Delta H}{RT^2} \tag{12.24}$$

or

$$\left[\frac{d(\ln P)}{dT}\right] = \frac{\Delta H}{RT^2} \tag{12.25}$$

This is an approximate form of the Clausius–Clapeyron equation which gives:

$$\int_{P_1}^{P_2} d(\ln P) = \Delta H \int_{T_1}^{T_2} \frac{1}{RT^2}\,dT \tag{12.26}$$

Because ΔH_m is constant this integrates simply to give:

$$\ln\left(\frac{P_2}{P_1}\right) = \frac{\Delta H}{RT}\left(\frac{1}{T_1} - \frac{1}{T_2}\right) \tag{12.27}$$

The value of ΔH for the vaporisation of any liquid is always positive. Therefore, **the vapour pressure of any liquid will increase with increasing temperature**. Figure 12.7 demonstrates this point for some simple liquids.

A simple application of Clapeyron's relation is to the effect of pressure on the temperature of a phase transition, or vice versa.

To calculate the change in the melting point of ice due to a pressure of 100 bar, we use data of the following form:

Latent heat of fusion of ice at $0\,^\circ\mathrm{C} = 333.88\ \mathrm{J\ g^{-1}}$

Volume of 1 g of ice at $0\,^\circ\mathrm{C} = 1.0907\ \mathrm{cm^3\ g^{-1}}$

Volume of 1 g of water at $0\,^\circ\mathrm{C} = 1.0001\ \mathrm{cm^3\ g^{-1}}$

Therefore, $\Delta V = -0.0906\ \mathrm{cm^3\ g^{-1}}$

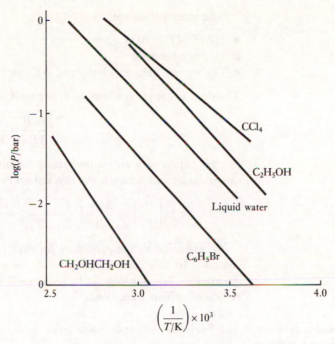

Figure 12.7   The log of the vapour pressure of several liquids as a function of the reciprocal of absolute temperature (data from Moore, 1972).

The Clausius equation (12.19) can be written as:

$$dT = T\frac{\Delta V}{\Delta H}dP \qquad (12.28)$$

Integrating, assuming ΔV and ΔH are constant, gives:

$$\int dT = T\frac{\Delta V}{\Delta H}\int dP \qquad (12.29)$$

$$\Delta T = T\frac{\Delta V}{\Delta H}\Delta P \qquad (12.30)$$

In order to use the above experimental data to predict ΔT it is clear that $\Delta V\Delta P$ must have the same dimensions as ΔH. Since $1\ cm^3\ g^{-1}$ equals $0.1\ J\ g^{-1}\ bar^{-1}$, multiplying ΔV in the above units by 0.1 puts ΔV in units of $J\ g^{-1}\ bar^{-1}$. Therefore,

$$\Delta T = -\left(\frac{0.0906 \times 0.1 \times 100}{333.88}\right) \times 273.15 = -0.74\,^{\circ}C \text{ per } 100 \text{ bar}$$

$$(12.31)$$

More sophisticated equations for liquid–vapour or solid–vapour transitions   The simplest improvement involves removing the restriction that ΔH is constant by expressing the measured temperature dependence of ΔH with empirical or theoretical equations. However, it is usually easier to measure heat capacities than enthalpies and simple formulae can be developed using Planck equations for the temperature dependence of ΔH. One useful equation is derived as follows.

The assumptions are:

- $d(\Delta H)/dT = \Delta C_P$.
- ΔC_P is a constant.
- The vapour behaves ideally and ΔV can be approximated by RT/P per mole.

The third assumption allows us to use the Clausius–Clapeyron equation written as:

$$\frac{d \ln P}{dT} = \frac{\Delta H}{RT^2} \tag{12.32}$$

In conjunction with an approximation for the temperature dependence of ΔH, this can be integrated between the two states (P_2, T_2) and (P_1, T_2) to give:

$$\ln\left(\frac{P_2}{P_1}\right) = \frac{\Delta H_{T_1}}{R}\left(\frac{1}{T_1} - \frac{1}{T_2}\right) + \frac{\Delta C_P}{R}\left[\frac{T_1}{T_2} + \ln\left(\frac{T_1}{T_2}\right) - 1\right] \tag{12.33}$$

An extension to the procedure involves removing the restriction that ΔC_P is a constant. In this case the ΔC_P would be expressed as a function of T prior to integration. The resulting final equation will still only be applicable to solid–vapour or liquid–vapour transitions.

The simplest equation for solid–solid or solid–liquid transitions

The following assumption was made:

- Both ΔH and ΔV are independent of temperature and pressure and are measurable.

Equation 12.19 rearranges to:

$$\Delta V_m \int_{P_1}^{P_2} dP = \int_{T_1}^{T_2} \frac{\Delta H}{T} dT \tag{12.34}$$

which integrates to give:

$$P_2 - P_1 = \frac{\Delta H}{\Delta V} \ln\left(\frac{T_2}{T_1}\right) \tag{12.35}$$

More accurate equations may be developed by removing the restrictions that ΔH and ΔV are constant. This can be performed by replacing both variables with empirical or theoretical expressions in T prior to integration.

12.3   The elevation of boiling point and the depression of freezing point

Any single phase such as pure anorthite, will melt at a fixed temperature and specified pressure to form a single-component melt. Conversely, it will freeze under the same conditions. If, however, the melt contains a second molten component (e.g. diopside), the mixture will freeze at a lower temperature than that for pure anorthite. Such phenomena are common in magmatic crystallizations. The thermodynamic rules which control the extent of the freezing point depression and the composition of the final solid are considered in this section. These phenomena are properties of mixtures. In view of their geological importance they will be considered separately from the general properties of mixtures. Section 18.4 shows those thermodynamic rules applied to relevant geological systems.

For mixtures containing an involatile solute and a volatile solvent, the boiling point of the mixture is higher than the boiling point of the solvent. Similarly, the freezing point of the mixture is lower than the freezing point of the solvent. These phenomena arise because the chemical potential of the solvent is lowered by the presence of a solute in accordance with Raoult's law, and occur with ideal as well as non-ideal solutions.

To explain this we recall that for all substances under all conditions the chemical potential decreases with increasing temperature. A phase transition will occur when changes in temperature equalise the chemical potentials of the solute in the two relevant phases (liquid–gas or liquid–solid). In order to compensate for the reduction in chemical potential of the solvent, due to the solute, a higher temperature is required to reach the chemical potential of the pure gas. Similarly, at the freezing point the chemical potentials of the pure solid and the liquid solute are equal. The freezing point must be reduced in order to accommodate the reduction in the chemical potential of the liquid solute. Figure 12.8 shows this graphically.

We note that the actual reduction in freezing point, or elevation of boiling point, is influenced by the slopes of chemical potential with temperature as well as by the lowering of the chemical potential of the solvent due to the solute.

An alternative way of viewing this phenomenon is to consider the partial pressure diagram for the system (figure 12.9). This shows that the vapour pressure of a mixture is lower than the vapour pressure of a pure solvent at any temperature. The freezing point for the mixture is depressed to the temperature at which the vapour pressure of the solvent equals the vapour pressure of the pure solid solvent. The boiling point is elevated to the point at which the vapour pressure equals the external pressure (atmospheric, for example).

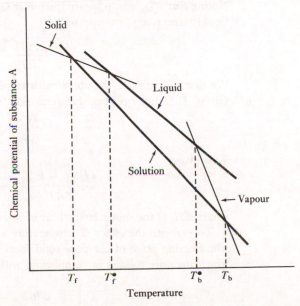

Figure 12.8   Elevation of boiling point and depression of freezing point. T_f temperature of freezing; T_b temperature of boiling; ● property of pure substance (after Everett, 1959).

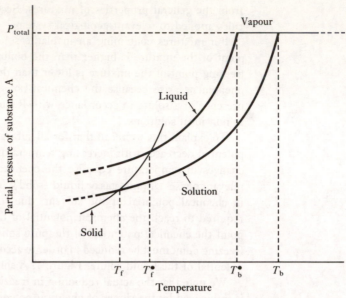

Figure 12.9   Elevation of boiling point and depression of freezing point. T_f temperature of freezing; T_b temperature of boiling, ● property of pure substance, P external pressure (after Everett, 1959).

12.3.1   The depression of freezing point

Equations for the depression of freezing point can be developed using the expression for chemical potential of a solvent A in an ideal mixture and equating this to the chemical potential of the pure solid:

$$\mu_{A(s)}^{\bullet} = \mu_{A(l)} = \mu_{A(l)}^{\bullet} + RT \ln x_A \tag{12.36}$$

Noting that $\mu_{A(s)}^{\bullet}$ and $\mu_{A(l)}^{\bullet}$ are the molar Gibbs energies for the pure solid and pure liquid respectively, we may write:

$$\frac{(\bar{G}_{A(s)}^{\bullet} - \bar{G}_{A(l)}^{\bullet})}{RT} = \ln x_A \tag{12.37}$$

We first differentiate the above with respect to temperature. We make use of the Gibbs–Helmholtz relationship (equation 9.109, section 9.2.3), defined as:

$$\frac{\partial (G/T)}{\partial T} = -\frac{H}{T^2} \tag{12.38}$$

This gives:

$$\frac{(\bar{H}_{A(l)}^{\bullet} - \bar{H}_{A(s)}^{\bullet})}{RT^2} = \frac{\partial \ln x_A}{\partial T} = \frac{\Delta H_f}{RT^2} \tag{12.39}$$

where ΔH_f is the molar latent heat of fusion of the **pure** components.

To evaluate the effect of temperature we must integrate the above between $T_f^{\bullet}$ (the freezing point of the pure solid from a pure liquid) to T_f (the temperature at which the pure solid is in equilibrium with a liquid of composition x_A), i.e.

$$\int_{x_A=1}^{x_A} d \ln x_A = \int_{T_f^{\bullet}}^{T_f} \frac{\Delta H_f}{RT^2} dT \tag{12.40}$$

The above relationship is thermodynamically exact for ideal solutions.

Simple expressions for the depression of freezing point

Assuming that ΔH_f is independent of temperature, integration of equation 12.40 gives:

$$\ln x_A = \frac{\Delta H_f}{R}\left(\frac{1}{T_f^\bullet} - \frac{1}{T_f}\right) \tag{12.41}$$

We observe a linear relationship between $\ln x_A$ and $1/T$ for ideal mixtures. For non-ideal mixtures it is necessary to replace x_A with $x_A\gamma_A$. If the integration path is a two-stage process from the pure solution of $T_f^\bullet$ to the pure solution at T_f and then from this state at constant temperature to the mole fraction of activity, a_A, we may write:

$$\int_{x_A=1}^{x_A} \mathrm{d}\ln a_A = \ln a_{A(T_f)} = \int_{T_f^\bullet}^{T_f} \frac{\Delta H_f}{RT^2}\,\mathrm{d}T \tag{12.42}$$

$$= \frac{\Delta H_f}{R}\left(\frac{1}{T_f^\bullet} - \frac{1}{T_f}\right) \tag{12.43}$$

The term $a_{A(T_f)}$ denotes an activity at the specified temperature and composition at which freezing occurs. It is clear that, if we know the activity (or mole fraction) of a solute and $T_f^\bullet$, the reduced temperature T_f can be calculated.

We may simplify the approach by defining the lowering of freezing point as $\Delta T_f = T_f^\bullet - T_f$ and making the approximation that $T_f^\bullet T_f = T_f^{\bullet 2}$. Then, for an ideal solution:

$$\ln a_{A(T_f)} = \ln x_A = -\frac{\Delta H_f}{RT^{\bullet 2}}\Delta T_f \tag{12.44}$$

The term $\ln x_A$ can be expanded as a Taylor series. Therefore, for a binary mixture where x_A tends to 1:

$$\ln x_A \sim \ln(1 - x_B) = -x_2 - \tfrac{1}{2}x_2^3 - \cdots \tag{12.45}$$

This truncates to give the simple expression:

$$\Delta T_f \sim \frac{RT_f^{\bullet 2}}{\Delta H_f}x_2 \tag{12.46}$$

Conversion to the molal scale, using the approximate expression at low concentrations (equation 2.20), leads to:

$$\Delta T_f \sim \frac{RT_f^{\bullet 2}W_B}{1000\Delta H_f}m_B \tag{12.47}$$

where W_B and m_B are the formula weight and molality of solute B respectively. The term $RT_f^{\bullet 2}W_B/1000\Delta H_f$ is a constant equal to K_f from which:

$$\Delta T_f = K_f m_B \tag{12.48}$$

The term K_f is called the freezing point depression constant or **cryoscopic constant**. Strictly, the value of K_f is only constant at zero concentration. If K_f at infinite dilution can be found, by extrapolation from measurements of ΔT at different concentrations, its value can be used to estimate ΔH_f.

Box 12.3   **More exact expressions for the depression of freezing point**

The above approach assumes the solid phase is pure, and not a solid solution of solute and solvent, and also that ΔH_f is constant.

To remove the first restriction we introduce the activity of A in the solid phase into equation 12.37. This step introduces an algebraic complexity but no new principles. The resulting equation is:

$$\ln\left(\frac{a_{A(l,T_f)}}{a_{A(s,T_f)}}\right) = \frac{\Delta H_f}{R}\left(\frac{1}{T_f^\bullet} - \frac{1}{T_f}\right) \tag{12.49}$$

where l and s subscripts denote the solid and liquid phase respectively. We note that the term on the left-hand side is equal to the equilibrium constant for the distribution of component A between the solid and liquid phases in accordance with the reaction:

$$A(s) \rightleftharpoons A(l) \tag{12.50}$$

To remove the second restriction it is necessary to know the temperature dependence of ΔC_P of fusion. As an example, we will consider the case when ΔC_P can be expressed as an empirical equation in temperature. A typical expression may be:

$$\Delta C_P = C_0 + C_1 T \tag{12.51}$$

where C_0 and C_1 are constants. The value of ΔH at any temperature is given by substituting equation 12.51 into equation 9.89 and integrating with respect to temperature. This gives:

$$\Delta H_{f(T_f)} = \Delta H_{f(T_f^\bullet)} + C_0(T - T_f^\bullet) + \frac{C_1}{2}(T^2 - T_f^\bullet) \tag{12.52}$$

If the constant terms involving $\Delta H_f^\bullet$, $T_f^\bullet$, C_0 and C_1 are collected together the above equation takes the simple form:

$$\Delta H_{f(T_f)} = D_0 + D_1 T + D_2 T^2 \tag{12.53}$$

where the Ds are constants. Equation 12.53 can be substituted into 12.48 and integrated to give:

$$R \ln a_A(T_f) = -\int_{T_f}^{T_f^\bullet} \frac{D_0 + D_1 T + D_2 T^2}{T^2}\, dT \tag{12.54}$$

$$= \left(\frac{D_0}{T} - D_1 \ln T - D_2 T\right)_{T_f}^{T_f^\bullet} \tag{12.55}$$

$$= \frac{D_0}{T_f^\bullet} - \frac{D_0}{T_f^\bullet} - D_1 \ln(T_f^\bullet/T_f) - D_2(T_f^\bullet - T_f) \tag{12.56}$$

The above equation can be easily extended to include more fitting parameters for ΔC_P if the additional flexibility is required.

A useful equation, based on the assumption that the value of ΔC_P is constant but non-zero, is:

$$\ln\left(\frac{a_{A(l,T_f)}}{a_{A(s,T_f)}}\right) = \frac{\Delta H_f}{R}\left(\frac{1}{T_f^\bullet} - \frac{1}{T_f}\right) + \frac{\Delta C_P}{R}\ln\left(\frac{T_f}{T_f^\bullet}\right) + \frac{\Delta C_P}{R}\left(\frac{T_f^\bullet}{T_f} - 1\right) \tag{12.57}$$

It is derived by a method described in section 12.26.

12.3.2   The elevation of boiling point

There exist equations analogous to 12.42 and 12.43 for the elevation of boiling point. These are:

$$\ln a_{A(T_v)} = \int_{T_v^\bullet}^{T_v} \frac{\Delta H_v}{R} \frac{1}{T^2} \, dT \tag{12.58}$$

and

$$\ln a_{A(T_v)} = -\frac{\Delta H_v}{R}\left(\frac{1}{T_v^\bullet} - \frac{1}{T}\right) \tag{12.59}$$

where $a_{A(T_v)}$ is the activity of component A in the liquid phase at the elevated boiling point T_v, ΔH_v is the molar latent heat of vaporisation defined as $\Delta H_v = H_{A(l)}^\bullet - H_{A(g)}^\bullet$ and $T_v^\bullet$ is the boiling point of the pure solvent. The negative sign before ΔH_v arises since the latent heat of vaporisation is defined as above, rather than as $\Delta H_v = H_{A(g)}^\bullet - H_{A(l)}^\bullet$. Similarly, we may write the more approximate equations:

$$\Delta T_v \sim \frac{RT_v^2}{\Delta H_v} x_A \simeq \frac{RT_v^{\bullet 2} W_B}{1000 \Delta H_v} m_A \tag{12.60}$$

and

$$\Delta T_v = K_v m_A \tag{12.61}$$

where $\Delta T_v = T_v - T_v^\bullet$ and K_v is called the boiling point elevation constant or **ebullioscopic constant**. Note that the difference in sign conventions for the definitions of ΔT_v and ΔT_f ensures that equations 12.48 and 12.61 have the same form.

12.4   General properties of multicomponent systems

Multicomponent systems will be considered with reference to binary liquid/vapour and liquid/liquid systems, which include liquid melts. A limited review of ternary systems is presented in this section.

The general rules for liquids and gases apply equally well to solids. However, systems containing solids have properties of interest to geochemists and will be considered separately in chapter 18.

For non-reacting binary systems, where $C = 2$, the phase rule becomes:

$$F = 2 + 2 - P \tag{12.62}$$

so that the number of degrees of freedom F can be 0, 1, 2 or 3, depending on the number of phases present. When only one phase exists ($P = 1$) the system is trivariant, i.e. three variables are necessary to define the system. If the three intensive properties are temperature, pressure and the mole fraction of one component, then all equilibrium states of the system can be described in three-dimensional $P-T-$composition space.

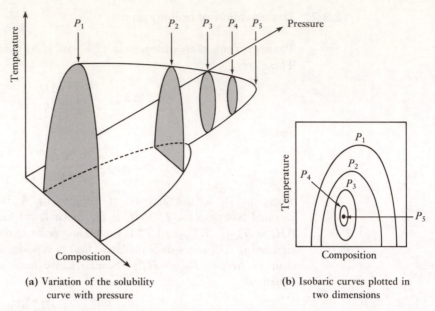

(a) Variation of the solubility
curve with pressure

(b) Isobaric curves plotted in
two dimensions

Figure 12.10   Three-dimensional and sectional temperature–pressure–composition plots for a bivariant system (after Ferguson and Jones, 1973).

If the two-component system has up to two phases in any region, the system is bivariant ($F = 2$). In this case, one region of P–T–composition space may contain a single mixed phase; another region may contain the two separate phases coexisting in equilibrium. The points of coexistence of two such regions define a surface in P–T–composition space. In three dimensions this is difficult to represent and it is common to set one variable constant, pressure for example, and examine the phase distributions as a function of the other two: figure 12.10 depicts this graphically.

By assigning a constant value to one variable, a degree of freedom is lost. The phase rule becomes:

$$P + F' = C + 2 - R \tag{12.63}$$

where R is the number of variables made constant and F' is the number of remaining degrees of freedom. If both temperature and pressure are fixed, the phase rule becomes $P + F' = C$.

12.4.1   Liquid/liquid and liquid/vapour systems

This section attempts to consolidate and extend the ideas on ideal and non-ideal liquid mixtures introduced in chapters 5 and 10.

Any two liquid components may be:

- completely miscible and separable by fractional distillation,
- completely miscible and not separable by fractional distillation,
- partially miscible, or
- immiscible.

These different types of behaviour result from different degrees of non-ideality in the systems.

The condition for spontaneous mixing of components is a negative value for ΔG_{mix}. The ideal solution, defined by Raoult's law, of course conforms with this condition and has zero values of the excess Gibbs energy at all compositions. For real systems we may think of ΔG_{mix} as having an ideal contribution and a non-ideal contribution (G^E). Molecular interactions between components lead to non-zero values for G^E. In a binary mixture of A and B there will be A–A, B–B and A–B interactions. If all interactions are similar in magnitude the value of G^E will be almost zero. Non-ideality results from A–B interaction being notably different from A–A and B–B. Positive deviation from Raoult's law is associated with weak A–B interactions, in which case the tendency of a substance to leave the liquid phase is increased (section 5.6.4). The activity of any component will therefore be greater than its mole fraction. This is consistent with activity coefficients greater than unity and therefore positive values for the excess Gibbs energy. Negative deviation from Raoult's law is associated with strong A–B interactions and activities lower than the respective mole fractions. This is equivalent to having activity coefficients lower than unity and therefore negative values for the excess Gibbs energy. We must remember that the strength of interactions may be composition-dependent and the sign of G^E may change with composition.

Moderate non-ideality leads to small non-zero values for the excess Gibbs energy which have a near-symmetric dependence on composition. In these cases case ΔG_{mix} will still be less than or equal to zero and will also show a near-symmetric dependence on composition. More significant non-ideality leads to a highly non-symmetric dependence of G^E on composition, in which case the composition dependence of ΔG_{mix} is more asymmetric and significantly different from the ideal case.

Completely miscible zeotropic systems

A **zeotropic** mixture is an ideal, or moderately non-ideal, mixture which can be separated by fractional distillation. The thermodynamic properties of such a mixture are linear, or moderately non-linear, combinations of the properties of the pure components. For example, in a mixture of liquids A and B we expect the vapour pressure of the mixture P to fall between the vapour pressures of the pure components at the same temperature as the mixture. For an ideal mixture, the vapour pressure can be calculated from Raoult's law using:

$$P = P_A + P_B = x_{A(1)}P_A^\bullet + x_{B(1)}P_B^\bullet \tag{12.64}$$

where $x_{A(1)}$ and $x_{B(1)}$ are the respective mole fractions in the liquid phase, P_A and P_B the respective partial pressures and $P_A^\bullet$ and $P_B^\bullet$ the vapour pressures of the pure components. A plot of P versus composition of the liquid phase is called the **liquids, or bubblepoint curve**. Figure 12.11(a) depicts a hypothetical liquidus. The slight curvature indicates a moderate degree of non-ideality which may be accommodated by replacing mole fractions in the above equation with activities ($a_i = x_{i(1)} \times \gamma_i$).

The activities may be calculated from a regular solution model using the expressions:

$$RT \ln \gamma_A = \alpha x_{B(1)} \quad \text{and} \quad RT \ln \gamma_B = \alpha x_{A(1)} \tag{12.65}$$

where α is a suitable interaction parameter. Equation 12.80 is sometimes called the Porter equation. Figure 12.11(a) is a phase diagram which shows that at any pressure P above the line the system will be a liquid: therefore $P = 1$ and $F = 3$. The system

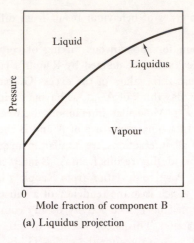

(a) Liquidus projection

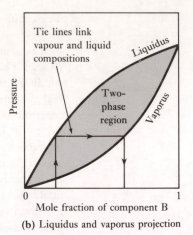

(b) Liquidus and vaporus projection

Figure 12.11   Pressure–composition plots.

is at constant temperature so we lose one degree of freedom ($F' = 2$). This means that two variables (pressure and composition) can be varied widely over the liquid region without encountering a phase change. At any pressure below the line, the complete fluid phase will be a mixed gas. For example, at any point on the line both gas and liquid phases exist so that $P = 2$ and the number of degrees of freedom at a fixed temperature is reduced to 1 – composition. Therefore, at any selected composition the vapour pressure must be fixed, and vice versa.

The vapour phase is likely to be ideal at low pressures. Therefore, the composition of the vapour phase can be derived from Dalton's law, which states that:

$$x_{A(g)} = P_A/P \quad \text{and} \quad x_{B(g)} = P_B/P \tag{12.66}$$

Combining the above equations gives expressions for the composition of the vapour phase in terms of the composition of the liquid phase, i.e.

$$x_{A(g)} = \left(\frac{x_{A(l)} P_A^{\bullet}}{\gamma_B P_B^{\bullet}} \right) + (\gamma_A P_A^{\bullet} - \gamma_B P_B^{\bullet}) x_A \quad \text{and} \quad x_{B(g)} = 1 - x_{A(g)} \tag{12.67}$$

A plot of vapour pressure versus vapour composition is called the **vaporus** or the **dewpoint curve**. Figure 12.11(b) shows the vaporus for the hypothetical system, using a small positive value for α. The curve is superimposed on the plot of the liquidus, so the composition axis represents the liquid phase for the liquidus and the vapour phase for the vaporus. Horizontal tie lines, at any temperature, can be used to link the composition of both phases via a single equilibrium vapour pressure. In this mixture component B is the more volatile component and the tie lines show that the vapour phase is rich in B compared with the liquid.

Similarly, figure 12.12 shows the hypothetical boiling point curve of this mixture as a function of temperature at a constant pressure of 1 bar. There is a clear elevation of three boiling point for mixtures richer in the less volatile component, A, although the dependence on composition is quite simple. As with the vapour pressure curves, tie lines show that at any boiling temperature the vapour is rich in the more volatile component, B.

The regions of temperature–composition space, or pressure–composition space,

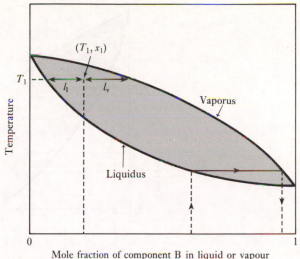

Figure 12.12   Temperature–composition plot.

between the liquidus and the vaporus define a two-phase region. Any mixture whose coordinates (T, x) or (P, x) fall in this region will spontaneously separate into a mixed-liquid and mixed-vapour phase. If the system is at constant pressure, a measurement of the vapour pressure defines the composition of both phases in this region. If the system is at fixed pressure, the composition of both phases is fixed by the temperature.

The relative proportions of liquid and vapour at a defined temperature or pressure and overall composition can be obtained from the **lever rule**. This states that **the relative amounts of two phases are inversely proportional to the distances of the phase lines from the overall composition point**. For example, in figure 12.12, the overall composition at point (T_1, x_1) has a vapour-to-liquid ratio proportional to the lengths l_v and l_l respectively. That is,

$$\frac{n_{liq}}{n_{vap}} = \frac{l_v}{l_l} \qquad (12.68)$$

where n denotes relative amounts in moles. Similar calculations can be made from vapour pressure plots if both liquidus and vaporus are plotted.

The ability to separate the mixture completely by distillation arises because the vapour phase contains a higher proportion of one component than the liquid phase at all compositions. Also, the vapour-phase concentration of this more volatile substance increases with temperature at any composition. Consider a fractionating column constructed from a flask, containing the boiling mixture, and a long outlet column over which there is a temperature gradient. Figure 12.13 shows that at temperature T_1 a boiling mixture, of overall composition Y, will have liquid composition l_1 and a vapour of composition v_1. As the vapour rises in the column it cools to temperature T_2. The tie line at this temperature shows that the liquid will separate into a gas of composition v_2 plus a liquid of composition l_2 which is richer in the less volatile substance. The liquid runs back into the flask, enriching the liquid with the less volatile substance and leaving a gas phase enriched with the more volatile liquid. This gas moves higher up the column and is cooled to temperature

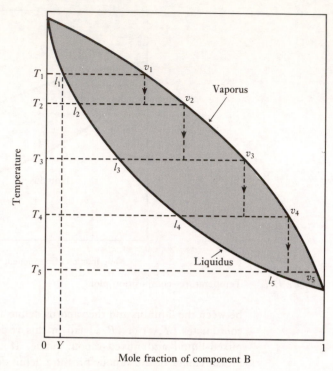

Figure 12.13   Composition changes in a fractioning column.

T_3, where a liquid of composition l_3 separates out, further enriching the gas phase. Continual repetition of this process leads to complete separation of the phases when the more volatile substance has been distilled off leaving the less volatile substance in the flask. Further heating will of course boil off the remaining pure liquid.

Completely miscible azeotropic mixtures

Figure 12.14 shows boiling-point diagrams for mixtures which exhibit significant positive and negative deviations from Raoult's law respectively. Deviation from ideality is so severe that the plots show either maxima or minima. The significant feature of these systems is that the turning points correspond to a state at which **the equilibrium composition of the liquid is equal to the equilibrium composition of the vapour.** Such a state is called an **azeotrope**.

Mixtures which exhibit azeotropes cannot be separated completely by fractional distillation. Attempts at fractionation lead to progressive partial separation of the phases until the azeotropic point is reached and the vapour distils with the same composition as the liquid. Further separation is not possible and the resulting mixture boils as if it were a pure liquid.

If positive deviation from ideality occurs, the vapour-pressure curve shows a maximum at the same composition as the minimum in the boiling-point curve. Similarly, if negative deviation from ideality is seen, the vapour-pressure curve shows a minimum at the same composition as the maximum in the boiling-point curve. Figure 12.15 shows the complementary vapour-pressure diagrams for the above two boiling-point diagrams.

The lever rule may be applied to any limb of the phase diagram for an azeotrope, provided both liquidus and vaporus are plotted.

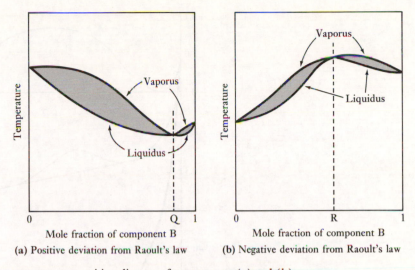

Figure 12.14   Temperature–composition diagrams for azeotropes (a) and (b).

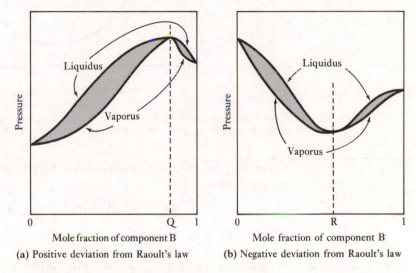

Figure 12.15   Pressure–composition diagrams for azeotropes (a) and (b).

Azeotropes reflect significant non–ideality but are sometimes given by regular solutions. In these cases we may use the composition of the azeotrope to calculate the parameters in simple Porter, Margules or van Laar equations. For example, Raoult's law gives $P_A = x_{A(l)}\gamma_A P^\bullet = x_{A(g)}P$ and $P_B = x_{B(l)}\gamma_B P^\bullet = x_{B(g)}P$. Noting that at the azeotrope $x_{(g)} = x_{(l)}$ for both components, we may write:

$$\gamma_A = (P/P_A^\bullet)\gamma_B \tag{12.69}$$

Therefore:

$$\left(\frac{\gamma_A}{\gamma_B}\right) = P_B^\bullet/P_A^\bullet \tag{12.70}$$

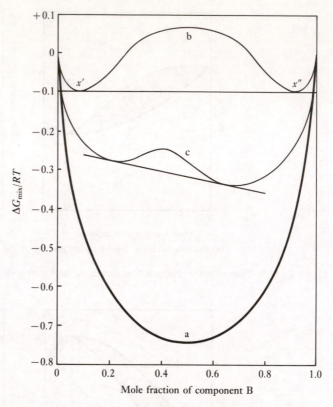

Figure 12.16   Phase separation and the Gibbs energy of mixing: a, one liquid phase; b, two liquid phases; c, two liquid phases with a more complex dependence of ΔG_{mix} on composition.

Substituting the Porter equation (equation 12.80) for both activity coefficients into the above gives:

$$RT \ln(\gamma_A/\gamma_B) = \alpha(x_{B(l)}^2 - x_{A(l)}^2) = P_B^{\bullet}/P_A^{\bullet} \qquad (12.71)$$

for which α can be calculated.

Partially immiscible   In some cases non-ideality is so severe that the plot of $\Delta G_{mix}/RT$ versus composition
liquid mixtures   may show more than one minimum, in which case spontaneous phase separation can occur. Figure 12.16 shows such a plot with two minima at compositions x' and x''. If a mixture is at any composition **between** x' and x'' the value of ΔG_{mix} can be reduced by the system separating into two different liquid phases of composition x' and x'' respectively. The relative amounts of the two phases will be determined by the initial composition of the system. The resulting phases have equilibrium compositions and conform with the requirement that the chemical potential of any substance has the same value in all phases.

   The point to remember is that molecular interactions result in non-zero values for G^E. In the hypothetical binary system, if A−B interactions are greater than A−A and B−B interactions, the values of G^E are negative, the resulting values of ΔG_{mix} become more negative and total miscibility is reinforced. If A−A interactions and B−B interactions are large and positive compared with A−B, then values of G^E are positive. This may increase the values of ΔG_{mix} in some composition ranges so that

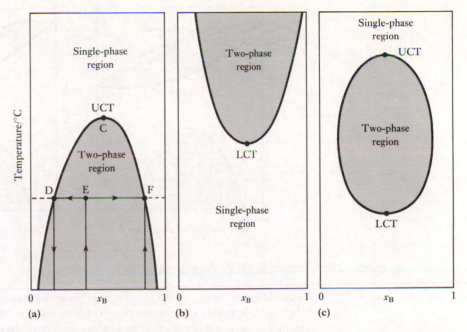

Figure 12.17 Phase equilibria for two partially immiscible liquids showing upper and lower consolute temperatures (UCT, LCT).

regions of immiscibility occur. Total immiscibility occurs when G^E is so large and positive that the value of ΔG_{mix} is positive at all compositions. Regular solutions may exhibit immiscibility.

The compositions at which points x' and x'' occur will change with temperature and possibly pressure, depending on the contributions which S^E and H^E make to the values of G^E. For many immiscible systems, changing the temperature will cause the values of x' and x'' to become closer together, finally coalescing to single point when the system becomes miscible at all compositions. This point is called a **critical miscibility** point and has the same meaning as the critical point for liquid/gas separation. This state of affairs is called critical mixing. The temperature at which critical mixing occurs is called the **consolute** temperature.

For a binary system ($C = 2$) with two coexisting liquid phases and a single vapour ($P = 3$), the phase rule indicates that there is only one degree of freedom. If temperature is fixed, for example, the compositions of both phases are fixed. A useful diagram can be constructed by plotting the compositions of these phases as a function of temperature. Commonly this would give a dome–shaped equilibrium line as in figure 12.17(a). The horizontal tie lines intersecting at points D and F indicate the compositions of the two equilibrium liquid phases formed by a mixture of overall composition E at the specified temperature. One liquid will be rich in component A, the other rich in component B. The point C is an upper consolute temperature (UCT).

Mixtures whose temperature/composition coordinates fall outside the dome will remain as a single homogeneous liquid phase.

Immiscibility is a feature of systems with a large and positive excess enthalpy and therefore a large positive value for ΔH_{mix}. If the entropy of mixing ΔS_{mix} is positive, then the values ΔG_{mix} will become more negative as temperatures increase

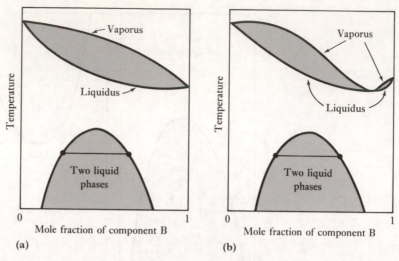

Figure 12.18   Immiscible liquids showing miscibility at the boiling points.

since $\Delta G_{mix} = \Delta H_{mix} - \Delta S_{mix}$. In this case increasing temperature promotes miscibility, it is this type of mixture which is indicated in figure 12.17(a). If the entropy of mixing is negative, increasing temperature will promote immiscibility and such a mixture will show a lower consolute temperature (LCT), as in figure 12.17(b). Some mixtures show both upper and lower consolute temperatures, as depicted in figure 12.17(c). This is rare at low temperatures and pressure but we may expect all systems with a lower consolute temperature to exhibit an upper consolute temperature over sufficiently large ranges of temperature. It is commonly observed that increasing pressure lowers the UCT and raises the LCT, giving rise to the type of variation depicted in figure 12.10.

Usually a consolute temperature is sensitive to impurities, which may be thought of as additional components. With each additional component there is another degree of freedom and the consolute temperature for the two-phase system is no longer fixed at fixed pressure. It varies with the amount of the additional component. This can be a way of testing for the presence of impurities.

The distillation of partially miscible liquids can be interpreted from the vapour-pressure and boiling-point diagrams for such systems. These diagrams must contain the phase boundary, the liquidus and the vaporus. Many different types of these diagrams have been observed. Their forms depend on whether the liquids are completely or partially miscible at the boiling point and whether or not the miscible regions exhibit azeotropy. If the liquid mixtures are miscible at their boiling points, the boiling-point diagrams may look like figure 12.18 with an immiscible region well below the boiling-point curve. In contrast, figure 12.19(a) shows a type of mixture which is similar to the mixture in figure 12.18(a) but with the upper consolute temperature raised above the boiling point. The variations in the vapour composition are denoted by lines ME and NE, whereas the variations in the liquid composition are denoted by MC and ND.

In figure 12.19 we observe the following areas:

● Area MEN unbounded in temperature: a homogeneous mixed vapour (denoted by V in figures 12.18 and 12.19) for which composition and temperature can be varied independently.

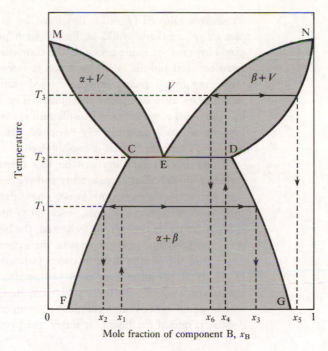

Figure 12.19   Immiscible liquids showing immiscibility at the boiling points.

- Area MCEM: a mixed liquid phase, rich in component A plus a mixed vapour phase.
- Area NEDN: a mixed liquid phase, rich in component B plus a mixed vapour phase.
- Area MCF(0)M: a single liquid phase α, variable in composition over a limited temperature range but rich in component A.
- Area NDG(1)N: a single liquid phase β, variable in composition over a limited temperature range at but rich in component B.
- Area FCDGF: two immiscible liquid phases $\alpha + \beta$. Both phases have compositions fixed by the temperature. The relative proportion of each phase is determined by the composition of the complete system.

Analysing the above figure is a good example of using tie lines. The rule is to select a composition and draw a vertical line to intersect with a horizontal line at the selected temperature. If the point of intersection is in a single-phase region, the system is stable. If the point of intersection falls in a two-phase region, the compositions of each phase can be determined by extrapolating horizontally in both directions until a boundary is reached. Vertical lines at these points define the composition of the respective phase. For example, at the overall composition x_1 and temperature T_1 the system has two liquid phases α and β, of composition x_2 and x_3 respectively. At the overall composition x_4 and temperature T_3 the system exists as a liquid phase α of composition x_5 and a vapour phase of composition x_6.

The significance of line CED and point E is of interest. If any mixture within the multiphase region is raised to temperature T_2, the system will spontaneously separate into three phases (two liquid phases α and β, and one vapour). The compositions of the liquid phases are defined by points C and D and the vapour phase by point E. The presence of three phases dictates that only one degree of freedom is allowed ($F = 1$); this must be pressure. In the full three-dimensional plot (T, P, x) there would be line of such points E extending along the pressure axis. However, by fixing pressure, the composition of all phases plus the temperature must be fixed and the system will boil to release a vapour of fixed composition. The remaining liquid phases will have fixed compositions but will change in relative proportions until one phase disappears. The liquid which remains depends on the initial overall composition. If distillation is continued, when one liquid has gone the vapour becomes progressively enriched with the more volatile component until complete phase separation is encountered.

There are several variations on this theme. Of interest is the case where the liquids approach total immiscibility and regions MCF(0)M and NDG(1)N become very narrow; an almost pure liquid phase exists below the line CED. We think of this as two separate liquids, each of which exerts its own vapour pressure independently. If the mixture is heated, the liquid boils when the sum of the vapour pressures (the total pressure) equals the external pressure. This temperature will be lower than the boiling point of either pure component since there is a contribution to the total pressure from each component. The distillation occurs at constant temperature while both the liquids remain. This is a way of distilling substances which would otherwise decompose at temperatures nearing their ordinary boiling points. If one of the liquids is water, the process is called steam distillation.

12.4.2 Ternary systems

As the number of components increases, the phase diagrams become difficult to construct and we need alternative graphical techniques. This section deals with the techniques for ternary liquid/liquid and liquid/vapour systems. Further details on the graphical techniques are given in chapter 18, which deals with systems containing solids.

Graphical representation   For a three-component system the phase rule is:

$$P + F = 5 \tag{12.72}$$

The maximum number of degrees of freedom is four, occurring when there is only one phase present. These variables are usually temperature, pressure and the concentrations of two of the components. The concentration of a third component is fixed if the mass fractions or mole fractions of the other two are known.

If pressure is held constant, the condensed phase rule is:

$$P + F' = 4 \tag{12.73}$$

Even with the loss of one degree of freedom it is not possible to plot all variables on a planar diagram. A complete representation for a system containing components A, B and C requires the intensive variables to be plotted on a three-dimensional graph having an equilateral triangle for its base, as in figure 12.20. The apices of the triangle denote a pure component (i.e. A = 100%, B = 100% or C = 100%), whereas the sides represent the binary limits. A horizontal section of the prism represents the overall composition at fixed temperature and pressure. Figure 12.21(a) shows such a projection.

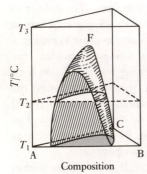

Figure 12.20 Three-dimensional equilibrium plot (after Ferguson and Jones, 1973).

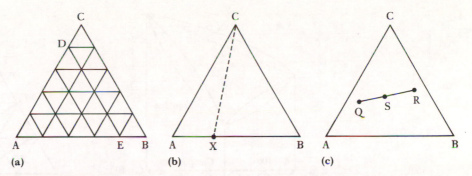

Figure 12.21   Ternary composition diagrams.

There are four important points to remember when using ternary plots.

- Any line parallel to one side of the triangle represents a fixed proportion of one component. Thus, in figure 12.21(a) all points on the line DE are for a mixture containing 20% A with varying amounts of B and C. The total amount of B + C on this line sums to 80%. To define the composition of any point on DE we draw a second line, parallel to one of the other sides of the triangle, through the selected point. For example, the proportion of C is denoted by the point at which a line parallel to AB intersects with either CA or CB.

- Any line through one apex represents mixtures for which two of the components are in fixed proportion. For example, any point on the broken line in figure 12.21(b) denotes a mixture with a fixed ratio of A:B equal to $A/B = BX/AX$. Along this line the amount of C changes from 100% to zero.

- If two ternary mixtures are themselves mixed, the final composition can be determined using a 'centre of gravity' principle. If mixtures of known compositions Q and R are mixed, the final composition S can be found by determining the ratio of lengths RS:QR on figure 12.21(c). This is given by:

$$\frac{RS}{QR} = \frac{\text{Mass of mixture Q}}{\text{Mass of mixture R}} \tag{12.74}$$

- As an alternative to percentage mass or mass fractions we may use mole fractions as composition units. In this case the apices represent $x_A = 1.0$, $x_B = 1.0$ and $x_C = 1.0$. The centre of gravity principle must be modified by replacing the mass of the mixture by the sum of the number of moles in the mixture.

Basic properties of ternary liquid mixtures   All the properties of ideal and non-ideal binary liquids mixtures are exhibited by ternary mixtures. Figure 12.22(a) shows a three-dimensional representation of composition–temperature space for a fully miscible liquid mixture. The liquidus and the vaporus enclose a space constituting the two-phase region of liquid and vapour. The position of the vaporus is changing with temperature and composition as expected. As with binary systems, tie lines can be used to link the compositions of coexisting phases at any temperature. No two tie lines at any one temperature may cross each other. Such isotherms may be depicted on a projected view of the space model. Figure 12.22(b) shows the vaporus projection for the hypothetical system. The contours are lines of constant temperature. Such a plot is equally valid

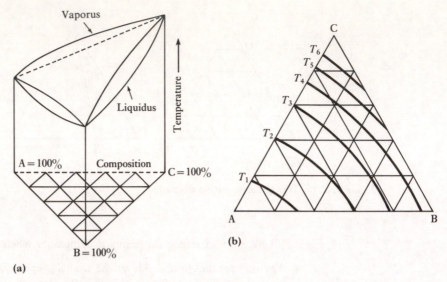

Figure 12.22   A ternary mixture with three miscible components.

for solid/liquid systems, in which case figure 12.22(b) would be for the liquidus rather than the vaporus.

In contrast to binary systems, few ternary liquid mixtures are completely miscible under all conditions. It is common to find one or more pairs of liquids showing immiscibility in some region of temperature–composition space.

Consider a three-component mixture where one pair of components A and B are partially miscible, but the third component C will dissolve in either of the other pure components or any mixture of A and B.

A ternary phase diagram, at fixed temperature and pressure, is constructed by mixing components A and B in immiscible proportions (no C at this stage) and plotting the compositions of the two immiscible phases. These would be points D and E on figure 12.23(a). If a small amount of C is added, the immiscible phases will now have the compositions D′ and E′. The tie lines linking D′ and E′ fall parallel to the AB baseline on the rare occasions when C is distributed equally between A and B. Progressively adding more C gives the linked points (D″, E″), (D‴, E‴), etc. These points define the **binodal curve**, within which all liquid mixtures form two phases and outside of which all mixtures are miscible. The binodal curve may be determined by starting with any initial mixture between the points D and E. The point F is a mixture at which the tie lines disappear and the two phases have identical compositions and merge into one. This is called the **plait point**. Experiments are required to establish the positions of the tie lines for individual systems. An example of this type of system is water and oil, as the immiscible components, with propylene glycol added to enhance the mutual solubilities of the oil/water mixture.

Figure 12.23 can be used to show two applications of this type of ternary diagram. Consider first a mixture at the initial composition G in figure 12.23(a). If the component C is progressively added, at a fixed ratio of A to B, the composition of C would progress along the line GC. At any point on this line the relative amounts of the two immiscible phases can be calculated if the position of the tie line at the

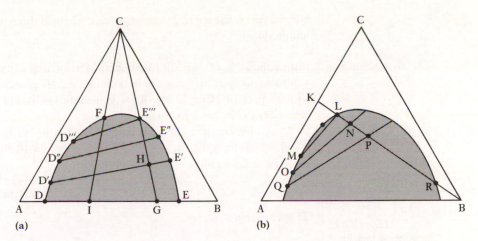

Figure 12.23 A ternary mixture with two immiscible components (after Ferguson and Jones, 1973).

point is known. For example, at the point H the relative amounts of the two phases is given by:

$$\frac{\text{Amount of liquid D}'}{\text{Amount of liquid E}'} = \frac{\text{Distance H}'\text{E}}{\text{Distance D}'\text{H}} \tag{12.75}$$

This is a useful application but it also emphasises the fact that the plait point can only be obtained exactly from a single starting point I. In view of this, the change in miscibility with increasing C may be quite different, depending on the initial starting composition. If the composition progresses along the line GC we will observe a gradual reduction in the amount of both phases and a progressive change in their compositions. However, one liquid phase will disappear 'faster' than the other. This is made clear since the distance between the line GC and the curve bounded by EE''' diminishes rapidly while the distance between the line GC and the curve bounded by DE''' diminishes slowly. At a infinitely small distance from E''', one phase will have disappeared, leaving a quantity of the second phase of composition D'''. Any further addition of C will lead to instant disappearance of the remaining phase. In contrast, if the starting composition is I, we observe both immiscible phases to remain right up to an infinitely small distance from F along the line IC. At this point we will have equal amounts of the two phases of almost identical composition. Any further addition leads to immediate mixing of the two phases.

A phenomenon known as **retrograde solubility** is exhibited by the above system. This is where the addition of solvent reduces solubility. Consider adding an immiscible component B to a mixture of A and C at the initial point K on figure 12.23(b). The mixture is miscible up to the point of intersection with the binodal line at point L. Here, a second phase of composition M precipitates out. As more B is added, a second phase will develop, progressing through points N and P, etc. The tie lines at these points show that the first phase to precipitate goes to points O and Q, i.e. this mixed liquid becomes progressively enriched in component A. This is the retrograde solubility effect. The effect continues until point R is reached, after which the two immiscible phases become miscible.

If more than one pair of mixtures is immiscible, there will be more than one binodal curve. Figure 12.24 shows the case when all three pairs of liquids are partially immiscible.

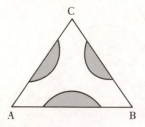

Figure 12.24
Composition plot for a ternary mixture of three immiscible pairs of liquids (after Ferguson and Jones, 1973).

Immiscibility is a feature of highly non-ideal mixing of two components. However, a third component may dissolve in both immiscible phases and behave almost ideally with respect to Henry's law if its concentration is low enough. This would be component C in our hypothetical example.

At equilibrium, the chemical potentials of C in the immiscible phases, denoted α and β, will be equal. In accordance with equation 10.48 (section 10.3.1), we may write for ideal mixing:

$$\mu^{\bullet}_{C(\alpha)} + RT \ln x_{C(\alpha)} = \mu^{\bullet}_{C(\beta)} + RT \ln x_{C(\beta)} \qquad (12.76)$$

This rearranges to give:

$$\ln\left(\frac{x_{C(\alpha)}}{x_{C(\beta)}}\right) = \left(\frac{\mu^{\bullet}_{C(\beta)} - \mu^{\bullet}_{C(\alpha)}}{RT}\right) \qquad (12.77)$$

The right-hand side of the above equation is a temperature- and pressure-dependent constant called the **partition coefficient** or **distribution coefficient**. The approximate relationship is called the **distribution law** or **partition law** and the ratio of mole fractions the **distribution ratio** or **partition ratio**.

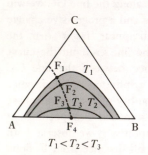

$$T_1 < T_2 < T_3$$

Figure 12.25
Changes in miscibility with temperature for a ternary mixture with two immiscible components (after Ferguson and Jones, 1973).

If ternary phase equilibria are measured at various temperatures it is possible to plot a family of binodal curves on the same axis. Figure 12.25 shows a series of curves which bound areas that are being reduced as temperature is increased. This is typical of a system where two components are immiscible and the immiscibility is decreasing with increasing temperature. Consequently, each temperature has its own plait point, denoted F_1, F_2, etc. The final plait point F_4 falls on the binary AB axis. The immiscibility finally disappears at the upper consolute temperature for the immiscible binary mixture which occurs at the plait point F_4. This is the upper consolute point for the binary AB mixture. In contrast, for any ternary system which shows a lower consolute temperature on the binary AB axis, the areas bounded by the binodal curve will increase with increasing temperature.

A second type of behaviour will be seen if the system exhibits a true ternary immiscibility rather than binary immiscibility in a ternary system. Figure 12.20 shows a hypothetical system where a ternary upper consolute point is observed at the plait point F. The 'solid' cone defines the immiscible region. The top of the cone is the upper consolute point. Figure 12.26 shows the planar representation of figure 12.20. The points F on both figures are corresponding upper consolute points. When a binodal curve on figure 12.26 gives a full unbroken contour in two dimensions there are two plait points at the specified temperature, e.g. points F_1 and F_4 at temperature T_2 on figure 12.26.

If a system shows a true ternary immiscibility and has a lower consolute point, the areas bounded by the binodal curves will increase with increasing temperature. The areas bounded by the binodal curves may grow so large that the curves intersect with a second axis, i.e. AC as well as AB. This situation is depicted in figure 12.27. At temperature T_0 the liquids A and C are now partially immiscible as well as liquids A and B.

There are two other situations commonly encountered. The first is a ternary

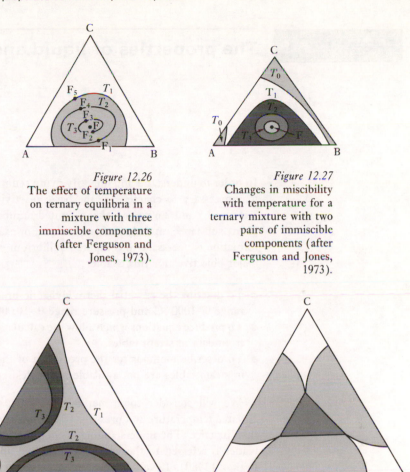

Figure 12.26
The effect of temperature on ternary equilibria in a mixture with three immiscible components (after Ferguson and Jones, 1973).

Figure 12.27
Changes in miscibility with temperature for a ternary mixture with two pairs of immiscible components (after Ferguson and Jones, 1973).

(a)

(b)

Figure 12.28   The effect of temperature on ternary equilibria in a mixture with two or three pairs of immiscible components (after Ferguson and Jones, 1973).

mixture with two immiscible pairs as in figure 12.28(a). If the immiscibility of such a system changes with temperature, the two binodal curves may merge, giving an immiscible band across the ternary diagram (temperature T_1). The second situation is that of a system where all three components show some immiscibility. If immiscibility changes with temperature, all three binodal curves may merge, as in figure 12.28(b). This gives three separate regions containing one miscible layer, three regions containing two immiscible layers and a third region, in the centre of the plot, where three immiscible layers exist.

Chapter 13

The properties of liquid and gaseous water

In order to understand and quantify many sedimentary and hydrothermal processes it is necessary to characterise the basic reactivity of liquid and gaseous water. In sections 2.5 and 2.6 we saw that a limited number of the thermodynamic properties of any substance can help in this exercise. For example, the standard Gibbs energy of formation is necessary to compute equilibrium constants. Therefore there are three major objectives for this chapter:

- To describe the essential thermodynamic properties of water in the temperature range 0–1000 °C and pressure range 0–10 000 bar.
- To produce equations which allow the calculation of apparent energies of formation from data in steam tables.
- To describe methods for the prediction of thermodynamic properties when data in steam tables are not available at the desired temperature and pressure.

We will consider some methods for calculating thermodynamic properties at elevated temperature and pressure. These are selected to emphasise principles rather than practice. The understanding of the properties of water is still evolving and the reader is referred to the works by Helgeson and Kirkham (1974a), Uematsu and Franck (1980), Levelt Sengers et al. (1983), Pitzer (1983), Haar et al. (1980), Johnson and Norton (1991) and Heger et al. (1980) for suitable working equations.

Some useful properties of liquid and gaseous water are as follows:

- **The molecular weight of water (molar mass):** This is 18.0153×10^{-3} kg, using the convention that the atomic weight of carbon-12 is exactly 12.0000×10^{-3} kg.
- **The triple point of water:** Temperature 273.16 K (0.01 °C), pressure $611.3\,\text{N m}^{-2}$ (0.006113 bar).
- **The critical point of water:** Temperature 647.286 K (374.136 °C), pressure $22088.0\,\text{kN m}^{-2}$ (220.88 bar).
- **The conventional reference state for pure water:** Temperature 298.15 K (25 °C), pressure $10^5\,\text{N m}^{-2}$ (1.000 bar).
- **The saturation line:** The saturation line defines the equilibrium conditions under which liquid water and steam coexist. The vapour pressure is the unique value required to maintain both liquid and gas in equilibrium at a specified temperature.

An accurate expression for the vapour pressure of water along the saturation curve is given by Keenan et al. (1969) as:

$$P_{\text{sat}} = P_{\text{c}} \exp\left[10^{-5}\tau(t_{\text{c}} - t) \sum_{i=1}^{8} F_i(0.65 - 0.01t)^{i-1} \right] \tag{13.1}$$

where t is temperature in °C, P_c and t_c are the values of temperature and pressure at the critical point (220.88 bar and 374.136 °C), τ is $1000/T$ (T in K) and F_i is a set of parameters with the following values: $F_1 = -741.9242$, $F_2 = -29.72100$, $F_3 = -11.55286$, $F_4 = -0.8685635$, $F_5 = 0.1094098$, $f_6 = 0.439993$, $F_7 = 0.2520658$ and $F_8 = 0.05218684$.

On the saturation line the pressure varies significantly with temperature, especially at elevated temperatures. For example, equation 13.1 shows that at 0 °C $P_{sat} = 0.006$ bar, at 100 °C $P_{sat} = 1.013$ bar, but at 350 °C $P_{sat} = 165.125$ bar.

13.1   Conventions and notation

The symbol ● is used to denote the property of pure water. In this chapter only, ● also indicates a specified reference temperature and pressure. Unless otherwise stated, P represents pressure in bars, T absolute temperature and t temperature in °C.

13.1.1   The standard state for water

The usual standard state for water in electrolyte solutions is the same as the reference state for dissolved ions (see section 14.1.2). This is the infinitely dilute solution at any temperature and pressure. This standard state ensures that the activity of **pure water** is unity under all conditions. It is used in the calculation of equilibrium constant reactions involving water (see Appendix 3 for further details).

Raoult's law is chosen as defining ideal behaviour, so the selected composition unit for water is the mole fraction, x_w. Even in concentrated solutions water is the major component; therefore x_w is usually close to unity.

To describe the non–ideal behaviour of water we use the practical osmotic coefficient, ϕ_w. The value of ϕ, as defined in section 10.3.10, is related to the activity of water a_w by:

$$\ln a_w = -\phi_w(18.0153 \sum_i m_i) \tag{13.2}$$

where m_i is the molal concentration of dissolved species in the solution, including protons and hydroxides but excluding water, and 18.0153 is the formula weight of water.

The link between the osmotic coefficient of water and the activities of other solution species (section 10.3.11) is:

$$\phi_w = 1 + \frac{1}{\sum_i m_i} \sum_i \int_0^{m_i} m_i \, \mathrm{d} \ln \gamma_i \tag{13.3}$$

This is a general equation valid in multicomponent electrolyte solutions and the summations include **all** ions in solution.

The activity of pure water   An **alternative standard state** is the **hypothetical ideal gas** at 1 bar pressure and any temperature. In this standard state the gas has unit fugacity ($f^{\ominus} = 1$); therefore the activity of water and the fugacity are equal at all temperatures and pressures.

This standard state is adopted in the following exercise to calculate apparent energies of formation.

A third standard state is the pure liquid at 1 bar pressure and any temperature. With this convention:

$$a = \frac{f}{f^{\ominus}} \qquad (13.4)$$

where $f^{\ominus}$ is equal to the fugacity of water at 1 bar and the specified temperature. The activity clearly equals unity in this standard state and no longer equals the fugacity, which has a non-unit value. With this standard state the activity of liquid water is dependent on pressure but independent of temperature. Water activities can be calculated at pressures other than 1 bar using:

$$\ln a = \frac{\Delta G_{(T, P)} - \Delta G^{\ominus}_{(T, P_r)}}{RT} \qquad (13.5)$$

where $\Delta G^{\ominus}_{(T, P_r)}$ equals the apparent Gibbs energy of formation at 1 bar and the specified temperature and $\Delta G^{\ominus}_{(T, P_r)}$ the complementary function at the specified pressure.

The point to note is that **we choose the standard state to suit our purpose and change it when appropriate**.

13.1.2   Apparent energies of formation for pure water

Summarising the essential features of section 11.1.4, we may state that the temperature and pressure dependence of the four energies of formation for water are given by:

$$\Delta G_{(T, P)} \equiv \Delta G^{\bullet}_f + [\bar{G}_{(T, P)} - \bar{G}_{(T_r, P_r)}] \qquad (13.6)$$

$$\Delta H_{(T, P)} \equiv \Delta H^{\bullet}_f + [\bar{H}_{(T, P)} - \bar{H}_{(T_r, P_r)}] \qquad (13.7)$$

$$\Delta A_{(T, P)} \equiv \Delta A^{\bullet}_f + [\bar{A}_{(T, P)} - \bar{A}_{(T_r, P_r)}] \qquad (13.8)$$

$$\Delta U_{(T, P)} \equiv \Delta U^{\bullet}_f + [\bar{U}_{(T, P)} - \bar{U}_{(T_r, P_r)}] \qquad (13.9)$$

In equation 13.6, $\Delta G_{(T, P)}$ is the **apparent** Gibbs energy of formation at a designated temperature and pressure (T, P). The function $\Delta G^{\bullet}_{f(T_r, P_r)}$ is the **true** standard Gibbs energy of formation under selected reference conditions (T_r, P_r), normally 298.15 K and 1 bar. The function $[\bar{G}_{(T, P)} - \bar{G}_{(T_r, P_r)}]$ is the difference between the Gibbs energy of the substance in the conditions (T_r, P_r) and (T, P). It is the property we must calculate in order to evaluate $\Delta G_{(T, P)}$. Similar definitions hold for H, A and U.

Note that in equations 13.6–13.9 the functions $\Delta G^{\bullet}_{f(T_r, P_r)}$, $\Delta H^{\bullet}_{f(T_r, P_r)}$, $\Delta A^{\bullet}_{f(T_r, P_r)}$ and $\Delta U^{\bullet}_{f(T_r, P_r)}$ denote properties at the specific standard state at 298.15 K and 1 bar. Since we are trying to calculate ΔG, ΔH, ΔA and ΔU at other temperatures and pressures, these functions do not carry any form of standard-state superscript. Once calculated, however, ΔG, ΔH, ΔA and ΔU may be employed in geochemical calculations using a standard state unconstrained in temperature and pressure. **In these applications** functions may carry a standard-state superscript, usually $\ominus$.

By definition the relationships between the true energies of formation are:

$$\Delta A^{\bullet}_{f(T_r, P_r)} = \Delta G^{\bullet}_{f(T_r, P_r)} - P_r \Delta V^{\bullet}_{f(T_r, P_r)} \qquad (13.10)$$

and

$$\Delta U^{\bullet}_{f(T_r, P_r)} = \Delta H^{\bullet}_{f(T_r, P_r)} - P_r \Delta V^{\bullet}_{f(T_r, P_r)} \qquad (13.11)$$

where P_r is the reference pressure.

13.1.3   The steam table convention

The thermodynamic properties of liquid and gaseous water are commonly available for the temperature range $0-1300\,°C$ and the pressure range $0-10\,000$ bar. Two comprehensive data-sets are those by Keenan *et al.* (1969) and also Burnham *et al.* (1969), who document the Gibbs energy, the entropy and the enthalpy at small temperature and pressure intervals.

In these tables entropies and enthalpies are based on the convention that $G_{triple} = S_{triple} = 0$, where the subscript 'triple' denote values at the triple point. This convention is inconvenient since it is inconsistent with other conventions used to tabulate thermodynamic data for minerals, gases and electrolytes. The actual values given in steam tables are equal to:

$$H_{ST} = (\bar{H}_{(T, P)} - \bar{H}_{triple}) \qquad (13.12)$$

$$G_{ST} = (\bar{G}_{(T, P)} - \bar{G}_{triple}) + \bar{S}_{triple}(T - T_{triple}) \qquad (13.13)$$

$$S_{ST} = (\bar{S}_{(T, P)} - \bar{S}_{triple}) \qquad (13.14)$$

where the subscript 'ST' denotes data in steam tables. The values of A and U are rarely given in steam tables. However, in this convention they would be equal to:

$$A_{ST} = (\bar{A}_{(T, P)} - \bar{A}_{triple}) + \bar{S}_{triple}(T - T_{triple}) \qquad (13.15)$$

$$U_{ST} = (\bar{U}_{(T, P)} - \bar{U}_{triple}) \qquad (13.16)$$

A second point is that these data are not energies of formation but can be used to calculate such properties. The following four equations will be found useful in such calculations:

$$(\bar{U}_{(T, P)} - \bar{U}_{triple}) = (\bar{H}_{(T, P)} - \bar{H}_{triple}) + (P\bar{V})_{(T, P)} - P_{triple}\bar{V}_{triple} \quad (13.17)$$

$$(\bar{G}_{(T, P)} - \bar{G}_{triple}) = (\bar{H}_{(T, P)} - \bar{H}_{triple}) - (T\bar{S})_{(T, P)} + T_{triple}\bar{S}_{triple} \quad (13.18)$$

$$(\bar{A}_{(T, P)} - \bar{A}_{triple}) = (\bar{U}_{(T, P)} - \bar{U}_{triple}) - (T\bar{S})_{(T, P)} + T_{triple}\bar{S}_{triple} \quad (13.19)$$

$$= (\bar{G}_{(T, P)} - \bar{G}_{triple}) - (P\bar{V})_{(T, P)} + P_{triple}\bar{V}_{triple} \quad (13.20)$$

All values on the left-hand side of equations 13.17–13.20 can be calculated directly from steam tables if data at the triple point are known. This is a key point.

13.1.4   The convention for evaluating triple-point data and energies of formation

Equations can be derived which allow us to calculate the apparent standard energies of formation directly from steam-table data. For example, the apparent enthalpy of formation can be expressed in terms of $(\bar{H} - \bar{H}_{triple})$ by reformulating equation 13.7

Table 13.1   **Triple-point and standard-state data (data from Helgeson and Kirkham, 1974a)**

Property	$J\,mol^{-1}$	$J\,g^{-1}$	$cal\,mol^{-1}$	Property	$J\,mol^{-1}$	$J\,g^{-1}$	$cal\,mol^{-1}$
$\Delta G^{\bullet}_{f(T_r, P_r)}$	-237178	-13165	-56687	$\Delta H^{\bullet}_{f(T_r, P_r)}$	-285830	-15866	-68315
ΔG_{triple}	-235517	-13073	-56290	ΔH_{triple}	-287729	-15971	-68767
$\bar{G}_{(T_r, P_r)} - \bar{G}_{triple}$	-1660.59	-92.18	-396.89	$\bar{H}_{(T_r, P_r)} - \bar{H}_{triple}$	1889.62	104.89	451.63
$\Delta A^{\bullet}_{f(T_r, P_r)}$	-233517	-12962	-55812	$\Delta U^{\bullet}_{f(T_r, P_r)}$	-282152	-15662	-67436
ΔA_{triple}	-231856	-12870	-55415	ΔU_{triple}	-284039	-15766	-67887
$\bar{A}_{(T_r, P_r)} - \bar{A}_{triple}$	-1587.07	-92.28	-397.32	$\bar{U}_{(T_r, P_r)} - \bar{U}_{triple}$	1887.82	104.79	451.20

Property	$J\,mol^{-1}\,K^{-1}$	$J\,g^{-1}\,K^{-1}$	$cal\,mol^{-1}\,K^{-1}$	Property	$J\,mol^{-1}$	$J\,g^{-1}$	$cal\,mol^{-1}$
$\bar{S}^{\bullet}_{(T_r, P_r)}$	69.9146	3.8808	16.71	$\bar{V}_{(T_r, P_r)}$	1.80690	0.100296	0.43186
$\bar{S}_{triple}$	63.1229	3.5144	15.132	$\bar{V}_{triple}$	1.80192	0.10002	0.43067
$\bar{S}_{(T_r, P_r)} - \bar{S}_{triple}$	6.60068	0.3664	1.5776	$\bar{V}_{(T_r, P_r)} - \bar{V}_{triple}$	0.004979	0.00028	0.00119

$\Delta G^{\bullet}_f$, $\Delta H^{\bullet}_f$ and $S^{\bullet}$ *are the apparent Gibbs energy of formation, the apparent enthalpy of formation and the molar third-law entropy respectively.*

to give:

$$\Delta H_{(T, P)} = \Delta H^{\bullet}_{f(T_r, P_r)} + (\bar{H}_{(T, P)} - \bar{H}_{triple}) - (\bar{H}_{(T_r, P_r)} - \bar{H}_{triple}) \quad (13.21)$$

Values of $(\bar{H}_{(T, P)} - \bar{H}_{triple})$, at any temperature and pressure including the reference conditions $(\bar{H}_{(T_r, P_r)} - \bar{H}_{triple})$, are given in steam tables and the value of $\Delta H^{\bullet}_{f(T_r, P_r)}$ is known from measurement. Therefore, ΔH under any conditions can be calculated. Other equations which can be used similarly are:

$$\Delta G = \Delta G^{\bullet}_{f(T_r, P_r)} + (\bar{G}_{(T, P)} - \bar{G}_{triple}) - (\bar{G}_{(T_r, P_r)} - \bar{G}_{triple}) \quad (13.22)$$

$$\Delta A = \Delta A^{\bullet}_{f(T_r, P_r)} + (\bar{A}_{(T, P)} - \bar{A}_{triple}) - (\bar{A}_{(T_r, P_r)} - \bar{A}_{triple}) \quad (13.23)$$

$$\Delta U = \Delta U^{\bullet}_{f(T_r, P_r)} + (\bar{U}_{(T, P)} - \bar{U}_{triple}) - (\bar{U}_{(T_r, P_r)} - \bar{U}_{triple}) \quad (13.24)$$

$$\bar{S} = \bar{S}^{\bullet}_{(T_r, P_r)} + (\bar{S}_{(T, P)} - \bar{S}_{triple}) - (\bar{S}_{(T_r, P_r)} - \bar{S}_{triple}) \quad (13.25)$$

where $\bar{S}^{\bullet}$ is the standard molal third-law entropy of **liquid** water at the reference temperature and pressure.

The values of $(\bar{H}_{(T_r, P_r)} - \bar{H}_{triple})$, $(\bar{G}_{(T_r, P_r)} - \bar{G}_{triple})$, $(\bar{U}_{(t_r, P_r)} - \bar{U}_{triple})$, $(\bar{A}_{(T_r, P_r)} - \bar{A}_{triple})$ and $(\bar{S}_{(T_r, P_r)})$ are given in table 13.1.

Box 13.1   **An example calculation using steam tables**

The values of ΔG, ΔH and S at 200 °C and 10 000 bar can be calculated using data from the steam tables. In this example we will retain the units in the original tables Burnham and Holloway *et al.*, 1969), which yields the following data:

$$S_{ST} = (\bar{S}_{(T, P)} - \bar{S}_{triple}) = 9.469 \text{ cal mol}^{-1}\,K^{-1}$$

$$H_{ST} = (\bar{H}_{(T, P)} - \bar{H}_{triple}) = 3889 \text{ cal mol}^{-1}$$

$$G_{ST} = (\bar{G}_{(T, P)} - \bar{G}_{triple}) = -604 \text{ cal mol}^{-1}$$

continued

Box 13.1
continued

Using equation 13.13 and data from table 13.1:

$$(\bar{G}_{(T,P)} - \bar{G}_{\text{triple}}) = -604 - 15.132 \times (473.15 - 273.16)$$

$$= -3630.25 \text{ cal mol}^{-1} \qquad (13.26)$$

Therefore, from equation 13.22,

$$\Delta G_{(T,P)} = -56\,687 - 3630.25 + 396.89 = -59\,920.36 \text{ cal mol}^{-1} (13.27)$$

$$= -250.707 \text{ kJ mol}^{-1}$$

Using equation 13.12 and data from table 13.1:

$$(\bar{H}_{(T,P)} - \bar{H}_{\text{triple}}) = 3889 \text{ cal mol}^{-1}$$

Therefore, from equation 13.21,

$$\Delta H_{(T,P)} = -68\,315 + 3889 - 451.63 = -64\,877.63 \text{ cal mol}^{-1} \quad (13.28)$$

$$= -271.448 \text{ kJ mol}^{-1}$$

Using equation 13.12 and data from table 13.1, $(\bar{S}_{(T,P)} - \bar{S}_{\text{triple}}) = 9.469 \text{ cal mol}^{-1} \text{ K}^{-1}$. Therefore, from equation 13.25,

$$\bar{S}_{(T,P)} = 16.71 + 9.469 - 1.5776 = 24.64 \text{ cal mol}^{-1} \text{ K}^{-1} \qquad (13.29)$$

$$= 103.09 \text{ J mol}^{-1}$$

Note: $1 \text{ cal mol}^{-1} = 41.8393 \text{ cm}^3 \text{ bar mol}^{-1}$.

13.2 Temperature and pressure dependence of the thermodynamic potentials

13.2.1 Molar volume

Table 13.2 and figures 13.1, 13.2 and 13.3 show the molar volume of water in the pressure range 0–10 000 bar and the temperature range 0–900 °C. We observe a hyperbolic surface in pressure–temperature–volume space.

There are dramatic changes in volume with temperature and pressure in some regions. Of particular note are volume changes on the saturation curve where, at low temperatures, the molar volume of water shows a minimum at 4 °C. This is the so-called anomalous behaviour of water. As pressure is increased the temperature of the minimum volume is lowered and the minimum becomes less distinct, finally disappearing at pressures above 2000 bar (figure 13.4). As temperature increases on the saturation curve, the molar volume of water increases almost exponentially as the substance approaches its critical point. Around the critical point all the thermodynamic properties of water vary significantly with temperature and pressure. Consequently, the mathematical expressions used to predict the molar volume of water in the region 0–1000 bar and 0–900 °C are quite complicated.

As the conditions of temperature and pressure progress away from the critical

Table 13.2   **Molar volume of water in units of $cm^3\,mol^{-1}$ (data from Helgeson and Kirkham, 1974a)**

t ($^\circ C$)	Sat.	0.5	1	2	3	4	5	6	7	8	9	10
						Pressure/kbar						
25	18.0681	17.6886	17.3547	16.84	16.41	16.05	15.72	15.45	15.22	15.01	14.79	14.61
50	18.2340	17.8610	17.5303	17.02	16.59	16.23	15.91	15.62	15.39	15.17	14.97	14.79
75	18.4820	18.0933	17.7526	17.22	16.18	16.42	16.09	15.80	15.55	15.34	15.13	14.96
100	18.7991	18.3778	18.0161	17.44	16.98	16.61	16.28	15.98	15.73	15.50	15.30	15.11
125	19.1851	18.7147	18.3211	17.70	17.21	16.82	16.48	16.17	15.91	15.67	15.46	15.27
150	19.6455	19.1075	18.6697	18.00	17.46	17.04	16.69	16.37	16.09	15.85	15.63	15.42
175	20.1901	19.5606	19.0638	18.33	17.73	17.28	16.91	16.57	16.28	16.03	15.79	15.58
200	20.8344	20.0807	19.5054	18.69	18.03	17.53	17.13	16.78	16.48	16.21	15.97	15.74
225	21.6039	20.6782	19.9976	19.08	18.34	17.80	17.37	17.00	16.68	16.40	16.14	15.90
250	22.5416	21.3703	20.5457	19.50	18.68	18.08	17.62	17.23	16.89	16.59	16.32	16.07
275	23.7224	22.1850	21.1593	19.95	19.05	18.38	17.88	17.46	17.10	16.79	16.51	16.25
300	25.2858	23.1669	21.8528	20.43	19.43	18.69	18.14	17.70	17.32	16.99	16.70	16.42
325	27.5307	24.3871	22.6469	20.95	19.83	19.02	18.43	17.95	17.54	17.20	16.89	16.60
350	31.3508	25.9614	23.5689	21.51	20.26	19.37	18.72	18.20	17.77	17.41	17.08	16.77
375	—	28.0920	24.6535	22.12	20.71	19.73	19.02	18.47	18.01	17.62	17.27	16.95
400	—	31.1824	25.9406	22.79	21.19	20.11	19.34	18.75	18.26	17.84	17.47	17.13
425	—	36.1590	27.4745	23.52	21.70	20.51	19.67	19.03	18.51	18.07	17.67	17.31
450	—	44.7894	29.3033	24.31	22.24	20.93	20.02	19.33	18.77	18.30	17.88	17.49
475	—	57.1878	31.4791	25.18	22.81	21.36	20.37	19.63	19.04	18.53	18.09	17.68
500	—	70.1236	34.0536	26.13	23.41	21.82	20.74	19.95	19.31	18.77	18.30	17.86
525	—	81.8168	37.0572	27.15	24.05	22.29	21.12	20.27	19.59	19.02	18.52	18.05
550	—	92.2038	40.4681	28.25	24.71	22.77	21.52	20.60	19.87	19.27	18.73	18.24
575	—	101.5571	44.1991	29.43	25.40	23.28	21.92	20.94	20.17	19.52	18.96	18.43
600	—	110.1153	48.1255	30.67	26.13	23.79	22.33	21.28	20.46	19.78	19.18	18.63
625	—	118.0515	61.1293	31.98	26.89	24.33	22.75	21.64	20.76	20.04	19.41	18.84
650	—	125.4894	56.1235	33.35	27.67	24.88	23.18	22.00	21.07	20.31	19.65	19.04
675	—	132.5197	60.0537	34.77	28.48	25.44	23.63	22.36	21.38	20.58	19.89	19.25
700	—	139.2110	63.8902	36.24	29.31	26.02	24.08	22.74	21.70	20.85	20.13	19.47
725	—	145.6166	67.6192	37.74	30.16	26.62	24.54	23.12	22.02	21.13	20.37	16.69
750	—	151.7790	71.2367	39.26	31.04	27.23	25.01	23.50	22.35	21.41	20.62	19.91
775	—	157.7331	74.7445	40.81	31.94	27.85	25.49	23.90	22.68	21.70	20.87	20.13
800	—	163.5079	78.1479	42.36	32.86	28.49	25.98	24.30	23.02	21.99	21.12	20.35
825	—	169.1279	81.4536	43.92	33.80	29.14	26.47	24.70	23.36	22.28	21.37	20.57
850	—	174.6138	84.6695	45.48	34.75	29.79	26.97	25.10	23.71	22.58	21.63	20.80
875	—	179.9838	87.8033	47.03	35.71	30.44	27.46	25.51	24.06	22.89	21.90	21.02
900	—	185.2535	90.8630	48.57	36.66	31.09	27.94	25.91	24.41	23.20	22.18	21.26

point, the molar volume of water changes less dramatically. At pressures higher than 1000 bar the effect of changing temperature becomes less pronounced. In these high-pressure regions the molar volume may be described with relatively simple empirical expressions.

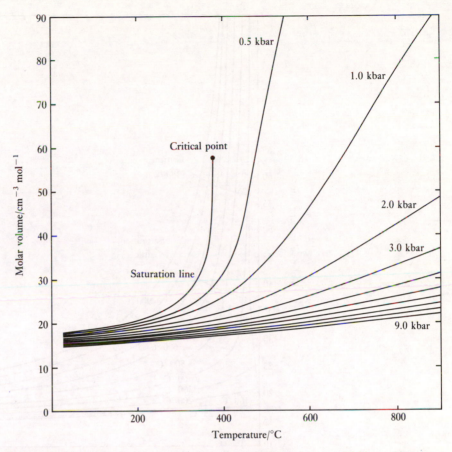

Figure 13.1   Molar volume of water versus temperature (data from Helgeson and Kirkham, 1974a).

13.2.2   Isothermal compression and isobaric expansion

The coefficient of isothermal compression and the coefficient of isobaric expansion were defined in section 9.1.5 as:

$$\kappa = -\frac{1}{V}\left(\frac{\partial V}{\partial P}\right)_T = -\left(\frac{\partial \ln V}{\partial P}\right)_T \tag{13.30}$$

and

$$\alpha = \frac{1}{V}\left(\frac{\partial V}{\partial T}\right)_P = \left(\frac{\partial \ln V}{\partial T}\right)_P \tag{13.31}$$

Both compressibility and expansibility show complex changes with temperature and pressure, especially in the 0–1000 bar region, as depicted in tables 13.3 and 13.4, and figures 13.5–13.10.

   The anomalous behaviour of water on the saturation curve is seen in figure 13.5, where the coefficient of isobaric expansion is negative below 4 °C and increases dramatically with temperature. The increase is most notable on the saturation line (figure 13.6).

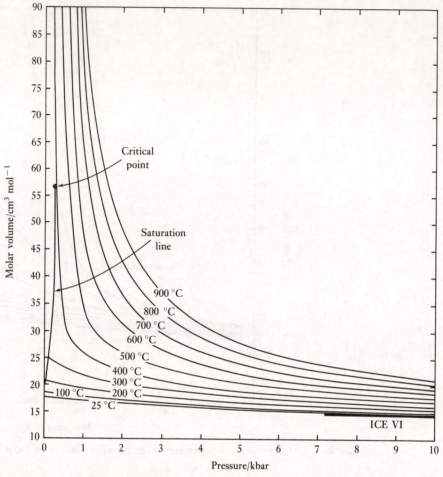

Figure 13.2   Molar volume of water versus pressure (data from Helgeson and Kirkham, 1974a).

On the saturation curve and at low pressures, the compressibility of water shows a minimum at around 46 °C and a dramatic increase with increasing temperature (figures 13.8–13.9). This property of water is a second type of anomalous behaviour, since the compressibilities of most liquids increase monotonically with increasing temperature. Inspection of table 13.4 and figure 13.10 shows that as pressure increases the minimum becomes less distinct, finally disappearing at pressures above 3000 bar.

13.2.3   The standard molar third-law entropy

By definition the third-law molar entropy equals zero at the absolute zero of temperature $[S = 0$ at $T(K) = 0]$. Values of the third-law entropy for water, consistent with this reference state, are given in table 13.5 and figures 13.11 and 13.12. An isobaric increase in temperature causes an increase in entropy at all pressures. The effect of temperature is most notable as conditions tend towards the critical point. The effect of pressure is less pronounced, especially in the liquid-phase

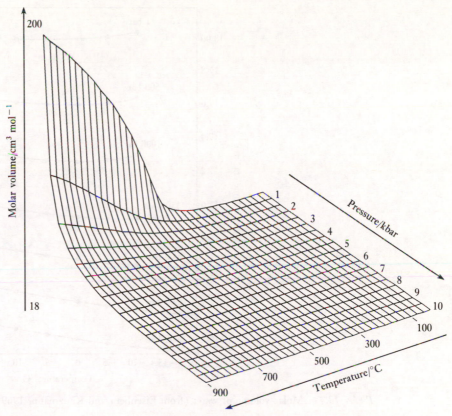

Figure 13.3   Molar volume of water versus temperature and pressure (data from Helgeson and Kirkham, 1974a).

regions where there is an almost linear reduction in entropy with increasing pressure. In contrast the entropy of steam decreases asymptotically with increasing pressure.

13.2.4   Third-law molar heat capacity

Changes in temperature and pressure have a marked effect on $\bar{C}_P$ but a less noticeable effect on $\bar{C}_V$, as shown in figures 13.13 and 13.14 and tables 13.6 and 13.7.

The temperature and pressure dependence of $\bar{C}_P$ and $\bar{C}_V$ show the same type of non-linearity associated with the molar volume, the isothermal compressibility and the isobaric expansivity. On the saturation curve $\bar{C}_P$ exhibits a minimum value at around 50 °C with a dramatic increase to infinity as the water approaches its critical point. At higher pressures the minimum in $\bar{C}_P$ disappears and the maximum becomes less pronounced and moves to higher temperatures. In contrast, $\bar{C}_V$ shows a more linear relationship with temperature and pressure, even close to the critical point.

13.2.5   The apparent Helmholtz and Gibbs energies of formation

The values of ΔA and ΔG do not change dramatically with temperature or pressure. Tables 13.8 and 13.9 with figures 13.15–13.18 show that both ΔA and ΔG decrease

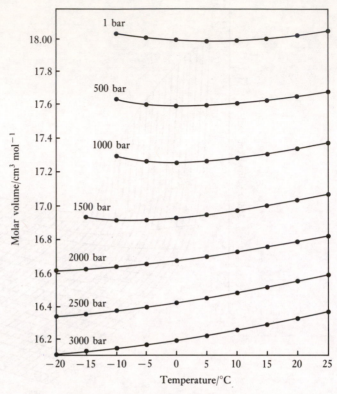

Figure 13.4   Molar volume of water (from Eisenberg and Kauzmann, 1969).

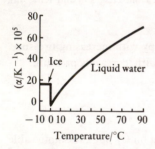

Figure 13.5 Isobaric expansivity of water at low temperatures (from Eisenberg and Kauzmann, 1969).

with increasing temperature at any pressure. Changes in pressure do not change the relative effect of temperature significantly.

In general, the effect of increasing pressure is to increase the values of both ΔA and ΔG at any temperature, although pressure has a smaller effect than temperature. The effect of pressure is greater on ΔG than on ΔA even at high temperatures (the steam region) where the effect of pressure on ΔA is greatest. At low pressures and close to the saturation curve, the values of ΔG and ΔA are similar.

At lower temperatures both functions vary almost linearly with pressure. There is greater pressure dependence at high temperatures and notable non-linearity, especially in the 0−1 bar region at high temperatures.

13.2.6   The apparent enthalpy and internal energy of formation

The effect of temperature on ΔH and ΔU is different from the effect on ΔA and ΔG. Tables 13.10 and 13.11 and figures 13.19 and 13.20 show that an isobaric increase in temperature increases the values of the ΔH and ΔU, in contrast to the decrease observed for ΔA and ΔG. The fact that pressure influences the relative effect of temperature can be seen in these figures, since the isobars fan out with changing pressure.

Table 13.3 **The coefficient of isobaric thermal expansion α in units of $K^{-1} \times 10^5$, consistent with equation 13.31 (data from Helgeson and Kirkham, 1974a)**

$T/°C$	Sat.	0.5	1	2	3	4	5	6	7
					Pressure/kbar				
25	25.53	30.99	34.3	38.4	40.6	42.3	43.4	43.9	44.2
50	46.24	45.8	45.76	44.5	44.6	44.8	44.7	44.5	44.3
75	61.39	57.21	54.79	50.4	48.8	47.7	46.7	45.8	45.0
100	74.86	67.55	63.03	56.1	52.9	50.6	48.8	47.2	46.0
125	88.36	77.84	71.27	61.6	56.7	53.5	50.9	48.7	47.0
150	102.71	8.36	79.51	67.0	60.4	56.1	52.8	50.1	48.0
175	118.7	99.24	87.61	72.4	63.9	58.5	54.5	51.4	48.8
200	137.62	110.87	95.60	77.8	67.2	60.7	56.1	52.4	49.6
225	161.75	124.03	103.8	83.1	70.4	62.7	57.4	53.3	50.1
250	195.13	139.93	112.71	88.3	73.5	64.6	58.6	54.1	50.6
275	245.37	160.3	123.0	93.5	76.5	66.4	59.8	54.8	51.0
300	329.48	187.56	135.41	98.7	79.4	68.2	60.9	55.5	51.4
325	499.05	225.16	150.64	103.9	82.4	70.0	62.0	56.2	51.8
350	1038.3	278.55	169.21	109.4	85.5	71.9	63.2	57.0	52.2
375	—	358.22	191.25	115.3	88.6	73.8	64.5	57.9	52.8
400	—	488.47	216.32	121.5	91.8	75.8	65.9	58.8	53.4
425	—	715.91	243.57	128.0	95.1	77.9	67.3	59.8	54.0
450	—	973.36	272.09	134.9	98.4	79.9	68.7	60.8	54.7
475	—	923.47	300.82	141.9	101.5	81.8	70.0	61.8	55.4
500	—	707.36	327.4	148.9	104.3	83.4	71.1	62.6	55.9
525	—	537.73	347.2	156.7	107.2	85.7	—	—	—
550	—	426.21	354.86	161.6	109.8	87.0	—	—	—
575	—	351.4	348.44	165.3	111.9	88.0	—	—	—
600	—	298.71	331.01	167.7	113.5	88.8	74.8	65.6	58.2
625	—	259.87	307.74	168.8	114.6	89.3	—	—	—
650	—	230.18	282.87	168.5	115.2	89.6	—	—	—
675	—	206.79	258.89	167.1	115.2	89.7	—	—	—
700	—	187.93	236.91	164.5	114.8	89.7	75.3	66.1	58.4
725	—	172.44	217.29	161.1	114.2	89.7	—	—	—
750	—	159.52	200.01	157.0	113.3	89.8	—	—	—
775	—	148.61	184.87	152.5	112.4	89.8	—	—	—
800	—	139.29	171.64	147.6	111.4	89.9	76.1	67.7	60.2
825	—	131.26	160.06	—	—	—	—	—	—
850	—	124.28	149.92	—	—	—	—	—	—
875	—	118.18	141.02	—	—	—	—	—	—
900	—	112.8	133.17	—	—	—	—	—	—

Figure 13.21 shows that an isothermal increase in pressure decreases the value of ΔU. The most pronounced effects are at high temperatures. An isothermal increase in pressure causes a similar effect on ΔH, although a minimum is seen in the isotherms at high temperature (figure 13.22). The points joining positions of the minima define the Joule–Thomson inversion curve. The Joule–Thomson coefficient μ_{JT}, defined

Table 13.4   **The coefficient of isothermal compression κ in units of bar^{-1}, consistent with equation 13.30 (data from Helgeson and Kirkham, 1974a)**

$T/°C$	Sat.	0.5	1	2	3	4	5	6	7	8
					Pressure/kbar					
25	45.60	39.61	36.57	30.0	25.3	21.6	18.8	16.7	15.2	14.1
50	44.36	38.95	36.07	29.8	25.1	21.6	18.8	16.7	15.2	14.1
75	45.89	39.86	36.44	30.0	25.2	21.7	18.9	16.8	15.3	14.2
100	49.50	42.11	37.70	30.8	25.6	22.0	19.2	17.1	15.5	14.4
125	55.12	45.60	39.82	32.0	26.4	22.6	19.7	17.5	15.9	14.6
150	63.09	50.41	42.83	34.0	27.5	23.3	20.3	18.0	16.3	15.0
175	74.24	56.82	46.82	36.6	29.0	24.3	21.0	18.6	16.8	15.4
200	90.09	65.35	52.02	39.7	30.9	25.5	21.9	19.3	17.3	15.8
225	113.48	76.87	58.73	43.6	33.1	27.0	23.0	20.1	18.0	16.4
250	149.96	92.75	67.45	48.1	35.7	28.6	24.1	21.0	18.7	17.0
275	211.54	115.26	78.84	53.3	38.6	30.4	25.4	22.0	19.5	17.7
300	329.06	148.27	93.85	59.3	41.9	32.5	26.9	23.1	20.4	18.4
325	607.65	198.70	113.72	66.3	45.6	34.8	28.5	24.3	21.4	19.2
350	1698.85	280.16	140.04	74.3	49.8	37.3	30.2	25.6	22.5	20.1
375	—	423.17	174.66	83.6	54.3	40.0	32.1	27.1	23.6	21.1
400	—	707.68	219.66	94.3	59.3	42.9	34.1	28.6	24.8	22.1
425	—	1349.14	227.43	106.7	64.8	46.1	36.2	30.2	26.1	23.2
450	—	2525.76	350.85	120.9	70.7	49.4	38.5	31.9	27.5	24.4
475	—	3289.09	442.71	137.1	77.1	53.0	40.9	33.7	29.0	25.6
500	—	3266.80	553.11	155.4	84.0	56.7	43.4	35.6	30.5	26.9
525	—	3043.18	674.90	—	—	—	—	—	—	—
550	—	2839.08	792.51	189.8	99.9	67.7	48.6	39.9	33.3	29.2
575	—	2680.23	889.54	—	—	—	—	—	—	—
600	—	2558.34	958.68	230.0	116.5	76.3	54.0	43.6	36.5	31.9
625	—	2463.48	1002.28	—	—	—	—	—	—	—
650	—	2388.27	1026.73	269.1	133.9	85.5	59.5	47.8	39.7	34.5
675	—	2327.61	1038.3	—	—	—	—	—	—	—
700	—	2277.93	1041.66	305.5	151.2	94.7	65.1	51.8	43.0	37.2
725	—	2236.74	1040.01	—	—	—	—	—	—	—
750	—	2202.21	1035.45	337.3	167.3	103.8	70.5	57.7	46.1	40.0
775	—	2173.01	1029.28	—	—	—	—	—	—	—
800	—	2148.12	1022.36	363.7	181.6	112.6	75.8	59.5	49.3	43.0
825	—	2126.74	1015.2	—	—	—	—	—	—	—
850	—	2108.27	1008.12	—	—	—	—	—	—	—
875	—	2092.2	1001.31	—	—	—	—	—	—	—
900	—	2078.18	994.87	—	—	—	—	—	—	—

in section 9.1.5, is a measure of temperature change on isenthalpic expansion. A positive value for μ_{JT} corresponds with cooling on expansion and a negative value corresponds with warming. For water, as pressure increases across the Joule–Thomson curve the value of μ_{JT} changes from negative to positive.

As with ΔA and ΔG, the most pronounced effect of pressure is in the 0–1000 bar region at high temperature.

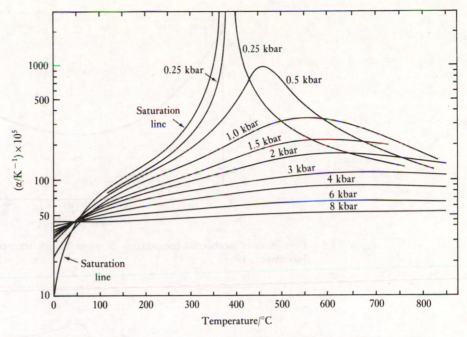

Figure 13.6   Isobaric expansivity of water versus temperature (data from Helgeson and Kirkham, 1974a).

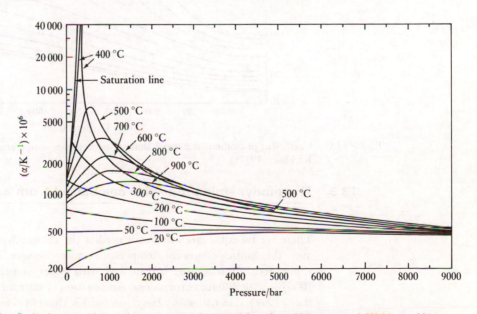

Figure 13.7   Isobaric expansivity of water versus pressure (data from Helgeson and Kirkham, 1974a).

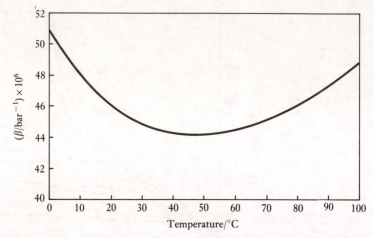

Figure 13.8   Coefficient of isothermal compression of water at low temperatures (from Eisenberg and Kauzmann, 1969).

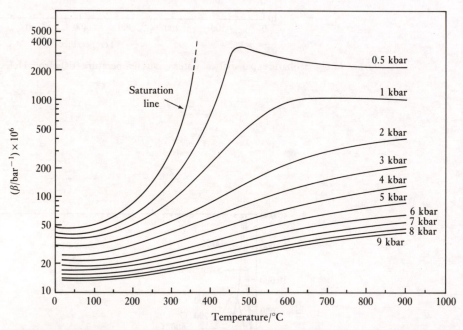

Figure 13.9   Coefficient of isothermal compression of water versus temperature (data from Helgeson and Kirkham, 1974a).

13.3   Thermodynamic properties calculated from an equation of state

There are no equations which can predict the thermodynamic properties of water over the entire ranges of temperature and pressure needed for geochemical applications. It is appropriate to deal with water at pressures above and below 1000 bar with different expressions. An equation of state for water derived by Keenan *et al.* (1969), summarized by Helgeson and Kirkham (1974a), can be used to compute the required thermodynamic properties of water at **pressures less than 1000 bar**

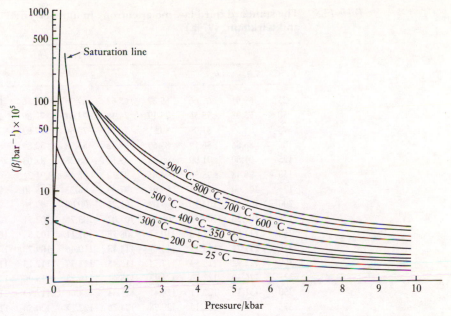

Figure 13.10   Coefficient of isothermal compression of water versus pressure (data from Helgeson and Kirkham, 1974a).

and temperatures up to 1300 °C. This equation is selected because it provides a good example of the use of equations of state within thermodynamics.

13.3.1   The equation of state for water

The Helmholtz energy for water (liquid or vapour) is given by the equation:

$$\psi = \psi_0 + RT[\ln \rho + \rho\xi] \tag{13.32}$$

where R is the gas constant (either $4.615 \, \text{cm}^3 \, \text{bar} \, \text{g}^{-1} \, \text{K}^{-1}$ or $0.4615 \, \text{J} \, \text{g}^{-1} \, \text{K}^{-1}$), ρ is density in $\text{g} \, \text{cm}^{-3}$ and ψ ($\text{J} \, \text{g}^{-1}$) is the Helmholtz energy defined in accordance with the steam table convention.

The other parameters are defined as follows.

The true Helmholtz energy $\bar{A}$ ($\text{J} \, \text{g}^{-1}$), at specified temperature and density, is related to ψ by:

$$\psi = (\bar{A}_{(T, P)} - \bar{A}_{\text{triple}}) + \bar{S}_{\text{triple}}(T - T_{\text{triple}}) \tag{13.33}$$

We see that A is a function of absolute temperature and density, which is consistent with the proper variables for A being temperature and volume (see section 19.2).

The temperature and density dependence of ψ_0 is given by:

$$\psi_0 = \left(\sum_{i=1}^{6} \frac{C_i}{\tau^{i-1}} \right) + C_7 \ln T + C_8 \ln(T/\tau) \tag{13.34}$$

where τ is $1000/T$. The function ξ has a temperature dependence given by:

$$\xi = (\tau - \tau_c) \sum_{j=1}^{7} (\tau - \tau_{aj})^{y-2} \left[\sum_{i=1}^{8} A_{ij}(\rho - \rho_{aj})^{i-1} + \exp(-4.8\rho) \sum_{i=9}^{10} A_{ij}\rho^{i-9} \right] \tag{13.35}$$

Table 13.5   **The standard third law molar entropy in units of $J\,mol^{-1}\,K^{-1}$ (data from Helgeson and Kirkham, 1974a)**

$T/°C$	Sat.	Pressure/kbar								
		0.5	1	2	3	4	5	6	7	8
25	69.91	69.66	69.37	68.62	67.78	67.36	66.53	66.11	65.27	64.85
50	75.98	75.56	75.19	74.48	73.64	72.80	71.96	71.55	70.71	70.29
75	81.59	81.09	80.58	79.50	78.66	77.82	77.40	76.57	75.73	74.89
100	86.86	86.19	85.60	84.52	83.68	82.84	82.01	81.17	80.33	79.50
125	91.80	91.00	90.33	89.12	88.28	87.03	86.19	85.35	84.52	84.10
150	96.48	95.56	94.81	93.30	92.47	91.21	90.37	89.54	88.70	87.86
175	101.00	99.91	99.04	97.49	96.23	95.40	94.14	93.30	92.47	91.63
200	105.31	104.10	103.09	101.25	100.00	99.16	97.91	97.07	96.23	95.40
225	109.50	108.11	106.94	105.02	103.76	102.51	101.25	100.42	99.58	98.74
250	113.64	111.96	110.67	108.78	107.11	105.86	104.60	103.76	102.93	102.09
275	117.74	115.77	114.27	112.13	110.46	109.20	107.95	107.11	105.86	105.02
300	121.92	119.54	117.74	115.48	113.39	112.13	110.88	110.04	109.20	107.95
325	126.36	123.26	121.13	118.41	116.73	115.06	113.80	112.97	111.71	110.88
350	131.38	127.11	124.47	121.34	119.66	117.99	116.73	115.48	114.64	113.80
375	—	131.13	127.78	124.26	122.17	120.50	119.24	117.99	117.15	116.32
400	—	135.44	131.08	127.19	125.10	123.43	121.75	120.92	119.66	118.83
425	—	140.29	134.35	130.12	127.61	125.94	124.26	123.01	122.17	121.34
450	—	145.98	137.65	132.63	130.12	128.45	126.78	125.52	124.26	123.43
475	—	151.75	140.88	135.56	132.63	130.54	128.87	127.61	126.36	125.52
500	—	156.48	144.10	138.07	134.72	132.63	130.96	129.70	128.45	127.61
525	—	160.25	147.28	140.16	137.24	134.72	133.05	131.80	130.54	129.29
550	—	163.26	150.37	142.67	139.33	136.82	135.14	133.89	132.63	131.38
575	—	165.85	153.34	144.77	141.00	138.49	136.82	135.14	133.89	133.05
600	—	168.11	156.11	147.28	143.09	140.58	138.91	137.24	135.98	135.14
625	—	170.16	158.70	149.37	145.18	142.67	141.00	139.33	138.07	136.82
650	—	172.00	161.08	151.46	147.28	144.77	142.67	141.42	139.75	138.91
675	—	173.76	163.30	153.97	149.37	146.86	144.77	143.09	141.84	140.58
700	—	175.35	165.35	155.64	151.46	148.53	146.44	144.77	143.51	142.26
725	—	176.86	167.28	157.74	153.13	150.21	148.11	146.44	145.18	143.93
750	—	178.28	169.08	159.41	154.81	151.88	149.79	148.11	146.86	145.60
775	—	179.66	170.75	161.08	156.48	153.55	151.46	149.79	148.11	146.86
800	—	180.92	172.30	162.76	158.16	155.23	152.72	151.04	149.79	148.53
825	—	182.17	173.80	—	—	—	—	—	—	—
850	—	183.34	175.18	—	—	—	—	—	—	—
875	—	184.51	176.52	—	—	—	—	—	—	—
900	—	185.60	177.82	—	—	—	—	—	—	—

where $\tau_c = 1000/T_{\mathrm{critical}} = 1.544912\,K^{-1}$, and

$$\tau_{aj} = \tau_c \text{ for } j = 1 \qquad \tau_{aj} = 2.5 \text{ for } j > 1$$

$$\rho_{aj} = 0.634 \text{ for } j = 1 \qquad \rho_{aj} = 1.0 \text{ for } j > 1$$

The variables A_{ij} and C_i are empirical coefficients given in table 13.12.

The advantage of a theoretical equation for any one characteristic function is that all other thermodynamic variables can be determined through the derivatives of the

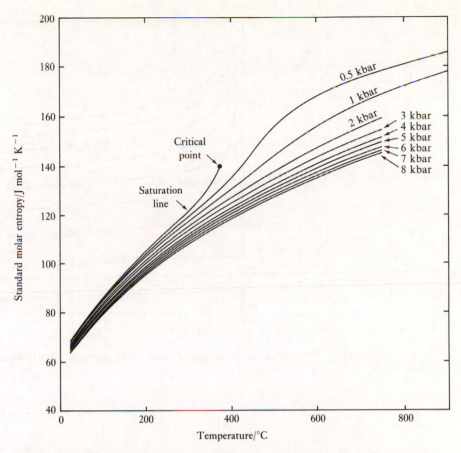

Figure 13.11   Third-law molar entropy of water versus temperature (after Helgeson and Kirkham, 1974a).

equation. For example, it follows from the fundamental equation for the Helmholtz energy $dA = -S\,dT - P\,dV$ that:

$$P = -\left(\frac{\partial \overline{A}}{\partial \overline{V}}\right)_T = \rho^2\left(\frac{\partial \overline{A}}{\partial \rho}\right)_\tau = \rho^2\left(\frac{\partial \psi}{\partial \rho}\right)_\tau \qquad (13.36)$$

and

$$(\overline{S}_{(T,\,P)} - \overline{S}_{\text{triple}}) = -\left(\frac{\partial \psi}{\partial T}\right)_\rho \qquad (13.37)$$

The range of partial derivatives at ψ and ξ necessary to calculate other thermodynamic variables from the equation of state are given by Helgeson and Kirkham (1974).

13.3.2   Molar volumes

In view of the relationship $P = \rho^2(\partial \psi/\partial \rho)_\tau$, the equation of state formulated by Keenan can be differentiated to give:

$$P = \rho RT\left[1 + \rho\xi + \rho^2\left(\frac{\partial \xi}{\partial \rho}\right)_T\right] \qquad (13.38)$$

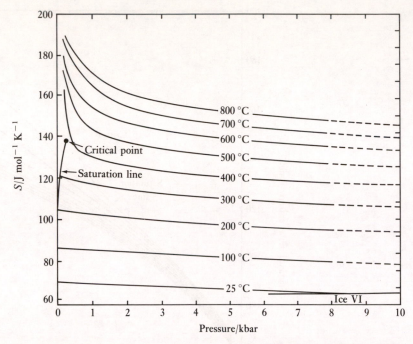

Figure 13.12   Third-law molar entropy of water versus pressure (after Helgeson and Kirkham, 1974a).

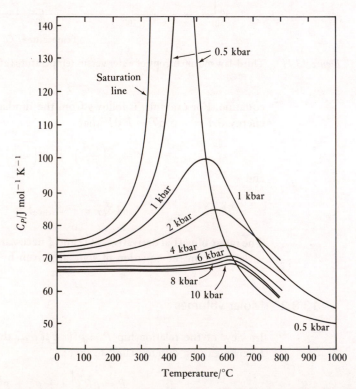

Figure 13.13   Heat capacity of water C_P versus temperature (after Helgeson and Kirkham, 1974a).

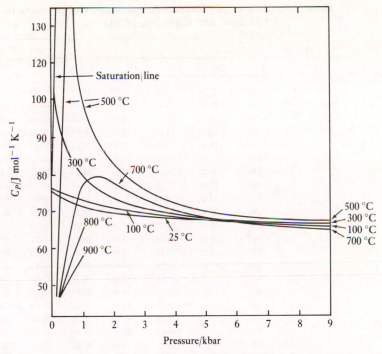

Figure 13.14   Heat capacity of water C_P versus pressure (after Helgeson and Kirkham, 1974a).

This equation gives pressure as a function of temperature and density, although a more common requirement would be the prediction of density (or molar volume) at known values of temperature and pressure. To perform such a calculation the known temperature would be inserted into equation 13.38 along with an estimate of density, to predict pressure. Density is then varied by successive approximation until the known pressure is predicted.

In order for pressure to be in bars, the value of the gas constant in equation 13.38 must be $4.6151\,\mathrm{cm^3\,bar\,g^{-1}\,K^{-1}}$.

13.3.3   Isothermal compression and isobaric expansion

In principle, α and κ can be obtained by differentiating any equation of state for volume with respect to temperature or pressure.

The equation of state for pressure (equation 13.38) can be differentiated with respect to pressure to yield:

$$\frac{1}{\kappa} = \rho\left(\frac{\partial P}{\partial \rho}\right)_T = P + \rho^2 RT\left[\xi + 3\rho\left(\frac{\partial \xi}{\partial \rho}\right)_T + \rho^2\left(\frac{\partial^2 \xi}{\partial \rho^2}\right)_T\right] \tag{13.39}$$

where ξ is as defined by equation 13.35. The partial derivatives of ζ are given by Helgeson and Kirkham (1974a). Once κ is known, α can be calculated at any temperature and pressure from the relationship $\alpha/\kappa = (\partial P/\partial T)_V$ derived in section 9.1.4, from which

$$\alpha = \frac{\kappa P}{T} + \rho RT\kappa\left\{\rho\left(\frac{\partial \xi}{\partial T}\right)_\rho + \rho^2\left[\frac{\partial(\partial \xi/\partial \rho)_T}{\partial T}\right]_\rho\right\} \tag{13.40}$$

The properties of liquid and gaseous water

Table 13.6 **Values of C_P versus temperature and pressure in units of $J\,mol^{-1}\,K^{-1}$ (data from Helgeson and Kirkham, 1974a)**

$T/°C$	Sat.	0.5	1	2	3	4	5	6	7	8
					Pressure/kbar					
25	75.35	73.05	71.50	69.87	69.04	68.20	67.78	67.78	67.78	67.78
50	75.27	73.47	72.13	70.29	69.04	68.20	67.78	67.36	67.36	66.94
75	75.52	73.85	72.59	70.71	69.45	68.62	67.78	67.36	66.94	66.53
100	75.94	74.14	72.76	70.71	69.45	68.20	67.37	66.94	66.53	66.11
125	76.65	74.60	73.05	70.71	69.45	68.20	67.36	66.94	66.53	66.11
150	77.70	75.27	73.51	71.13	69.45	68.20	67.36	66.94	66.53	66.11
175	79.08	76.15	74.14	71.55	69.87	68.62	67.78	66.94	66.53	66.11
200	80.96	77.19	74.85	71.55	69.87	68.62	67.78	66.94	66.53	66.11
225	83.55	78.49	75.60	71.96	69.87	68.62	67.78	67.36	66.53	66.53
250	87.40	80.17	76.44	72.38	70.29	68.62	67.78	67.36	66.94	66.53
275	93.47	82.51	77.49	72.80	70.29	69.04	67.78	67.36	66.53	66.53
300	103.68	85.81	78.83	73.22	70.71	69.04	67.78	67.36	66.53	66.53
325	123.51	90.58	80.58	74.06	71.13	69.04	68.20	67.36	66.94	66.53
350	181.79	97.28	82.80	74.89	71.55	69.45	68.20	67.36	66.94	66.53
375	—	106.98	85.48	75.73	71.96	69.87	68.62	67.36	66.94	66.53
400	—	122.09	88.41	76.57	72.38	69.87	68.20	67.36	66.53	66.11
425	—	147.07	91.46	77.82	73.64	70.29	68.62	67.36	66.94	66.11
450	—	172.38	94.39	79.08	74.48	70.71	68.62	67.36	66.94	66.53
475	—	160.08	97.03	80.33	75.31	71.13	69.04	67.78	67.36	66.53
500	—	129.62	99.16	82.01	76.15	71.55	69.45	67.78	67.36	66.94
525	—	106.52	100.25	83.26	77.40	71.96	70.29	68.20	67.78	67.36
550	—	91.59	99.70	84.10	78.66	72.80	71.13	68.62	68.20	67.78
575	—	81.67	97.32	84.52	79.50	73.22	71.55	69.04	68.62	68.20
600	—	74.81	93.64	82.42	76.15	73.22	71.55	70.29	69.45	69.04
625	—	69.79	89.37	82.01	76.15	73.22	71.55	70.29	69.45	69.04
650	—	65.98	85.02	80.75	75.31	72.38	70.71	69.87	69.04	68.62
675	—	69.97	80.88	78.66	74.06	71.55	69.87	68.62	67.78	67.36
700	—	60.54	77.11	76.57	72.80	70.29	68.62	67.36	66.11	65.27
725	—	58.58	73.76	—	—	—	—	—	—	—
750	—	56.90	70.71	—	—	—	—	—	—	—
775	—	55.48	68.03	—	—	—	—	—	—	—
800	—	54.27	65.65	—	—	—	—	—	—	—
825	—	53.22	63.47	—	—	—	—	—	—	—
850	—	52.30	61.55	—	—	—	—	—	—	—
875	—	51.55	59.83	—	—	—	—	—	—	—
900	—	50.88	58.28	—	—	—	—	—	—	—

13.3.4 Standard molar third-law entropy

The third-law entropy can be obtained if equation 13.33 is differentiated with respect to temperature at constant density. This gives:

$$-\left(\frac{\partial \psi}{\partial T}\right)_{\rho} = -\left(\frac{\partial \bar{A}}{\partial T}\right)_{\rho} - \bar{S}_{\text{triple}} = \bar{S}_{(T, P)} - \bar{S}_{\text{triple}} \qquad (13.41)$$

Substituting equation 13.32 gives:

$$\bar{S}_{(T, P)} = \bar{S}_{\text{triple}} - \left(\frac{\partial \psi_0}{\partial T}\right)_{\rho} - R(\ln \rho + \rho \xi) - RT\rho\left(\frac{\partial \xi}{\partial T}\right)_{\rho} \qquad (13.42)$$

Table 13.7 **Values of C_V versus temperature and pressure in units of $J \, mol^{-1} \, K^{-1}$ (data from Helgeson and Kirkham, 1974a)**

					Pressure/kbar						
$T/°C$	*Sat.*	*0.1*	*0.2*	*0.3*	*0.4*	*0.5*	*0.6*	*0.7*	*0.8*	*0.9*	*1.0*
25	74.60	73.93	73.35	72.80	72.26	71.80	71.34	70.92	69.83	70.17	70.54
50	72.43	71.96	71.50	71.13	70.71	70.37	70.04	69.71	68.87	69.12	69.41
75	70.21	69.87	69.54	69.25	68.96	68.66	68.41	68.16	67.49	67.70	67.95
100	67.99	67.70	67.45	67.20	66.94	66.73	66.53	66.32	65.69	65.90	66.11
125	65.86	65.56	65.35	65.10	64.89	64.68	64.52	64.31	63.72	63.93	64.14
150	63.81	63.55	63.35	63.14	62.93	62.76	62.59	62.38	61.84	62.05	62.22
175	61.92	61.67	61.46	61.30	61.13	60.96	60.79	60.63	60.12	60.29	60.46
200	60.21	60.00	59.79	59.62	59.45	59.33	59.20	59.08	58.62	58.79	58.91
225	58.74	58.53	58.28	58.12	57.99	57.86	57.78	57.66	57.32	57.45	57.57
250	57.45	57.24	56.99	56.78	56.65	56.57	56.48	56.44	56.23	56.27	56.36
275	56.48	56.23	55.86	55.61	55.48	55.40	55.35	55.31	55.23	55.27	55.27
300	55.86	55.73	54.98	54.60	54.43	54.31	54.31	54.31	54.35	54.35	54.31
325	56.02	56.61	54.64	53.85	53.47	53.35	53.30	53.35	53.56	53.47	53.39
350	57.82	41.84	55.81	53.56	52.76	52.47	52.38	52.43	52.76	52.63	52.51
375	—	38.16	61.30	54.64	52.47	51.76	51.55	51.55	52.01	51.84	51.67
400	—	36.02	48.62	62.13	52.97	51.30	50.79	50.71	51.21	51.00	50.84
425	—	34.73	43.89	54.02	54.56	51.17	50.17	49.92	50.46	50.17	50.00
450	—	33.97	39.75	46.74	52.01	50.88	49.58	49.16	49.66	49.37	49.16
475	—	33.56	37.95	42.84	47.32	49.12	48.74	48.37	48.87	48.58	48.37
500	—	33.30	36.82	40.58	44.14	46.57	47.40	47.49	48.12	47.82	47.57
525	—	33.22	36.19	39.20	42.13	44.48	45.86	46.48	47.40	47.07	46.78
550	—	33.18	35.77	38.37	40.88	43.01	44.56	45.48	46.78	46.40	45.98
575	—	33.22	35.52	37.82	40.04	42.01	43.51	44.60	46.15	45.77	45.27
600	—	33.30	35.40	37.49	39.46	41.25	42.76	43.85	45.61	45.19	44.64
625	—	33.43	35.31	37.20	39.04	40.71	42.13	43.26	45.10	44.64	44.06
650	—	33.56	35.27	37.03	38.70	40.29	41.63	42.72	44.60	44.14	43.51
675	—	33.68	35.27	36.86	38.45	39.92	41.17	42.22	44.14	43.64	43.01
700	—	33.85	35.27	36.78	38.20	39.54	40.75	41.76	43.64	43.14	42.55
725	—	34.02	35.31	36.65	37.99	39.25	40.33	41.30	43.10	42.63	42.05
750	—	34.23	35.35	36.61	37.78	38.95	39.96	40.84	42.55	42.13	41.55
775	—	34.39	35.44	36.53	37.61	38.66	39.58	40.42	42.01	41.59	41.05
800	—	34.60	35.52	36.48	37.49	38.41	39.25	39.96	41.46	41.09	40.58
825	—	34.81	35.61	36.44	37.32	38.16	38.91	39.58	40.92	40.54	40.12
850	—	35.02	35.69	36.44	37.20	37.95	38.58	39.16	40.33	40.04	39.66
875	—	35.23	35.82	36.44	37.11	37.74	38.28	38.79	39.83	39.54	39.20
900	—	35.44	35.94	36.48	37.03	37.53	38.03	38.45	39.29	39.08	38.79

The above equation can be used with the value of $\overline{S}_{\text{triple}}$ in table 13.1 to calculate the third-law entropy of water at pressures at or below 1000 bar and temperature up to 1300 °C. Entropies calculated by this method are included in table 13.5.

13.3.5 The apparent Helmholtz energy and the Gibbs energy of formation

The value of ΔA can be calculated directly from the equation of state since:

$$(\overline{A}_{(T, P)} - \overline{A}_{\text{triple}}) = \psi - \overline{S}_{\text{triple}}(T - T_{\text{triple}}) \qquad (13.43)$$

Therefore, equation 13.23 can be used directly with data from table 13.1 to predict

Table 13.8 **The apparent Gibbs energy of formation in units of kJ mol^{-1} (data from Helgeson and Kirkham, 1974a)**

$T/°C$	Sat.	0.001	0.5	1	2	3	4	5	6	7	8	9
25	−237.178	−237.18	−236.28	−235.40	−233.68	−232.04	−230.41	−228.82	−227.27	−225.73	−224.22	−222.71
50	−238.998	−239.00	−238.10	−237.21	−235.48	−233.80	−232.17	−230.54	−228.99	−227.44	−225.89	−224.39
75	−240.969	−240.97	−240.06	−239.16	−237.40	−235.73	−234.05	−232.42	−230.83	−229.82	−227.74	−226.19
100	−243.074	−243.11	−242.15	−241.24	−239.45	−237.73	−236.06	−234.43	−232.43	−231.21	−229.66	−228.11
125	−245.308	−248.04	−244.36	−243.44	−241.63	−239.91	−238.20	−236.52	−234.89	−233.30	−231.71	−230.16
150	−247.655	−253.03	−246.70	−245.75	−243.93	−242.13	−240.41	−238.74	−237.07	−235.48	−233.47	−232.30
175	−250.115	−258.06	−249.14	−248.18	−246.31	−244.51	−242.76	−241.04	−239.37	−237.73	−236.10	−234.51
200	−252.680	−263.15	−251.69	−250.70	−248.78	−246.94	−245.18	−243.47	−241.75	−240.08	−238.45	−236.86
225	−255.345	−100.93	−254.35	−253.33	−251.37	−249.49	−247.49	−245.94	−244.22	−242.55	−240.87	−239.24
250	−258.103	−273.46	−257.09	−256.05	−254.05	−252.13	−250.29	−248.53	−246.77	−245.06	−243.38	−241.75
275	−260.952	−278.69	−259.94	−258.86	−256.81	−254.85	−253.01	−251.17	−249.41	−247.69	−245.98	−244.35
300	−263.881	−283.95	−262.88	−261.76	−259.66	−257.65	−255.77	−253.93	−252.13	−250.37	−248.66	−246.98
325	−266.893	−289.25	−265.92	−264.75	−262.59	−260.54	−258.61	−256.73	−254.89	−251.25	−251.42	−249.70
350	−269.981	−295.59	−269.05	−267.82	−265.56	−263.51	−261.50	−259.62	−257.78	−255.98	−254.22	−252.46
375	—	−299.97	−272.28	−270.97	−268.65	−263.88	−264.51	−262.55	−260.66	−258.85	−257.06	−255.35
400	—	−305.38	−275.61	−274.21	−271.79	−269.92	−267.52	−265.56	−263.68	−261.83	−259.99	−258.24
425	—	−310.83	−279.05	−277.52	−275.01	−272.75	−270.66	−268.65	−266.73	−264.85	−263.01	−261.21
450	—	−316.31	−282.63	−280.93	−278.28	−275.98	−273.84	−271.79	−269.83	−267.90	−266.06	−264.26
475	—	−321.82	−286.35	−284.41	−281.67	−279.24	−277.06	−274.97	−272.96	−271.04	−269.16	−267.32
500	—	−327.37	−290.21	−287.97	−285.06	−282.59	−280.33	−278.24	−276.19	−274.22	−272.34	−270.45
525	—	−332.95	−294.17	−291.61	−288.53	−285.98	−283.68	−281.50	−279.45	−277.44	−275.52	−273.73
550	—	−338.56	−298.21	−295.34	−292.09	−289.45	−287.06	−284.89	−282.75	−280.75	−278.78	−276.90
575	—	−344.19	−302.33	−299.13	−295.68	−292.96	−290.54	−288.28	−286.14	−284.09	−282.09	−280.16
600	—	−349.86	−306.50	−303.00	−299.32	−296.52	−294.05	−291.75	−289.57	−287.48	−285.47	−283.51
625	—	−355.55	−310.73	−306.93	−303.05	−300.12	−297.61	−295.22	−294.14	−290.91	−288.86	−286.90
650	—	−361.28	−315.01	−310.93	−306.81	−303.80	−301.21	−298.78	−296.52	−294.39	−292.29	−290.33
675	—	−367.02	−319.33	−314.99	−310.62	−307.52	−304.85	−302.38	−300.08	−297.90	−295.81	−293.80
700	—	−372.80	−323.70	−319.10	−314.51	−311.29	−308.53	−306.02	−303.67	−301.46	−299.32	−297.32
725	—	−378.06	−328.10	−323.26	−318.44	−315.10	−312.25	−309.70	−307.36	−305.10	−302.92	−300.82
750	—	−381.50	−332.54	−327.46	−322.38	−318.95	−316.06	−313.42	−311.04	−308.74	−306.56	−304.43
775	—	−390.28	−337.01	−331.71	−326.39	−322.84	−320.24	−317.19	−314.76	−312.42	−310.20	−308.03
800	—	−396.16	−341.52	−335.99	−330.45	−326.77	−323.72	−321.04	−318.49	−316.14	−313.88	−311.75
825	—	−402.07	−346.06	−340.32	—	—	—	—	—	—	—	—
850	—	−407.99	−350.63	−344.68	—	—	—	—	—	—	—	—
875	—	−413.94	−355.23	−349.08	—	—	—	—	—	—	—	—
900	—	−419.92	−359.85	−353.51	—	—	—	—	—	—	—	—

ΔA, i.e.

$$\Delta A_{(T, P)} = \Delta A_{f(T_r, P_r)}^{\bullet} + (\bar{A}_{(T, P)} - \bar{A}_{\text{triple}}) - (\bar{A}_{(T_r, P_r)} - \bar{G}_{\text{triple}}) \quad (13.44)$$

The simplest way to calculate ΔG is with equation 13.20 written as:

$$(\bar{G}_{(T, P)} - \bar{G}_{\text{triple}}) = (\bar{A}_{(T, P)} - \bar{A}_{\text{triple}}) + (P\bar{V})_{(T, P)} - (P_{\text{triple}}\bar{V}_{\text{triple}}) \, (13.45)$$

where $(\bar{A} - \bar{A}_{\text{triple}})$ is calculated as above, $\bar{V}$ is calculated from the density given by equation 13.38 and $P_{\text{triple}}V_{\text{triple}}$ is derived from table 13.1. The value of ΔG can be obtained from equation 13.22 and data from table 13.1, i.e.

$$\Delta G_{(T, P)} = \Delta G_{f(T_r, P_r)}^{\bullet} + (\bar{G}_{(T, P)} - \bar{G}_{\text{triple}}) - (\bar{G}_{(T_r, P_r)} - \bar{G}_{\text{triple}}) \quad (13.46)$$

Table 13.9 **The apparent Helmholtz energy of formation in units of kJ mol^{-1} (data from Helgeson and Kirkham, 1974a)**

$T/^\circ C$	Sat.	0.5	1	2	3	4	5	6	7	8	9
						Pressure/kbar					
25	−233.517	−233.50	−233.48	−233.38	−233.30	−233.17	−233.01	−232.88	−232.71	−232.55	−232.38
50	−235.342	−235.33	−235.31	−235.22	−235.10	−234.97	−234.85	−234.68	−234.56	−234.39	−234.22
75	−237.312	−237.30	−237.28	−237.19	−237.07	−236.94	−236.81	−236.65	−236.48	−236.31	−236.14
100	−239.417	−239.41	−239.38	−239.28	−239.20	−239.03	−238.91	−238.74	−238.57	−238.40	−238.24
125	−241.651	−241.64	−241.61	−241.50	−241.37	−241.25	−241.12	−240.96	−240.75	−240.58	−240.41
150	−244.007	−243.99	−243.96	−243.84	−243.72	−243.59	−243.43	−243.26	−243.05	−242.88	−242.71
175	−246.475	−246.46	−246.43	−246.31	−246.14	−246.02	−245.85	−245.64	−245.48	−245.27	−245.10
200	−249.057	−249.04	−248.99	−248.86	−248.70	−248.53	−248.36	−248.15	−247.94	−247.78	−247.57
225	−251.743	−251.72	−251.67	−251.54	−251.33	−251.17	−250.96	−250.75	−250.54	−250.33	−250.12
250	−254.534	−254.50	−254.45	−254.30	−254.09	−253.89	−253.68	−253.47	−253.22	−253.01	−252.80
275	−257.433	−257.39	−257.32	−257.15	−256.90	−256.69	−256.44	−256.23	−255.98	−255.77	−255.52
300	−260.437	−260.38	−260.29	−260.08	−259.83	−259.58	−259.32	−259.07	−258.82	−258.61	−258.32
325	−263.567	−263.48	−263.35	−263.09	−262.84	−262.55	−262.29	−262.00	−261.75	−261.50	−261.25
350	−266.839	−266.69	−266.52	−266.23	−265.89	−265.60	−265.31	−265.01	−264.76	−264.47	−264.18
376	—	−270.02	−269.78	−269.41	−269.07	−268.74	−268.40	−268.11	−267.82	−267.52	−267.23
400	—	−273.51	−273.14	−272.67	−272.29	−271.92	−271.58	−271.25	−270.91	−270.62	−270.29
425	—	−277.20	−276.61	−276.06	−275.60	−275.18	−274.81	−274.39	−274.14	−273.80	−273.47
450	—	−281.21	−280.20	−279.49	−278.99	−278.53	−278.11	−277.73	−277.40	−277.02	−276.69
475	—	−285.55	−283.90	−283.01	−282.42	−281.92	−281.50	−281.08	−280.70	−280.33	−280.29
500	—	−290.06	−287.72	−286.60	−285.98	−285.39	−284.93	−284.47	−284.09	−283.68	−283.26
525	—	−294.60	−291.66	−290.29	−289.53	−288.95	−288.40	−287.94	−287.52	−287.06	−286.65
550	—	−299.16	−295.72	−294.05	−293.21	−292.55	−291.96	−291.46	−291.00	−290.54	−290.08
575	—	−303.75	−299.89	−297.90	−296.94	−296.19	−295.60	−295.06	−294.55	−294.05	−293.59
600	—	−308.85	−304.15	−301.83	−300.70	−299.91	−299.24	−298.70	−298.15	−297.65	−297.15
625	—	−312.98	−308.49	−305.77	−304.55	−303.67	−302.96	−302.38	−301.79	−301.25	−300.70
650	—	−317.62	−312.62	−309.83	−308.44	−307.48	−306.73	−301.10	−305.47	−304.93	−304.34
675	—	−322.30	−317.34	−313.93	−312.43	−311.37	−310.54	−309.87	−309.24	−308.61	−308.83
700	—	−327.00	−321.82	−318.11	−316.44	−315.31	−314.43	−313.67	−313.01	−312.38	−311.75
725	—	−331.72	−326.36	−322.34	−320.49	−319.24	−318.32	−317.57	−316.85	−316.18	−315.52
750	—	−336.47	−330.92	−326.60	−324.59	−323.30	−322.29	−321.46	−320.70	−320.03	−319.32
775	—	−341.24	−335.52	−330.91	−328.74	−327.36	−326.31	−325.43	−324.64	−323.88	−313.17
800	—	−346.04	−340.15	−335.26	−332.96	−331.46	−330.33	−329.41	−328.61	−327.82	−327.06
825	—	−350.86	−344.81	—	—	—	—	—	—	—	—
850	—	−355.70	−349.49	—	—	—	—	—	—	—	—
875	—	−360.56	−354.20	—	—	—	—	—	—	—	—
900	—	−365.46	−358.94	—	—	—	—	—	—	—	—

13.3.6 The apparent enthalpy and internal energy of formation

Values of ΔU and ΔH are calculated from:

$$\Delta U_{(T, P)} = \Delta U_{\text{triple}} + (\Delta A_{(T, P)} - \Delta A_{\text{triple}}) + (T\bar{S})_{(T, P)} - T_{\text{triple}}\bar{S}_{\text{triple}} \quad (13.47)$$

and

$$\Delta H_{(T, P)} = \Delta H_{\text{triple}} + (\Delta E_{(T, P)} - \Delta E_{\text{triple}}) + P\bar{V} - P_{\text{triple}}\bar{V}_{\text{triple}} \quad (13.48)$$

using data from table 13.1, the value of ΔA from equation 13.44 and volume (via density) calculated from equation 13.38.

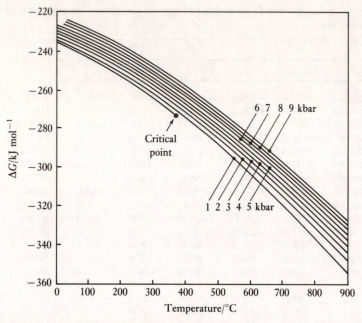

Figure 13.15   Apparent Gibbs energy of formation of water versus temperature (after Helgeson and Kirkham, 1974a).

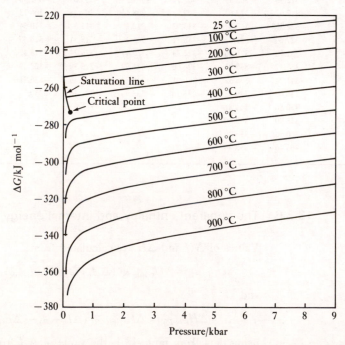

Figure 13.16   Apparent Gibbs energy of formation of water versus pressure (after Helgeson and Kirkham, 1974a).

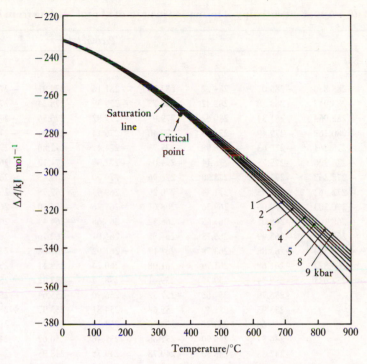

Figure 13.17   Apparent Helmholtz energy of formation of water versus temperature (after Helgeson and Kirkham, 1974a).

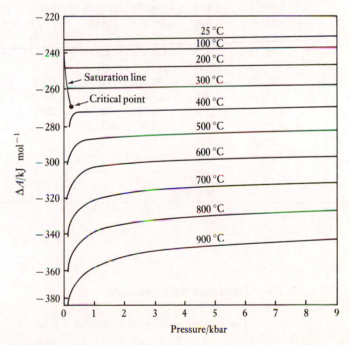

Figure 13.18   Apparent Helmholtz energy of formation of water versus pressure (after Helgeson and Kirkham, 1974a).

Table 13.10   **Apparent enthalpy of formation in units of kJ mol^{-1} (data from Helgeson and Kirkham, 1974a)**

$T/°C$	Sat.	0.5	1	2	3	4	5	6	7	8
					Pressure/kbar					
25	−285.830	−285.01	−284.22	−282.84	−281.16	−279.91	−278.24	−276.98	−275.73	−274.47
50	−283.947	−283.18	−282.42	−280.75	−279.49	−278.24	−276.56	−275.31	−274.05	−272.80
75	−282.064	−281.34	−280.62	−279.07	−277.82	−276.56	−274.89	−273.63	−272.38	−271.12
100	−280.169	−279.49	−278.80	−277.40	−276.14	−274.89	−273.22	−271.96	−270.70	−269.45
125	−278.261	−277.63	−276.98	−275.73	−274.47	−272.80	−271.54	−270.29	−269.03	−267.78
150	−276.328	−275.75	−275.14	−274.05	−272.38	−271.12	−269.87	−268.61	−267.36	−266.10
175	−272.362	−273.86	−273.30	−271.96	−270.70	−269.45	−268.19	−266.94	−265.68	−264.43
200	−272.362	−271.95	−271.44	−270.29	−269.03	−267.78	−266.52	−265.27	−264.01	−262.76
225	−270.303	−270.00	−269.56	−268.61	−267.36	−266.10	−264.85	−263.59	−263.24	−261.08
250	−268.165	−268.02	−267.65	−266.52	−265.68	−264.43	−263.17	−261.92	−260.66	−259.41
275	−265.918	−265.99	−265.73	−264.85	−264.01	−262.76	−261.50	−260.24	−258.99	−257.73
300	−263.504	−263.88	−263.78	−263.17	−261.92	−261.08	−259.83	−258.57	−257.32	−256.06
325	−260.822	−261.68	−261.79	−261.08	−260.24	−259.41	−258.15	−256.90	−255.64	−254.39
350	−257.621	−259.34	−259.75	−259.41	−258.57	−257.32	−256.48	−255.22	−253.97	−252.71
375	—	−256.79	−257.65	−257.32	−256.90	−255.64	−254.81	−253.55	−252.30	−251.04
400	—	−253.95	−255.47	−255.64	−254.81	−253.97	−253.13	−251.88	−250.62	−249.37
425	—	−250.60	−253.22	−253.55	−253.13	−252.30	−251.46	−250.20	−248.95	−248.11
450	—	−246.57	−250.90	−251.88	−251.46	−250.62	−249.78	−248.53	−247.27	−246.44
475	—	−242.33	−248.50	−249.78	−249.78	−248.95	−248.11	−246.86	−246.02	−244.76
500	—	−238.71	−246.05	−248.11	−247.69	−247.27	−246.44	−245.18	−244.35	−243.09
525	—	−235.79	−243.55	−246.02	−246.02	−245.60	−244.76	−243.93	−242.67	−241.84
550	—	−233.32	−241.05	−243.93	−244.35	−243.93	−243.09	−242.25	−241.00	−240.15
575	—	−231.16	−238.59	−242.25	−242.67	−242.67	−241.84	−241.00	−239.74	−238.91
600	—	−229.21	−236.20	−240.58	−241.00	−240.58	−239.74	−238.91	−238.07	−237.65
625	—	−227.41	−233.91	−238.49	−238.91	−238.91	−238.49	−237.23	−236.40	−235.56
650	—	−225.71	−231.73	−236.40	−237.23	−236.81	−236.40	−235.56	−234.72	−233.89
675	—	−224.10	−271.50	−234.30	−235.14	−235.14	−234.72	−233.89	−233.05	−231.79
700	—	−222.56	−227.68	−232.21	−233.47	−233.47	−233.05	−232.21	−231.38	−230.12
725	—	−221.07	−225.80	−230.54	−231.79	−231.79	−231.38	−230.54	−229.70	−228.86
750	—	−219.63	−223.99	−228.45	−229.70	−230.12	−229.70	−228.86	−228.03	−227.19
775	—	−218.22	−222.26	−226.77	−228.03	−228.45	−228.03	−227.19	−226.77	−225.52
800	—	−216.85	−220.59	−225.10	−226.35	−226.77	−226.35	−225.94	−225.10	−224.26
825	—	−215.51	−219.15	—	—	—	—	—	—	—
850	—	−214.19	−217.41	—	—	—	—	—	—	—
875	—	−212.89	−215.90	—	—	—	—	—	—	—
900	—	−211.61	−214.42	—	—	—	—	—	—	—

13.3.7   Third-law heat capacity

The fundamental relationship for $\bar{C}_V$ is:

$$\bar{C}_V = T\left(\frac{\partial \bar{S}}{\partial T}\right)_\rho = -T\left(\frac{\partial^2 \psi}{\partial t^2}\right)_\rho \tag{13.49}$$

Table 13.11   **Apparent internal energy of formation in units of kJ mol^{-1} (data from Helgeson and Kirkham, 1974a)**

$T/°C$	Sat.	0.5	1	2	3	4	5	6	7	8
					Pressure/kbar					
25	−282.152	−282.22	−282.28	−282.42	−282.42	−282.42	−282.42	−282.84	−282.84	−282.84
50	−280.269	−280.39	−280.50	−280.75	−280.75	−280.75	−281.16	−281.16	−281.16	−281.16
75	−278.387	−278.57	−278.71	−279.07	−279.07	−279.49	−279.49	−279.49	−279.49	−279.49
100	−276.495	−276.73	−276.92	−277.40	−277.40	−277.40	−277.82	−277.82	−278.24	−278.24
125	−274.588	−274.89	−275.13	−275.31	−275.73	−276.14	−276.14	−276.56	−276.56	−276.56
150	−272.663	−273.04	−273.34	−273.63	−274.05	−274.47	−274.47	−274.89	−274.89	−275.31
175	−270.709	−271.17	−271.53	−271.96	−272.38	−272.80	−273.22	−273.22	−273.63	−273.63
200	−268.717	−269.27	−269.71	−270.29	−270.70	−271.12	−271.54	−271.54	−271.96	−271.96
225	−266.680	−267.36	−267.88	−268.61	−269.03	−269.45	−269.87	−270.29	−270.29	−270.70
250	−264.579	−265.41	−266.04	−266.94	−267.36	−267.78	−268.19	−268.61	−269.03	−269.03
275	−258.199	−263.42	−264.17	−265.27	−265.68	−266.52	−266.94	−266.94	−267.36	−267.78
300	−260.044	−261.37	−262.29	−263.59	−264.01	−264.85	−265.27	−265.68	−266.10	−266.10
325	−257.479	−259.22	−260.37	−261.92	−262.76	−263.17	−263.59	−264.01	−264.43	−264.43
350	−254.463	−256.96	−258.43	−259.83	−261.08	−261.50	−262.34	−262.34	−262.76	−263.17
375	—	−254.52	−256.43	−258.15	−259.41	−259.83	−260.66	−261.08	−261.50	−261.50
400	—	−251.83	−254.39	−256.48	−257.73	−258.57	−258.99	−259.41	−259.83	−260.24
425	—	−248.74	−252.30	−254.81	−256.06	−256.90	−257.32	−258.15	−258.57	−288.57
450	—	−245.14	−250.15	−253.13	−254.39	−255.22	−256.06	−256.48	−256.90	−257.32
475	—	−241.52	−247.98	−251.04	−252.71	−253.55	−254.39	−255.22	−255.64	−256.06
500	—	−238.54	−245.78	−249.37	−251.04	−252.30	−253.13	−253.55	−253.97	−254.39
525	—	−236.20	−243.58	−247.69	−249.78	−250.62	−251.46	−252.30	−252.71	−253.13
550	—	−234.25	−241.43	−246.02	−248.11	−249.37	−250.20	−250.62	−251.46	−251.88
575	—	−232.56	−239.33	−244.76	−246.86	−248.11	−248.95	−249.78	−250.20	−250.62
600	—	−231.04	−237.34	−242.67	−245.18	−246.44	−247.27	−248.11	−248.53	−249.37
625	—	−229.63	−235.45	−241.00	−243.51	−244.76	−246.02	−246.44	−247.27	−247.69
650	—	−228.31	−233.67	−239.32	−241.84	−243.09	−244.35	−245.18	−245.60	−246.44
675	—	−227.05	−231.99	−237.65	−240.16	−241.84	−242.67	−243.51	−244.35	−244.76
700	—	−225.84	−230.40	−235.98	−238.49	−240.16	−241.42	−242.25	−242.67	−243.51
725	—	−224.67	−228.88	−234.30	−237.23	−238.49	−239.74	−240.58	−241.42	−241.84
750	—	−223.54	−227.44	−232.63	−235.56	−237.23	−238.49	−239.32	−240.16	−240.58
775	—	−222.43	−226.06	−231.38	−233.89	−235.98	−237.23	−238.07	−238.49	−239.32
800	—	−221.35	−224.73	−229.70	−232.63	−234.30	−235.56	−236.40	−237.65	−238.07
825	—	−220.29	−223.45	—	—	—	—	—	—	—
850	—	−219.25	−222.20	—	—	—	—	—	—	—
875	—	−218.22	−221.00	—	—	—	—	—	—	—
900	—	−217.20	−219.83	—	—	—	—	—	—	—

Substituting equation 13.32 into 13.49 gives:

$$\bar{V}_V = -T\left(\frac{\partial^2 \psi_0}{\partial T^2}\right)_\rho - 2RT\rho\left(\frac{\partial \xi}{\partial T}\right)_\rho - RT^2\rho\left(\frac{\partial^2 \xi}{\partial T^2}\right)_\rho \qquad (13.50)$$

Values of $\bar{C}_P$ can be calculated using the fundamental relationship between $\bar{C}_V$ and $\bar{C}_P$ given in section 9.2.7, with computed values of $\bar{C}_V$ and κ, i.e.

$$\bar{C}_P = \bar{C}_V + \frac{\alpha^2 \bar{V} T}{\kappa} \qquad (13.51)$$

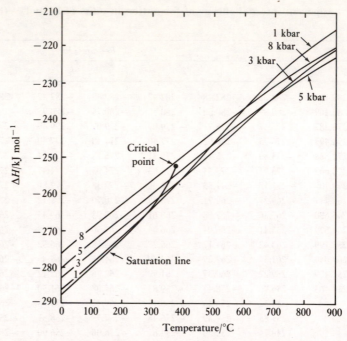

Figure 13.19   Apparent enthalpy of formation of water versus temperature (after Helgeson and Kirkham, 1974a).

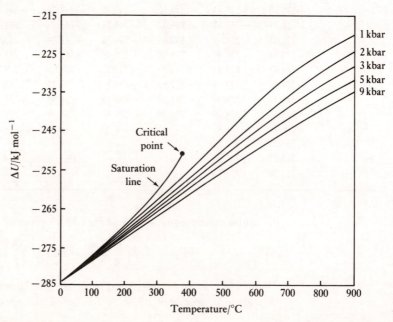

Figure 13.20   Apparent internal energy of formation of water versus temperature (after Helgeson and Kirkham, 1974a).

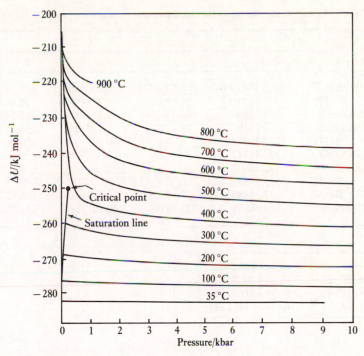

Figure 13.21   Apparent internal energy of formation of water versus pressure (after Helgeson and Kirkham, 1974a).

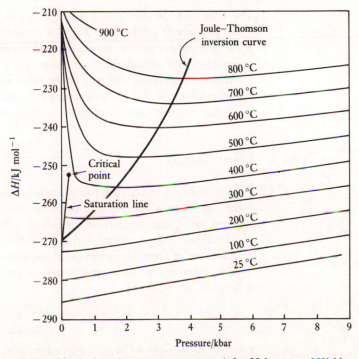

Figure 13.22   Apparent enthalpy of formation of water versus pressure (after Helgeson and Kirkham, 1974a).

Table 13.12 A_{ij} and C_i coefficients for the equation of state for water (data from Helgeson and Kirkham, 1974a)

i	A_{ij}			j			
	1	2	3	4	5		7
1	29.492937	−5.1985860	6.8335354	−0.1564104	−6.3972405	−3.9661401	−0.69048554
2	−132.13917	7.7779182	−26.149751	−0.72546108	26.409282	15.453016	2.7407416
3	274.64632	−33.301902	65.326396	−9.2734389	−47.740374	−29.142470	−5.1028070
4	−360.93828	−16.254622	−26.181978	4.3125840	56.323130	29.568796	3.963085
5	342.18431	−177.31074	0	0	0	0	0
6	−244.50042	127.48742	0	0	0	0	0
7	155.18535	137.46153	0	0	0	0	0
8	5.9728487	155.97836	0	0	0	0	0
9	−410.30848	337.31180	−137.46618	6.7874983	136.87317	79.847970	13.041253
10	−416.05860	−209.88866	−733.96848	10.401717	645.81880	399.17570	71.531353

$C_1 = 1857.065$	$C_5 = -20.5516$
$C_2 = 3229.12$	$C_6 = 4.85233$
$C_3 = -419.465$	$C_7 = 46.0$
$C_4 = 36.6649$	$C_8 = -1011.249$

13.4   Working expressions for thermodynamic potentials for pressures above 1000 bar

The 0.001 bar–10 kbar pressure range can be found in various conditions at the Earth's surface to a lithostatic depth of around 35 km. Many geochemical processes occur at pressures greater than 1 kbar, including hydrothermal ore deposition, and sedimentary and metamorphic processes.

At pressures above 1000 bar, in the temperature range 0–1000 °C, there are no suitable theoretical equations of state for the thermodynamic properties of water. In this region we must resort to empirical expressions to fit measured or estimated data.

13.4.1   Molar volumes

Even at pressures above 1000 bar it is convenient to subdivide pressure–temperature space into two distinct regions as shown in figure 13.23. Each region requires a different expression representing pressure–temperature–volume data. Burnham *et al.* (1969) introduced a third region at high temperatures, which is not addressed here because of uncertainties in the molar volumes in this region.

The molar volume of water (in units of cm^3 per mole) is given by a high-order polynomial of the form:

$$\bar{V} = 18.0153 \times \sum_{i=0}^{N} \sum_{j=0}^{N-i} a_{ij} t^i P^{[(r \times j)-1]} \tag{13.52}$$

where P is pressure in bars, t is temperature in °C and a_{ij} is an array of fitting coefficients given in table 13.13. The coefficient r is a parameter with the value 1

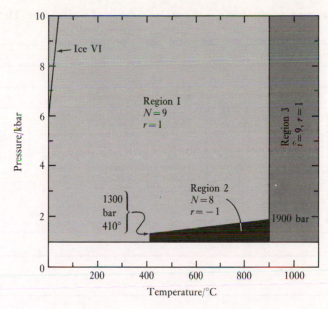

Figure 13.23 Alternative regions of temperature and pressure for molar volume fitting equations (after Helgeson and Kirkham, 1974a).

or -1 depending on the specific region of $P-T$ space, and N is the order of the polynomial. The coefficients were obtained by regression of volume data.

13.4.2 Isothermal compression and isobaric expansion

In principle, equation 13.52 for molar volume could be differentiated to give both α and κ. However, the uncertainty in experimental data gives residual errors in the fitting coefficients of this equation. These errors prevent the accurate prediction of compressibility and expansibility because the functions $(\partial \bar{V}/\partial T)_P$ and $(\partial \bar{V}/\partial P)_T$ show a sensitive dependence on temperature and pressure and are therefore sensitive to errors in the fitting coefficients.

In acknowledging this point, Helgeson and Kirkham (1974a) proposed a different approach. This involves calculating the partial derivatives $(\partial \bar{V}/\partial T)_P$ and $(\partial \bar{V}/\partial P)_T$ by finite difference. This finite difference technique involves using equation 13.52 to calculate the value of $\bar{V}$ at the points (T, P) and $(T, P + \Delta P)$ where ΔP is a small but finite increment in P. The partial derivative of V with respect to P at constant temperature can be estimated from:

$$\left(\frac{\partial \bar{V}}{\partial P}\right)_T \simeq \left(\frac{\bar{V}_{(T, P)} - \bar{V}_{(T, P+\Delta P)}}{\Delta P}\right)_T \qquad (13.53)$$

Similarly:

$$\left(\frac{\partial \bar{V}}{\partial T}\right)_P \simeq \left(\frac{\bar{V}_{(T, P)} - \bar{V}_{(T+\Delta T, P)}}{\Delta T}\right)_P \qquad (13.54)$$

The increments ΔT and ΔP must be carefully selected to be small, yet larger than uncertainties in the regression equations. The selection procedure is difficult especially if the volume varies strongly with temperature and pressure.

Table 13.13   **Fitting coefficients for the molar volume of water (from Helgeson and Kirkham, 1974a)**

$$a_{ij} = \hat{a}_{ij} \times 10^{a_{ij}^*}$$

Region 1: N = 9, r = 1 $\hat{a}_{ij}$

i	0	1	2	3	4	5	6	7	8	9
0	-3.8106590	1.0564665	-7.6454683	7.0956600	1.0454935	-4.4887845	9.0586408	-9.8313225	5.5147402	-1.2547995
1	6.6431473	-4.2654409	8.2193439	-4.6017503	1.5484386	-3.0809442	3.4462611	-1.9805288	4.5366798	
2	-8.9413365	8.9222030	1.9719748	-4.8796036	1.9017140	-2.7155343	1.5518783	-2.9266747		
3	9.2076993	4.3805877	3.1779963	-1.4111684	1.9087901	-3.3016835	-2.0239777			
4	-4.7887922	-6.1891487	1.9043321	-1.1153384	-1.2650068	6.0544695				
5	1.3798565	-1.3233046	-1.4942777	2.5569347	-3.4096098					
6	-1.9886789	2.3912344	-7.5721292	-6.0823080						
7	1.5064451	-1.2791592	5.7844948							
8	-6.0211303	1.7501650								
9	1.0682110									

a_{ij}^*

i	0	1	2	3	4	5	6	7	8	9
0	1	0	-5	-10	-11	-15	-19	-23	-27	
1	-1	-4	-7	-10	-13	-17	-21	-25	-30	
2	-3	-7	-10	-13	-16	-20	-24	-29		
3	-5	-9	-13	-15	-19	-24	-28			
4	-7	-12	-14	-18	-22	-27				
5	-9	-13	-17	-21	-26					
6	-12	-16	-21	-25						
7	-15	-19	-24							
8	-19	-23	-22							

$$a_{ij} = a_{ij} \times 10^{a_{ij}^*}$$

Region 2: N = 8, r = -1 $\hat{a}_{ij}$

i	0	1	2	3	4	5	6	7	8
0	6.2333394	2.7062946	-1.0740139	2.5322920	-4.8726693	6.4561026	-5.4093771	2.5488337	-5.10200454
1	-1.8568559	-1.0956591	2.0430020	1.9631668	-5.9663828	6.7298320	-3.6953264	7.9719676	
2	1.0560104	3.9724858	-1.0855790	5.1510625	-3.3120122	1.4949128	-2.3568834		
3	-3.0872023	-5.4077509	1.17693190	-2.2749747	1.4365951	-1.6687313			
4	4.7086592	2.8059817	-1.9688673	1.3446174	2.8271453				
5	-3.5684092	2.7903289	1.173349	-6.4912023					
6	4.5184760	-4.2896650	-2.7139198						
7	1.0666197	1.5083195							
8	-4.8594179								

a_{ij}^*

i	0	1	2	3	4	5	6	7	8
0	3	8	12	15	18	21	24	27	29
1	2	6	9	12	15	18	21	23	
2	0	3	7	9	12	15	17		
3	-3	0	4	6	8	11			
4	-6	-3	1	3	5				
5	-9	-6	-2	-1					
6	-13	-9	-6						
7	-15	-12							
8	-19								

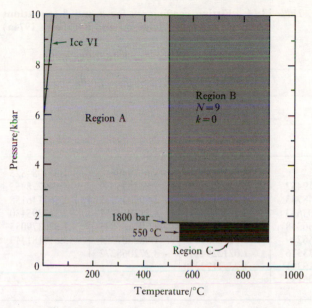

Figure 13.24 Alternative regions of temperature and pressure for α and κ fitting equations (after Helgeson and Kirkham, 1974a).

Once the partial derivatives are estimated at the point (T, P), the values of α and κ can be calculated from equations 13.30 and 13.31.

Using this procedure it is possible to map out α and κ as a function of $P-T$ space. The estimated values can then be regressed against polynomial equations in temperature and pressure to give simple working equations. As in the case of volume, no single expression can predict α and κ over the whole of $P-T$ space and it is convenient to subdivide the system into different $P-T$ regions, as in figure 13.24.

The form of the equations which fit the estimated data are as follows.

Region A

$$\kappa^{-1} = \sum_{j=0}^{5} \sum_{i=0}^{8-j} jB_{ij}T^i\rho^j \tag{13.55}$$

and

$$\alpha = \kappa \sum_{j=0}^{5} \sum_{i=0}^{8-j} iB_{ij}T^{i-1}\rho^j \tag{13.56}$$

where ρ is density calculated from a suitable expression for molar volume and B_{ij} is an array of coefficients given in table 13.14.

Regions B and C The fitted polynomial expressions for α and κ are:

$$-\kappa\bar{V} = \sum_{i=0}^{N} \sum_{j=0}^{N+k-i} jD_{ij}T^iP^{j-1} \tag{13.57}$$

and

$$\alpha\bar{V} = \sum_{i=0}^{N} \sum_{j=0}^{N+k-i} iD_{ij}T^{i-1}P^j \tag{13.58}$$

where $\bar{V}$, the molar volume, is calculated from a suitable expression such as equation 13.52 and D_{ij} is an array of coefficients given in table 13.15. The difference between

Table 13.14   Fitting coefficients for α and β (equations 13.55 and 13.56) for region A of figure 13.24 (from Helgeson and Kirkham, 1974a)

$$B_{ij} = \hat{B}_{ij} \times 10^{B_{ij}^*}$$

$\hat{B}_{ij}$

i	j					
	0	1	2	3	4	5
0	0	5.541654961	−5.64794994	1.391084525	9.314051406	−4.224085781
1	2.001619452	−6.480330829	7.804635059	−4.149778955	8.284235229	−3.17327205
2	−8.445248783	2.599081363	−2.997059355	1.567146158	−3.437390375	1.864672542
3	1.849203613	−5.126544377	5.215914752	−2.327295430	4.123082012	−1.650950925
4	−2.344156348	5.544129599	−4.597168845	1.520379055	−1.562138539	
5	1.782166279	−3.321663469	1.982307354	−3.619181443		
6	−8.063473576	1.021229414	−3.272068272			
7	2.064415341	−1.2371172220				
8	−2.498805152					

B_{ij}^*

i	j					
	0	1	2	3	4	5
0	0	6	6	6	5	5
1	4	4	4	4	3	1
2	1	2	2	2	1	0
3	−1	−1	−1	−1	−2	−3
4	−4	−4	−4	−4	−5	
5	−7	−7	−7	−8		
6	−11	−10	−11			
7	−14	−14				
8	−18					

regions B and C is in the choice of N and k. For region B, $N = 9$ and $k = 0$. For region C, $N = 5$ and $k = 1$.

13.4.3   The standard molar third-law entropy

The entropy value at 1000 bar pressure and any specified temperature ($\overline{S}_{T, 1000\,bar}$) can be used as a **reference** value to calculate entropies at the **same** temperature and any other pressure. This requires using the relationship:

$$\overline{S}_{(T, P)} = \overline{S}_{T, 1000\,bar} - \int_{1000\,bar}^{P} \overline{V}\alpha \, dP \qquad (13.59)$$

which is taken from equation 9.86 with the constraint of constant temperature. Values of $\overline{S}_{T, 1000\,bar}$ may be calculated from equation 13.42 and the integral evaluated graphically or numerically. Helgeson and Kirkham (1974a) used a numerical

Table 13.15 **Coefficients for α and β equations 13.57 and 13.58 in regions B and C of figure 13.24 ($N = 9$ and $k = 0$ for region B, and $N = 5$ and $k = 1$ for region C) (from Helgeson and Kirkham, 1974a)**

$$D_{ij} = \hat{D}_{ij} \times 10_{ij}^{D*}$$

$\hat{D}_{ij}$ for region B

i	j									
	0	1	2	3	4	5	6	7	8	9
0	0	−1.506919791	1.507973492	−7.507671325	2.471604532	−5.205698231	7.023757340	−5.865613363	2.757432199	−5.572807869
1	−1.787558824	2.810937616	7.031625811	−5.954625622	2.223890468	−4.458179301	4.957412121	−2.879352320	6.815947672	
2	4.503625642	−7.193725721	5.156729178	−1.997843646	4.203016232	−4.702839663	2.617250953	−5.673263000		
3	1.986537032	−1.133016027	5.705056876	−1.410074499	1.247849426	−2.017582894	−1.207488032			
4	−5.376132969	−1.172316191	3.305071187	3.521599490	−9.298387117	3.676616808				
5	1.484589030	−3.488262960	−6.138922413	7.391640366	−1.171172984					
6	−1.126923624	7.599081341	−1.272723651	−1.546104985						
7	−8.992177755	−4.338459618	1.368240982							
8	1.459740213	6.914686939								
9	−4.903210863									

D_{ij}^{*} for region B

i	j									
	0	1	2	3	4	5	6	7	8	9
0	0	−3	−6	−10	−13	−17	−21	−25	−29	−34
1	−4	−8	−10	−13	−16	−20	−24	−28	−33	
2	−6	−9	−12	−15	−19	−23	−27	−32		
3	−8	−11	−15	−18	−22	−27	−31			
4	−11	−15	−17	−22	−26	−30				
5	−13	−17	−21	−25	−29					
6	−16	−20	−24	−28						
7	−20	−23	−27							
8	−22	−27								
9	−26									

$\hat{D}_{ij}$ for region C

i	j						
	0	1	2	3	4	5	6
0	0	4.668287584	−7.085868594	2.111356062	7.954938710	−4.642576017	6.17798632
1	−7.041518081	2.955659758	2.779240901	−1.768464908	3.383614741	−2.22308253	
2	2.151504359	−1.525948185	4.132847534	9.211139144	−1.082504726		
3	−2.773357282	2.067817826	−2.526689074	−1.019479759			
4	1.650848239	−1.137352755	1.051203720				
5	−3.731274546	2.162263875					

D_{ij}^{*} for region C

i	j						
	0	1	2	3	4	5	6
0	0	−3	−5	−8	−12	−15	−19
1	−1	−4	−7	−10	−14	−18	
2	−3	−6	−11	−14	−17		
3	−6	−9	−13	−17			
4	−9	−12	−16				
5	−13	−16					

technique to evaluate the integral in region A on figure 13.24 using data for $\bar{V}$ and α calculated by the methods described above.

For regions B and C, equation 13.59 was combined with equation 13.58 and integrated with respect to pressure to yield:

$$\bar{S}_{(T, P)} = \bar{S}_{T, 1000\,\text{bar}} - \sum_{i=0}^{5} \sum_{j=0}^{6-i} i\dot{D}_{ij} T^{i-1} \left(\frac{1800^{j+1} - 1000^{j+1}}{j+1} \right)$$

$$- \sum_{i=0}^{9} \sum_{j=0}^{9-i} iD_{ij} T^{i-1} \left(\frac{P^{j+1} - 1800^{j+1}}{j+1} \right) \tag{13.60}$$

The constants $\dot{D}_{ij}$ and D_{ij} are the coefficients given in table 13.15 for regions B and C respectively. Equation 13.60 is valid for region B for temperatures above 550 °C. For temperatures less than or equal to 550 °C in region B, the equation is modified by setting all values of $\dot{D}_{ij}$ to zero and replacing $\bar{S}_{T, 1000\,\text{bar}}$ with $\bar{S}_{T, 1800\,\text{bar}}$ from region A. For predictions in the temperature and pressure region C, the coefficients D_{ij} are set to zero and the number 1800 becomes a variable equal to pressure P in bars.

Entropies calculated by this method are included in table 13.5.

13.4.4 The apparent energies of formation

Apparent Gibbs energies of formation can be evaluated at elevated temperatures and pressures by using values of ΔG at 1000 bar pressure and any temperature as reference values. The fundamental relationship is:

$$\Delta G_{(T, P)} = \Delta G_{T, 1000\,\text{bar}} + \int_{1000\,\text{bar}}^{P} \bar{V}\,dP \tag{13.61}$$

which can be evaluated using molar volume data calculated from equation 13.52.

A general equation valid for regions 1 and 2 in figure 13.23 can be obtained by substituting equation 13.52 into equation 13.61 and integrating with respect to pressure. Thus:

$$\Delta G_{(T, P)} = \Delta G_{T, 1000\,\text{bar}} + \sum_{i=0}^{8} \left[\dot{a}_{i0} t^i \ln(P^*/1000) - \sum_{j=1}^{8-i} \dot{a}_{ij} t^i \frac{(P^{*-j} - 1000^{-j})}{j} \right]$$

$$+ \sum_{i=0}^{9} \left[a_{i0} t^i (P/P^*) + \sum_{j=1}^{9-i} a_{ij} t^i \frac{(P^j - P^{*j})}{j} \right] \tag{13.62}$$

where $\dot{a}_{ij}$ and a_{ij} are the coefficients given in table 13.13 for regions 1 and 2 respectively.

To calculate ΔG in region 1, all the coefficients $\dot{a}_{ij}$ and a_{ij} are used and P^* is set equal to 1000 bar for temperatures less than or equal to 410 °C, but is a variable given by $P^* = 1300 + 122\,499(t - 410)$ for temperatures between 410 and 800 °C.

To calculate ΔG in region 2, the coefficients a_{ij} should be set to zero and P^* replaced by pressure P in bars.

Once values of ΔG have been evaluated, values of ΔA can be calculated simply from:

$$\Delta A_{T, P} = \Delta A_{T, 1000\,\text{bar}} + (\Delta G_{T, P} - \Delta G_{T, 1000\,\text{bar}}) - P\bar{V}_{P, T} + 1000\bar{V}_{T, 1000\,\text{bar}} \tag{13.63}$$

where $\bar{V}$ and $\bar{V}_{T, 1000\,bar}$ are the values of molar volume at the specified temperature and pressure and the specified temperature and 1000 bar respectively, which can be calculated from equation 13.52.

The apparent enthalpy and internal energy of formation can be calculated using equations 13.47 and 13.48, which are the same equations for pressures below 1000 bar.

13.4.5 Third-law heat capacity

At pressures above 1000 bar the value of $\bar{C}_P$, at any specified temperature, can be calculated using $\bar{C}_P$ at 1000 bar as reference value from:

$$\bar{C}_{P,\,T} = \bar{C}_{1000\,bar,\,T} - T\left[\int_{1000\,bar}^{P} \left(\frac{\partial^2 \bar{V}}{\partial T^2}\right)_P dP\right]_T \qquad (13.64)$$

Partial differentiation of equation 13.64 gives:

$$\bar{C}_{P,\,T} = \bar{C}_{1000\,bar,\,T} - T\left\{\int_{1000\,bar}^{P} \bar{V}\left[\alpha^2 + \left(\frac{\partial \alpha}{\partial T}\right)_P\right] dP\right\}_T \qquad (13.65)$$

For data in region A of figure 13.24, the integral on the right-hand side of equation 13.65 can be evaluated numerically with the aid of equation 13.52 and the temperature derivative of equation 13.56.

A general expression for regions B and C can be generated since the function $(\partial^2 \bar{V}/\partial T^2)_P$ can be determined from the second differential of equation 13.58. The resulting expressions for $(\partial^2 \bar{V}/\partial T^2)_P$ can be substituted into equation 13.64 to give:

$$\bar{C}_{P,\,T} = \bar{C}_{1000\,bar,\,T} - T\left[\sum_{i=0}^{5} \sum_{j=0}^{6-i} i(i-1)\dot{D}_{ij} T^{i-2}\left(\frac{1800^{j+1} - 1000^{j+1}}{j+1}\right)\right.$$
$$\left. + \sum_{i=0}^{9} \sum_{j=0}^{9-i} i(i-1)D_{ij} T^{i-2}\left(\frac{P^{j+1} - 1800^{j+1}}{j+1}\right)\right] \qquad (13.66)$$

The values of $\dot{D}_{ij}$ and D_{ij} are the coefficients given in table 13.15 for regions B and C in figure 13.24.

To calculate $\bar{C}_P$ in region B, all the coefficients should be used for temperatures below 550 °C. At temperatures above 550 °C the coefficients $\dot{D}_{ij}$ should be set to zero and the value of $\bar{C}_{1000\,bar,\,T}$ replaced by the value of $\bar{C}_{1800\,bar,\,T}$.

To calculate $\bar{C}_P$ in region C, the values of D_{ij} should be set to zero and the number 1800 should be replaced by pressure P in bars.

In view of the sensitive dependence of $\bar{C}_V$ and $\bar{C}_P$ on T and P, the errors in the coefficients $\dot{D}_{ij}$ and D_{ij} prevent calculation of $\bar{C}_V$ from $\bar{C}_P$ at these elevated pressures and temperatures.

13.5 Fugacity and the fugacity coefficient

The fugacity of water, f, is related to the pressure by:

$$f = \chi P \qquad (13.67)$$

Table 13.16   **The fugacity of water (from Helgeson and Kirkham, 1974a)**

$T/°C$	Sat.	0.001	0.5	1	2	3	4	5	6	7	8	9
										Pressure/kbar		
25	0.9995	0.0331	0.0001	0.0001	0.000	0.000	0.000	0.000	0.000	0.000	0.001	0.001
50	0.9968	0.1270	0.0003	0.0002	0.000	0.000	0.000	0.001	0.001	0.001	0.002	0.003
75	0.9925	0.3912	0.0010	0.0007	0.001	0.001	0.001	0.001	0.002	0.003	0.005	0.007
100	0.9838	0.9851	0.0027	0.0018	0.002	0.002	0.002	0.003	0.004	0.006	0.009	0.014
125	0.9732	0.9888	0.0060	0.0040	0.003	0.004	0.005	0.006	0.009	0.012	0.017	0.024
150	0.9587	0.9915	0.0120	0.0078	0.007	0.007	0.009	0.011	0.015	0.021	0.029	0.040
175	0.9397	0.9934	0.0218	0.0141	0.012	0.013	0.015	0.019	0.025	0.033	0.045	0.061
200	0.9165	0.9950	0.0366	0.0236	0.019	0.020	0.024	0.030	0.038	0.050	0.066	0.089
225	0.8889	0.9969	0.0577	0.0369	0.029	0.031	0.037	0.044	0.055	0.071	0.093	0.122
250	0.8572	0.9966	0.0859	0.0546	0.043	0.045	0.051	0.062	0.077	0.097	0.125	0.162
275	0.8221	0.9971	0.1219	0.0773	0.061	0.062	0.070	0.083	0.102	0.128	0.163	0.208
300	0.7840	0.9975	0.1659	0.1050	0.082	0.083	0.092	0.109	0.132	0.173	0.205	0.259
325	0.7431	0.9978	0.2176	0.1377	0.106	0.107	0.118	0.138	0.166	0.203	0.252	0.315
350	0.6990	0.9982	0.2763	0.1752	0.135	0.135	0.148	0.171	0.203	0.246	0.303	0.375
375	—	0.9984	0.3406	0.2169	0.167	0.165	0.180	0.207	0.244	0.293	0.357	0.439
400	—	0.9987	0.4083	0.2622	0.202	0.199	0.216	0.245	0.287	0.343	0.414	0.504
425	—	0.9989	0.4766	0.3100	0.239	0.253	0.253	0.286	0.333	0.394	0.473	0.571
450	—	0.9991	0.5416	0.3594	0.278	0.273	0.293	0.329	0.380	0.477	0.533	0.640
475	—	0.9992	0.5990	0.4094	0.319	0.312	0.334	0.373	0.429	0.501	0.593	0.708
500	—	0.9993	0.6481	0.4589	0.361	0.353	0.376	0.418	0.478	0.556	0.654	0.776
525	—	0.9994	0.6899	0.5070	0.403	0.394	0.418	0.464	0.528	0.611	0.715	0.843
550	—	0.9995	0.7259	0.5528	0.445	0.435	0.461	0.510	0.577	0.665	0.775	0.909
575	—	1.0000	0.7575	0.5960	0.486	0.476	0.504	0.555	0.627	0.719	0.833	0.973
600	—	1.0000	0.7848	0.6358	0.527	0.517	0.546	0.600	0.675	0.771	0.890	1.034
625	—	1.0000	0.8087	0.6724	0.566	0.557	0.587	0.644	0.722	0.821	0.945	1.093
650	—	1.0000	0.8297	0.7057	0.603	0.595	0.628	0.686	0.767	0.870	0.997	1.150
675	—	1.0000	0.8482	0.7359	0.639	0.632	0.667	0.727	0.818	0.917	1.047	1.203
700	—	1.0000	0.8646	0.7633	0.673	0.668	0.704	0.767	0.753	0.962	1.095	1.254
725	—	1.0000	0.8791	0.7880	0.705	0.703	0.740	0.805	0.894	1.005	1.142	1.302
750	—	1.0000	0.8920	0.8103	0.735	0.735	0.775	0.842	0.932	1.046	1.184	1.347
775	—	1.0000	0.9034	0.8305	0.763	0.766	0.808	0.877	0.969	1.085	1.225	1.390
800	—	1.0000	0.9137	0.8487	0.789	0.795	0.839	0.910	1.004	1.122	1.263	1.429
825	—	1.0000	0.9228	0.8651	—	—	—	—	—	—	—	—
850	—	1.0000	0.9311	0.8799	—	—	—	—	—	—	—	—
875	—	1.0000	0.9385	0.8934	—	—	—	—	—	—	—	—
900	—	1.0000	0.9452	0.9056	—	—	—	—	—	—	—	—

where χ is the fugacity coefficient, which is a measure of the deviation from ideal gas behaviour.

Table 13.16 and figures 13.25 and 13.26 show the effect of temperature and pressure on fugacity. It can be seen that water exhibits a marked negative deviation from ideal behaviour at low temperatures and pressures. At higher temperatures and pressures water shows a positive deviation from ideality which becomes more positive with increasing temperature and pressure.

In order to evaluate the fugacity coefficient it is helpful to recall equation 10.24 which defines the activity of water as:

$$a = \frac{f}{f^\ominus} \tag{13.68}$$

where $f^\ominus$ is an arbitrarily chosen standard state.

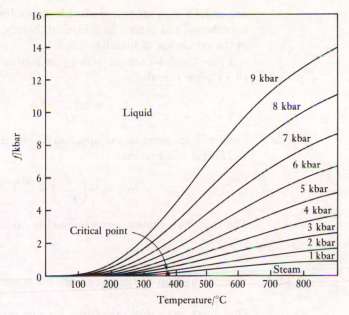

Figure 13.25   Fugacity of water versus temperature (after Helgeson and Kirkham, 1974a).

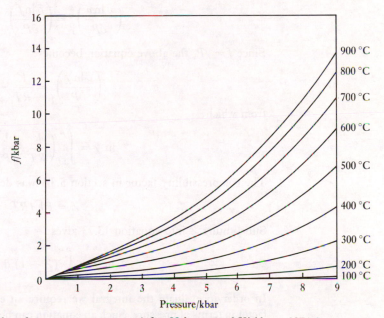

Figure 13.26   Fugacity of water versus pressure (after Helgeson and Kirkham, 1974a).

A convenient standard state for water is that of the **hypothetical ideal gas** at 1 bar pressure and any temperature. In this standard state the gas has unit fugacity ($f^\ominus = 1$).

This hypothetical standard state can be assigned a Gibbs energy and an apparent Gibbs energy of formation, denoted by $\bar{G}^\ominus_{(T, 1\,\text{bar})}$ and $\Delta G^\ominus_{(T, 1\,\text{bar})}$ respectively. The standard state chosen ensures that the values of $\bar{G}^\ominus_{(T, 1\,\text{bar})}$ and $\Delta G^\ominus_{(T, 1\,\text{bar})}$ are

dependent on temperature but independent of pressure. The actual values are hypothetical and cannot be measured directly. They serve only as reference points in the calculation of fugacity.

From the definition of activity in section 10.1.2 and equation 10.62 (section 10.3.5), it is true that:

$$\ln a = \ln\left(\frac{f}{f^{\ominus}}\right) = \frac{\bar{G}_{(T, P)} - \bar{G}^{\ominus}_{(T, \text{1 bar})}}{RT} \tag{13.69}$$

From the definition of the apparent Gibbs energy of formation (ΔG in equation 13.22) it is also true that:

$$\ln a = \ln\left(\frac{f}{f^{\ominus}}\right) = \frac{\Delta G_{(T, P)} - \Delta G^{\ominus}_{(T, \text{1 bar})}}{RT} \tag{13.70}$$

The above two equations can be used to calculate fugacity by one of the following methods.

13.5.1   The calculation of fugacity from an equation of state

One direct method of calculating fugacity involves using equations 13.68 or 13.69 with equation 10.18 which defines fugacity. From them we can write:

$$\left(\frac{\partial \ln a}{\partial P}\right)_T = \left(\frac{\partial \ln f}{\partial P}\right)_T = \frac{\bar{V}}{RT} \tag{13.71}$$

Since $f = \chi P$, the above equation becomes:

$$\left(\frac{\partial \ln \chi}{\partial P}\right)_T = \frac{\bar{V}}{RT} - \frac{1}{P} \tag{13.72}$$

from which:

$$\ln \chi = \int_0^P \left(\frac{\bar{V}}{RT} - \frac{1}{P}\right) dP \tag{13.73}$$

The compressibility factor in section 5.3.1 was defined as

$$Z = P\bar{V}/RT \tag{13.74}$$

Substituting Z into equation 13.73 gives:

$$\ln \chi = \int_0^P (Z - 1)\, d\ln P \tag{13.75}$$

In order to evaluate the integral we require an expression for the compressibility factor in terms of pressure. Such an equation can be obtained by rearranging equation 13.36 derived from the equation of state by Keenan *et al.* (1969), i.e.

$$Z - 1 = \rho\xi + \rho^2\left(\frac{\partial \xi}{\partial \rho}\right)_T \tag{13.76}$$

Substituting the above equation into equation 13.75 gives:

$$\ln \chi = \int_0^P \left[\rho\xi + \rho^2\left(\frac{\partial \xi}{\partial \rho}\right)_T\right] d\ln P \tag{13.77}$$

The integrated form of this equation gives χ at any temperature and pressure at which the equation of state is valid. (In this case at any temperature below 1300 °C and pressures below 1000 bar.)

The lower limit of this integral is indeterminate; however, the integral can be evaluated numerically or graphically by adopting a small but finite value of P at the lower limit.

13.5.2   The calculation of fugacity at elevated pressures

At pressures above 1000 bar the equation of state is not valid. However, it is possible to calculate χ and f at any temperature and pressure using equation 13.70. The requirement is a value of $\Delta G^{\ominus}_{(T, P)}$, at the temperature and pressure of interest, plus the value of $\Delta G^{\ominus}_{(T, 1\,\text{bar})}$ at the temperature of interest. Values of $\Delta G^{\ominus}_{(T, P)}$ can be taken from steam tables or by equations given previously. Values of $\Delta G^{\ominus}_{(T, 1\,\text{bar})}$, despite being hypothetical, can be calculated from a previously determined value of $\chi_{(T, 1\,\text{bar})}$ using the relationship:

$$\Delta G^{\ominus}_{(T, 1\,\text{bar})} = \Delta G_{(T, 1\,\text{bar})} - RT \ln \chi_{(T, 1\,\text{bar})} \qquad (13.78)$$

Here, $\Delta G_{(T, 1\,\text{bar})}$ and $\chi_{(T, 1\,\text{bar})}$ are the apparent Gibbs energy of formation and the fugacity coefficient at 1 bar pressure and the temperature of interest.

Equation 13.77 can be used to calculate $\chi_{(T, 1\,\text{bar})}$, from which $\Delta G^{\ominus}_{(T, 1\,\text{bar})}$ can be calculated using equation 13.78.

The combination of equations 13.70 and 13.78 can be used to calculate χ at any temperature and pressure, including low pressures where the equation of state is valid. Indeed this is the simplest way of calculating fugacity once the values of $\chi_{T, 1\,\text{bar}}$ are known.

Table 13.17   **Coefficients for the relative permittivity of water (equation 13.79) (from Helgeson and Kirkham, 1974a)**

$$e_{ij} = \hat{e}_{ij} \times 10^{e^{*}_{ij}}$$

			$\hat{e}_{ij}$		
			i		
j					
	0	*1*	*2*	*3*	*4*
0	4.39109592	−2.33277456	4.61662109	−4.03643333	1.31604037
1	−2.18995148	1.00498361	−1.35650709	5.94046919	
2	1.82898246	−2.08896146	1.60491325		
3	1.54886800	−6.13941874			
4	−6.13542375				

			e^{*}_{ij}		
			i		
j					
	0	*1*	*2*	*3*	*4*
0	2	0	−3	−6	−9
1	2	0	−3	−7	
2	1	−1	−4		
3	2	−2			
4	1				

Table 13.18   **The relative permittivity of water**

$T/°C$	Sat.	0.5	1	1.5	2	2.5	3	3.5	4	4.5	5
						Pressure/kbar					
25	78.47	80.20	81.78	83.05	84.38	85.56	86.63	87.62	88.57	89.48	90.35
50	69.96	71.59	73.09	74.23	75.51	76.64	77.65	78.58	79.48	80.34	81.18
75	62.30	63.91	65.38	66.56	67.78	68.87	69.84	70.72	71.57	72.38	73.17
100	55.47	57.10	58.55	59.78	60.98	62.06	63.00	63.85	64.65	65.43	66.17
125	49.37	51.05	52.53	53.76	54.97	56.04	56.98	57.82	58.60	59.34	60.05
150	43.91	45.68	47.19	48.42	49.64	50.73	51.68	52.52	53.28	54.00	54.68
175	39.02	40.90	42.46	43.69	44.92	46.03	47.00	47.85	48.61	49.31	49.96
200	32.60	36.63	38.27	39.51	40.75	41.88	42.87	43.73	44.48	45.17	45.81
225	30.58	32.79	34.54	35.82	37.06	38.20	39.21	40.07	40.83	41.51	42.13
250	26.87	29.31	31.20	32.55	33.79	34.94	35.95	36.83	37.58	38.25	38.86
275	23.38	26.12	28.20	29.63	30.88	32.03	33.04	33.92	34.67	35.33	35.92
300	19.99	23.15	25.46	27.00	28.27	29.41	30.43	31.30	32.05	32.70	33.27
325	16.58	20.32	22.94	24.61	25.90	27.05	28.05	28.92	29.66	30.29	30.85
350	12.87	17.57	20.58	22.40	23.74	24.88	25.88	26.74	27.47	28.09	28.63
375	—	14.86	18.37	20.35	21.74	22.89	23.89	24.74	25.45	26.06	26.58
400	—	12.13	16.27	18.42	19.88	21.06	22.05	22.88	23.59	24.18	24.68
425	—	9.38	14.30	16.63	18.16	19.35	20.34	21.17	21.86	22.44	22.92
450	—	6.80	12.46	14.95	16.56	17.78	18.77	19.59	20.27	20.84	21.31
475	—	4.98	10.78	13.41	15.08	16.33	17.34	18.16	18.83	19.39	19.85
500	—	3.94	9.27	12.01	13.75	15.03	16.05	16.88	17.56	18.11	18.56
525	—	3.30	7.97	10.78	12.57	13.90	14.94	15.78	16.46	17.02	17.48
550	—	2.87	6.89	9.74	11.58	12.95	14.03	14.90	15.59	16.16	16.63
575	—	2.56	6.08	8.91	10.81	12.24	13.36	14.26	14.99	15.58	16.06
600	—	2.39	5.53	8.33	10.30	11.80	12.98	13.93	14.70	15.32	15.83

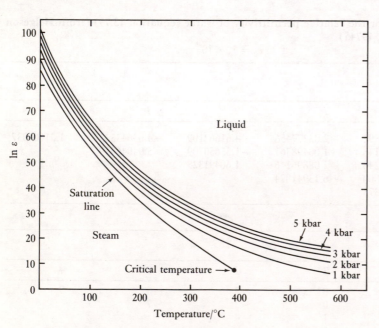

Figure 13.27   Relative permittivity of water versus temperature (after Helgeson and Kirkham, 1974a).

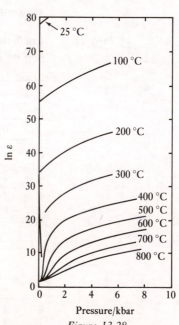

Figure 13.28
Relative permittivity of water versus pressure (after Helgeson and Kirkham, 1974a).

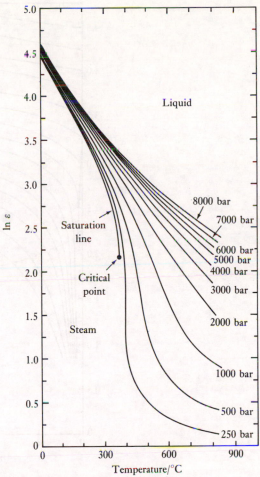

Figure 13.29   Natural log (relative permittivity) of water versus temperature (after Helgeson and Kirkham, 1974a).

13.6   Relative permittivity of pure water

The relative permittivity $\varepsilon_{r(s)}$ is a property of liquid water that influences the ion–water and ion–ion interactions in aqueous electrolytes. In particular, the tendency for ions to associate increases as the relative permittivity of water decreases. It is therefore important to characterise the changes in $\varepsilon_{r(s)}$ as a function of temperature and pressure. Detailed measurements of the relative permittivity of water have been made at pressures below 5000 bar and over the temperature range 0–600 °C. At pressures above 5000 bar and temperatures above 600 °C the data are less comprehensive. The available data within this T–P range was consolidated by Helgeson and Kirkham (1974a), who proposed a simple fourth-degree polynomial equation for the dependence of $\varepsilon_{r(s)}$ on temperature and density:

$$\varepsilon_{r(s)} = \sum_{i=0}^{4} \sum_{j=0}^{4-i} e_{ij} T^i \rho^j \qquad (13.79)$$

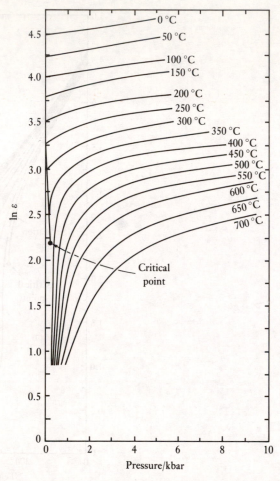

Figure 13.30 Natural log (relative permittivity) of water versus pressure (after Helgeson and Kirkham, 1974a).

where T is temperature (in K), ρ is density (in g cm^{-3}) and e_{ij} is an array of coefficients given in table 13.17. The density must be calculated from a suitable equation of state for volume, or taken from data such as those in table 13.2.

Table 13.18 and figures 13.27 and 13.28 show the relative permittivity of water as a function of temperature and pressure. Pressure has some effect on the values of ε. However, the effect of temperature is dramatic, causing a decrease of almost an order of magnitude as temperature is increased from 0 to 600 °C.

Figure 13.29 shows a near-linear dependence of ln $\varepsilon_{r(s)}$ on temperature, for all temperatures less than ~ 250 °C, at most pressures. The greatest degree of non-linearity with temperature occurs at high temperature and low pressures. Figure 13.30 shows a near-linear dependence of ln $\varepsilon_{r(s)}$ on pressure at all temperatures below 100 °C. At elevated temperatures linearity with pressure is only approached at high pressure. In all cases the greatest variability with temperature and pressure is at pressures below 1 bar and under conditions approaching the critical point.

Figure 13.31 shows the relationship between the relative permittivity of water

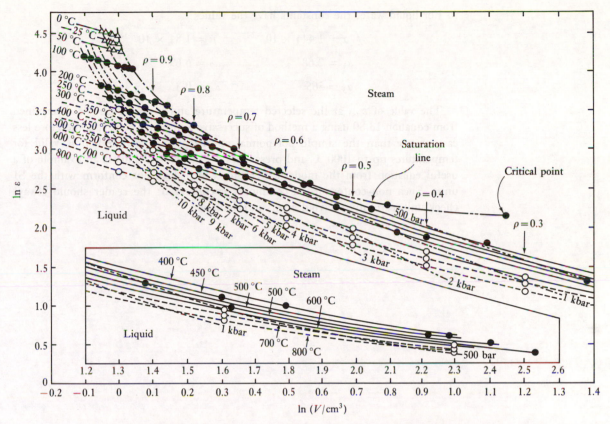

Figure 13.31   Relative permittivity versus density of water (data from ● Heger, 1969; ○ Quist and Marshall, 1965; △ Owen *et al.*, 1961). (From Helgeson and Kirkham, 1974a.)

and its molar volume. As molar volume increases and density decreases, the relative permittivity of water also decreases. We may therefore expect increasing ion association to occur as the density of water decreases, which occurs at high temperatures and low pressures.

An alternative approach to predicting the relative permittivity of any liquid containing associating dipoles was given by Kirkwood (1939). His expression has the form:

$$\frac{(\varepsilon_{r(s)} - 1)(2\varepsilon_{r(s)} + 1)}{9\varepsilon_{r(s)}} = \frac{4\pi L \rho^{\bullet}}{3W_{(s)}}\left(\alpha + \frac{\mu^2 g}{3kT}\right) \tag{13.80}$$

where L is Avogadro's constant, ρ g cm^{-3} the density of the pure solvent, g the Kirkwood correlation factor, α/cm g s the molecular polarisability, μ/cm g is the dipole moment of the solvent, W_s/g mol^{-1} the formula weight of the solvent and k is Boltzmann's constant, which, to be consistent, is 1.3086×10^{-16} erg K^{-1}.

The coefficient g varies with absolute temperature in accordance with a formulation given by Pitzer (1983), i.e.

$$g = 1 + a_1\rho^{\bullet} + a_2\rho^{\bullet 5}[(a_3/T)^{a_4} - 1] \tag{13.81}$$

For liquid water the constants have the values:

$$\alpha = 1.444 \times 10^{-24} \qquad \mu = 1.84 \times 10^{-18}$$

$$a_1 = 2.68 \qquad a_2 = 6.69$$

$$a_3 = 565 \qquad a_4 = 0.3$$

The value of $\varepsilon_{r(s)}$ at the selected temperature and pressure can be determined from equation 13.80 using a method of successive approximation. It is therefore less easy to use than the simple polynomial approach, but the method is valid for temperatures up to 1000 °C and pressures up to 5000 bar. This is an example of a useful equation from the modern literature which does not conform with the SI units. Such non-conformity is commonly encountered: the reader should always check units carefully to avoid errors in calculation.

Chapter 14

The thermodynamic properties of aqueous electrolytes

The reactivities of dissolved ions and salts play a central role in aqueous sedimentary and hydrothermal processes. In order to understand these processes we must measure or predict the thermodynamic properties of ions as well as those for liquid water (section 2.6). This chapter is therefore designed to complement chapter 13 on the properties of water.

The apparent Gibbs energies of formation for dissolved species are of special importance because they are used to calculate equilibrium constants for reactions at specified temperatures and pressures. However, a full characterisation of the properties of electrolytes involves calculating all apparent energies of formation. In this chapter, the effect of temperature and pressure on these quantities will be explored. In some cases pertinent features of working equations will be discussed. Most equations quoted here are valid to pressures up to 5 kbar, which is somewhat less than the 10 kbar level normally associated with maximum conditions in the Earth's crust.

For species at elevated concentrations, the temperature, pressure and concentration dependence of activity coefficients will also be discussed.

Throughout this chapter the superscript $\circ$ will denote the infinitely dilute solution, $\ominus$ the standard state and $\bullet$ pure water.

14.1 Some basic considerations

For a species to be characterised thermodynamically it must be an independent variable in a system. In principle, this restriction prevents us from considering single ions as components, since to vary the concentration of an ion independently would require the input of enormous energy to overcome the interactions between electric charges. In contrast, neutral salts can be true independent variables and their thermodynamic properties are measurable. Despite the restriction, we do use ions as components. For example, the precipitation of gypsum from water is described by the ionic reaction:

$$Ca^{2+} + SO_4^{2-} + 2H_2O \rightleftharpoons CaSO_4 . 2H_2O(s) \qquad (14.1)$$

The equilibrium constant has the form:

$$K_{eq} = \frac{a_{CaSO_4 . 2H_2O(s)}}{a_{Ca^{2+}} \times a_{SO_4^{2-}} \times a_{H_2O}^2} \qquad (14.2)$$

An alternative reaction, in terms of the dissolved salt $CaSO_4$, is:

$$CaSO_4 + 2H_2O \rightleftharpoons CaSO_4 . 2H_2O(s) \qquad (14.3)$$

for which the equilibrium constant is:

$$K_{eq} = \frac{a_{CaSO_4.2H_2O(s)}}{a_{CaSO_4} \times a_{H_2O}^2} \qquad (14.4)$$

Both conventions are equivalent. The values of the equilibrium constants are identical in both cases, as are the standard Gibbs energy changes. The only choice we make is between using the activities of the neutral salt or the activities of the single ions. In electrolyte solutions containing a **single salt** both options are equally convenient. However, in multicomponent electrolyte solutions, where any one ionic species may be involved in several different reactions, it is easier to treat ions as components.

14.1.1 The relationship between the properties of ions and the properties of salts

Since the properties of ions are not directly measurable we must adopt some convention which allows ions to be used in the formalism while maintaining consistency with the measured properties of the neutral salts.

Basic relationships Consider 1 mole of a neutral salt composed of v_A moles of anion A^{z_A-} and v_B moles of cation B^{z_B+}, where z_A and z_B represent the valence charges on the ions A and B, respectively. If the salt is dissolved in 1 kg of water to a molality of m_{AB}, the concentrations of the ions (m_A and m_B) are given by

$$m_A = v_A m_{AB} \quad \text{and} \quad m_B = v_B m_{AB} \qquad (14.5)$$

For example, the molal concentration of calcium chloride (with the formula $CaCl_2$) is related to the molal concentrations of the ions by:

$$m_{CaCl_2} = [m_{Ca^{2+}}] + [2 \times m_{Cl^-}] \qquad (14.6)$$

Any thermodynamic property of a salt, Ξ_{AB}, is given by:

$$\Xi_{AB} = [v_A \times \Xi_A] + [v_B \times \Xi_B] \qquad (14.7)$$

where Ξ_A and Ξ_B are the complementary properties of the ions A and B respectively. The properties are any we choose to consider, such as partial molar volumes, partial molar heat capacities or apparent energies of formation. For the calcium chloride example, the partial molar volume of the neutral salt is $\bar{V}_{CaCl_2}$. This is related to the partial molar volumes of calcium $\bar{V}_{Ca^{2+}}$ and chloride $\bar{V}_{Cl^-}$ by:

$$\bar{V}_{CaCl_2} = [\bar{V}_{Ca^{2+}}] + [2 \times \bar{V}_{Cl^-}] \qquad (14.8)$$

Similarly, the apparent standard Gibbs energy of formation of the salt is related to the corresponding Gibbs energy of the ions by:

$$\Delta G^{\ominus}_{CaCl_2} = [\Delta G^{\ominus}_{Ca^{2+}}] + [2 \times \Delta G^{\ominus}_{Cl^-}] \qquad (14.9)$$

A generalised notation For multicomponent electrolytes we may define the kth salt in a system to contain v_{ik} moles of the ith ion in its accepted formula. A general expression for the relationship between the properties of the ions and the properties of the salt is then:

$$\Xi_k = \sum_j [v_{ik} \times \Xi_i] \qquad (14.10)$$

where Ξ_k is an extensive property of the kth salt and Ξ_i is the property of the ith ion. This formalism is more general than the one where cations and anions are specified, and is preferred by some authors. The convention will sometimes be used in this text.

In this convention a commonly encountered variable is the **sum of the number of moles of each ion in the specified salt**, i.e.

$$v_k = \sum_i v_{ik} \qquad (14.11)$$

A thermodynamic convention

If a property can be measured for a salt, it is possible to assign the values for the ions by convention. To do this we measure the property of the salt and define the property of one ion to have a fixed value. The value of the property for the second ion can then be obtained by subtraction. In principle, it is necessary to define the properties of only one ion, from which the properties of all other ions can be determined successively.

The usual convention is that:

THE PARTIAL MOLAR PROPERTIES AND APPARENT ENERGIES OF FORMATION FOR THE PROTON ARE ZERO UNDER ALL CONDITIONS.

This rule applies to **all** the extensive properties of the proton.

14.1.2 The standard states for dissolved species

The usual composition unit for ions dissolved in aqueous solutions is molality (mol kg^{-1}). **The standard state for an ion is unit activity in a hypothetical 1-molal solution at any temperature and pressure.**

Ideal behaviour is defined in accordance with Henry's law extrapolated from the infinitely dilute solution. Accordingly, for the ith ion,

$$a_i = \frac{m_i \gamma_i}{m_i^\ominus} \qquad (14.12)$$

where γ_i is the molal single-ion activity coefficient and $m_i^\ominus$ is the molal concentration in the standard state, which is of course 1 mol kg^{-1}.

The activity and the activity coeffficient are dimensionless and vary with temperature and pressure. The activity coefficient tends to unity as the solution tends to infinite dilution at any temperature and pressure. (See Appendix 3 for further comments on standard states.)

14.1.3 Single-ion and stoichiometric activity coefficients

Basic formalism

When m_{AB} moles of a neutral salt, composed of v_A moles of anion A^{z_A-} and v_B moles of B^{z_B+}, dissolve in 1 kg of water there are m_A moles of anion and m_B moles of cation released. The molal concentrations of ions and salts are related by $m_A = v_A m_{AB}$ and $m_B = v_B m_{AB}$.

The chemical potential of the salt μ_{AB} can be related to the chemical potentials for the single ions by:

$$m_{AB}\mu_{AB} = m_A\mu_A + m_B\mu_B \tag{14.13}$$

Eliminating single-ion molalities and dividing by m_{AB} gives:

$$\mu_{AB} = \nu_A\mu_A + \nu_B\mu_B \tag{14.14}$$

Substituting $\mu_i = \mu_i^\ominus + RT \ln a_i$ for both anion and cation into equation 14.14 gives an equation for the chemical potential of a neutral salt:

$$\mu_{AB} = (\nu_A\mu_A^\ominus + \nu_B\mu_B^\ominus) + \nu_A RT \ln(m_A\gamma_A) + \nu_B RT \ln(m_B\gamma_B) \tag{14.15}$$

Collecting standard chemical potential terms and simplifying yields:

$$\mu_{AB} = \mu_{AB}^\ominus + \nu_{AB}RT \ln[m_{AB}(\nu_A^{\nu_A}\nu_B^{\nu_B})^{1/\nu_{AB}} \times (\gamma_A^{\nu_A}\gamma_B^{\nu_B})^{1/\nu_{AB}}] \tag{14.16}$$

where

$$\nu_{AB} = \nu_A + \nu_B \tag{14.17}$$

and

$$\mu_{AB}^\ominus = \nu_A\mu_A^\ominus + \nu_B\mu_B^\ominus \tag{14.18}$$

The above three equations allow us to define the following two terms.
The **mean molal stoichiometric activity coefficient** for the neutral salt is:

$$\gamma_{\pm AB} \equiv (\gamma_A^{\nu_A}\gamma_B^{\nu_B})^{1/\nu_{AB}} \tag{14.19}$$

The **mean molality** of the neutral salt is:

$$m_{\pm AB} \equiv m_{AB}(\nu_A^{\nu_A}\nu_B^{\nu_B})^{1/\nu_{AB}} \tag{14.20}$$

The **fundamental equation for the chemical potential of a salt** is therefore:

$$\mu_{AB} = \mu_{AB}^\ominus + \nu_{AB}RT \ln(m_{\pm AB}\gamma_{\pm AB}) \tag{14.21}$$

The activity of the salt can now be defined by:

$$\ln a_{AB} = \nu_{AB}RT \ln[m_{\pm AB}\gamma_{\pm AB}] \tag{14.22}$$

For example, the mean stoichiometric activity coefficient and mean molality of aqueous calcium chloride and aqueous aluminium sulphate are respectively:

$$\gamma_{\pm,CaCl_2} = (\gamma_{Ca^{2+}}\gamma_{Cl^-}^2)^{1/3} \quad \text{and} \quad m_{\pm,CaCl_2} = m_{CaCl_2} \times 4^{1/3}$$

$$\gamma_{\pm,Al_2(SO_4)_3} = (\gamma_{Al^{3+}}^2\gamma_{SO_4^{2-}}^3)^{1/5} \quad \text{and} \quad m_{\pm,Al_2(SO_4)_3} = m_{Al_2(SO_4)_3} \times 108^{1/5}$$

An alternative expression for the chemical potential of a salt in terms of its molality is:

$$\mu_{AB} = \mu_{AB}^\ominus + RT[\nu_{AB} \ln(m_{AB}\gamma_{\pm AB}) + \nu_A \ln \nu_A + \nu_B \ln \nu_B] \tag{14.23}$$

A function commonly referred to in the geochemical literature is the Gibbs energy of mixing, $\bar{G}_{AB} - \bar{G}_{AB}^\ominus$. This is given by:

$$\bar{G}_{AB} - \bar{G}_{AB}^\ominus = \mu_{AB} - \mu_{AB}^\ominus$$

$$= RT[\nu_{AB} \ln(m_{AB}\gamma_{\pm AB}) + \nu_A \ln \nu_A + \nu_B \ln \nu_B] \tag{14.24}$$

Similarly, for the single ions:

$$\bar{G}_A - \bar{G}_A^\ominus = \mu_A - \mu_A^\ominus = RT(\ln m_A \gamma_A)$$

and

$$\bar{G}_B - \bar{G}_B^\ominus = \mu_B - \mu_B^\ominus = RT(\ln m_B \gamma_B) \tag{14.25}$$

Activity coefficients on
different concentration
scales

The notation for activity coefficients is:

- Mole fraction scale (x_i):   Activity coefficients f_i
- Molar scale ($m_i/\text{mol kg}^{-1}$):   Activity coefficients γ_i or $\gamma_\pm$
- Molar scale ($c_i/\text{mol dm}^{-3}$):   Activity coefficients y_i or $y_\pm$

The inter-relationships are:

$$f_i = \gamma_i \left(1 + 0.001 W_{H_2O} \sum_i m_i \right) \tag{14.26}$$

$$f_i = y_i \left[\frac{\rho + 0.001(W_{H_2O} \sum_i c_i - \sum_i c_i W_i)}{\rho^\bullet} \right] \tag{14.27}$$

$$\gamma_i = y_i \left(\frac{\rho - 0.001 \sum_i c_i W_i}{\rho^\bullet} \right) = y_i \frac{c_i}{m_i \rho^\bullet} \tag{14.28}$$

$$y_i = \gamma_i \left(1 + 0.001 \sum_i m_i W_i \right) \frac{\rho^\bullet}{\rho} = \gamma_i \frac{m_i \rho^\bullet}{c_i} \tag{14.29}$$

where W_i is the formula weight of the subscripted species in grams, ρ is the density of the **solution** and $\rho^\bullet$ the density of **pure water** in g cm^{-3}.

14.1.4   Partial molar and apparent molar quantities of electrolytes

True partial molar
properties

The value of any extensive property of a mixture is equal to the sum of the moles of each species multiplied by its partial molar property. The governing equation from section 9.1.5 is:

$$\Xi = \sum n_i \bar{\Xi}_i \tag{14.30}$$

where any partial molar property $\bar{\Xi}_i$ of the ith species is:

$$\bar{\Xi}_i = \left(\frac{\partial \Xi}{\partial n_i} \right)_{T, P, n_j} \tag{14.31}$$

and Ξ is the corresponding property of the mixture.

For example, the volume V of a mixture is related to the partial molar volumes of its constituents by:

$$V = \sum_i n_i \bar{V}_i \tag{14.32}$$

where n_i is the number of moles of the ith substance and $\bar{V}_i$ its partial molar volume. In a neutral aqueous solution, containing n_A moles of anion A^{z_A-} and n_B moles of cation B^{z_B+}, this would be

$$V = n_w \bar{V}_w + (n_A \bar{V}_A) + (n_B \bar{V}_B) \tag{14.33}$$

where the subscript w denotes a property of water. Alternatively, the total volume could be expressed in terms of the molar volume and amount of the neutral salt AB.

$$V = n_w \bar{V}_w + n_{AB} \bar{V}_{AB} \tag{14.34}$$

Apparent molar properties

The volume of a solution containing a single salt AB plus water is expressed in terms of the partial molar properties of the components. However, the only measurements we can actually make are the total volume of the system and the composition. The partial molar volumes remain inaccessible.

The total volume of the system will not be a linear function of composition since the partial molar volumes themselves change with composition. Unfortunately, we cannot say to what extent the partial molar volume of an individual component changes.

To help tabulate data we assume that the partial molar volume of water does not change with composition. Consequently, all the deviations from ideal behaviour are attributed to the partial molar volume of the salt. This allows us to use the partial molar volume of pure water ($\bar{V}_w^{\bullet}$) in equation 14.34 rather than its true partial molar volume.

The volume of the solution can then be written:

$$V = n_w \bar{V}_w^{\bullet} + n_{AB} \Phi_V \tag{14.35}$$

where Φ_V is defined as the **apparent molar volume of the electrolyte**.

The definition for the apparent molar volume of an aqueous electrolyte in a solution containing any number of salts is:

$$\Phi_V = \frac{V - 55.51 \bar{V}_w^{\bullet}}{\sum_k m_k} \tag{14.36}$$

where m_k is the molal concentration of the kth salt and 55.51 is the number of moles of water in 1 kg of water. Equation 14.35 becomes:

$$V = n_w \bar{V}_w^{\bullet} + \sum_k n_k \Phi_V \tag{14.37}$$

Note that there is only one value for the apparent molar volume of the mixed electrolyte. We do not define an apparent molar volume for every salt in the mixture which retains Φ_V as a single measurable property of all solutions. For a solution containing only one solute species at infinite dilution, any apparent molar property is equal to the true partial molar property.

The apparent molar volume of any single electrolyte solution of total volume V containing m moles of solute per kg of water is given by:

$$\Phi_V = \frac{1}{m} \left[\left(\frac{1000 + m_k W_k}{\rho} \right) - \frac{1000}{\rho_w^{\bullet}} \right] \tag{14.38}$$

where $\rho_w^{\bullet}$ is the density of pure water (g cm^{-3}), ρ the density of the solution and W_k the formula weight of the solute in g mol^{-1}.

Any partial molar property such as heat capacity, compressibility or expansibility can have an apparent molar counterpart. A general expression, valid for any solvent, is therefore:

$$\Phi_\Xi = \frac{\Xi - (1000/W_s)\bar{\Xi}_w^\bullet}{\sum_k m_k} \qquad (14.39)$$

where Φ_Ξ represents an apparent molar property, $\bar{\Xi}_w^\bullet$ represents the corresponding standard molar property of pure solvent, Ξ is the property of the amount of solution containing 1 kg of solvent and W_s is the formula weight of the solvent in g mol^{-1}.

14.1.5 Relative partial molar properties

For any salt the partial molar volume is dependent on concentration through the stoichiometric activity coefficient for the salt. This is also true for the partial molar heat capacity, partial molar compressibility and partial molar expansibility.

To deduce the important conclusion above we must consider the general expression for the partial molar Gibbs energy of the kth salt taken from equation 14.24:

$$\bar{G}_k = \bar{G}_k^\ominus + v_k RT \ln(m_k \gamma_\pm) + RT \sum_j \ln v_{jk}^{v_{jk}} \qquad (14.40)$$

where, from equation 14.11, $v_k = \sum_j v_{jk}$. We see that $\bar{G}_k$ is equal to $\bar{G}_k^\ominus$ when the salt is in its standard state and under no other conditions. The compositional dependence of $\bar{G}_k$ is through molality as well as the stoichiometric activity coefficient.

Other partial molar properties, at any specified concentration, are related to $\bar{G}_k$ through the identities defined in sections 9.1.9, 9.1.10 and (later) 14.2.4. These are:

$$\bar{H}_k = -T^2 \left[\frac{\partial(\bar{G}_k/T)}{\partial T} \right]_{P,n} \qquad (14.41)$$

$$\bar{C}_{P,k} = \left(\frac{\partial \bar{H}_k}{\partial T} \right)_{P,n} \qquad (14.42)$$

$$\bar{V}_k = \left(\frac{\partial \bar{G}_k}{\partial P} \right)_{T,n} \qquad (14.43)$$

$$\bar{E}_k = \left(\frac{\partial \bar{V}_k}{\partial T} \right)_{P,n} \qquad (14.44)$$

$$\bar{B}_k = -\left(\frac{\partial \bar{V}_k}{\partial P} \right)_{T,n} \qquad (14.45)$$

Clearly, the temperature and pressure dependence of $\bar{G}_k$ plays a central role in determining all other partial molar properties. For example, the partial molar enthalpy of the salt in equation 14.41 is given by:

$$-T^2 \frac{\partial}{\partial T}\left(\frac{\bar{G}_k}{T} \right)_{P,n} = -T^2 \frac{\partial}{\partial T}\left(\frac{\bar{G}_k^\ominus}{T} \right)_{P,n} - v_k RT^2 \left(\frac{\partial \ln \gamma_{\pm k}}{\partial T} \right) \qquad (14.46)$$

Since

$$\bar{H}_k^\ominus \equiv -T^2\left[\frac{\partial}{\partial T}\left(\frac{\bar{G}_k^\ominus}{T}\right)_{P,n}\right]$$

we may write:

$$\bar{H}_k = \bar{H}_k^\ominus - v_k RT^2\left(\frac{\partial \ln \gamma_{\pm k}}{\partial T}\right) \tag{14.47}$$

The compositional dependence of enthalpy, through the temperature dependence of the activity coefficient at the specified composition, is apparent.

At any temperature and pressure, the value of $\gamma_{\pm k}$ in the standard state is unity. The temperature derivative of $\ln \gamma_{\pm k}$ in the standard state is therefore zero and the partial molar enthalpy at infinite dilution has the same value as in its standard state, i.e.

$$\bar{H}_k^\circ = \bar{H}_k^\ominus \tag{14.48}$$

Partial molar enthalpies are usually measured relative to infinite dilution. These are called **relative partial molar enthalpies** and given the symbol $\bar{L}_k$. We therefore have:

$$\bar{L}_k = \bar{H}_k - \bar{H}_k^\ominus = \bar{H}_k - \bar{H}_k^\circ \tag{14.49}$$

from which:

$$\bar{L}_k = -v_k RT^2\left(\frac{\partial \ln \gamma_{\pm k}}{\partial T}\right)_{P,n} \tag{14.50}$$

Consider now the partial molar heat capacity given by:

$$\bar{C}_{P,k} = \bar{C}_{P,k}^\ominus - v_k\left(T^2\frac{\partial^2 \ln \gamma_{\pm k}}{\partial T^2} + 2T\frac{\partial \ln \gamma_{\pm k}}{\partial T}\right)_{P,n} \tag{14.51}$$

We may deduce that:

$$\bar{C}_{P,k}^\circ = \bar{C}_{P,k}^\ominus \tag{14.52}$$

In a similar way to the definition of $\bar{L}_k$, we may define the relative partial molar heat capacity $\bar{J}_k$. This is

$$\bar{J}_k = \bar{C}_{P,k} - \bar{C}_{P,k}^\ominus = \bar{C}_{P,k} - \bar{C}_{P,k}^\circ \tag{14.53}$$

The other partial molar properties for salts have a compositional dependence through the relationships:

$$\bar{V}_k = \bar{V}_k^\ominus + v_k RT\left(\frac{\partial \ln \gamma_{\pm k}}{\partial P}\right)_{T,n} \tag{14.54}$$

$$\bar{E}_k = \bar{E}_k^\ominus + v_R T\left[\frac{\partial(\partial \ln \gamma_{\pm k}/\partial P)_{T,n}}{\partial T}\right]_{P,n} - \frac{\bar{V}_k - \bar{V}_k^\ominus}{T} \tag{14.55}$$

$$\bar{B}_k = \bar{B}_k^\ominus + v_k RT\left(\frac{\partial^2 \ln \gamma_{\pm k}}{\partial T}\right)_{P,n} \tag{14.56}$$

and

$$\bar{S}_k = \bar{S}_k^\ominus + \frac{(\bar{H}_k - \bar{H}_k^\ominus) - (\bar{G}_k - \bar{G}_k^\ominus)}{T} \tag{14.57}$$

From equations 14.54 to 14.57 we may also conclude that:

$$\bar{V}^\circ = \bar{V}^\ominus \tag{14.58}$$

$$\bar{B}^\circ = \bar{B}^\ominus \tag{14.59}$$

$$\bar{E}^\circ = \bar{E}^\ominus \tag{14.60}$$

We may not make similar claims about $G_k^\ominus$ and $S_k^\ominus$ since the compositional dependence of G_k and S_k through modality shows that:

$$S^\circ = + \infty \text{ and } G^\circ = - \infty \tag{14.61}$$

These rather abstract conclusions will be used in section 14.2 to help predict the temperature and pressure dependence of energies of formation.

The above expressions are consistent with the properties of the single ions through the relationships:

$$\bar{G}_k - \bar{G}_k^\ominus = \sum_j v_{jk}(\bar{G}_j - \bar{G}_j^\ominus) \tag{14.62}$$

$$\bar{H}_k - \bar{H}_k^\ominus = \sum_j v_{jk}\bar{L}_j = \sum_j v_{jk}(\bar{H}_j - \bar{H}_j^\ominus) \tag{14.63}$$

$$\bar{V}_k - \bar{V}_k^\ominus = \sum_j v_{jk}(\bar{V}_j - \bar{V}_j^\ominus) \tag{14.64}$$

$$\bar{E}_k - \bar{E}_k^\ominus = \sum_j v_{jk}(\bar{E}_j - \bar{E}_j^\ominus) \tag{14.65}$$

$$\bar{B}_k - \bar{B}_k^\ominus = \sum_j v_{jk}(\bar{B}_j - \bar{B}_j^\ominus) \tag{14.66}$$

$$\bar{S}_k - \bar{S}_k^\ominus = \sum_j v_{jk}(\bar{S}_j - \bar{S}_j^\ominus) \tag{14.67}$$

Relative partial molar properties of single ions can only be evaluated by convention.

14.2 Properties of the infinitely dilute solution and apparent energies of formation

14.2.1 Measuring partial molar properties at infinite dilution

It is impossible to measure apparent molar properties at infinite dilution. This problem is overcome by measuring the apparent molar property as a function of concentration and back-extrapolating to zero concentration. For example, concentration dependence of the apparent molar volumes of sodium chloride solutions at 1 bar is shown in figure 14.1(a). It is clear that the partial molar volume at any temperature varies with concentration, although back-extrapolation by eye is inaccurate.

Theoretical equations for the compositional dependence of apparent partial molar properties are useful for estimating properties at infinite dilution. For volume

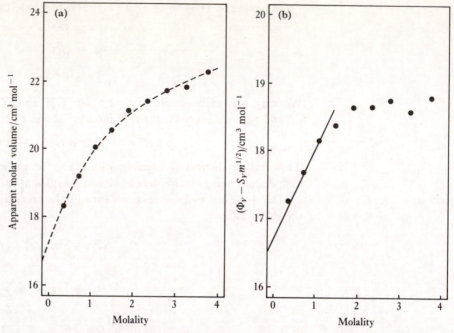

Figure 14.1   Apparent molar volume of sodium chloride versus molality (data from Helgeson and Kirkham, 1976).

measurements a modification of the Redlich–Maier equation describes the concentration dependence of Φ_V for a solution containing a single salt composed of v_A moles of anion A of charge z_A and v_B moles of cation B of charge z_B, i.e.

$$\Phi_V = \bar{V}_{AB}^{\circ} + S_V m_{AB}^{1/2} + b_{AB} m_{AB} \qquad (14.68)$$

where $\bar{V}_{AB}^{\circ}$ is the partial molar volume of the salt at infinite dilution. Also,

$$S_V = \frac{2A_V}{3}\left(\frac{v_A + v_B}{2}|z_A z_B|\right)^{3/2} \qquad (14.69)$$

The variable A_V is a temperature- and pressure-dependent coefficient characteristic of the solvent and equal to 2.75 cm^2 kg$^{1/2}$ mol$^{-3/2}$ at 298.15 K and 1 bar for water. The coefficient b_{AB} is characteristic of the salt. This equation is valid for most electrolyte solutions for concentrations up to 0.5 mol kg^{-1}. The values of S_V for water are known; therefore a plot of $\Phi_V m_{AB}^{1/2}$ versus m_{AB} should have a slope of b_{AB} and an intercept of $\bar{V}_{AB}^{\circ}$.

Figure 14.1(b) shows data for sodium chloride at 298.15 K and 1 bar. The linearity at low concentrations is apparent and the values of $\bar{V}_{AB}^{\circ}$ easily determined. We note that the linearity in figure 14.1(b) disappears at high concentrations.

14.2.2   Partial molar volumes of ions and salts

The temperature and pressure dependence of the standard molar volume of sodium chloride (figures 14.2–14.4) is typical for inorganic electrolytes.

The effect of temperature is especially pronounced on the saturation line where

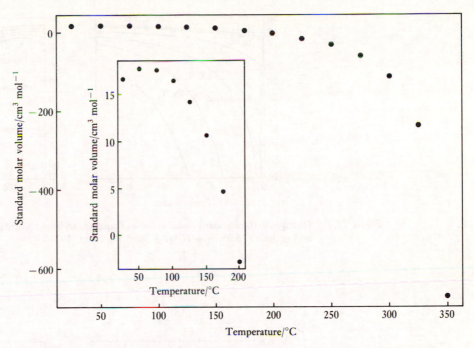

Figure 14.2   Standard molar volume of aqueous sodium chloride on the saturation line (data from Helgeson and Kirkham, 1976).

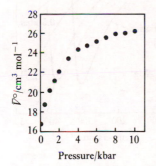

Figure 14.3
Standard molar volume of
aqueous sodium chloride
as a function of pressure
at 298 K (data from
Adams, 1931).

the molar volume first increases with increasing temperature to a maximum at around 70 °C (figure 14.2). The molar volume becomes negative above 200 °C and decreases dramatically as the temperature approaches the critical point.

A plot of molar volumes versus pressure, as in figure 14.3, indicates that increasing pressure invariably increases the molar volume. The effect is less pronounced, however, than that due to temperature.

The general effects of temperature and pressure can be seen in figure 14.4. This shows that the greatest sensitivity to temperature is in the pressure region 0–1 kbar, and as pressure is increased the effect of temperature becomes less. Increasing pressure causes a slight drop in the temperature of the initial maximum at around 70 °C, which becomes less apparent, finally disappearing at pressures above ~3500 bar. In contrast, the greatest sensitivity to pressure is at the higher temperatures.

Many inorganic salts and single ions exhibit the same general features as sodium chloride. Figure 14.5 shows data for Mg^{2+} and SO_4^{2-}. This emphasises the similarities but suggests that the values of the molar volumes and their precise temperature and pressure dependencies are characteristic of the individual ions.

14.2.3   A simple interpretation of the trends

Liquid water has an open framework structure (section 4.1.2). In the presence of an ion, a proportion of the water molecules are removed from the structure to become part of the ion's solvation shell. This phenomenon leads to a slight collapse of the

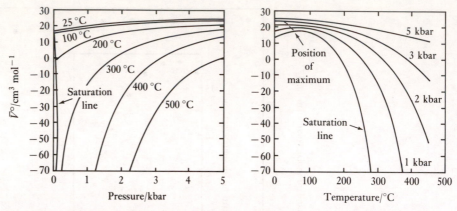

Figure 14.4   Changes in the standard molar volume of aqueous sodium chloride as a function of temperature and pressure (data from Helgeson and Kirkham, 1976).

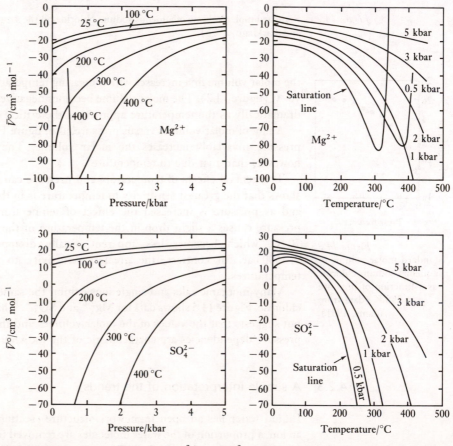

Figure 14.5   Standard molar volume of aqueous Mg^{2+} and SO_4^{2-} (data from Helgeson and Kirkham, 1976).

solvent structure, especially in the region between the solvation shell and the bulk solvent, which causes an overall reduction in volume. A second contribution to the volume arises because water molecules in the solvation shell are aligned and packed more closely than in the bulk. This leads to a local increase in the solvent density in the near-vicinity of an ion. When these effects are large enough for the volume of the solution to be less than the original volume of the pure water before mixing, the partial molar volumes of these ions become negative.

The term **electrostriction** (section 4.4.3) is given to the compression of a material due to the presence of an electric field. In principle, the factors contributing to electrostriction cannot be properly separated. However, it is instructive to consider the standard molar volume of an ion as having three additive contributions. These are:

- An intrinsic volume characteristic of the specific ion, V_{int}. The magnitude of this term is thought to be proportional to the molar volume of its crystallographic counterpart and essentially independent of temperature and pressure.
- An electrostriction volume loss due to solvation, ΔV_{solv}. This term is related to the relative permittivity of water, which influences the extent to which electrostatic ion–solvent interactions can promote solvation.
- An electrostriction volume loss due to local collapse of the solvent structure, ΔV_{coll}. This contribution changes asymptotically with increasing temperature and increases linearly with pressure.

The three terms combine to give:

$$\bar{V}^{\circ} = V_{int} + \Delta V_{solv} + \Delta V_{coll} \tag{14.70}$$

Expressions developed by Helgeson and Kirkham (1976) and Helgeson *et al.* (1981) can give some insight into the relative contributions from the solvation and solvent collapse terms. These equations will be used to demonstrate how equations of state can be employed; their use does not imply that they are the most suitable ones available to date.

In section 4.4.4 it was shown how the energy of interaction due to the solvation of 1 mole of ions could be calculated from a simple theory due to Born. This energy can be set equal to the standard partial molal Gibbs energy of solvation at infinite dilution (ΔG_s°). Then, for any ion of charge z and apparent electrostatic radius r_e (see section 4.4.4):

$$\Delta G_s^{\circ} = L \frac{(ze)^2}{2r_e}\left(\frac{1}{\varepsilon_{r(s)}} - 1\right) \tag{14.71}$$

where L is Avogadro's constant, e is the charge on one electron and $\varepsilon_{r(s)}$ is the relative permittivity of water. The term $L(ze)^2/2r_e$ is given the symbol ω^{abs} and named the Born coefficient. Therefore,

$$\Delta G_s^{\circ} = \omega^{abs}\left(\frac{1}{\varepsilon_{r(s)}} - 1\right) \tag{14.72}$$

The relative electrostatic radii can be estimated from the crystallographic radii of the ions (r_c) using:

$$r_e = r_c + z \times \Gamma \tag{14.73}$$

where Γ is a constant equal to 0.094 nm for cations and zero for anions.

The basic thermodynamic identity, $(\partial G/\partial P)_T = V$, allows the calculation of ΔV_{solv}, since:

$$\Delta V_{\text{solv}} = \left(\frac{\partial \Delta G_s^{\circ}}{\partial P}\right)_T \tag{14.74}$$

This differentiates giving:

$$\Delta V_{\text{solv}} = -\omega^{\text{abs}}Q \tag{14.75}$$

where

$$Q = \frac{1}{\varepsilon_{\text{r(s)}}}\left(\frac{\partial \ln \varepsilon_{\text{r(s)}}}{\partial P}\right)_T \tag{14.76}$$

The term Q is a **Born function** which must be calculated from the temperature and pressure dependence of the relative permittivity of pure water. There are several Born functions which play a role in the continuum properties of aqueous solutions.

Equation 14.75 can be used to estimate ΔV_{solv} for any ion. The corresponding property for any salt can be calculated from the values for appropriate ions.

The second electrostriction term, due to local solvent collapse, is approximated by an equation of the form:

$$\Delta V_{\text{coll}} = a_1 P + \frac{(a_2 + a_3 P)T}{(T - \theta)} \tag{14.77}$$

where a_1, a_2, a_3 and θ are ion-specific fitting coefficients, pressure P is in bars and temperature T in kelvin. The form of this empirical equation has been deduced from experiments but has a theoretical basis described by Bernal and Fowler (1933). The variable θ is called the **structural temperature**.

The two electrostriction terms can be combined in equation 14.70 to give an equation of state for ions which has the form:

$$\bar{V}^{\circ} = V_{\text{int}} + a_1 P + \frac{(a_2 + a_3 P)T}{(T - \theta)} - \omega^{\text{abs}}Q \tag{14.78}$$

Once an equation of this nature is formulated, the unknown fitting coefficients can be determined by regression against experimental data. Helgeson and Kirkham (1976) were able to evaluate these parameters for an extensive range of ions and salts. The term Γ, used to calculate the relative electrostatic radii, was also treated as a fitting parameter and optimised accordingly.

Figures 14.6–14.9 show values of molar volumes and the contributions from each component for selected ions. At all temperatures and pressures the electrostriction terms contribute to the molar volumes of ions, especially on the saturation line and at pressures less than 1000 bar. The tendencies of all ions are similar although each ion has its own individual characteristics.

The solvent collapse term is generally smaller than the solvation term and varies less with temperature and pressure. However, the two electrostriction terms give different contributions in different regions of $P-T$ space. At temperatures below 150 °C and pressures below 1000 bar the two terms are of similar magnitude. The solvent collapse term increases with temperature and dominates initially. At temperatures above 50 °C the solvation term, which decreases with temperature, starts to play a role leading to a maximum and the subsequent decrease with

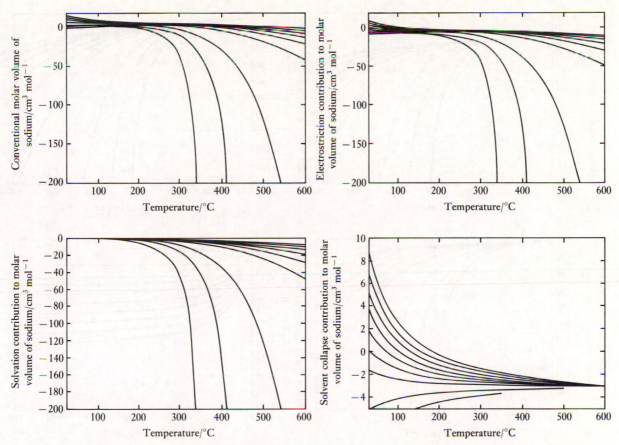

Figure 14.6   Specific contributions to the standard molar volume of aqueous sodium (data from Helgeson and Kirkham, 1976).

temperature. At higher pressures the solvent collapse term decreases with increasing temperature, accounting for the loss in the maximum.

On the saturation line at temperatures above 150 °C, and at pressures below 1000 bar, the solvation term dominates but it varies dramatically with temperature. This is a direct consequence of the temperature dependence of the relative permittivity of water in this region.

As pressures increase to above 2000 bar, and temperatures to above 300 °C, the solvation term becomes smaller and the solvent collapse term once more becomes influential.

Clearly, the characterisation of such behaviour gives serious problems to experimentalists and theoreticians alike. Equation 14.78 affords a good representation of the standard molar volume of dissolved ions at pressures lower than 3000 bar and temperatures below 250 °C with the exception of temperatures and pressures close to the critical point for water. This failure of the model is attributed to the assumption that the Born coefficient ω is temperature- and pressure-independent. A more realistic model is given by Tanger and Helgeson (1988).

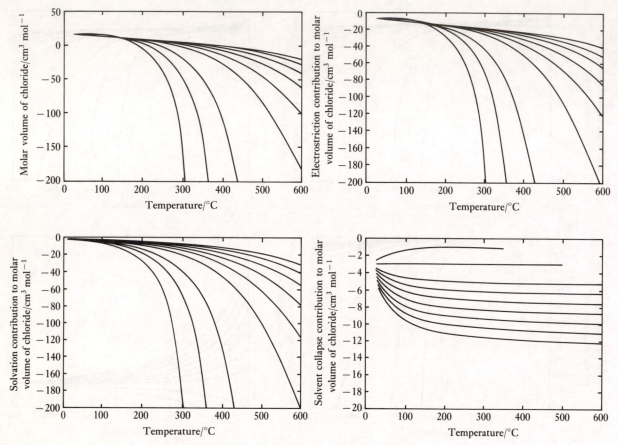

Figure 14.7   Specific contributions to the standard molar volume of aqueous chloride (data from Helgeson and Kirkham, 1976).

14.2.4   Isothermal compressibilities and expansibilities

The expansion of a mixture with increasing temperature and the compression with increasing pressure can be characterised with the following two properties:

Expansibility:

$$E = \left(\frac{\partial V}{\partial T}\right)_P \tag{14.79}$$

Compressibility:

$$B = -\left(\frac{\partial V}{\partial P}\right)_T \tag{14.80}$$

These are related to the coefficient of isobaric expansion and the coefficient of isothermal compression by:

$$E = \alpha V \quad \text{and} \quad B = \kappa V \tag{14.81}$$

The advantage of using E and B is that they are each equal to the sum of the

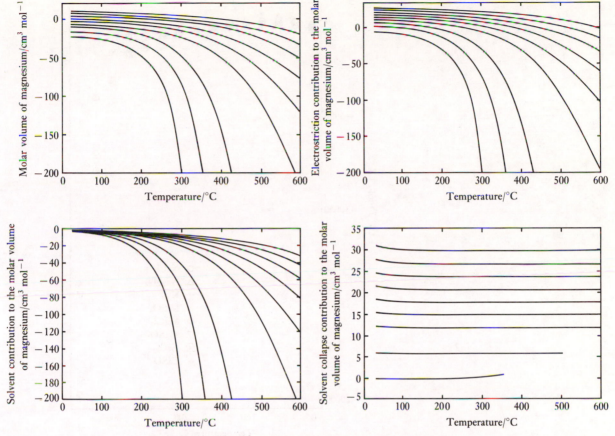

Figure 14.8   Specific contributions to the standard molar volume of aqueous sulphate (data from Helgeson and Kirkham, 1976).

corresponding partial molar properties of the components in the solution. In accordance with the definitions of partial molar properties in section 9.1.5, these are:

$$\bar{E}_i = \left(\frac{\partial E}{\partial n_i}\right)_{P,nj} = \left(\frac{\partial \bar{V}_i}{\partial T}\right)_P \tag{14.82}$$

and

$$\bar{B}_i = \left(\frac{\partial B}{\partial n_i}\right)_{T,nj} = -\left(\frac{\partial \bar{V}_i}{\partial P}\right)_T \tag{14.83}$$

where i denotes any solution component including water. From this,

$$E = \sum_i n_i \bar{E}_i = \sum_i n_i \left(\frac{\partial E}{\partial n_i}\right)_{P,nj} \tag{14.84}$$

and

$$B = \sum_i n_i \bar{B}_i = \sum_i n_i \left(\frac{\partial B}{\partial n_i}\right)_{P,nj} \tag{14.85}$$

Compressibilities and expansibilities vary with temperature, pressure and

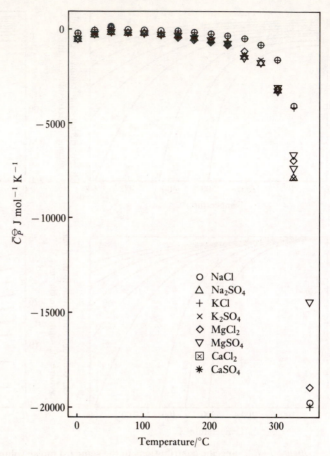

Figure 14.9   Standard molar heat capacities for several ions (data from Helgeson and Kirkham, 1976).

composition. In the standard state, and at infinite dilution, the corresponding properties are

$$\bar{E}_i^{\ominus} = \bar{E}_i^{\circ} = \left(\frac{\partial \bar{V}_i^{\circ}}{\partial T}\right)_P \tag{14.86}$$

and

$$\bar{B}_i^{\ominus} = \bar{B}_i^{\circ} = -\left(\frac{\partial \bar{V}_i^{\circ}}{\partial P}\right)_T \tag{14.87}$$

For ions and salts $\bar{E}_i^{\circ}$ and $\bar{B}_i^{\circ}$ are difficult to measure. The simplest way of determining their values at any temperature and pressure is to establish an equation of state for the standard molar volume of a species and differentiate this with respect to temperatures and pressure respectively. For example, equation 14.78 can be differentiated in accordance with equations 14.86 and 14.87 to give ion compressibilities and expansibilities respectively. Dropping the subscript i, since we are always referring to a specified ion, the resulting equations are:

$$\bar{E}^{\circ} = -\frac{(a_2 + a_3 P)\theta}{(T - \theta)^2} - \omega_{\text{abs}} U \tag{14.88}$$

and

$$-\bar{B}^\circ = a_1 + \frac{a_2 T}{(T - \theta)} - \omega_{\mathrm{abs}} N \tag{14.89}$$

The variables U and N are two more Born functions defined as:

$$U = \frac{1}{\varepsilon_{\mathrm{r(s)}}} \left[\left(\frac{\partial (\partial \ln \varepsilon_{\mathrm{r(s)}} / \partial P)_T}{\partial T} \right)_P - \left(\frac{\partial \ln \varepsilon_{\mathrm{r(s)}}}{\partial T} \right)_P \left(\frac{\partial \ln \varepsilon_{\mathrm{r(s)}}}{\partial P} \right)_T \right] \tag{14.90}$$

and

$$N = \frac{1}{\varepsilon_{\mathrm{r(s)}}} \left[\left(\frac{\partial^2 \ln \varepsilon_{\mathrm{r(s)}}}{\partial P^2} \right)_T - \left(\frac{\partial \ln \varepsilon_{\mathrm{r(s)}}}{\partial P} \right)_T^2 \right] \tag{14.91}$$

Since the intrinsic volume contribution to $\bar{V}^\circ$ is supposedly independent of temperature and pressure, the values of $\bar{E}^\circ$ and $\bar{B}^\circ$ for any ion are dominated by electrostriction. Consequently, $\bar{B}^\circ$ and $\bar{E}^\circ$ show the same type of complicated dependence on temperature and pressure dependencies as do standard molar volumes.

14.2.5   Partial molar heat capacities and standard partial molar entropies

Partial molar heat capacities of salts at infinite dilution, $\bar{C}_P^\circ$, can be evaluated calorimetrically. Their determination requires the extrapolation of the apparent partial molar heat capacities for concentrated electrolytes back to zero concentration. It is notoriously difficult to make these calorimetric determinations at elevated pressure, and most data for salt solutions are at low pressure but over a range of temperatures.

Figure 14.10 shows the standard partial molar heat capacity of other ions over a range of temperatures on the saturation curve. A dramatic increase with increasing temperature is observed.

Using such data it is possible to deduce expressions for the temperature dependence of the standard molar heat capacity of ions at 1 bar pressure.

The extension of such expressions to elevated pressures requires the use of a Maxwell equation (section 19.1.1) and is worth considering in detail.

We start by assuming that the standard molar heat capacity of an ion, like the standard molar volume, can be broken down into separate contributions from an intrinsic term, a solvation term and a local solvent collapse term. An approximate relationship for the temperature dependence of an aqueous species becomes:

$$\bar{C}_P^\circ = \bar{C}_{P,\mathrm{int}} + \frac{c_1 T}{T - \theta} + \omega^{\mathrm{abs}} T X \tag{14.92}$$

where X is another Born function, which has the form:

$$X = \frac{1}{\varepsilon_{\mathrm{r(s)}}} \left[\left(\frac{\partial^2 \ln \varepsilon_{\mathrm{r(s)}}}{\partial T^2} \right)_P - \left(\frac{\partial \ln \varepsilon_{\mathrm{r(s)}}}{\partial T} \right)_T^2 \right] \tag{14.93}$$

In equation 14.92 the term $\omega^{\mathrm{abs}} T X$ is the solvation contribution and the term $c_1 T / (T - \theta)$ is the contribution from local solvent collapse.

Equation 14.92 is limited to the reference pressure. To extend it to higher pressures we recall from section 8.1.1 that an isothermal change in the standard partial molar

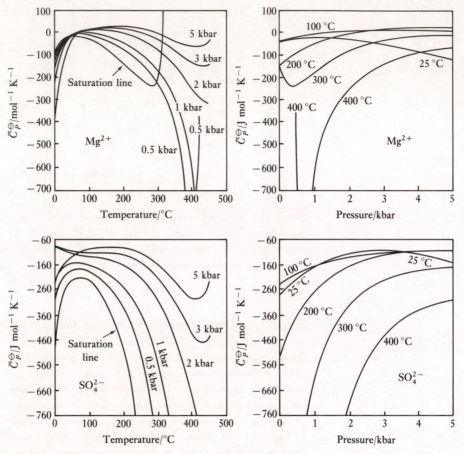

Figure 14.10   Third-law molar heat capacity for Mg^{2+} and SO_4^{2-} versus temperature and pressure (data from Tanger and Helgeson, 1988).

heat capacity, resulting from an infinitesimal change in pressure, is given by:

$$d\bar{C}_P^\circ = \left(\frac{\partial \bar{C}_P^\circ}{\partial P}\right)_T dP \qquad (14.94)$$

Therefore, the change in the standard molar heat capacity as pressure goes from the reference pressure P_r to P is given by:

$$\int_{P_r}^{P} d\bar{C}_P^\circ = \bar{C}_{P(T,P)}^\circ - \bar{C}_{P(T,P_r)}^\circ = \int_{P_r}^{P} \left(\frac{\partial \bar{C}_P}{\partial P}\right)_T dP \qquad (14.95)$$

The value of $\bar{C}_{P(T,P_r)}^\circ$ is given by equation 14.92 at the reference pressure of 1 bar and temperature T. Clearly, we must obtain an expression for $(\partial \bar{C}_P/\partial P)_T$ in order to evaluate the pressure integral. This can be obtained using the Maxwell equation $(\partial S/\partial P)_T = -(\partial V/\partial T)_P$ and the basic definition of heat capacity, which is:

$$\bar{C}_P = T\left(\frac{\partial \bar{S}}{\partial T}\right)_P \qquad (14.96)$$

From equation 14.96 we can write:

$$\left(\frac{\partial \bar{C}_P^\circ}{\partial P}\right)_T = T\frac{\partial}{\partial P}\left(\frac{\partial \bar{S}^\circ}{\partial T}\right)_P = T\frac{\partial}{\partial T}\left(\frac{\partial \bar{S}^\circ}{\partial P}\right)_T \qquad (14.97)$$

Substituting the above Maxwell equation into equation 14.97 leads to:

$$\left(\frac{\partial \bar{C}_P^\circ}{\partial P}\right) = -T\left(\frac{\partial \bar{V}^\circ}{\partial T^2}\right)_P \qquad (14.98)$$

This gives a way of evaluating $(\partial \bar{C}_P^\circ/\partial P)_T$ from the equation of state for the standard partial molar volume of an ion. Thus, if equation 14.78 is differentiated twice with respect to temperature, and the resulting equation substituted into equation 14.95, the necessary integration can then be performed. This gives the full equation for the standard molar heat capacity of an ion at any temperature and pressure as:

$$\bar{C}_P^\circ = \bar{C}_{P,\text{int}} + \frac{c_1 T}{T-\theta} - \frac{\theta T[a_2(P-P_r)+a_3(P^2-P_r^2)]}{(T-\theta)^3} + \omega^{\text{abs}}TX \quad (14.99)$$

Once an equation of state for $\bar{C}_P^\circ$ is established, the generation of an equation for the standard partial molar entropy $\bar{S}^\ominus$ is a simple procedure. Recalling that $\bar{C}_P^\ominus = \bar{C}_P^\circ$, integration of equation 14.99 with respect to temperature at constant pressure gives:

$$\bar{S}^\ominus = \left(\frac{\partial \bar{C}_P^\ominus}{\partial T}\right)_P = \left(\frac{\partial \bar{C}_P^\circ}{\partial T}\right)_P$$

$$= \bar{S}_{(T_r, P_r)}^\ominus + \bar{C}_{P,\text{int}}\ln(T/T_r) + c_1\ln\left(\frac{T-\theta}{T_r-\theta}\right)$$

$$+ \frac{\theta[2a_2(P-P_r)+a_3(P^2-P_r^2)]}{2(T-\theta)^2} + \omega^{\text{abs}}(Y - Y_{P_r,T_r}) \quad (14.100)$$

where Y is a Born function defined as:

$$Y = \frac{1}{\varepsilon_{r(s)}}\left(\frac{\partial \ln \varepsilon_{r(s)}}{\partial T}\right)_P \qquad (14.101)$$

Both $\bar{S}^\ominus$ and $\bar{C}_P^\circ$ have a complex dependence on temperature and pressure, as indicated in figures 14.10 and 14.11 for selected ions. As with most other partial molar properties the greatest sensitivity to temperature is in the $0-1$ kbar region and around the critical point. The sensitivity to pressure is maximised at high temperatures. As the system approaches the critical point the values of $\bar{C}_P^\circ$ and $\bar{S}^\ominus$ for any ion in the aqueous phase tend to $-\infty$ (in the gas phase to $+\infty$). This makes measurement and theoretical prediction difficult. The general trends in temperature and pressure are influenced strongly by the solvation contribution to electrostriction. This is shown by figure 14.12 which depicts changes in the Born function X at very high temperatures. The Born function shows an oscillation with changing temperature which is damped out by increasing pressure. These trends are followed closely by the trends in $\bar{C}_P^\circ$ as calculated by equation 14.99.

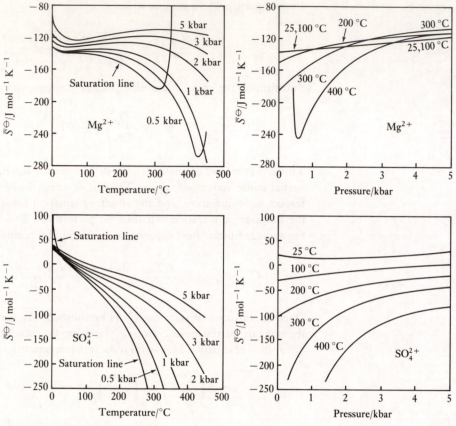

Figure 14.11   Third-law molar entropies for Mg^{2+} and SO_4^{2-} versus temperature and pressure (data from Tanger and Helgeson, 1988).

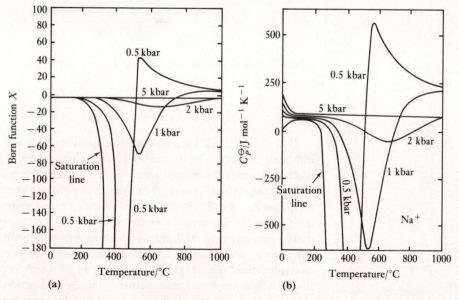

Figure 14.12   The Born function X and standard molar heat capacity of aqueous Na^+ versus temperature and pressure (after Helgeson, 1985).

14.2.6   Apparent standard energies of formation

The standard state properties of ionic species

Following the definitions in section 11.1.4, the four apparent standard energies of formation of **ions** are: $\Delta G^\ominus$, $\Delta H^\ominus$, $\Delta A^\ominus$ and $\Delta U^\ominus$. The corresponding values for neutral salts can be obtained from the relationships in section 14.1.1.

The temperature and pressure dependence of the apparent standard Gibbs energy of formation of a species may be calculated from the change in the standard partial molar Gibbs energy for the species. The working equation is:

$$\Delta G^\ominus_{(T, P)} = \Delta G^\ominus_{f(T_r, P_r)} + (\bar{G}^\ominus_{(T, P)} - \bar{G}^\ominus_{(T_r, P_r)}) \tag{14.102}$$

where $\Delta G^\ominus_{f(T_r, P_r)}$ is the true standard Gibbs energy of formation of the species at a reference temperature and pressure. Similarly for the apparent standard enthalpy of formation we have

$$\Delta H^\ominus_{(T, P)} = \Delta H^\ominus_{f(T_r, P_r)} + (\bar{H}^\ominus_{(T, P)} - \bar{H}^\ominus_{(T_r, P_r)}) \tag{14.103}$$

The objective, therefore, is to calculate $(\bar{G}^\ominus_{(T, P)} - \bar{G}^\ominus_{(T_r, P_r)})$ and $(\bar{H}^\ominus_{(T, P)} - \bar{H}^\ominus_{(T_r, P_r)})$. For this purpose we use the following fundamental relationships:

$$\left[\frac{\mathrm{d}(\Delta G^\ominus)}{\mathrm{d}T}\right]_P = -\bar{S}^\ominus \tag{14.104}$$

$$\left[\frac{\mathrm{d}(\Delta G^\ominus)}{\mathrm{d}P}\right]_T = \bar{V}^\ominus = \bar{V}^\circ \tag{14.105}$$

$$\left[\frac{\mathrm{d}(\Delta H^\ominus)}{\mathrm{d}T}\right]_P = \bar{C}_P^\ominus = \bar{C}_P^\circ \tag{14.106}$$

$$\left[\frac{\mathrm{d}(\Delta H^\ominus)}{\mathrm{d}P}\right]_T = \bar{V}^\ominus(1 - \bar{\alpha}^\ominus T) = \bar{V}^\circ(1 - \bar{\alpha}^\circ T) \tag{14.107}$$

These relationships combine to give:

$$(\bar{G}^\ominus_{(T, P)} - \bar{G}^\ominus_{(T_r, P_r)}) = -S^\ominus_{(T_r, P_r)}(T - T_r) + \int_{T_r}^{T} \bar{C}_P^\circ \, \mathrm{d}T$$
$$- T \int_{T_r}^{T} \bar{C}_P^\circ \, \mathrm{d}\ln T + \int_{P_r}^{P} \bar{V}^\circ \, \mathrm{d}P \tag{14.108}$$

and

$$(\bar{H}^\ominus_{(T, P)} - \bar{H}^\ominus_{(T_r, P_r)}) = \int_{T_r}^{T} \bar{C}_P^\circ \, \mathrm{d}T + \int_{P_r}^{P} \left[\bar{V}^\circ - T\left(\frac{\partial \bar{V}^\circ}{\partial T}\right)_p\right]_T \mathrm{d}P \tag{14.109}$$

Suitable equations of state for the temperature and pressure dependence of $\bar{V}^\circ$ and $\bar{C}_P^\circ$, plus a value for $\bar{S}^\ominus_{T_r, P_r}$, may be substituted into the above two equations and integrated to give $(\bar{G}^\ominus_{(T, P)} - \bar{G}^\ominus_{(T_r, P_r)})$ and $(\bar{H}^\ominus_{(T_r, P_r)})$.

Box 14.1   **Helgeson's equation of state**

As an example, we will integrate the equation of state by Helgeson *et al.*
(1981) to produce:

$$(\bar{G}^{\ominus}_{(T,P)} - \bar{G}^{\ominus}_{(T_r,P_r)}) = -S^{\ominus}_{(T_r,P_r)}(T - T_r) - \bar{C}_{P,\text{int}}[T \ln(T/T_r) - T + T_r]$$

$$+ c_1\left[T - T_r - (T - \theta)\ln\left(\frac{T - \theta}{T_r - \theta}\right)\right]$$

$$+ \frac{2(\bar{V}_{\text{int}}(T - \theta) + a_2 T)(P - P_r) + (a_1(T - \theta) + a_4 T)(P^2 - P_r^2)}{2(T - \theta)}$$

$$- \omega^{\text{abs}}[Z_{T,P} - Z_{T_r,P_r} - Y_{T_r,P_r}(T - T_r)] \qquad (14.110)$$

and

$$(\bar{H}^{\ominus}_{(T,P)} - \bar{H}^{\ominus}_{(T_r,P_r)}) = (\bar{C}^{\circ}_{P,\text{int}} + c_1)(T - T_r) + c_1\theta \ln\left(\frac{T - \theta}{T_r - \theta}\right)$$

$$+ \frac{2\bar{V}^{\circ}_{\text{int}}(P - P_r) + a_1(P^2 - P_r^2)}{2}$$

$$+ \frac{a_2 T(P - P_r) + a_3 T(P^2 - P_r^2)}{2(T - \theta)}$$

$$+ \frac{2\theta a_2 T(P - P_r) + \theta a_3 T(p^2 - p_r^2)}{2(T - \theta)^2}$$

$$+ \omega^{\text{abs}}{}_j(TY_{T,P} - T_r Y_{T_r,P_r} - Z_{T,P} - Z_{T_r,P_r}) \quad (14.111)$$

where

$$Z_{T,P} = -\frac{1}{\varepsilon_{P,T}} \qquad (14.112)$$

The algebraic complexity of equations 14.110 and 14.111 suggests that $\Delta G^{\ominus}$
and $\Delta H^{\ominus}$ may have a complex dependence on temperature and pressure. This is
indeed true.

Figures 14.13 and 14.14 show trends in the temperature and pressure dependence
of $\Delta G^{\ominus}$ for selected ions (computed using equation 14.110 and necessary coefficients
calculated by Helgeson *et al.*, 1981). Typically, changes in pressure influence $\Delta G^{\ominus}$
less than changes in temperature. The values of $\Delta G^{\ominus}$ for all ions show a characteristic
sensitivity to temperature and pressure as the systems approach the critical point.
This sensitivity is a consequence of electrostriction. However, we cannot make simple
qualitative generalisations about the effect of temperature and pressure on $\Delta G^{\ominus}$.
For example the value of $\Delta G^{\ominus}$ for sodium decreases with increasing temperature,
whereas for magnesium the value of $\Delta G^{\ominus}$ decreases. Greater complexities are seen
with chloride and sulphate ions, which show temperature-dependent minima. The
effect of pressure is also not easily generalised since it has little effect on $\Delta G^{\ominus}$ for
sodium but a notable effect for magnesium. Typically, the values of $\Delta G^{\ominus}$ for anions
are more sensitive to pressure changes than those for cations.

The solubilities of minerals are directly related to the apparent standard Gibbs
energy of formation. We would therefore expect the stabilities of minerals to be

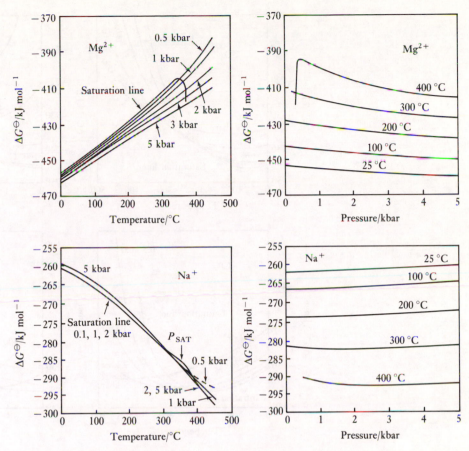

Figure 14.13   Apparent Gibbs energies of formation for Mg^{2+} and Na^+ cations (data from Tanger and Helgeson, 1988).

complex functions of temperature and pressure. This underscores the need to make precise measurements and have reliable equations of state in order to understand mineral transformations under geological conditions.

The standard state properties of neutral species   In principle, neutral, non-polar, species do not exhibit electrostriction. However, many neutral species, including ion pairs and neutral molecules, are naturally dipolar or can be polarised under the influence of an electric field. They are therefore expected to solvate to some extent.

Aqueous silica is a good example since it both solvates and polymerises in aqueous solutions. A general formula for the resulting molecule is:

$$(SiO_2)_n . mH_2O \qquad (14.113)$$

where n and m denote the degree of polymerisation and solvation. These two parameters have been the subject of much debate and the formula $(SiO_2) . 2H_2O$ is often assumed, i.e. monomeric silica with a solvation number of two. An alternative simple formula can be obtained by setting n equal to 1 and m equal to zero, i.e. for a monomeric species with no bound water.

The assumption of explicit solvation numbers causes difficulties since the choice

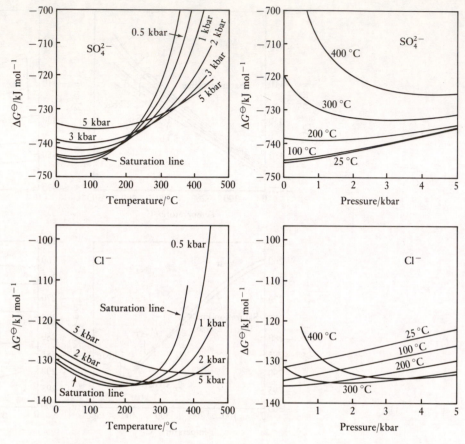

Figure 14.14 Apparent Gibbs energies of formation for SO_4^{2-} and Cl^- anions (data from Tanger and Helgeson, 1988).

of formula dictates the stoichiometry of any subsequent reaction of the species. For example, the precipitation of quartz could be written:

$$(SiO_2) \cdot 2H_2O \rightleftharpoons (SiO_2)_{quartz} + 2H_2O \qquad (14.114)$$

or alternatively,

$$(SiO_2)_{aq} \rightleftharpoons (SiO_2)_{quartz} \qquad (14.115)$$

if the simple formula was chosen. In one convention it is necessary to include a water activity term in the equilibrium constant; in the other convention it is not. This affects the prediction of solubilities in concentrated solutions. The situation becomes even more complicated since the solvation number actually changes with temperature and pressure. The easiest way to deal with this second difficulty is to avoid specifying the solvation numbers in reaction equations and to include solvation implicitly within the standard partial molar properties. We use exactly this procedure for inorganic ions, which we know solvate strongly.

Figure 14.15 shows the standard partial molar volume of aqueous silica along with that of a strong electrolyte for comparison. The molecular formula of aqueous silica is assumed to contain no water, in which case we might expect solvation and

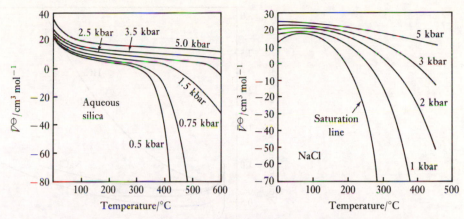

Figure 14.15 Standard molar volume of aqueous silica (data from Walther and Helgeson, 1977).

electrostriction. Indeed, $\bar{V}^{\circ}$ for aqueous silica shows similar trends to that of the strong electrolyte. The major difference is that the decrease in $\bar{V}^{\circ}$ with temperature in the region of $P-T$ space approaching the critical point is proportionally less. This implies that the solvation contribution to electrostriction is less pronounced. An additional missing feature is the maximum with temperature at low pressures. This implies that the local solvent collapse contribution to electrostriction is either negative or much less than the solvation term. A smaller local collapse term is to be expected in view of the small charge that must be on the molecule.

The general implication of figure 14.15 is that an equation of state such as equation 14.78 can be used to describe the properties of polar neutral molecules with a suitable choice of fitting parameters.

This approach avoids the complexity of explicit reference to solvation numbers. It is not explicitly wrong to use solvation numbers; indeed many properties of aqueous solutions can be explained in these terms.

14.3   The properties of concentrated aqueous electrolyte solutions

14.3.1   General trends for stoichiometric activity coefficients

One factor controlling the value of $\gamma_{\pm}$ is the **ionic strength** of the solution (I). Ionic strength is a measure of the charge density in the solution phase and is defined as:

$$I \equiv \frac{1}{2} \sum_i m_i z_i^2 \tag{14.116}$$

where z_i is positive for cations and negative for anions. The summation is over the molal concentrations all aqueous **solute** species. For example, a neutral solution containing 0.5 mol kg^{-1} chloride, 0.25 mol kg^{-1} sulphate, 0.4 mol kg^{-1} potassium and 0.3 mol kg^{-1} calcium has an ionic strength of:

$$I = \tfrac{1}{2}(0.5 \times 1^2 + 0.25 \times 2^2 + 0.4 \times 1^2 + 0.3 \times 2^2) = 1.05/\text{mol kg}^{-1} \tag{14.117}$$

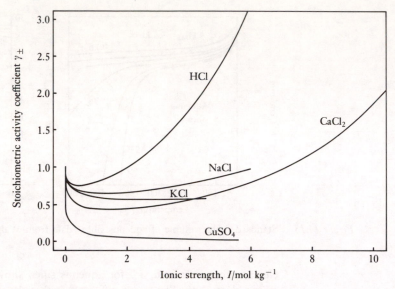

Figure 14.16   Ionic strength dependence of the stoichiometric activity coefficient $\gamma_\pm$ (data from Robinson and Stokes, 1959).

Figure 14.16 shows the ionic strength dependence of values of $\gamma_\pm$, for selected salts at 298.15 K.

The activity coefficients are constrained to unity in the infinitely dilute solution and generally show an initial decrease with increasing ionic strength. At ionic strengths in excess of 1 mol kg^{-1} there is usually a minimum in the value of $\gamma_\pm$ followed by a notable increase. Activity coefficients for salts composed of highly charged ions show the most dramatic decrease at low ionic strengths. These trends are observed for most salts at low temperatures and pressures, although there are exceptions, as demonstrated by the data for potassium sulphate.

Increasing the temperature, at constant ionic strength, generally decreases the value of the activity coefficient. This observation is demonstrated in figure 14.17 showing data for sodium chloride on the saturation line at different temperatures. In contrast, increasing pressure generally increases the value of the activity coefficient. The effect due to pressure is most marked at high temperatures and high ionic strengths (figure 14.18).

The minimum in the ionic strength dependence of $\gamma_\pm$ suggests there are opposing forces at work. These forces fall into three major categories which strictly are not separable. However, it is common to separate the mean activity coefficient for the kth salt into three terms, i.e.

$$\log \gamma_{\pm k} = \log \gamma_{\pm k}^{\text{el}} + \log \gamma_{\pm k}^{\text{solv}} + \log \gamma_{\pm k}^{\text{assoc}} \qquad (14.118)$$

where $\log \gamma_{\pm k}^{\text{el}}$ is an electrostatic contribution, $\log \gamma_{\pm k}^{\text{solv}}$ is a solvation contribution and $\log \gamma_{\pm k}^{\text{assoc}}$ is a contribution due to ion association. This is consistent with the relationship:

$$G_k^{\text{E}} = \Delta G_k^{\text{el}} + \Delta G_k^{\text{solv}} + \Delta G_k^{\text{assoc}} \qquad (14.119)$$

where G_k^{E} is the excess Gibbs energy for one mole of the salt.

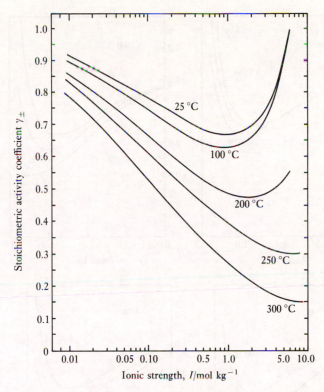

Figure 14.17   The stoichiometric activity coefficient for sodium chloride versus ionic strength at various temperatures (data from Helgeson *et al.*, 1981).

Generally, the electrostatic contribution to the activity coefficients progressively decreases with ionic strength, as does the ion association term. In contrast, the solvation contribution increases with ionic strength.

14.3.2   Deviation from ideality due to electrostatic forces

The electrostatic contribution to activity coefficients is usually calculated from Debye–Hückel theory. This approach yields simple equations for activity coefficients in dilute solutions. Two useful expressions are as follows.

The Debye–Hückel limiting law   The **Debye–Hückel limiting law** links the ionic strength to the electrostatic contribution to the **rational activity coefficient** of a salt. The expression for the kth binary salt composed of cation i of charge z_i+ and anion j of charge z_j- is:

$$\ln f_{\pm k} = -|z_i z_j| A_\gamma I^{1/2} \qquad (14.120)$$

Theoretical studies and experiments have shown that this law represents behaviour to which all ionic species tend as the ionic strength drops below $10^{-5}\,\mathrm{mol\,kg^{-1}}$. The relationship is a measure of the long-range electrostatic forces between ions.

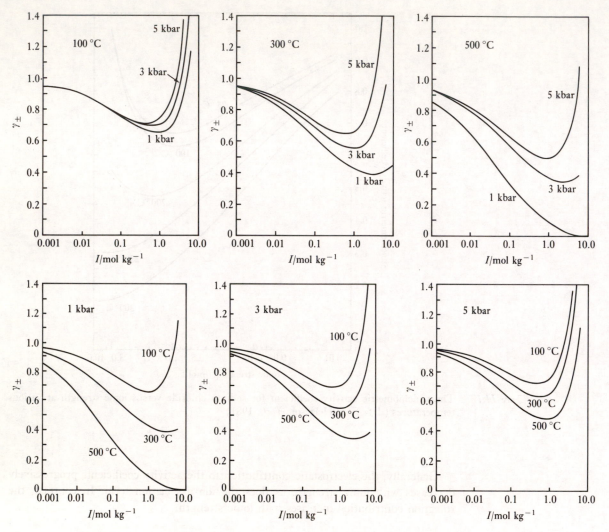

Figure 14.18   General temperature and pressure dependence of $\gamma_\pm$ for NaCl (from Helgeson *et al.*, 1981).

The extended Debye–Hückel equation

For more concentrated solutions we use the expression:

$$\log f_{\pm k} = -|z_i z_j| \left(\frac{A_\gamma I^{1/2}}{1 + B_\gamma \mathring{a} I^{1/2}} \right) \qquad (14.121)$$

This is the **extended Debye–Hückel equation**. The numerator of the right-hand side of equation 14.121 is equal to the long-range electrostatic contribution to non-ideality assuming that ions are point charges. The denominator in equation 14.121, i.e. $(1 + B \mathring{a} I^{1/2})$, accommodates the fact that ions are not point charges. The symbol $\mathring{a}$ is the distance of closest approach. It may be thought of as the diameter of the ion, although a clear physical significance cannot be truly ascribed to this distance. This term is a major weakness in the Debye–Hückel equation since it assumes ions to be non-deformable spheres of equal radii. This is hopelessly idealised; however, theoretical studies have suggested that with a suitable value for $\mathring{a}$ the basic equation may be valid for completely dissociated monovalent electrolyte solutions at concentrations up to $1 \, \text{mol dm}^{-3}$. Consequently, the extended Debye–Hückel equation is used in most semi-empirical expressions for activity coefficients.

Box 14.2

Deriving the Debye–Hückel limiting law

To dervie an expression for $\gamma_{\pm}$ we must consider the rational activity coefficient f_j^{el} of a single ion of charge z_j. This is defined by:

$$\mu_j = \mu_j^{\ominus} + RT \ln x_j + RT \ln f_j^{el} \tag{14.122}$$

where $RT \ln f_j^{el}$ is an **excess energy per mole**. Equation 14.122 represents a situation where the only contribution to non-ideality is electrostatic.

The excess energy per ion is:

$$kT \ln f_j^{el} \tag{14.123}$$

and is the non-ideal contribution to the chemical potential.

The critical step is to equate $kT \ln f_j^{el}$ to an expression for the non-ideal contribution to mixing of ions, ΔG^{el}. This expression should give the total energy of electrostatic interaction between ion j and all other ions in the solution in terms of solution composition. An equation valid at extreme dilution is (section 4.5):

$$\Delta G^{el} \equiv -\frac{z_j^2 e^2 \kappa}{8\pi\varepsilon_0\varepsilon_r} \tag{14.124}$$

where e is the charge on one electron, ε_0 is the permittivity of a vacuum and ε_r the relative permittivity of the fluid. The variable κ has the dimensions of reciprocal length and is called the **Debye inverse length**. It is defined as:

$$\kappa^2 = \frac{2Le^2\rho^{\bullet}}{\varepsilon_0\varepsilon_r kT} I \tag{14.125}$$

where L is Avogadro's constant and k is the Boltzmann constant.

We may therefore write:

$$kT \ln f_j^{el} = \Delta G^{el} = -\frac{z_j^2 e^2 \kappa}{8\pi\varepsilon_0\varepsilon_r} \tag{14.126}$$

Substituting for κ in equation 14.126 to give:

$$\ln f_j^{el} = -z_j^2 \left\{ \left[\frac{e^3(2L)^{1/2}}{8\pi(\varepsilon_0 k)^{3/2}} \right] \left(\frac{\rho^{\bullet}}{\varepsilon_r^3 T^3} \right)^{1/2} \right\} I^{1/2} \tag{14.127}$$

The term in square brackets is grouped this way to distinguish the constant terms from the temperature- and pressure-dependent term $(\rho^{\bullet}/\varepsilon_r^3 t^3)^{1/2}$.

It is usual to express equation 14.127 in decadic logarithms. Thus, collecting the constant terms and performing the log transformation gives:

$$\log f_j^{el} = -A_\gamma z_j^2 I^{1/2} \tag{14.128}$$

continued

Box 14.2
(c continued

where

$$A_\gamma = \left[\frac{e^3(2L)^{1/2}}{8\pi(\varepsilon_0 k)^{3/2}}\right]\left(\frac{\rho^\bullet}{\varepsilon_r^3 T^3}\right)^{1/2}$$

$$= 1.824\,829 \times 10^6 \left(\frac{\rho^\bullet}{\varepsilon_r^3 T^3}\right)^{1/2} \tag{14.129}$$

The units of $\rho^\bullet$ are $cm^3\ g^{-1}$ and the units of A_γ are $mol^{-1/2}\ kg^{1/2}\ K^{1/2}$.

Equation 14.128 is for a single ion. It can be applied to the kth neutral electrolyte composed of v_{ik} moles of cation i and v_{jk} moles of anion j to give:

$$(v_{ik} + v_{jk})\log f_{\pm k}^{el} = v_{ik}\log f_i^{el} + v_{jk}\log f_j^{el} \tag{14.130}$$

Substituting equation 14.128 for both ions into equation 14.130 gives:

$$(v_{ik} + v_{jk})\log f_{\pm k} = v_{ik}(-z_i^2 A_\gamma I^{1/2}) + v_{jk}(-z_j^2 A_\gamma I^{1/2}) \tag{14.131}$$

The condition of electroneutrality imposes the restriction that:

$$|v_{ik}z_i| = |v_{jk}z_j| \tag{14.132}$$

Therefore, the terms v_{ik} and v_{ik} and v_{ik} can be eliminated from equation 14.131 to give:

$$\ln f_{\pm k} = -|z_i z_j| A_\gamma I^{1/2} \tag{14.133}$$

Deriving the extended Debye–Hückel equation

To derive this relationship we equate $kT \ln f_j^{el}$ to ΔG^{el} as given by section 4.5:

$$\Delta G^{el} \equiv -\frac{z_j^2 e^2 \kappa}{8\pi\varepsilon_0\varepsilon_r}\left(\frac{\kappa}{1 + \kappa \mathring{a}}\right)$$

That is,

$$\ln f_j^{el} = -z_j^2\left(\frac{e^2}{8\pi\varepsilon_0\varepsilon_r kT}\right)\left(\frac{\kappa}{1 + \kappa\mathring{a}}\right) \tag{14.134}$$

Collecting the constant terms and converting to decadic logarithms gives:

$$\log f_j^{el} = -z_j^2\left(\frac{A_\gamma I^{1/2}}{1 + B_\gamma \mathring{a} I^{1/2}}\right) \tag{14.135}$$

where B_γ is another Debye–Hückel parameter defined as:

$$B_\gamma = \left(\frac{2Le^2}{\varepsilon_0 k}\right)^{1/2}\left(\frac{\rho^\bullet}{\varepsilon_r T}\right)^{1/2} \tag{14.136}$$

$$= 50.291\,587 \times 10^8 \left(\frac{\rho^\bullet}{\varepsilon_r T}\right)^{1/2} \tag{14.137}$$

If the units of $\rho^\bullet$ are $cm^3\ g^{-1}$, then B_γ has units of $cm^{-1}\ mol^{-1/2}\ kg^{1/2}\ K^{1/2}$ and $\mathring{a}$ has units of cm.

The value of B_γ is temperature- and pressure-dependent through the term $(\rho^\bullet/\varepsilon_r T)^{1/2}$. The corresponding equation for the stoichiometric activity coefficient is:

$$\log f_{\pm k} = -|z_i z_j|\left(\frac{A_\gamma I^{1/2}}{1 + B_\gamma \mathring{a} I^{1/2}}\right) \tag{14.138}$$

Practical activity coefficients

For most applications we are interested in determining the electrostatic contribution to the **practical activity coefficient** of a single ion or neutral salt. This is defined by the fundamental equation for chemical potential:

$$\mu_j = \mu_j^{\ominus} + RT \ln m_j + RT \ln \gamma_j \tag{14.139}$$

At low concentrations the rational activity coefficient is approximately equal to the practical activity coefficient. However, as the ionic strength increases it becomes necessary to make the correction for the different concentration scales described in section 14.1.3. That is:

$$\log \gamma_j^{\text{el}} = \log f_j^{\text{el}} - \log\left(1 + 0.018 \sum_i m_i\right) \tag{14.140}$$

Defining the function Γ as

$$\Gamma = -\log\left(1 + 0.018 \sum_i m_i\right) \tag{14.141}$$

we have

$$\log \gamma_j^{\text{el}} = -z_j^2\left(\frac{A_\gamma I^{1/2}}{1 + B_\gamma \mathring{a} I^{1/2}}\right) + \Gamma \tag{14.142}$$

and

$$\log \gamma_{\pm k}^{\text{el}} = -|z_i z_j|\left(\frac{A_\gamma I^{1/2}}{1 + B_\gamma \mathring{a} I^{1/2}}\right) + \Gamma \tag{14.143}$$

For a solution containing a single salt, where the molality of the cation $m_i = v_{ik}m_k$ and for the anion $m_j = v_{jk}m_k$, we have:

$$\log \gamma_{\pm k}^{\text{el}} = -|z_i z_j|\left(\frac{A_\gamma I^{1/2}}{1 + B_\gamma \mathring{a} I^{1/2}}\right) - \log(1 + 0.018 v_k m_k) \tag{14.144}$$

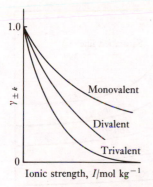

Figure 14.19 The electrostatic contribution to $\gamma_{\pm k}$ (data from Helgeson *et al.*, 1981).

Figure 14.19 shows some predicted values of $\gamma_{\pm k}^{\text{el}}$ for idealised salts at 298.15 K and 1 bar.

If the only contribution to non-ideality was electrostatic, then $\log \gamma_{\pm k}^{\text{el}}$ would be equal to the stoichiometric activity coefficient.

The effect of temperature and pressure

The effect of temperature and pressure on the electrostatic contribution to the activity coefficient is contained entirely within the temperature and pressure dependence of A_γ and B_γ. Values of A_γ and B_γ can be calculated using methods described in sections 13.2.1 and 13.6. Figures 14.20 and 14.21 show data for A_γ and B_γ at elevated temperature and pressure: it can be seen that both A_γ and B_γ decrease with increasing pressure. The relative effect of pressure becomes less at pressures above 3000 bar. There is a general increase with increasing temperature below 500 °C, at which point the density contribution becomes progressively more important than the relative permittivity, giving rise to extrema.

The effect of changes in A_γ and B_γ on activity coefficients are difficult to infer from these figures. In general, increasing the temperature at any pressure tends to lower the activity coefficient, whereas increasing the pressure at any fixed temperature will increase the value of the activity coefficient.

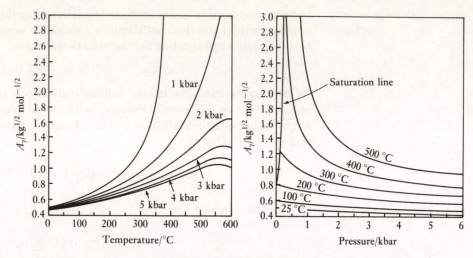

Figure 14.20 A_γ versus temperature and pressure (from Helgeson and Kirkham, 1974b).

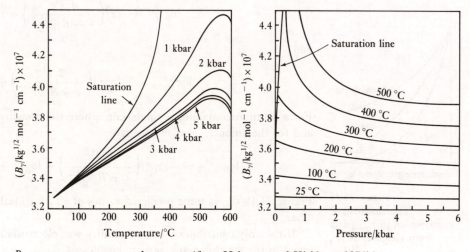

Figure 14.21 B_γ versus temperature and pressure (from Helgeson and Kirkham, 1974b).

The effects due to pressure become more significant at elevated temperatures, and temperature effects are most pronounced in the 0–1 kbar region. These two observations are common for any property of an aqueous species which is influenced by the electrostatic properties of water.

14.3.3   Deviation from ideality due to ion solvation

Forces of attraction between a charged ion and the permanent dipoles on water molecules cause a realignment of water molecules in the near-vicinity of the ion. Ions therefore exist within a sphere of water molecules closely bound to the ion. These water molecules have lost their independent translational motion and move around with the ion as a single entity. The bound molecules contribute to the size of the mobile species, which has an effective radius greater than its bare crystallographic radius.

The effects due to changes in concentration   Solvation has a direct effect on the activity coefficient of a neutral salt or single ion. Qualitatively, this is equivalent to reducing the water content of the system by the amount bound to the ion. The effect is to increase the apparent concentration of the ion and therefore increase its activity coefficient.

An approximate quantitative treatment by Robinson and Stokes (1959) emphasises this point well. This approach involves separating the mean stoichiometric activity coefficient into a simple electrostatic term log $\gamma^{el}_{\pm k}$ plus a solvation term log $\gamma^{solv}_{\pm k}$. The equation for $\gamma_{\pm k}$ has the form:

$$\log \gamma_{\pm k} = \log \gamma^{el}_{\pm k} - \frac{h}{v_k} \log a_w - \log(1 - 0.018 h m_k) \qquad (14.145)$$

The solvation term log $\gamma^{solv}_{\pm k}$ is equal to $[-(h/v_k)\log a_w - \log(1 - 0.018 h m_k)]$ where h is the number of moles of solvent bound to each mole of the salt and a_w is the activity of water in the solution.

At all concentrations the additional terms contributing to log $\gamma^{solv}_{\pm k}$ are positive, which leads to the expected increase in log $\gamma_{\pm k}$ over and above the electrostatic term.

The effect of solvation is significant. Figure 14.22(a) shows the activity coefficient for a hypothetical 1:1 salt, calculated with and without the solvent terms. To make the calculation simple the water activity is estimated by $a_w = 1 - 0.04m$ and h set to 4, which is not unreasonable for sodium chloride at 298 K. At concentrations above 1 mol kg^{-1} the solvent terms start to dominate, leading to activity coefficients

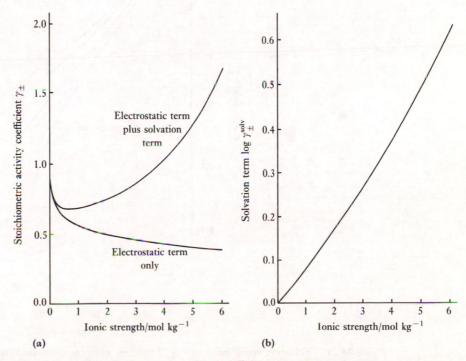

Figure 14.22   The general effect of solvation on activity coefficients (calculated using equation 14.145).

much greater than unity at high concentrations. Inadequacies in the calculations arise from the simplistic expression for water activity and the assumption that h is independent of the concentration level.

Figure 14.22(b) shows just the solvation term, calculated for the hypothetical salt. The term shows an almost linear dependence on ionic strength. The linearity of the solvation term was proposed by Hückel (1925). This idea has been recently reinforced by Helgeson *et al.* (1981), who deduced the relationship from an analysis of the relative permittivities of electrolytes. From their analysis (which is an alternative to the approach of Robinson and Stokes) we may write for a single salt:

$$\log \gamma_{\pm k}^{solv} = \omega_k^{abs} b_k I \tag{14.146}$$

where ω_k^{abs} is the Born parameter for the salt, defined as $\omega_k^{abs} = \sum_j v_{jk} \omega_j^{abs}$ where ω_j^{abs} is the Born parameter for ion j, and b_k a temperature- and pressure-dependent parameter for the salt.

This approach to the solvation correction is informative because the parameters ω_k^{abs} and b_k have been evaluated for sodium chloride over a large range of temperatures and pressures. Figure 14.23 shows the temperature and pressure dependence of log

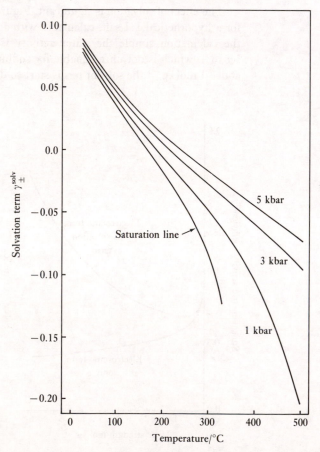

Figure 14.23   Solvation term for $1 \, \text{mol} \, \text{kg}^{-1}$ sodium chloride solution versus temperature at different pressures (data from Helgeson *et al.*, 1981).

$\gamma_{\pm k}^{\text{solv}}$ for NaCl(aq) at 1 mol kg^{-1} using such data. It gives some insight into the effects we might expect in electrolytes generally. We observe that an isobaric increase in temperature results in a reduction in the solvation term. This contributes to the generally observed trend of lower activity coefficients at higher temperatures.

14.3.4   Deviations from ideality due to ion association

General features   The ability of ions in aqueous solutions to associate is well established. However, the concept of an ion-pair of complex is nebulous. Bjerrum in 1926 introduced the idea of an ion-pair to compensate for any ions that may be closer together than the minimum distance of approach required by the extended Debye–Hückel equation. Ion-pairs in this class were treated as a single ion carrying a charge equal to the sum of the charges on the ions from which they are derived. The quantity of this ion-pair was calculated from mass action, although this formalism was not meant to imply that the ion-pair was a stable molecular entity. However, for weak electrolytes and many strong electrolytes such as sulphates, carbonates and salts of transition metals, a stable molecular complex is apparent from conductivity and freezing point determinations. Regardless of the nature of the species, ion association has an effect on stoichiometric or single ion activity coefficients.

One effect is on the electrostatic interaction directly. This is easily accommodated by including the associated species in the equation for ionic strength. The ionic strength defined this way is called the 'true' ionic strength and is given the symbol $\bar{I}$.

For example, in a neutral solution of calcium chloride at the concentration m_{CaCl_2}, the total concentration of the ions are:

$$m_{\text{Ca}^{2+}} = m_{\text{CaCl}_2} \quad \text{and} \quad m_{\text{Cl}^-} = 2 \times m_{\text{CaCl}_2} \tag{14.147}$$

The stoichiometric ionic strength is:

$$I = \tfrac{1}{2}(m_{\text{Ca}^{2+}} \times 2^2 + 2 \times m_{\text{Cl}^-}) \tag{14.148}$$

However, if an ion-pair is formed by the reaction:

$$\text{Ca}^{2+} + \text{Cl}^- \rightleftharpoons \text{CaCl}^+ \tag{14.149}$$

the true concentrations of calcium and chloride would be lower than that implied by the molal concentration m_{CaCl_2} and equal to $\bar{m}_{\text{Ca}^{2+}}$ and $\bar{m}_{\text{Cl}^-}$. If the concentration of the ion-pair is $\bar{m}_{\text{CaCl}^+}$, the true ionic strength is:

$$\bar{I} = \tfrac{1}{2}(\bar{m}_{\text{Ca}^{2+}} \times 2^2 + 2 \times \bar{m}_{\text{Cl}^-} + \bar{m}_{\text{ClCl}^+}) \tag{14.150}$$

In general, for a multicomponent electrolyte solution that contains N ions and N^{c} associated species, the true ionic strength is given by:

$$\bar{I} = \sum_{i=1}^{N} \bar{m}_l z_l^2 + \sum_{p=1}^{N^{\text{c}}} \bar{m}_p^{\text{c}} z_p^2 \tag{14.151}$$

where $\bar{m}_l$ is the true concentration of the lth ion and $\bar{m}_p^{\text{c}}$ is the concentration of the pth complex or ion-pair.

The value of γ_k^{el}, the electrostatic contribution to the activity coefficient for the salt, is then calculated from:

$$\log \gamma_k^{\text{el}} = -|z_i z_j| \left(\frac{A_\gamma \bar{I}^{1/2}}{1 + B_\gamma \mathring{a} \bar{I}^{1/2}} \right) - \log \left[1 + 0.018 \left(\sum_l \bar{m}_l + \sum_p \bar{m}_p^{\text{c}} \right) \right] \tag{14.152}$$

The effect is to **increase** slightly the value of the activity coefficient, since the ionic strength is reduced.

A second, and major, effect takes into account the actual reduction in the ion concentrations due to ion association. The equation which describes the effect on the mean molal stoichiometric activity coefficient of a salt is:

$$\log \gamma_{\pm k} = (\log \gamma_{\pm k}^{\text{el}} + \log \gamma_{\pm k}^{\text{solv}}) + \frac{1}{\nu_k} \log\left(\frac{\bar{m}_i^{\nu_{ik}} \bar{m}_j^{\nu_{jk}}}{m_i^{\nu_{ik}} m_{jk}^{\nu_{jk}}} \right) \qquad (14.153)$$

We may collect the electrostatic and solvation contributions into a single term, $\bar{\gamma}_{\pm k}$. This term is often called the **true mean activity coefficient** of a salt. There is an analogous term for the single-ion activity coefficient. The derivation of equation 14.153 is in Appendix 1.

By inspection of equation 14.153 we can see that the term

$$\frac{1}{\nu_k} \log\left(\frac{\bar{m}_i^{\nu_{ik}} \bar{m}_j^{\nu_{jk}}}{m_i^{\nu_{ik}} m_j^{\nu_{jk}}} \right)$$

will always be negative when ion association occurs because the true ion concentrations are less than the total ion concentrations. Therefore, the major effect is to **reduce** the value of the stoichiometric activity coefficient from that given by the combined electrostatic and solvation terms ($\bar{\gamma}_{\pm k}$).

It is possible to subdivide the true mean activity coefficient further into the true single ion activity coefficients $\bar{\gamma}_i$ and $\bar{\gamma}_j$ using:

$$\bar{\gamma}_{\pm k} \equiv (\bar{\gamma}_i^{\nu_{ik}} \bar{\gamma}_k^{\nu_{jk}})^{1/\nu_k} \qquad (14.154)$$

Equations 14.153 and 14.154 lead to the conclusion that the following relationships must hold:

$$a_i = m_i \gamma_i = \bar{m}_i \bar{\gamma}_i \qquad (14.155)$$

and

$$a_j = m_j \gamma_j = \bar{m}_j \bar{\gamma}_j \qquad (14.156)$$

Figure 14.24 shows the magnitude of the effect for a hypothetical system with varying degrees of ion association. In the infinitely dilute solution the proportion of ions (α) converted to an ion-pair is zero. At elevated concentrations the value of α commonly increases with concentration. Data in figure 14.24 are calculated assuming that α increases linearly with concentration to a maximum value $[\alpha_{(\text{max})}]$ at the highest concentration. This assumption of linearity is for demonstration purposes only and may not occur in real systems.

In addition to showing the effect of ion association on the stoichiometric activity coefficient, equation 14.153 is an equation of constraint. If the actual species concentrations are known, then the values of the true single-ion activity coefficients are constrained by the known values of the stoichiometric activity coefficients. In this way any model proposing to describe the true single-ion activity coefficients can be validated. The reverse is also true: the extent of ion association can be calculated from the stoichiometric activity coefficients if a suitable model for $\bar{\gamma}_{\pm k}$ is adopted.

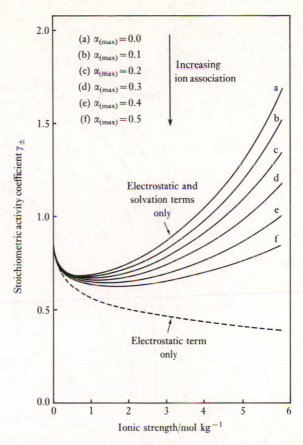

Figure 14.24   The effect of ion association on activity coefficients (calculated using equation 14.153).

The effects of temperature, pressure and ionic strength on ion association

In general, increasing temperature promotes ion association, whereas increasing pressure promotes dissociation of species. Typically, changes in temperature directly influence electrostatic ion−ion interactions, whereas the effect of pressure is to change the equilibrium constant by making subtle modifications to the ion−solvent interactions. The reasoning is as follows.

High temperatures and low water densities promote ion association. This is mainly a response to the changes in the relative permittivity of water. As the relative permittivity **decreases**, the strength of electrostatic interactions **increases**. The relative permittivity of water decreases with increasing temperature, but generally increases with increasing pressure. Figure 14.25 shows the formation constant for the sodium chloride ion-pair, and emphasises the expected trend of association increasing with increasing temperature but decreasing with increasing pressure. We should expect increased thermal motion at elevated temperatures to reduce ion association, but the effect due to decreasing relative permittivity wins out.

Along the saturation line the effect is especially noticeable. Here, as temperature increases water progresses into a low-density region near the critical point and the association constant changes by orders of magnitude.

The effect of temperature is dominated by ion−ion interactions. However, it is known that the strength of ion−solvent interactions increases with increasing

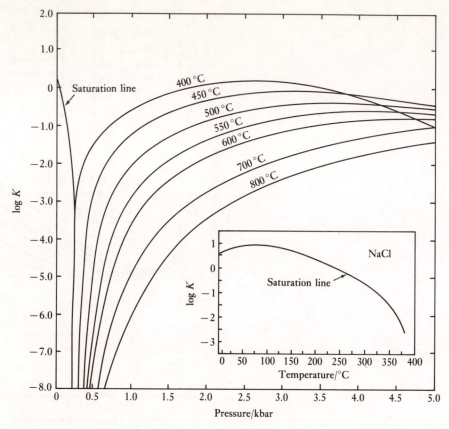

Figure 14.25　The association constants for sodium chloride (data from Helgeson *et al.*, 1981; Seward, 1981).

temperature. Thus, at high temperatures, the probability that an ion can form an inner-sphere complex by displacing a solvated water molecule from a neighbouring ion becomes less. This tendency is to reduce the equilibrium constant with increasing temperature and is most pronounced for associated species which contain several anions displacing many water molecules. Thus, the stability of high-coordination complexes such as $AgCl_3^{2-}$, $AgCl_4^{3-}$, $FeCl_3^0$ and $FeCl_4^-$ decreases with increasing temperature.

Additionally, the relative permittivity of water decreases with increasing ionic strength. This is a direct consequence of the breakdown in the structure of water in concentrated electrolytes. This increases further the tendency of ions to associate (figure 14.26).

The effect of pressure on $\varepsilon_{r(s)}$ is slight compared with the effect of temperature. Typically, a change in pressure of 1000 bar leads to a change in the value of $\varepsilon_{r(s)}$ of only a few per cent. From this we expect the effect of pressure on ion association to be slight. In fact, the effect of pressure is generally not negligible and we must look to ion–solvent interactions to explain this effect.

The effect of pressure on an equilibrium constant is given by:

$$\Delta V^{\ominus} = -RT\left(\frac{\partial \ln K}{\partial P}\right)_T \tag{14.157}$$

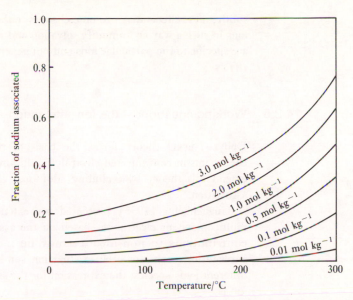

Figure 14.26 The extent of NaCl association versus temperature (NaCl concentrations are shown on the curves).

where $\Delta V^{\ominus}$ is the standard molar volume change for the reaction equal to the sum of the molar volumes of the products minus that for the reactants. In section 14.2.3 it was proposed that the standard molar volume of an ionic species is the sum of an intrinsic component and an electrostriction component, i.e.

$$\bar{V}^{\ominus} = V_{int} + \Delta V_{elect} \tag{14.158}$$

The electrostriction term is itself composed of two terms arising from solvation and local solvent collapse.

For ions and complexes, V_{int} is relatively independent of temperature and pressure. In contrast, the value of ΔV_{elect} is dependent on ionic charge, temperature and pressure and is generally negative. It is more negative for ions of high valence charge.

In a reaction to form an associated species from ions there is a neutralisation of charge. Thus, the value of ΔV_{elect} for the associated species is likely to be less negative than the sum of the electrostriction terms for the free ions. That is, the value of $\Delta V^{\ominus}$ for the overall reaction will be positive. Equation 14.157 shows that this leads to a decrease in the equilibrium constant with increasing pressure. Therefore, **increasing pressure favours dissociation of associated species**.

There is a need for further modification to this argument since inner-sphere complexes are expected to have a more effective neutralisation of ion charge than outer-sphere complexes because the ions are in closer contact. Consequently, the value of ΔV_{elect} for an inner-sphere complex is expected to be larger (less negative) than for a similar outer-sphere complex. This leads to greater values of $\Delta V^{\ominus}$ for inner-sphere complexes than for outer-sphere complexes. The result is that increasing pressure causes inner-sphere complexes to dissociate more than similar outer-sphere complexes.

Other factors play a role in ion association. In particular, the electronic structure of transition-metal complexes may be modified by changes in pressure. Although subtle, these effects may promote changes to the coordination chemistry of the metal ions in such a way as to modify solvation and ion association. These types of effects are specific to the particular ions and not generalised as easily as simple electrostatic effects.

14.3.5   Working equations – the ion-interaction approach

Debye–Hückel theory forms the basis of most working equations for activity coefficients in concentrated electrolyte solutions. Many attempts have been made to extend this theory by including additional contributions from solvation and ion association.

Guggenheim (1935) presented an extension of Debye–Hückel theory for multicomponent electrolytes based on the assumption that **all non-electrostatic contributions to non-ideality may be incorporated into a single ion–ion interaction term** for each existing ion combination. For a system containing N different ionic species, the expression for the single-ion activity coefficient of the jth ion is:

$$\log \gamma_j = \frac{-z_j^2 A_\gamma I^{1/2}}{1 + I^{1/2}} + 2 \sum_{i=1}^{N} \hat{\beta}_{jl} m_l + \Gamma \tag{14.159}$$

where $\hat{\beta}_{jl}$ is a coefficient accommodating interactions between the ion j and all other ions denoted l. These coefficients are defined to be zero for ions of like charge. This makes the equation consistent with Brønsted's theory of specific interaction, which proposes that only ions of like charge interact. The ionic-strength term is the stoichiometric ionic strength rather than the true ionic strength since there is no attempt to account explicitly for ion association.* Guggenheim (1935) acknowledged that the interaction terms cannot be attributed to any single factor and they compensate for all contributions, including ion association, solvation and inadequacies in the electrostatic term. The Brønsted restriction is only valid if the terms represent ion association. However, it is retained to simplify the application of equation 14.159.

For a neutral salt the stoichiometric activity coefficient is:

$$\log \gamma_{\pm k} = \frac{-z_i z_j A_\gamma I^{1/2}}{1 + I^{1/2}} + \frac{2z_j}{z_i + z_j} \sum_{l=1} \hat{\beta}_{jl} m_l + \frac{2z_i}{z_i + z_j} \sum_{l=1} \hat{\beta}_{il} m_l + \Gamma \tag{14.160}$$

which simplifies for a single-salt solution to:

$$\log \gamma_{\pm k} = \frac{-z_i z_j A_\gamma I^{1/2}}{1 + I^{1/2}} + \frac{4z_i z_j}{z_i + z_j} \hat{\beta}_k m_k + \Gamma \tag{14.161}$$

where $\hat{\beta}_k = \hat{\beta}_{ij} = \hat{\beta}_{ji}$.

Despite its simplicity, equation 14.159 may still be the most useful equation available for electrolyte mixtures at ionic strengths below 0.1 mol dm^{-3}. In practice,

---

* Equation 14.159 omits $B_\gamma \mathring{a}$ from the denominator in the electrostatic term. This is because $\mathring{a}$ is different for different ion combinations, which cannot be consistent with the Gibbs–Duhem equation. Consequently, in equation 14.159 it is assumed that $B_\gamma \mathring{a}$ equals 1. This restricts the validity of the electrostatic term but ensures the equation is thermodynamically consistent for multicomponent solutions, whereas equation 14.121 is strictly applicable in single-salt solutions only.

the procedure would be to evaluate $\hat{\beta}_{il}$ for various ion combinations by analysis of activity coefficient data for pure electrolytes. The fact that only one interaction parameter is required for each stoichiometric activity coefficient makes this a simple procedure. However, for more concentrated solutions it must be acknowledged that $\hat{\beta}_{jl}$ varies with ionic strength.

There have been many treatments of electrolyte solutions designed to extend the Guggenheim approach or develop similar techniques. One study of particular note is that by Pitzer (1973), who accommodated the compositional dependence of the ion–ion terms in the Guggenheim approach with a more generalised expression for the excess Gibbs energy. This involved a comprehensive array of ion–ion interaction terms which included terms for ions of like charge.

The general equation for the excess Gibbs energy was:

$$\frac{G^{\mathrm{E}}}{RT} = n_{\mathrm{w}}f(I) + \frac{1}{n_{\mathrm{w}}}\sum_i\sum_j \lambda_{ij}(I)n_in_j + \frac{1}{n_{\mathrm{w}}^2}\sum_i\sum_j\sum_k \mu_{ijk}n_in_jn_k \quad (14.162)$$

where n_{w} is the number of moles of water and n_i, etc., the number of moles of the subscripted ion (either cation or anion). The term $f(I)$ is an ionic strength-dependent parameter accommodating all electrostatic contributions, $\lambda_{ij}(I)$ is an interaction coefficient for the subscripted ions which is dependent on ionic strength, and finally μ_{ijk} is a third-order interaction coefficient which is assumed to be independent of ionic strength.

The activity coefficient and osmotic coefficient can be derived from the following derivatives of G^{E}:

$$\ln \gamma_i = \frac{1}{RT}\left(\frac{\partial G^{\mathrm{E}}}{\partial n_i}\right) \quad (14.163)$$

and

$$\phi_{\mathrm{w}} - 1 = \frac{1}{RT}\left(\frac{\partial G^{\mathrm{E}}}{\partial n_{\mathrm{w}}}\right) \quad (14.164)$$

The differentiations give rise to a range of ion–ion terms which are not determinable independently. Consequently, interaction terms of similar type were grouped together into a limited set of virial coefficients to give an equation for a single-ion activity coefficient of the form:

$$\ln \gamma_j = z_j^2 f^\gamma + \sum_{i=1}^N m_i\left(2B_{ji} + C_{ji}\sum_{k=1}^N z_km_k\right) + \sum_{i=1}^N m_i\left(2\Theta_{ji} + \sum_{k=1}^N m_k\psi_{jik}\right)$$
$$+ \sum_{i=1}^N\sum_{k=1}^N m_im_k(z_j^2 B'_{ik} + z_jC_{ik}) + \frac{1}{2}\sum_{i=1}^N\sum_{k=1}^N m_im_k\psi_{jik} \quad (14.165)$$

The terms B_{ji}, C_{ji}, Θ_{ij}, ψ_{jik} and B'_{jk} are virial coefficients for the interactions between the subscripted ions. The terms B_{ji}, B'_{jk} and C_{ji} are major parameters in the equation representing interactions between ions of different charge, i.e. they are zero for ions of like charge, whereas terms Θ_{ji} and ψ_{jik} are second-order terms for binary and triplet interactions respectively. The parameters B_{ji} and B' have compositional dependencies given by:

$$B_{ji} = 2\beta(0)_{ji} + \frac{2\beta(1)_{ji}}{\alpha(1)_{ji}I}\{1 - [1 + \alpha(1)_{ji}I^{1/2}]\exp[-\alpha(1)_{ji}I^{1/2}]\}$$

$$+ \frac{2\beta(2)_{ji}}{\alpha(2)_{ji}I}\{1 - [1 + \alpha(2)_{ji}I^{1/2}]\exp[-\alpha(2)_{ji}I^{1/2}]\} \qquad (14.166)$$

and

$$B'_{ik} = \frac{2\beta(1)_{ik}}{\alpha(1)_{ik}I}\{-1 + [1 + \alpha(1)_{ik}I^{1/2}]\exp[-\alpha(1)_{ik}I^{1/2}]\}$$

$$+ \frac{2\beta(2)_{ik}}{\alpha(2)_{ik}I}\{-1 + [1 + \alpha(2)_{ik}I^{1/2}]\exp[-\alpha(2)_{ik}I^{1/2}]\} \quad (14.167)$$

where $\alpha(1)_{ji}$ and $\alpha(2)_{ji}$ have the value 2 for 1:1 and 1:2 electrolytes and 1.4 for higher-valence electrolytes; $\alpha(2)_{ji}$ has the value 12.0. The term C_{ji} is related to a compositionally independent parameter C_{ji}^{ϕ} via:

$$C_{ji} = \frac{C_{ji}^{\phi}}{2(z_j z_i)^{1/2}} \qquad (14.168)$$

The other major term in equation 14.165 is the electrostatic term (f^{γ}), produced from the derivative of $f(I)$. For this term Pitzer substituted the pressure equation of statistical mechanics. This equation, which is a modern equivalent of the extended Debye–Hückel equation, has the form:

$$f^{\gamma} = -A^{\phi}\left[\frac{I^{1/2}}{1 + bI^{1/2}} + \frac{2}{b}\ln\left(1 + bI^{1/2}\right)\right] \qquad (14.169)$$

The parameter b represents a mean value for $B_{\gamma}\hat{a}$ having the value 1.2 for all solutions, and $A^{\phi} = 2.303A_{\gamma}/3$.

The Pitzer equation for the osmotic coefficient is:

$$(\phi_w - 1) = \frac{1}{\sum_{j=1}^{N}m_j}\left[2If^{\phi} + 2\sum_{i=1}^{N}\sum_{k=1}^{N}m_j m_i\left(B_{ji}^{\phi} + C_{ji}^{\phi}\sum_{l=1}^{N}z_l m_l/(z_i z_k)\right)\right.$$

$$\left. + \sum_{i=1}^{N}\sum_{k=1}^{N}\left(\Theta_{ik} + \sum_{l=1}^{N}m_l\psi_{jkl}\right)\right] \qquad (14.170)$$

where

$$B_{ji}^{\phi} = \beta(0) + \beta(1)_{ji}\exp[-\alpha(1)_{ji}I^{1/2}] + \beta(2)_{ji}\exp[-\alpha(2)_{ji}I^{1/2}] \qquad (14.171)$$

and

$$f^{\phi} = \frac{-A^{\phi}I^{1/2}}{1 + bI^{1/2}} \qquad (14.172)$$

Although algebraically complex, the Pitzer equation requires only the parameters $\beta(0)_{ji}$, $\beta(1)_{ji}$, $\beta(2)_{ji}$ and C_{ji}^{ϕ} for most salts, with the inclusion of the higher-order terms for only a limited number of situations.

Table 14.1 shows a range of measured activity coefficients compared with those calculated from the Pitzer equation using data in Appendix 2.

Table 14.1 Activity coefficients $\gamma\pm$, measured and predicted using the Pitzer equation

Conc'n/mol kg⁻¹	HCl	Pitzer	NaCl	Pitzer	CaCl₂	Pitzer	MgSO₄	Pitzer	HCl (50°C)	Pitzer
0.1	0.796	0.79567	0.778	0.77150	0.518	0.52032	0.150	0.16646	0.785	0.78391
0.5	0.757	0.75918	0.681	0.68007	0.448	0.44914	0.0675	0.07634	0.734	0.73873
1.0	0.809	0.81219	0.657	0.65609	0.500	0.50251	0.0485	0.05498	0.770	0.78478
2.0	1.009	1.01203	0.668	0.66804	0.792	0.80476	0.0417	0.04684	0.933	0.97014
2.5	1.147	1.15125	0.688	0.68778	1.063	1.07778	0.0439	0.04929		
3.0	1.316	1.31967	0.714	0.71391	1.483	1.47410	0.0492	0.05503		
3.5	1.518	1.52170	0.746	0.74577	2.08	2.04655				

A virtue of this approach is that it is supported by an extensive collection of values for the necessary parameters for over 100 salts at 25 °C with, in many cases, temperature derivatives valid up to ~ 50 °C (Pitzer, 1979). The technique has been tested extensively and shown to be valid for both major and trace components in concentrated electrolytes. Indeed, Harvie *et al.* (1982) reported remarkable success in the application to salt evaporites.

This virial coefficient approach is a semi-empirical fitting procedure and is valid over the ranges of experimental data used to determine the coefficients. It has no ability to predict activity coefficients outside these ranges where the fitting coefficients are not constrained.

In general the Pitzer equation is accurate at concentrations up to 2 or 3 mol dm^{-3} for solutions containing divalent ions and up to 6 mol dm^{-3}, or higher, for solutions dominated by monovalent ions. Comprehensive tables of Pitzer coefficients are given by Grant and Fletcher (1993). The Pitzer equation does not contain enough parameters to account for interactions between ions of like sign at high concentrations. Not does it account for ion—water interactions and water—water interactions under these conditions. For very concentrated solutions other expressions may be more appropriate although there is no single approach that is adequate for all salts over all concentrations. Other expressions are reviewed by Zemaitis *et al.* (1986) and Horvath (1985).

An equation commonly applied to concentrated electrolytes by the National Bureau of Standards is:

$$\log \gamma_{\pm k} = \frac{-A_\gamma |z_i z_j| I^{0.5}}{1 + B_\gamma \mathring{a} I^{0.5}} + C^{0.5} + DI^2 \ldots \tag{14.173}$$

where the terms C and D are empirical coefficients which are fitted to data but have no unambiguous theoretical significance (see Lietzke and Stoughton, 1962; Hammer and Wu, 1972).

14.3.6   Working equations – the ion association approach

The Pitzer equation for stoichiometric activity coefficients is a compact single equation requiring only the total concentrations of ions in solution and the necessary fitting coefficients. Its weakness is in the application to strongly associating and weak electrolytes. In these cases there is no reason to suppose that the excess Gibbs energy equation is suitable.

For many years geochemists have recognised the existence of associated ionic species and their influence on mineral solubilities. Within this field the practice has been to calculate **explicitly** the concentrations of the associated species, rather than absorb their effects in virial coefficients. This is simply an alternative way of approaching the same problem. However, the algebraic procedure is different since it involves using the equilibrium constants for ion association within a numerical scheme.

The numerical requirements   To perform calculations accommodating complex formation requires a procedure to calculate the species concentrations from the total ion concentrations in solution. For a solution containing N ions and N^c complexes, it must be postulated that the total concentration of the jth ion, m_j (either anion or cation), is equal to the concentration of the free uncomplexed ion plus the sum of the concentrations of the ion in each attendant complex. This can be described with the following mass balance equation:

$$m_j = \bar{m}_j + \sum_{i=1}^{N^c} q_{ji} \bar{m}_i^c \qquad (14.174)$$

where $\bar{m}_i^c$ is the molal concentration of the ith complex and q_{ji} is the number of moles of the jth ion in the ith complex. Here, we have N equations but $N + N^c$ unknowns (the concentrations of the free ions and the associated species). Clearly, we need more equations. To acquire these we acknowledge that the concentrations of the associated species are not independent but related to the concentrations of the unbound species by the overall formation constants for the reactions.

It is conventional to describe ion association with an equilibrium constant for the formation of the species from primary aqueous species. Primary aqueous species are those thought to arise from the dissociation of an electrolyte, although the distinction between primary species and associated species is arbitrary.

For each ion association reaction we have one equilibrium constant of the form:

$$K_i = \frac{\bar{m}_i^c \bar{\gamma}_i^c}{\prod_{j=1}^{N} (\bar{m}_j \bar{\gamma}_j)^{q_{ji}}}. \qquad (14.175)$$

There are therefore N^c equilibrium constants in total. Then, **for the situation when the activity coefficients are known**, there are $N + N^c$ soluble equations with $N + N^c$ unknowns. The species concentrations can now be determined using any one of a range of conventional techniques for solving simultaneous equations.

The procedure is simple. There is, however, a complication, since single-ion activity coefficients are evaluated using an equation involving **known** species concentrations. It is these concentrations we are trying to calculate. To solve this problem a common practice is to:

- estimate activity coefficients;
- solve for species concentrations using the estimates; and
- make an improved estimate of activity coefficients from which a better estimate of species concentrations can be made.

The process is repeated until a specified convergence criterion is met.

Using the example of an aluminium chloride solution in which the species $AlCl^{2+}$, $AlCl_2^+$, $AlCl_3^0$ and $AlCl_4^-$ are formed, mass balance equations become:

$$m_{Al} = \bar{m}_{Al^{3+}} + \bar{m}_{AlCl^{2+}} + \bar{m}_{AlCl_2^+} + \bar{m}_{AlCl_3^0} + \bar{m}_{AlCl_4^-} \qquad (14.176)$$

and

$$m_{Cl} = \bar{m}_{Cl} + \bar{m}_{AlCl} + 2\bar{m}_{AlCl_2} + 3\bar{m}_{AlCl_3} + 4\bar{m}_{AlCl_4} \qquad (14.177)$$

There are six unknowns. The other four equations are the equilibrium constants:

$$K_1 = \frac{\bar{m}_{AlCl^{2+}} \bar{\gamma}_{AlCl^{2+}}}{\bar{m}_{Al^{3+}} \bar{\gamma}_{Al^{3+}} \times \bar{m}_{Cl^-} \bar{\gamma}_{Cl^-}}, \qquad (14.178)$$

$$K_2 = \frac{\bar{m}_{AlCl_2^+} \bar{\gamma}_{AlCl_2^+}}{\bar{m}_{Al^{3+}} \bar{\gamma}_{Al^{3+}} \times (\bar{m}_{Cl^-} \bar{\gamma}_{Cl^-})^2} \qquad (14.179)$$

$$K_3 = \frac{\bar{m}_{\text{AlCl}_2^0}\,\bar{\gamma}_{\text{AlCl}_2^0}}{\bar{m}_{\text{Al}^{3+}}\bar{\gamma}_{\text{Al}^{3+}} \times (\bar{m}_{\text{Cl}^-}\bar{\gamma}_{\text{Cl}^-})^3} \tag{14.180}$$

$$K_4 = \frac{\bar{m}_{\text{AlCl}_4^-}\,\bar{\gamma}_{\text{AlCl}_4^-}}{\bar{m}_{\text{Al}^{3+}}\bar{\gamma}_{\text{Al}^{3+}} \times (\bar{m}_{\text{Cl}^-}\bar{\gamma}_{\text{Cl}^-})^4} \tag{14.181}$$

To make the necessary activity corrections we need a suitable equation for the activity coefficients of all the species including the associated species. Such an equation must contain terms for electrostatic and solvation effects but not ion association. For this purpose, an approximate equation for any species i was given by Davies (1962).

$$\log \bar{\gamma}_i = -z_i^2 A_\gamma \left(\frac{\bar{I}^{0.5}}{1 + \bar{I}^{0.5}} - 0.3\bar{I} \right) \tag{14.182}$$

where the $\bar{I}$ is the true ionic strength, and the term $-0.3\bar{I}$ compensates for solvation and inadequacies in the electrostatic term.

This equation may be useful for ionic strengths less than 0.5 mol dm^{-3}, at 298 K and 1 bar, if used with equilibrium constants for complex formation such as those compiled by Smith and Martell (1976) or Hogfeldt (1982).

A comprehensive
expression for activity
coefficients

For higher ionic strengths Helgeson *et al.* (1981) developed an expression similar to the Davis equation.

The characteristic feature of this treatment is that the major contributions to the single-ion activity coefficients are treated explicitly rather than with non-specific fitting coefficients. It is argued that the resulting equation has a sound molecular basis from which **extrapolations** may be made, rather than being an equation for **interpolations** over a range of fit, which is the distinguishing feature of the Pitzer approach. The basic equation for the ith ion in a solution containing k salts is:

$$\log \bar{\gamma}_i = \frac{-A_\gamma z_i^2 \bar{I}^{0.5}}{1 + B_\gamma \mathring{a} \bar{I}^{0.5}} + \Gamma + \omega_i^{\text{abs}} \sum_{n=1}^{K} b_n y_n \bar{I} + \sum_{l=1}^{N} \beta_{il} \bar{m}_l \tag{14.183}$$

The first electrostatic term contains $\mathring{a}$ estimated from the relative electrostatic radii (r_l^{el}) of the ions using:

$$\mathring{a} = \frac{2 \sum_{l=1}^{N} m_l r_l^{\text{el}}}{\sum_{l=1}^{N} m_l} \tag{14.184}$$

The term $\omega_i^{\text{abs}} \sum_{n=1}^{K} b_n y_n \bar{I}$ is an ion-solvation term written as **a summation over all salts** (denoted n). Within this term b_n is the coefficient for the nth salt and y_n the proportional contribution of the nth salt to the total ionic strength, i.e.

$$y_n = \frac{\sum_l v_{ln}(z_l^2/2)m_l}{I} \tag{14.185}$$

where v_{ln} is the number of moles of the lth ion in the nth salt and m_n the total molality of the designated salt. Note that y_n is in terms of the stoichiometric ionic strength and the total concentrations of ions in solution. The variable β_{il} is a short-range electrostatic parameter to account for ion–ion interactions not specifically accommodated in ion-association equilibria. The necessary coefficients for a range of ions including complexes are given by Helgeson *et al.* (1981). We note that the solution composition must be defined in terms of salts rather than single ions or complexes, e.g. a solution containing the ions Na^+, K^+, Cl^- and NO_3^- with their attendant complexes must be re-defined in terms of the equivalent quantities of the neutral salts $NaCl$, KCl, $NaNO_3$ and KNO_3. This is a purely formal procedure.

The mean ionic activity coefficient for the kth salt composed of ions i and j is therefore:

$$\log \bar{\gamma}_{\pm k} = \frac{-A_\gamma |z_i z_j| \bar{I}^{0.5}}{1 + B_\gamma \mathring{a} \bar{I}^{0.5}} + \Gamma + \frac{\omega_k^{abs}}{\nu_k} \sum_{n=1}^K b_n y_n \bar{I} + \frac{1}{\nu_k} \sum_{l=1}^N \nu_{ik} \beta_{il} \bar{m}_l + \frac{1}{\nu_k} \sum_{l=1}^N \nu_{jk} \beta_{jl} \bar{m}_l$$

(14.186)

The mean stoichiometric activity coefficient is:

$$\log \gamma_k = \log \bar{\gamma}_k + \frac{1}{\nu_k} \log \left(\frac{\bar{m}_i^{\nu_{ik}} \bar{m}_j^{\nu_{jk}}}{m_i^{\nu_{ik}} m_j^{\nu_{jk}}} \right)$$

(14.187)

The expression for the osmotic coefficient can be derived from the activity coefficients of the salts using a variation of equation 10.98 written in terms of the true ionic strength. This gives:

$$\phi_w = \frac{-2.303}{\sum_{j=1}^N m_j} \sum_{j=1}^N m_j \left[\frac{A_\gamma z_i^2 \bar{I}^{0.5} \sigma}{3} + \frac{\Gamma}{0.018 \sum_{j=1}^N m_j} - \frac{1}{2} \left(\omega_j^{abs} \sum_{k=1}^N b_k y_k + \sum_{i=1}^N \beta_{ij} \right) \right]$$

(14.188)

where

$$\sigma = \frac{3}{\mathring{a} B_\gamma^3 \bar{I}^{3/2}} \left(\Lambda - \frac{1}{\Lambda} + 2 \ln \Lambda \right)$$

(14.189)

and

$$\Lambda = 1 + B_\gamma \mathring{a} \bar{I}^{0.5}$$

(14.190)

Appendix 2 gives the fitting parameters for some selected salts at 298.15 K and 1 bar. Under these conditions, and only these, the salt parameters b_k are not strictly fitting parameters but are set equal to:

$$b_k = \left[\frac{\sum_{l=1}^N \nu_{lk} \omega_l^{abs}}{\sum_{l=1}^N (Z_l^2/2)} \right] \times 10^{-11}$$

(14.191)

where the summations are over all the ions in the solution but ν_{lk} will only be non-zero when the lth ions is in the kth salt. For example, the values of ω_l^{abs} for Ca^{2+} and Cl^- are 2.314×10^5 and 0.9173×10^5 cal mol^{-1} respectively, giving b_k a value of 1.3829×10^{-6} kg mol^{-1} for the salt $CaCl_2$. This approximation allows equations 14.186 and 14.188 to be expressed solely in terms of the ions in solution using:

$$\log \bar{\gamma}_j = \frac{-A_\gamma z_j^2 \bar{I}^{0.5}}{1 + B_\gamma \mathring{a} \bar{I}^{0.5}} + \Gamma + \omega_j^{abs} \sum_{i=1}^N Q_i \bar{m}_i + \sum_{i=1}^N \beta_{ij} \bar{m}_i$$

(14.192)

and

$$\phi_w = \frac{-2.303}{\sum_{j=1}^N n_j} \sum_{j=1}^N m_j \left[\frac{A_\gamma z_i^2 \bar{I}^{0.5} \sigma}{3} + \frac{\Gamma}{0.018 \sum_j m_j} - \frac{1}{2} \left(\omega_j^{abs} \sum_{k=1}^N Q_{imj} + \sum_{k=1}^N \beta_{jk} \right) \right]$$

(14.193)

Q_i is an ion–solvent interaction term which at 25 °C can be estimated from $Q_i = \omega_i^{abs} \times 10^{-11}$. At temperatures and pressures other than 298.15 K and 1 bar equations 14.192 and 14.193 are not valid since the approximation for of b_k does not hold and b_k and β_{ij} must both be determined.

Simplified expressions for activity coefficients

It is especially difficult to evaluate the individual parameters in equation 14.186; a simplified version can be produced by merging the solvation and the short-range ion–ion interactions terms into one term for each salt. The single-ion activity coefficient becomes:

$$\log \bar{\gamma}_i = \frac{-A_\gamma z_j^2 \bar{I}^{0.5}}{1 + B_\gamma \mathring{a} \bar{I}^{0.5}} + \Gamma + \omega_i^{\mathrm{abs}} \sum_{n=1}^{N} \beta_n y_n \bar{I} \qquad (14.194)$$

where the extended term parameter β_k has subsumed the original two terms. For a neutral salt in a mixed electrolyte, the above equation becomes:

$$\log \bar{\gamma}_{\pm k} = \frac{-A |z_i z_j| \bar{I}^{0.5}}{1 + B_\gamma \mathring{a} \bar{I}^{0.5}} + \Gamma + \frac{\omega_k^{\mathrm{abs}}}{v_k} \sum_{n=1}^{K} \beta_n y_n \bar{I} \qquad (14.195)$$

In order to generalise we have adopted a slightly different notation from that in the original work and equation 14.194 is an approximation to the original equation. The generalisation restricts the above equations to completely dissociated salt solutions or those where the salts form 1:1 complexes only. For a single-salt solution, equation 14.196 becomes:

$$\log \bar{\gamma}_{\pm k} = \frac{-A_\gamma |Z_i z_j| \bar{I}^{0.5}}{1 + B_\gamma \mathring{a} \bar{I}^{0.5}} + \Gamma + b_{\gamma k} \bar{I} \qquad (14.196)$$

where $b_{\gamma k} = \omega_k^{\mathrm{abs}} \beta_k / v_k$. Helgeson *et al.* (1981) have evaluated the extended term parameter $b_{\gamma k}$ for a range of salts at different temperatures and pressures. It is proposed that β_k implicitly incorporates the consequences of electrostriction, in which case the temperature and pressure dependence may be described by analogues of the equations for the temperature and pressure dependence of the standard partial molar properties. The appropriate expressions are given in Appendix 2. The osmotic coefficient consistent with equation 14.194 is:

$$\phi_{\mathrm{w}} = \frac{-2.303}{\sum_{j=1}^{N} m_j} \sum_{j=1}^{N} m_j \left[\frac{A_\gamma z_i^2 \bar{I}^{0.5} \sigma}{3} + \frac{\Gamma}{0.018 \sum_{j=1}^{N} m_j} - \frac{1}{2} \left(\omega_j^{\mathrm{abs}} \sum_{*k=1}^{N} \beta_k y_k \right) \right] \qquad (14.197)$$

Some comments on the explicit ion association approach

In principle there is no conflict between the use of virial coefficients to describe non-ideal behaviour and the explicit use of ion-association equilibria, since equation 14.153 can be used as an equation of self-consistency to link the two approaches.

In practice, the use of any equation for true single-ion activities at high concentrations requires precise knowledge of equilibrium constants and activity coefficients for complexes. Extensive data sources do exist but there is still no completely self-consistent dataset for this purpose.

However, for ions that associate insignificantly, equations 14.186, 14.188, 14.192 and 14.193 become simple equations for stoichiometric activity coefficients. In these cases, the equations predict the properties of pure electrolytes at high concentrations and at temperatures up to 3000 °C on the water saturation curve. The ability of the ion association technique to predict the properties of trace components in concentrated solutions still requires investigation.

*Other simplified equations which approximate the composition dependence of γ_i over extensive ranges of temperature and pressure are given by Helgeson *et al.* (1981).

Chapter 15

The thermodynamics of ion-exchange reactions

The reaction chemistry of minerals is strongly influenced by subtle reactions occurring at the mineral/solution interface. This is particularly true of the various types of mineral dissolution. If these minor reactions are extensive, they may modify the bulk composition of solutions as well as affecting the interfacial region; they may have a notable effect on other reactions occurring through the solution phase. Surface chemistry is clearly important and this chapter is concerned with ion-exchange, a major type of surface reaction.

The surfaces of many solid phases, including clays, zeolites and numerous oxides, bear electric charge. This charge is neutralised by loosely bound ions existing in a solution phase close to the solid surface. These ions are not fixed in space, but in dynamic equilibrium with ions in the bulk of the solution phase far from the solid–liquid interface. If the relative proportion of ions in the bulk liquid phase is changed, the relative proportion of ions in the interfacial region will change accordingly. For convenience we think of the system as a two-phase system. One phase is the bulk liquid phase, the other is the solid phase including the ions and water molecules in the interfacial region.

The process of **ion exchange** is defined as **the stoichiometric replacement of ions in one phase with ions of the same charge from a second phase.**

Ion exchange is often accompanied by a simultaneous change in the water content of the interfacial region and the adsorption of neutral salts. The chemical properties of many sedimentary and diagenetic processes are influenced by such reactions. Consequently, the measurement and thermodynamic characterisation of ion exchange is important.

15.1 Some general comments

The thermodynamic properties of many ion-exchange reactions have been studied under ambient conditions. Unfortunately, there are fewer data at higher temperatures and pressures and no data in the supercritical region. To add to this, the theoretical description of electrified interfaces is a formidable task. It is likely that the affinity of an ion for a charged surface is controlled by the overall surface charge density, the way the charge is distributed over particle surfaces, the charges and radii of the ions and the solvation properties of the ions. However, currently we do not have the means to predict thermodynamic data for ion-exchange reactions under conditions of geological interest. If standard entropy and enthalpy changes are measured for reactions at 25 °C we may estimate equilibrium constants at temperatures up to 50 °C or so, but little higher. In view of this situation, this chapter is devoted to the techniques used to measure equilibrium constants and activity coefficients for ion-exchange reactions. A limited discussion of equations for evaluation of activity coefficients in multicomponent ion-exchange systems is also presented.

15.2   The exchange capacity

The **cation exchange capacity (CEC) is the number of moles of exchangeable cations per unit mass of exchanger**.

CEC usually has the units of equivalents per gram, where 1 equivalent is equal to the charge on 1 mole of any **monovalent** ion. For example, the clay montmorillonite has a CEC of around 1×10^{-3} equivalents per gram. This could be equal to 1×10^{-3} moles of sodium or 0.5×10^{-3} moles of calcium. If the sodium ions on such a clay were progressively replaced by calcium, the total number of moles of cations on the clay would be reduced although the CEC remains fixed.

It is usual to determine CECs by extracting cations from the interfacial region with a solution containing an excess of a cation which is preferred by the solid surface. In principle this places all the charge-neutralising ions plus adsorbed salts in the extracting solution. Chemical analysis of the resulting solution gives the number of moles of ions extracted. The difference between the number of moles of extracted cations and anions is a measure of the CEC. For some cation exchangers, such as zeolites, there is little salt adsorption if electrolyte concentrations are less than $1 \; mol \; kg^{-1}$, and the number of moles of replaced cations is a direct measure of the CEC. For clays, however, salt adsorption is often observed, even in dilute solutions, and neglecting the extracted anions leads to values of CEC which are higher than the true exchangeable charge. Protons should also be considered as charge-neutralising cations. Neglecting these ions in CEC calculations may lead to the conclusion that a CEC is pH-dependent when it may not be.

Exchanger charge can arise from several sources. In clays and zeolites the charge arises predominantly from a net imbalance of ionic charge **within the structure of the solid**. In composite substances there may be molecules present on or near particle surfaces which can yield charge. For example, in soils or organic-rich shales, charge can arise from the dissociation of acidic or basic groups on the organic substances. If carboxylic acid groups are present dissociation may occur via the reaction:

$$—COOH \rightleftharpoons H^+ + —COO^- \qquad (15.1)$$

At low pH the carboxylic acid will be fully associated. At higher pH dissociation may occur, yielding negative charge available for neutralisation by other cations, e.g.

$$—COO^- + Na^+ \rightleftharpoons —COONa \qquad (15.2)$$

It is usual to define CEC operationally. For example, if it is convenient **not** to determine the number of protons bound to a composite solid, the apparent CEC will have a pH dependence. Alternatively, if bound protons are measured, the CEC will be constant and equal to the sum of all charge contributions.

In clay/water systems a pH-dependent CEC is observed because of hydrolysis of the Si–O groups at the edges of particles in accordance with the reaction $-SiO + H_2O \rightleftharpoons -SiOH^- + H^+$. This leads to a residual negative charge on the edges which progressively increases with pH. The edge charge may be as high as 0.1×10^{-3} mol g^{-1} and contributes up to 10% of the total clay charge. At pH values around 8 or 9 the edge groups are fully hydrolysed. As conditions become more acidic the negative charges are neutralised by protons; generally neutralisation is complete at pH values below 4. At lower pH a residual positive charge may even develop due to dissociation of —AlOH groups in accordance with the reaction $-AlOH \rightleftharpoons -Al^+ + OH^-$. This imparts a pH-dependent anion-exchange capacity to the clay edges. At all pH values the residual charge on clay surfaces remains negative. Hydrolysis and dissociation reactions at the edges of particles are common with clays, metal oxides and silicate minerals. They are not observed to any significant extent with zeolites in the pH range 4–10. In addition to modifying the bulk solution composition, the pH-dependent charge of minerals such as fosterite is known to influence the rate of dissolution of the mineral in an aqueous solution (Blum and Lasaga, 1988).

15.3 Ion-exchange isotherms

The conventional way of depicting an ion-exchange reaction between two ions is with an ion-exchange isotherm. This is constructed by measuring the concentrations of exchangeable ions in both the bulk solution and on the exchanger phase, for a series of solutions. Different solutions must contain different ratios of exchanging ions. The composition of any phase is quantified by the equivalent fraction or mole fraction of exchanging ions in the phase and the isotherm is a plot of the equivalent fraction of one ion in one phase against the equivalent fraction of the same ion in the second phase.

The equivalent fraction of the jth ion in any phase is given by:

$$E_j = \frac{z_j n_j}{\sum_i z_i n_i} \qquad (15.3)$$

where n_i, n_j are the numbers of moles of the ith, jth ion, etc. The mole fraction of exchangeable ions in any phase is given by:

$$x_j = \frac{n_j}{\sum_i n_i} \qquad (15.4)$$

Note that **only the exchanging ions are included in the summations.** For example, if a solution, in contact with a cation exchanger, contained sodium, calcium and chloride at pH 7 it would be conventional to neglect the protons and include only sodium and calcium in the calculations for equivalent fraction. If the concentrations of ions on the exchanger were determined to be 0.1×10^{-3} mol Na$^+$ g^{-1}, 0.2×10^{-3} mol Ca^{2+} g^{-1} and 0.5×10^{-3} mol Cl$^-$ g^{-1},

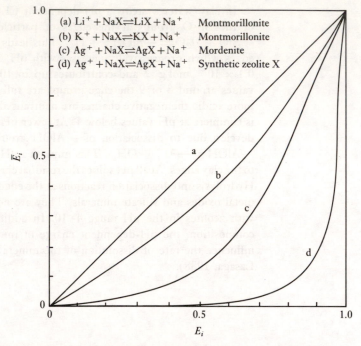

The legend inside the figure reads:

(a) $Li^+ + NaX \rightleftharpoons LiX + Na^+$ Montmorillonite
(b) $K^+ + NaX \rightleftharpoons KX + Na^+$ Montmorillonite
(c) $Ag^+ + NaX \rightleftharpoons AgX + Na^+$ Mordenite
(d) $Ag^+ + NaX \rightleftharpoons AgX + Na^+$ Synthetic zeolite X

Figure 15.1   Ion exchange isotherms (data from Gast, 1969; Fletcher and Townsend, 1980).

then the exchange-phase equivalent fractions of sodium and calcium are:

$$\left.\begin{array}{l} E_{Na} = \dfrac{0.1 \times 10^{-3}}{0.1 \times 10^{-3} + 2 \times 0.2 \times 10^{-3}} \\[3mm] E_{Ca} = \dfrac{2 \times 0.2 \times 10^{-3}}{0.1 \times 10^{-3} + 2 \times 0.2 \times 10^{-3}} \end{array}\right\} \tag{15.5}$$

An alternative way of calculating equivalent fractions is from mole fractions, using:

$$E_j = \frac{z_j x_j}{\sum_i z_i x_i} \tag{15.6}$$

Equivalent fractions of exchangeable ions in the solution phase can be calculated similarly if the exchanging ions are determined in molal units. The equivalent fraction of the ith ion in the solution phase will be denoted $\bar{E}_i$ and in the solid phase E_i.

Cation-exchange isotherms are usually determined using a range of solutions where the concentration of cations is varied but the molal concentration of the anionic background is kept constant. Typical experimental ion-exchange isotherms for the replacement of sodium from clays and zeolites are shown in figure 15.1.

15.4   Thermodynamics of cation-exchange reactions

The thermodynamic treatment of ion-exchange reactions first requires the definition of **reaction stoichiometry**. Once stoichiometry is defined it is possible to define the properties of an ideal ion-exchange reaction. The final state is to characterise the deviation of any real system from the ideal behaviour, using activity coefficients. The methods are as follows.

15.4.1   Defining ideal ion-exchange behaviour

The reaction stoichiometry   The ion-exchange reaction can be described by the following general equation:

$$z_i A_j^{z_j+}(e) + z_j A_i^{z_i+}(aq) \rightleftharpoons z_i A_j^{z_j+}(aq) + z_j A_i^{z_i+}(e) \qquad (15.7)$$

where z_i and z_j represent the valence charges on the ions $A_i^{z_i+}$ and $A_j^{z_j+}$, and abbreviations (aq) and (e) define ions to be in either the aqueous phase or the exchanger phase respectively. For example, the replacement of sodium ions on an exchanger by calcium ions can be written:

$$2Na^+(e) + Ca^{2+}(aq) \rightleftharpoons 2Na^+(aq) + Ca^{2+}(e) \qquad (15.8)$$

Equation 15.7 is correct. However, a reaction between ions disguises the fact that a thermodynamic treatment requires reactions to be between components which are independent variables in the system. Thus, thermodynamic components should be neutral species rather than single ions if a convention of the kind used for aqueous ions is to be avoided.

In section 14.3 we saw how the properties of single ions in an aqueous phase can be characterised. In contrast, for the exchanger phase, it is conventional to use components which are composed of a quantity of the anionic exchanger material in neutral combination with a quantity of cationic charge. Historically, two components with different formulae have been used. In this text we will be concerned with only one of these components.

Consider an ion $A_i^{z_i+}$ and a hypothetical exchanger species X^- carrying one mole of anionic charge. The exchange component has the composition $A_i X_{z_i}$. This represents 1 mole of cation i in association with a quantity of exchanger carrying z_i moles of negative charge.

For this type of component the ion-exchange reaction can be written:

$$z_i[A_j X_{z_j}] + z_j A_i^{z_i+}(aq) \rightleftharpoons z_i A_j^{z_j+}(aq) + z_j[A_i X_{z_i}] \qquad (15.9)$$

For example, the replacement of sodium ions from an exchanger by calcium ions is:

$$2[NaX] + Ca^{2+}(aq) \rightleftharpoons 2Na^+(aq) + [CaX_2] \qquad (15.10)$$

This equation is entirely compatible with equation 15.7 although the nature of ion-exchange components is made clearer with equation 15.9.

The equilibrium constant   The equilibrium constant for the general ion-exchange reaction is constructed as follows.

At equilibrium the sum of the Gibbs energies for the products is equal to the sum of the Gibbs energies for the reactants. This leads to the following equilibrium relationship:

$$[z_i \mu_j(aq) + z_j \mu_i(e)] - [z_i \mu_j(e) + z_j \mu_i(aq)] = 0 \qquad (15.11)$$

The basic chemical potential equation ($\mu = \mu^\ominus + RT \ln a$) for ions in both phases can be substituted into equation 15.11. This gives the fundamental equation for the condition of equilibrium as:

$$[z_i\mu_j^{\ominus}(\text{aq}) + z_i RT \ln a_j(\text{aq}) + z_j\mu_i^{\ominus}(\text{e}) + z_j RT \ln a_i(\text{e})]$$
$$- [z_i\mu_j^{\ominus}(\text{e}) + z_i RT \ln a_j(\text{e}) + z_j\mu_i^{\ominus}(\text{aq}) + z_j RT \ln a_i(\text{aq})]$$
$$= 0 \tag{15.12}$$

where $\mu_i^{\ominus}(\text{e})$ and $a_i(\text{e})$, etc., are standard chemical potentials and activities of the subscripted component in the exchanger. Similarly, $\mu_i^{\ominus}(\text{aq})$ and $a_i(\text{aq})$, etc., are the standard chemical potentials and single-ion activities for the ith ions in solution. Collecting the standard chemical potentials and logarithmic terms yields:

$$[z_j\mu_i^{\ominus}(\text{e}) + z_i\mu_j^{\ominus}(\text{aq})] - [z_i\mu_j^{\ominus}(\text{aq}) + z_j\mu_i^{\ominus}(\text{e})] = -RT \ln\left[\frac{a_j(\text{aq})^{z_i}a_i(\text{e})^{z_j}}{a_i(\text{aq})^{z_j}a_j(\text{e})^{z_i}}\right] \tag{15.13}$$

The logarithmic function contains the exchange equilibrium constant, defined as:

$$_j^i K \equiv \left[\frac{a_j^{z_i}(\text{aq}) \times a_i^{z_j}(\text{e})}{a_i^{z_j}(\text{aq}) \times a_j^{z_i}(\text{e})}\right] \tag{15.14}$$

The sum of the standard chemical potentials is the standard molar change of Gibbs energy for the cation-exchange reaction; therefore

$$_j^i\Delta G_A^{\ominus} = -RT \ln {}_j^i K \tag{15.15}$$

The convention used here is that the thermodynamic functions $_j^i\Delta G^{\ominus}$ and $_j^i K$ carry a superscript to denote the solid-phase 'product' ion and a subscript to denote the solid-phase 'reactant' ion. With this notation the reaction stoichiometry can be deduced from the symbols for the constants.

Ideal behaviour With the cation-exchange reaction stoichiometry and its attendant equilibrium constants established, ideal behaviour can now be defined.

The activity of a single ion i in the aqueous phase is defined in terms of its molal concentration m_i by $a_i(\text{aq}) = m_i \times \gamma_i$. Using this definition the equilibrium constant becomes:

$$_j^i K = \left[\frac{(m_j\gamma_j)^{z_i} \times a_i^{z_j}(\text{e})}{(m_i\gamma_i)^{z_j} \times a_j^{z_i}(\text{e})}\right] \tag{15.16}$$

To define the activity of components in the exchanger phase we must define standard states for the components plus ideal reference behaviour.

The most widely used reference behaviour for ion-exchange reactions is Raoult's law. This states that activity of a component is equal to its mole fraction in the exchanger phase ($a_i = x_i$). Strictly, we refer to the mole fraction of the neutral component rather than the mole fraction of the exchanging ion. With this choice of exchanger component, the mole fraction of the ion equals the mole fraction of the neutral component.

For non-ideal systems we define the rational activity coefficient f_i, which describes the deviation from the chosen reference behaviour, thus:

$$f_i \equiv \frac{a_i}{x_i} \tag{15.17}$$

This yields the usual equation for the chemical potential of a component:

$$\mu_i(e) = \mu_i^{\ominus}(e) + RT \ln(x_i f_i) \tag{15.18}$$

The most common standard state for a Raoult's law component is the pure component at any temperature and pressure. In this standard state the mole fraction and the activity coefficient of a component are unity.

A 'pure' exchange component is a cation exchanger containing a single exchangeable species. Unfortunately, for ion exchangers which adsorb water the conventional standard state is ambiguous since an exchanger could be in its pure form in the presence a pure salt solution of any concentration, and the amount of adsorbed water is therefore variable. Accordingly, the composition of the liquid phase must be specified to complete the definition of the solid-phase standard state. Consequently, the usual standard state for an ion-exchange component is **the homoionic form in equilibrium with an infinitely dilute solution of the exchanger ion at any temperature and pressure**. This is convenient since it is consistent with the conventional standard state for water.

Using this standard state and reference behaviour, the equilibrium constant becomes:

$$_j^i K = \frac{(m_j \gamma_j)^{z_i} \times (x_i f_i)^{z_j}}{(m_i \gamma_i)^{z_j} \times (x_j f_j)^{z_i}} \tag{15.19}$$

For convenience we define an experimentally measurable mass action coefficient called the Vanselow coefficient, i.e.

$$_j^i k_V = \frac{(m_j \gamma_j)^{z_i} \times (x_i)^{z_j}}{(m_i \gamma_i)^{z_j} \times (x_j)^{z_i}} \tag{15.20}$$

Therefore,

$$_j^i K = {}_j^i k_V \left(\frac{f_i^{z_j}}{f_j^{z_i}}\right) \tag{15.21}$$

Ideal behaviour is observed if the activity coefficients f_i and f_j are unity over the **entire range of an ion-exchange isotherm**, in which case the function $_j^i k_V$ is a constant. For most ion-exchange systems the value of $_j^i k_V$ will vary over all or part of the isotherm and the activity coefficients will not be unity. The selection of a standard state is arbitrary and other standard states may be appropriate. One common standard state is the hypothetical ideal molal solution of the exchanger, with a reference state being the infinitely dilute solution. With this convention there is no difference between the standard states for any solid-phase component and the exchange equilibrium constant approaches unity. The measured solid-phase activity coefficients therefore become the measure of the selectivity of the exchanger for given ions.

In view of the need to plot isotherms in terms of equivalent fractions, it is sometimes convenient to replace the mole fraction terms in the Vanselow coefficient with the complementary equivalent fractions. For binary ion exchange only, mole fractions can be expressed in terms of equivalent fractions with:

$$x_i = \frac{z_j E_i}{z_i + (z_j - z_i) E_i} \quad \text{and} \quad x_j = \frac{z_i E_j}{z_j + (z_i - z_j) E_j} \tag{15.22}$$

Therefore, we may eliminate x_i and x_j from equation 15.20 to produce:

$$_j^i k_V = \left(\frac{z_i^{z_i}}{z_j^{z_j}} \right) \left[\frac{a_j(aq)^{z_i} E_i^{z_j}}{a_i(aq)^{z_j} E_j^{z_i}} \right] (z_i E_j + z_j E_i)^{(z_i - z_j)} \qquad (15.23)$$

15.4.2 Selectivity and isotherm shape

The shape of any ion-exchange isotherm is controlled by three major factors. The first is the relative affinities the ions have for the exchanger phase. This is directly related to the magnitude of the equilibrium constant for the reaction. Figure 15.1 shows two isotherms for sodium/potassium and sodium/lithium exchange on montmorillonite. The equilibrium constant for the replacement of sodium by lithium is 0.95, i.e. little preference is shown. In contrast, potassium is quite strongly preferred over sodium and the curved isotherm shows a relatively high fraction of potassium on the clay, even when the fraction of potassium in solution is low. The third isotherm is for sodium/silver exchange on the synthetic mineral mordenite, where a massive preference for silver is exhibited.

The second controlling factor is the ionic charges. We can see this effect by inspecting the equation for $_j^i k_V$. To simplify the argument we will assume that all ions in both phases are behaving ideally. The value of $_j^i k_V$ is therefore a constant given by:

$$_j^i k_V = \left(\frac{m_j^{z_i} \times x_i^{z_j}}{m_i^{z_j} \times x_j^{z_i}} \right) \qquad (15.24)$$

If the exchange is heterovalent ($z_i \neq z_j$), the shape of the isotherm is dependent on the total concentration of ions in the solution phase. We deduced this from the fact that for given values of m_i and m_j the value of the ratio $m_j^{z_i}/m_i^{z_j}$ is fixed; therefore the value of $x_j^{z_i}/x_i^{z_j}$ is also fixed. If the actual values of m_j and m_i are increased by the same proportions, the fractions of the ions in the solution phase remain as before; however, the value of $m_j^{z_i}/m_i^{z_j}$ changes because the charges on the ions are different. This in turn leads to a different value for $x_j^{z_i}/x_i^{z_j}$, i.e. a different fraction of ions on the exchanger. This is called the **concentration valence effect**.

In general, as the total concentration of ions in the aqueous phase decreases, the relative proportion of the most highly charged ion increases on the exchanger. For homovalent ion-exchange this effect is not observed.

Figure 15.2 shows an example of sodium/calcium exchange on montmorillonite at different total ion concentrations.

The final factor is non-ideal behaviour where the activity coefficients vary with exchanger phase composition. This effect may lead to significant curvature of an exchange isotherm, especially for zeolitic ion exchangers which show marked deviation from ideality (Fletcher and Townsend, 1985; Barrer and Klinowski, 1974).

15.4.3 Characterising non-ideal behaviour

The objective of this section is to show how measurements of $_j^i k_V$ can be used to obtain the equilibrium constant and the activity coefficients. The procedure starts

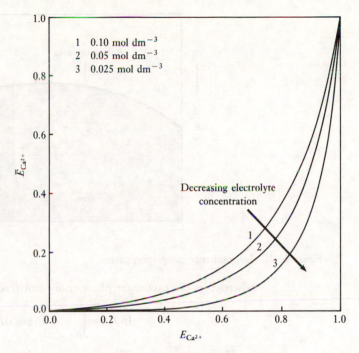

Figure 15.2   Sodium/calcium isotherms (data from Fletcher and Sposito, 1989).

with equation 15.21, re-written as:

$$\ln {}_j^i K = \ln {}_j^i k_V + \ln f_i^{z_i} - \ln f_j^{z_i} \qquad (15.25)$$

The above equation can be transformed into an equation of constraint by forming the differential of $\ln {}_j^i k_V$, i.e.

$$0 = d \ln {}_j^i K = d \ln {}_j^i k_V + z_j\, d \ln f_i - z_i\, d \ln f_j \qquad (15.26)$$

A second equation of constraint is the Gibbs–Duhem equation, which for any closed system is:

$$\sum_k n_k\, d\, \mu_k = 0 \qquad (15.27)$$

where n_k is the number of moles of the kth component in the system. For the simple binary mixture under consideration the Gibbs–Duhem equation becomes:

$$n_i\, d \ln(x_i f_i) + n_j\, d \ln(x_j f_j) = 0 \qquad (15.28)$$

Equations 15.26 and 15.28 constitute two equations with two unknowns (f_i and f_j). It is possible to solve the differential equations 15.26 and 15.28 to yield two equations which can be used to evaluate activity coefficients from isotherm data. The solutions to equations 15.26 and 15.28 are given in Grant and Fletcher (1993) and are:

$$\ln f_i^{z_j} = \int_{0(E_i=1)}^{E_j} \ln {}_j^i k_V\, dE_j - E_j \ln {}_j^i k_V \qquad (15.29)$$

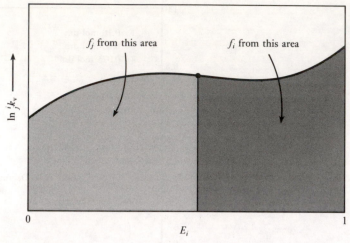

f_j from this area f_i from this area

Figure 15.3   Ion-exchange integration paths.

where E_i is the exchanger phase equivalent fraction of the ith ion. Similarly for f_j,

$$\ln f_j^{z_i} = \int_{0(E_j=1)}^{E_j} \ln {}_i^j k_V \, dE_i - E_i \ln {}_i^j k_V \tag{15.30}$$

The exchange equilibrium constant is obtained by eliminating the activity coefficients from equation 15.25 using equations 15.29 and 15.30, i.e.

$$\ln {}_j^i K = \int_0^1 \ln {}_j^i k_V \, dE_j \tag{15.31}$$

Equation 15.31 dictates the experimental measurements necessary to evaluate the thermodynamic properties. The requirement is a series of equilibrium measurements to determine values for ${}_j^i k_V$ over the **complete** isotherm range.

Figure 15.3 shows the two integration paths necessary to calculate activity coefficients at the specified composition.

15.4.4   The effect of water activity and salt adsorption

Water activity   For many ion exchanges, the amount of adsorbed water changes as one ion is progressively replaced by another.

To accommodate this effect it is necessary to include a term for the adsorbed water in the Gibbs–Duhem equation. That is,

$$n_i \, d\ln(x_i f_i) + n_j \, d\ln(x_j f_j) + n_w \, d\ln a_w = 0 \tag{15.32}$$

where n_w and a_w are the number of moles of bound water and the adsorbed water activity respectively.

Including this term gives a modified version of equations 15.29 and 15.30, i.e.

$$\ln f_i^{z_j} = \int_{0(E_i=1)}^{E_j} \ln {}_j^i k_V \, dE_j - E_j \ln {}_j^i k_V$$

$$- z_i z_j \left[\int_{0(E_i=1)}^{\ln a_w(\phi_i)} \bar{n}_w^{\phi_i} \, d\ln a_w + \int_{\ln a_w(\phi_i)}^{\ln a_w} \bar{n}_w \, d\ln a_w \right] \tag{15.33}$$

where $\bar{n}_w$ is number of moles of bound water per mole of exchanger-phase charge and is in principle measurable at any exchanger-phase composition. The symbol ϕ_i denotes a homoionic state where the ion i is the only ion in the system but the solution phase is at the concentration at which the experiment is being performed. Consequently, $\bar{n}_w^{\phi_i}$ is the number of moles of water (per mole of exchanger) which are bound to the exchanger in this state.

Similarly for f_j,

$$\ln f_j^{z_i} = \int_{0(E_j=1)}^{E_j} \ln {}_j^i k_V \, dE_i - E_i \ln {}_j^i k_V$$
$$- z_i z_j \left[\int_{0(E_j=1)}^{\ln a_w(\phi_j)} \bar{n}_w^{\phi_j} \, d \ln a_w + \int_{\ln a_w(\phi_j)}^{\ln a_w} \bar{n}_w \, d \ln a_w \right] \quad (15.34)$$

The exchange equilibrium constant becomes:

$$\ln {}_j^i K = \int_0^1 \ln {}_j^i k_V \, dE_j + z_i z_j \left[\int_{0(E_i=1)}^{\ln a_w(\phi_i)} \bar{n}_w \, d \ln a_w \right.$$
$$\left. - \int_{0(E_j=1)}^{\ln a_w(\phi_j)} \bar{n}_w \, d \ln a_w + \int_{\ln a_w(\phi_i)}^{\ln a_w(\phi_i)} \bar{n}_w \, d \ln a_w \right] \quad (15.35)$$

In view of the chosen standard states the activity of water in both phases will be identical and can be calculated from the composition of the aqueous phase if the quantity of bound water is known. Thus, all integrals in equations 15.33, 15.34 and 15.35 can be evaluated graphically or numerically. The integration steps for the water contributions to activity coefficients are:

1. from the standard state to the state (ϕ_i);
2. from the final state of step 1 to a state characterised by the chosen exchanger composition, with water activity being that in the attendant mixed electrolyte solution.

The water activity terms which contribute to the activity coefficients are positive in sign; therefore their neglect will lead to an underestimation of the activity coefficients. The effect of the water activity terms on the values of the activity coefficients will increase as the ionic strength of the electrolyte solution increases. Even for exchangers that do not swell in water, neglecting the water terms may lead to errors in the activity coefficient as high as 20% for monovalent ions and 30% for divalent ions when the ionic strength is around $2 \, mol \, kg^{-1}$. In contrast, the error in the value $\Delta G^\ominus$ for the reaction may be less than $100 \, J \, mol^{-1}$ at the same ionic strength because the water contributions in equation 15.35 cancel each other to some extent.

It is common to neglect water activity terms. However, for the most precise measurements it may be desirable to perform the water activity corrections for exchangers which show significant water uptake, such as clay minerals. The water corrections will be most pronounced for exchange between ions with markedly different hydration characteristics. For example, it is established that large monovalent cations such as potassium and caesium ions can exclude water from montmorillonite particles. In these cases the progressive change in water content accompanying ion exchange with sodium may require explicit thermodynamic treatment even at low electrolyte concentrations.

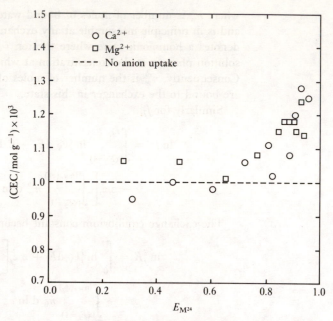

Figure 15.4   The effect of salt adsorption on CEC (data from Fletcher and Sposito, 1989).

General reviews of the thermodynamics of ion exchange are given by Gaines and Thomas (1951), Argersinger *et al.* 1950) and Sposito (1981).

Salt adsorption   For zeolites there is little adsorption of neutral salts until electrolyte concentrations are greater than $1 \, mol \, kg^{-1}$. However, for **dilute suspensions** of some clays, the replacement of monovalent cations with divalent cations gives rise to an increase in the total number of equivalents of cations on the clay surface. Figure 15.4 shows this for some heterovalent ion-exchange reactions on montmorillonite.

Since the clay phase must be electrically neutral it is presumed that the excess cations are accompanied by an equivalent of anions. This is consistent with the simultaneous adsorption of divalent-ion salt accompanying ion exchange. It suggests that highly charged cations promote salt adsorption whereas monovalent ions do not. This effect is seen in the presence of small anions such as chloride, nitrate or sulphate, but is less apparent with large anions such as perchlorate or organic polymers. With divalent cations the effect is significant, even in dilute electrolyte solutions.

A detailed review of the methods for characterising salt adsorption is beyond the scope of this book. However, as an approximation, salt adsorption may be viewed as the formation of a surface complex. For example, the replacement of sodium by calcium on montmorillonite (reaction equation 15.10) is accompanied by the reaction:

$$[NaX] + Ca^{2+} + Cl^- \rightleftharpoons Na^+ + [CaClX] \qquad (15.36)$$

The surface complex is the species CaClX and is akin to the formation of the ion-pair CaCl close to the surface of the clay. It should not be thought of as a molecule but rather a formal way of describing a quantity of chloride ions at the clay–water interface. This method is convenient since it is compatible with the mathematical techniques used to model multicomponent equilibria described in chapter 19. The complete mathematical details for the use of this convention are given elsewhere (Fletcher and Sposito, 1989).

The properties of suspensions containing high concentrations of clays have distinct features, including anion exclusion, which are not discussed here.

15.4.5 Multisite ion exchange

In many inorganic ion exchangers, the charge-neutralising cations can be distributed through a range of chemically distinct sites. Materials such as these are termed **polyfunctional**. This is most notable in zeolites, where several energetically distinct sites can be detected crystallographically, and soils, where the CEC can arise from oxide surfaces, clays and organic functional groups.

Polyfunctionality is a direct cause of non-ideality in exchange reactions. In order to understand this phenomenon we will consider polyfunctional materials as if they were mixtures of chemically distinct exchange materials. For example, if a substance is known to have K distinct cation sites, it will be considered as a mixture of K different exchanger materials, each with its own single-valued Vanselow coefficient designated $^{i}_{j}k_{V,k}$ for the kth site. This implies that ions in each site are at equilibrium with the solution phase.

Consider the case of binary exchange between ions i and j. For the kth site we may define p_k as the proportion of the total charge which resides on this site. This is the ratio of the number of equivalents of charge composing the site to the total number of equivalents of charge on all the sites. The value of p_k is given by:

$$p_k = \frac{z_i m_{i,k} + z_j m_{j,k}}{\sum_q z_i m_{i,q} + \sum_q z_j m_{j,q}} \tag{15.37}$$

where $m_{i,k}$, $m_{j,k}$, etc., are the numbers of moles of the ith, jth (etc.) ion in the kth site per unit mass of exchanger. The values of p_k are fractions constrained by:

$$\sum_k p_k = 1 \tag{15.38}$$

The equivalent fractions of ions i and j in the kth site ($E_{i,k}$ and $E_{j,k}$) are defined by:

$$E_{i,k} = \frac{z_i m_{i,k}}{z_i m_{i,k} + z_i m_{j,k}} \quad \text{and} \quad E_{j,k} = \frac{z_j m_{j,k}}{z_i m_{i,k} + z_i m_{j,k}} \tag{15.39}$$

The equivalent fractions for each site are constrained by the relationship:

$$E_{i,k} + E_{j,k} = 1 \tag{15.40}$$

We now define the Vanselow coefficient for the kth site as:

$$^{i}_{j}k_{V,k} = \left[\frac{a_j(\text{aq})^{z_i} \times x_{i,k}^{z_j}}{a_i(\text{aq})^{z_j} \times x_{j,k}^{z_i}} \right] \tag{15.41}$$

where $a_i(\text{aq})$ and $a_j(\text{aq})$ are the solution-phase activities of the designated ions. It is convenient to express the Vanselow coefficient in equivalent fractions rather than mole fractions, as in equation 15.23, i.e.

$$^{i}_{j}k_{V,k} = \left(\frac{z_i^{z_i}}{z_j^{z_j}} \right) \left[\frac{a_j(\text{aq})^{z_i} E_{i,k}^{z_j}}{a_i(\text{eq})^{z_j} E_{j,k}^{z_i}} \right] (z_i E_{j,k} + z_j E_{i,k})^{(z_i - z_j)} \tag{15.42}$$

Consider now **total** equivalent fractions E_i and E_j. These are the equivalent fractions of ions i and j on the entire polyfunctional exchanger and are related to their site counterparts by:

$$E_i = \sum_k p_k E_{i,k} \quad \text{and} \quad E_j = \sum_k p_k E_{j,k} \tag{15.43}$$

The conventional Vanselow coefficient, defined in terms of the total equivalent fractions, is given by equation 15.23:

$$_j^i k_V = \left(\frac{z_i^{z_i}}{z_j^{z_j}} \right) \left[\frac{a_j(aq)^{z_i} E_i^{z_j}}{a_i(aq)^{z_j} E_j^{z_i}} \right] (z_i E_j + z_j E_i)^{(z_i - z_j)} \tag{15.44}$$

We are interested in whether or not the presence of multiple sites causes the value of $_j^i k_V$ to be compositionally dependent. To assess this possibility it is convenient to simplify the equations by considering the case of heterovalent exchange between monovalent ions ($z_i = z_j = 1$). In this case:

$$_j^i k_V = \frac{a_j(aq) \times E_i}{a_i(aq) \times E_j} \tag{15.45}$$

By eliminating the total equivalent fractions in the above expression, using E_i and E_j from equation 15.43, we have:

$$_j^i k_V = \frac{a_j(aq) \times (\sum_k p_k E_{i,k})}{a_i(aq) \times (\sum_k p_k E_{j,k})} \tag{15.46}$$

The expressions for the Vanselow coefficients for the individual sites (equation 15.42) can be substituted into the above expression to give:

$$_j^i k_V = \frac{\sum_k p_k E_{j,k} \, _j^i k_{V,k}}{E_j} \tag{15.47}$$

This shows that the value of $_j^i k_V$ is a compositionally weighted mean of the values of $_j^i k_{V,k}$ for each site. As the total mole fraction of ion j on the mixed exchanger changes, the individual mole fractions of this ion in all sites also changes. Consequently, **the value of $_j^i k_V$ will be a function of total exchanger composition even if the individual sites are behaving ideally and have compositionally independent values of $_j^i k_{V,k}$**.

This conclusion is valid for exchange between ions of any charge, not only monovalent ones. The general expression for multisite heterovalent binary exchange has the form:

$$_j^i k_V = \left[\left(\frac{z_j^{z_j}}{z_i^{z_i}} \right) (z_i E_j + z_j E_i)^{(z_i - z_j)} \right]$$
$$\times \frac{\left\{ \sum_k p_k E_{j,k}^{z_i/z_j} [k_{V,k} (z_i^{z_i} z_j^{z_j})(z_i E_{j,k} + z_j E_{j,k})^{(z_j - z_i)}]^{1/z_j} \right\}^{z_j}}{E_j^{z_i}} \tag{15.48}$$

The above expression suggests that plots of $_j^i k_V$ versus total mole fraction of ion j may be complex. Indeed, the above expression can exhibit maxima, minima, inflexion points and sigmoidal shapes, depending on the number of sites, the total charge per site and the magnitude of the Vanselow coefficient for each site. Such shapes are consistent with the exchange properties of zeolites. Clays usually exhibit simpler exchanger behaviour.

The foregoing conclusions are based on the assumption of ideal mixing on the individual sites. However, it is presumed that in many exchangers non–ideal site mixing occurs. This introduces a compositional dependence to the values of ${}_j^ik_{V,k}$ which in turn introduces greater complexity into the plots of ${}_j^ik_V$ versus total composition. The effect is particularly noticeable in zeolites. There is a further complexity for zeolites arising from the fact that the population of sites is known to change as the overall composition changes. In this case values of p_k are no longer constant.

15.4.6   Multicomponent ion exchange

Thermodynamic measurements

In naturally occurring ion-exchange systems it is unusual to find only two exchanging cations. It is therefore necessary to understand how the activity coefficients of ion-exchange components behave in multicomponent systems.

One way is to measure the exchanger activity coefficients. This is possible since the thermodynamic description of multicomponent ion exchange has a greater algebraic complexity than the simple binary treatment but uses no new principles (Chu and Sposito, 1981). The general reaction equation (equation 15.7) applies to any binary ion combination in a multi-ion system. For a system containing N exchangeable cations, there are $\frac{1}{2}(N^2 - N)$ independent reactions. A total of $N^2 - N$ reactions can be formulated, of which half will be the reverse reactions for the other half.

The general equilibrium constant has the same form as the binary constant, i.e.

$$
{}_j^iK = \left[\frac{a_j^{z_i}(\mathrm{aq})x_i^{z_j}}{a_i^{z_j}(\mathrm{aq})x_j^{z_i}} \right] \frac{f_i^{z_j}}{f_j^{z_i}}
\tag{15.49}
$$

where the mole fraction terms x_k are constrained by:

$$
\sum_{k=1}^{N} x_k = 1
\tag{15.50}
$$

These are sometimes called pseudobinary exchange equilibrium constants. The value of any constant will be **independent** of whether it is determined in a binary or in a multicomponent system. The difficulty is in calculating the activity coefficients in multicomponent systems.

Such calculations can be made since the differentials of a minimum of $N - 1$ exchange equilibrium constants (equation 15.49), combined with a general Gibbs–Duhem equation, formalised as

$$
\sum_{k=1}^{N} x_k \, \mathrm{d} \ln f_k = 0
\tag{15.51}
$$

gives a minimum of N equations with N values of f_k as unknowns. These can be solved in exactly the same way as for a binary system. For example, a ternary system containing the ions 1, 2 and 3, yields three differential equations derived from the equilibrium constants:

$$
\mathrm{d} \ln {}_2^1k_V = z_1 \, \mathrm{d} \ln f_2 - z_2 \, \mathrm{d} \ln f_1
\tag{15.52}
$$

$$
\mathrm{d} \ln {}_3^1k_V = z_1 \, \mathrm{d} \ln f_3 - z_3 \, \mathrm{d} \ln f_1
\tag{15.53}
$$

and

$$d \ln {}^3_2 k_V = z_3 \, d \ln f_2 - z_2 \, d \ln f_3 \tag{15.54}$$

Combining these with equation 15.51, and following a similar integration to that for equations 15.26–30, yields a general equation for the kth activity coefficient in the set, i.e.

$$z_i z_j \ln f_k = z_j \left(E_i \ln {}^k_i k_V - \int_{E_k=1}^{E_j, E_k} \ln {}^k_i k_V \, dE_i \right)$$

$$+ z_i \left(E_j \ln {}^k_j k_V - \int_{E_k=1}^{E_j, E_k} \ln {}^k_i k_V \, dE_j \right) \tag{15.55}$$

where k, i and j represent any permutation of the three ions. The general equation for a designated exchange equilibrium constant is:

$$\ln {}^i_j K = \int_{E_i=1}^{E_i, E_j} \left(\ln {}^i_j k_V \, dE_j + \frac{z_j}{z_k} \ln {}^i_k k_V \, dE_k \right) \tag{15.56}$$

which can be evaluated by integration over any chosen path, including the binary limit, in a set of ternary ion-exchange experiments.

It is possible to collect enough data to characterise a ternary system fully. However, the amount of data required to characterise a system containing more than three cations is almost prohibitive. It is convenient, therefore, if a method can be devised whereby the activity coefficients in a multicomponent system can be **predicted** from data on binary systems. To do this we use an approach based on excess Gibbs-energy models.

15.5   Excess Gibbs-energy models for ion exchange

We recall from section 10.5 that the excess Gibbs energy is a measure of the total deviation from ideal behaviour of an entire mixture. It is a single composition-dependent variable which, for an N-component system, is related to the activity coefficients of the components by:

$$G^E = RT \sum_i^N x_i \ln f_i \tag{15.57}$$

where x_i denotes the mole fractions of the ith components. If the compositional variation of G^E is known, then the activity coefficient of the nth component can be obtained from equation 10.141 written in terms of f_n for the nth component, i.e.

$$RT \ln f_k = G^E + \sum_{p=1}^N (\delta_{kp} - x_p) \left(\frac{\partial G^E}{\partial x_p} \right)_{x_{j,j \neq k,n}} \tag{15.58}$$

where δ_{kp} is the Kronecker delta symbol, which is zero when $k \neq p$ and unity where $k = p$.

The application involves first measuring the compositional dependence of the activity coefficients, and therefore values of G^E, for every binary combination of ions

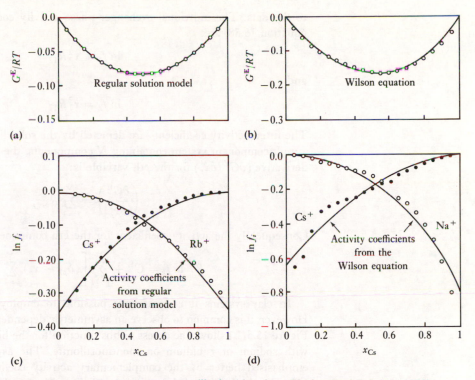

Figure 15.5   Excess functions for ions on montmorillonite (data from Elprince and Babcock, 1975).

relevant to a particular multi–ion system. We then select a suitable general expression
for the compositional dependence of G^E which contains **only binary ion-interaction
coefficients**, simplify this to the binary case and fit the resulting expression to the
binary G^E data. This process involves estimating the fitting coefficients so that the
fit to the binary G^E data is optimised. The extracted fitting coefficients may then
be used then to predict the value of G^E or the activity coefficients of any
multicomponent ion exchanger of fixed composition.

For example, if values of G^E for binary systems show a symmetric dependence
on composition, as in figure 15.5(a), a regular solution theory may be employed (see
section 10.5.2). The generalisation of the regular solution model for a solution with
N components has the form:

$$\frac{G^E}{RT} = \sum_{i=1}^{N-1} \sum_{j>i}^{N} R_{ij}x_i x_j \qquad (15.59)$$

where R_{ij} is an empirical fitting parameter for the ith and jth ions. The resulting
expression for a binary (1,2) mixture is:

$$\frac{G^E}{RT} = R_{12}x_1 x_2 \qquad (15.60)$$

In a system containing four ions there would be similar expressions for combinations
(1,3), (1,4), (2,3), (2,4) and (3,4). (Regular solution theory assumes the coefficient
matrix is symmetric, i.e. $R_{ij} = R_{ji}$).

The solid line on figure 15.5(a) shows a fit of the binary regular solution model to the data. The individual exchanger–phase activity coefficients, calculated using equation 15.58, are:

$$\ln f_1 = x_2^2 R_{12} \tag{15.61}$$

and

$$\ln f_2 = x_1^2 R_{12} \tag{15.62}$$

The fitted activity coefficients are depicted by the solid lines in figure 15.5(c). For a multicomponent system containing N components, the general expression for the derivative $(\partial G^E / \partial x_p)$ for the pth variable is:

$$\left(\frac{\partial G^E}{\partial x_p}\right) = \sum_{j=1}^{N} R_{pj} x_j \tag{15.63}$$

Consequently the activity coefficient for the kth component is given by:

$$RT \ln f_k = G^E + \sum_{p=1}^{N} \left[(\delta_{pk} - x_p) \sum_{j=1}^{N} R_{pj} x_j \right] \tag{15.64}$$

In clay systems it is sometimes possible to employ regular solution theory. However, it is common to observe an asymmetric dependence of G^E on composition. Figure 15.5(b) shows the excess Gibbs function for the binary exchange of caesium with sodium or rubidium on montmorillonite. The asymmetry is slight and is emphasised better by the complementary activity coefficients in figure 15.5(d). However, both G^E and the activity coefficients show a dependence on composition which cannot be modelled accurately by the simple regular solution equation.

A model which can account for moderate asymmetry is the Wilson equation (see section 10.5.2, equations 10.159–10.161). For the binary case this takes the form

$$\frac{G^E}{RT} = -x_1 \ln(1 - x_2 W_{21}) - x_2 \ln(1 - x_1 W_{12}) \tag{15.65}$$

where W_{12} is an empirical coefficient. For any system it is assumed that W_{12} does not equal W_{21}, in which case there are two parameters for every binary system. If both parameters are set equal, a plot of the excess Gibbs energy versus composition will show a maximum or a minimum at the 50:50 composition. The model is then similar to the regular solution model. Unequal values of the fitting parameters move the position of the turning point slightly, giving the required flexibility. The solid lines on figures 15.5(b) and (d) show the fit of the Wilson equation and the activity coefficients as predicted by the model. This model has been successfully applied to a number of binary exchange reactions on clays but has yet to be validated for multicomponent systems.

In order that this general approach may work, it is essential that the values of G^E in multicomponent systems are dominated by the binary ion-interaction terms. If experimental data suggest that higher-order terms are significant, it is necessary to evaluate these terms from an analysis of measured activity coefficients in ternary or higher-component systems.

As a note of caution, figure 15.6 shows some excess Gibbs function for reactions on zeolites that are far from symmetrical with composition. Such behaviour is usually attributed to the polyfunctional nature of zeolites. In figure 15.6(a) the fit of the Wilson equation is seen to be adequate. However, the compositional dependence of

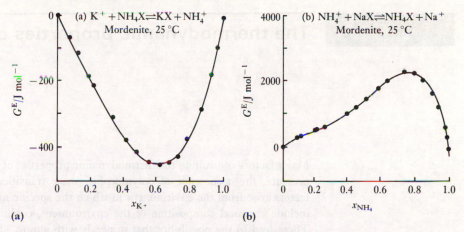

Figure 15.6   Extensively asymmetric excess functions (data from Fletcher *et al.*, 1984).

G^E in figure 15.6(b) cannot be modelled by such an expression. For such cases the general polynomial expressions such as the Margules equation should be used (section 10.5.2). These polynomial models contain an array of fitting parameters and are very flexible. However, for any specific system, their application to multicomponent systems should be validated prior to use. Their applicability should not be accepted uncritically (see Chu and Sposito, 1981).

Chapter 16

The thermodynamic properties of solids

Major factors controlling the thermodynamic properties of minerals are temperature, pressure, the occurrence of polymorphic phase transitions and impurities. Other factors arise from the environment in which the specific mineral was formed. These include chemical composition of the environment, cooling rate and strain history. These lead to the possibility that minerals with almost identical compositions may have different thermodynamic properties due to order/disorder effects, differences in crystal defect state or differences in particle size and crystallinity. Generally, there are enough data in the literature to understand the major factors and predict their consequences. In contrast, the secondary factors are less well characterised, although these factors may be the main reason for discrepancies between laboratory measurements and the properties of minerals in natural systems. Indeed, it has been estimated that the presence of crystal defects may increase the Gibbs energy of a mineral by up to $10 \, \text{kJ mol}^{-1}$ or more. The magnitude of order/disorder effects is similar to this although such effects are usually small at ambient temperatures, progressing to around $10 \, \text{kJ mol}^{-1}$ at temperatures in excess of $1000\,°\text{C}$. Provision for these secondary factors should be made for the most precise calculations (see Helgeson *et al.*, 1978).

16.1 General properties of minerals

16.1.1 The standard state for minerals

The standard state is the pure component at any specified temperature and pressure which is consistent with that for water and dissolved species. In this state the mole fraction of the component is 1 and the activity equals 1. There are two major consequences to this choice of standard state.

1. The equilibrium constant for any reaction involving a solid phase will have a temperature and pressure dependence.
2. Most solids will have an activity of 1 at all temperatures and pressures. Exceptions are those minerals containing high levels of impurities, where the mole fraction of the mineral is less than 1.

The superscript $\ominus$ is used to denote the standard state. However, since the standard state is unconstrained in temperature and pressure, all pure minerals are in their standard states and some authors prefer not to use the superscript.

Further comments on the use of this standard state are given in Appendix 3.

16.1.2   The solubility product

The reaction for the **precipitation** of a single mineral M(s) can be written as:

$$\sum_j v_j A_j = M(s) \tag{16.1}$$

where the summation is over the solution species A_j, including water, and v_j is the stoichiometric number of moles of the subscripted species in the reaction. The equilibrium constant is given by:

$$\ln K_{eq} = \ln a_{M(s)} + \sum_j v_j \ln a_{A_j} \tag{16.2}$$

When the activity of the mineral, $a_{M(s)}$, is unity, the equilibrium constant becomes:

$$\ln K_{eq} = \sum_j v_j \ln a_{A_j} \tag{16.3}$$

That is, **at equilibrium the product of the activities of the aqueous species, each raised to the power of the number of moles in the reaction, is a constant** (note that v_j is negative for reactants). From this, the **solubility product** K_s is defined as:

$$\ln K_s = - \sum_j v_j \ln a_{A_j} \tag{16.4}$$

For example:

$$Ag^+ + Cl^- \rightleftharpoons AgCl(s) \qquad K_s = a_{Ag^+} \times a_{Cl^-}$$

$$Ca^{2+} + SO_4^{2-} + 2H_2O \rightleftharpoons CaSO_4 . 2H_2O(s) \qquad K_s = a_{Ca^{2+}} \times a_{SO_4^{2-}} \times a_{H_2O}^2$$

If the activity product for ions in equilibrium with a solid is constant, then increasing the concentration of an ion common to the reaction will result in more solid being precipitated to re-establish the equilibrium. This explains the observation that any salt is **less soluble** in a solution containing a common ion than in pure water.

16.1.3   The saturation index

Any natural solution may or may not be at equilibrium with a given mineral even if all the necessary species are present. For example, a dissolving mineral may not have reached equilibrium, or a mineral which should precipitate may remain in solution on account of slow nucleation. **At any given instance**, the product of the activities of the dissolved species, each raised to the power of the number of moles in a given precipitation reaction, can be measured. This is called the **ionic activity product** (IAP). Thus in a solution containing Na^+, Ca^{2+}, Cl^- and SO_4^{2-}, we can calculate the IAPs for sodium chloride, gypsum, anhydrite, sodium sulphate heptahydrate or any other mineral that may be formed from this combination of ions. For example,

$$IAP(gypsum) = a_{Ca^{2+}} \times a_{SO_4^{2-}} \times a_{H_2O}^2$$

$$IAP(NaCl) = a_{Na^+} \times a_{Cl^-}$$

A **saturation index** (SI) is defined as:

$$SI = \log\left(\frac{IAP}{K_s}\right) \tag{16.5}$$

This variable gives the same information as the chemical affinity (section 11.2.4).

If the SI is **zero**, then $IAP = K_s$ and the solution is at **equilibrium** with the mineral.

If the SI is **negative** the solution is **undersaturated** with respect to the mineral and the mineral may dissolve.

If the SI is **positive** the solution is **oversaturated** with respect to the mineral and the mineral may precipitate.

16.1.4   Pressure solution

The solubility of any mineral is influenced by the factors controlling the chemical potentials of the relevant species. Pressure is usually the least important factor except when a mineral/fluid assembly has a high solid-to-liquid ratio and the mineral particles are in contact.

Consider two spherical particles, in contact, bathed in a solution with which they are at equilibrium (figure 16.1a). If the particles experience an axial force through their point of contact there will be a large pressure exerted at this point. In principle the pressure is infinite at a point contact, but real particles will have a finite area of contact and experience a finite pressure. In the local region at the contact point the Gibbs energy for the particles will increase, since:

$$dG = V\,dP \tag{16.6}$$

and V is positive. The increase in the Gibbs energy for the particles leads to a greater tendency for the particles to dissolve and the result is a net flux of dissolved particles away from the contact point. This phenomenon is called **pressure solution**. It is a local effect since the axial force is supported by the particles and the solvent does not experience the same pressure as the particles. Once a macroscopic quantity

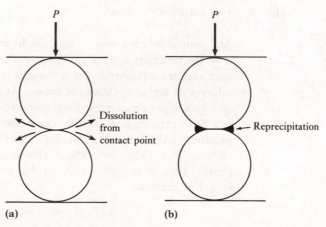

Figure 16.1   Pressure solution and creep.

of ions from the dissolving particles have detached they are oversaturated with respect to the solution phase and will re-precipitate close to the contact point. The general effect is that consolidated particles, supporting an overburden, will deform and 'creep' through the solution phase (figure 16.1(b)).

16.1 Basic thermodynamic relationships

At temperatures T_2 and pressure P_2 the apparent standard Gibbs energy of formation and apparent standard enthalpy of formation are related to the true standard energy of formation ($\Delta G^{\ominus}_{f(T_r,P_r)}$, and $\Delta H^{\ominus}_{f(T_r,P_r)}$) by:

$$\Delta G^{\ominus}_{(T_2,P_2)} = \Delta G^{\ominus}_{f(T_r,P_r)} + [\bar{G}^{\ominus}_{(T_2,P_2)} - \bar{G}^{\ominus}_{(T_r,P_r)}] \qquad [16.7]$$

and

$$\Delta H^{\ominus}_{(T_2,P_2)} = \Delta H^{\ominus}_{(T_r,P_r)} + [\bar{H}^{\ominus}_{f(T_2,P_2)} - \bar{H}^{\ominus}_{(T_r,P_r)}] \qquad (16.8)$$

where r denotes the reference condition, from which:

$$[\bar{G}^{\ominus}_{(T_2,P_2)} - \bar{G}^{\ominus}_{(T_r,P_r)}] = \int_{P_r}^{P_2} \bar{V}^{\ominus}_{(T_2,P)} \, dP - \bar{S}^{\ominus}_{(T_r,P_r)}(T - T_r)$$

$$+ \int_{T_r}^{T_2} \bar{C}^{\ominus}_{P(T,P_r)}\left(1 - \frac{T_2}{T}\right) dT \qquad (16.9)$$

and

$$[\bar{H}^{\ominus}_{(T_2,P_2)} - \bar{H}^{\ominus}_{(T_r,P_r)}] = \int_{T_r}^{T_2} \bar{C}^{\ominus}_{P(T,P_r)} \, dT + \int_{P_r}^{P_2} \bar{V}^{\ominus}_{(T_2,P)}(1 - \alpha_{(T_2,P)}) \, dP$$

$$(16.10)$$

where α is isothermal compressibility.

The third-law entropy is given by:

$$[\bar{S}^{\ominus}_{(T_2,P_2)} - \bar{S}^{\ominus}_{(T_r,P_r)}] = \int_{T_r}^{T_2} \frac{\bar{C}^{\ominus}_{P(T,P_r)}}{T} dT - \int_{P_r}^{P_2} \bar{V}^{\ominus}_{(T_2,P)}\alpha_{(T_2,P)} \, dP \qquad (16.11)$$

In calculating the standard-state properties of minerals at elevated temperatures and pressures the following points must be noted.

- The values of $\Delta G^{\ominus}_{f(T_r,P_r)}$, $\Delta H^{\ominus}_{f(T_r,P_r)}$ and $\bar{S}^{\ominus}_{(T_r,P_r)}$ must be known for the reference temperature and pressure, which is usually 298.15 K and 1 bar.
- The integration path is from temperature T_r to temperature T_2 at constant pressure P_r and then from pressure P_r to P_2 at constant temperature T_2.
- The integration path dictates that the necessary information includes temperature dependence of $\bar{C}_P$ at P_r in addition to the pressure dependence of molar volume V and isothermal compressibility α at T_2. This path is selected because $\bar{C}_P$ is more easily measured at low pressures.

To produce suitable retrieval equations we must select suitable empirical expressions for the temperature dependence of $\bar{C}_P$ at 1 bar and of the pressure

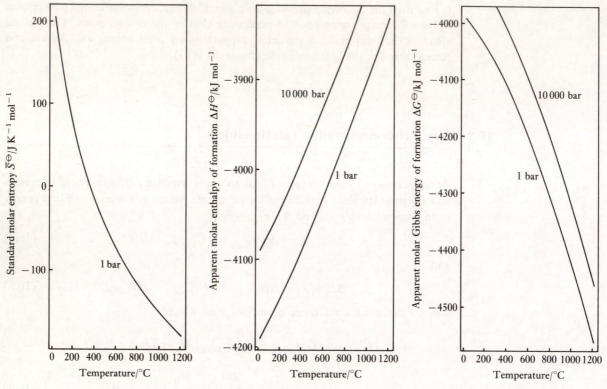

Figure 16.2   Thermodynamic data of anorthite (data calculated using the Mayer–Kelly equation, after Helgeson *et al.*, 1978).

dependence of $\bar{V}$ at the higher temperature, and then perform the necessary integrations.

Figure 16.2 shows the effect of temperature and pressure on the standard entropy, apparent Gibbs energy and apparent enthalpy of formation for the mineral anorthite. The effects are typical of minerals and less complex than those observed for aqueous electrolytes. Consequently, relatively simple retrieval equations are used for prediction. We will see later that the apparent lack of complexity reflects the relatively simple dependence of $\bar{C}_{P,1\text{ bar}}$ on temperature, and the insensitivity of the molar volume to changes in pressure. A major complexity, not depicted in figure 16.2, arises from the occurrence of second-order phase transitions.

16.2.1   Thermodynamic data at the reference point

For many minerals, values of $\Delta G^{\ominus}_{f(T_r, P_r)}$, $\Delta H^{\ominus}_{f(T_r, P_r)}$, $\bar{S}^{\ominus}_{(T_r, P_r)}$, and the appropriate temperature or pressure dependence of $\Delta C^{\ominus}_{P(T, P_r)}$ and $\bar{V}^{\ominus}_{(T_2, P)}$ have been calculated from solubility data, calorimetric measurements and density data. For other minerals it may be necessary to use estimation techniques to produce data as follows.

Heat capacity, C_P   The molal heat capacity of a mineral is approximately equal to the sum of the molal heat capacities of the constituent elements. This is Kopp's rule of additivity.

$$\overline{C}_{P(\text{min})} = \sum v_i \overline{C}_{P,i} \tag{16.12}$$

where v_i is the mean number of moles of the ith element in the mineral and $\overline{C}_{P,i}$ its molar heat capacity. For example, the measured molar heat capacity of acanthite ($Ag_2S(s)$) at 298.18 K and 1 bar is 75.36 J mol^{-1}, which compares well with the value of 73.57 J mol^{-1} calculated from:

$$\overline{C}_{P(Ag_2S)} = 2\overline{C}_{P(Ag^+)} + \overline{C}_{P(S^{2-})} = 2 \times 25.4 + 22.77 = 73.57 \text{ J mol}^{-1} \tag{16.13}$$

This technique is poor for minerals composed of elements which are gaseous under ambient conditions.

Standard molar   An analogous equation to 16.12 may be used for entropies.
entropy, S

$$\overline{S}_{\text{min}} = \sum v_i \overline{S}_i \tag{16.14}$$

This equation is less reliable than that for heat capacity. An improvement was made by Kubaschewski and Alcock (1979), who proposed that the entropies for anions should be corrected by a simple factor proportional to the nominal charge on the cation.

A more reliable approach involves expressing the formula of a mineral in terms of its oxide components and then equating the standard entropy of the mineral to the sum of the standard entropies of the oxides. The procedure involves writing a reaction for the mineral in terms of reactions between oxides.

$$\overline{S}_{\text{min}} = \sum v_{\text{oxides}} \overline{S}_{\text{oxides}} \tag{16.15}$$

For example, the mineral akermanite is written:

$$Ca_2MgSi_2O_7(s) = CaO(s) + MgO(s) + 2\alpha\text{-}SiO_2(s) \tag{16.16}$$

The standard entropy at 298.15 K and 1 bar is:

$$\overline{S}_{\text{aker}} = 2\overline{S}_{CaO} + \overline{S}_{MgO} + 2\overline{S}_{\alpha\text{-quartz}} \tag{16.17}$$

This approach is equivalent to assuming that the standard entropy change for the reaction is zero.

If the molar volume change for the reaction ΔV_r is known, an improved estimate can be made using the equation:

$$\overline{S}_{\text{min}} = \sum v_{\text{oxides}} \overline{S}_{\text{oxides}} + \overline{Q}\Delta V_r \tag{16.18}$$

where $\overline{Q}$ is an empirical coefficient with a value for silicates of around 2.5 if the volume change is in cm^3 mol^{-1}. In practice, $\overline{Q}$ should be determined using minerals from the same structural class as the unknown. The example of akermanite formation would then be:

$$Ca_2MgSi_2O_7(s) + \alpha\text{-}Al_2O_3(s) = Ca_2Al_2SiO_7(s) + MgO(s) + \alpha\text{-}SiO_2(s) \tag{16.19}$$

from which:

$$\overline{S}_{\text{aker}} + = \overline{S}_{\text{gehl}} + \overline{S}_{\text{per}} + \overline{S}_{\alpha\text{-quartz}} - \overline{S}_{\text{cor}} + \overline{Q}\Delta V_r \tag{16.20}$$

The predicted value using this equation is very close to the measured value.

An equation recommended by Helgeson *et al.* (1978) for non-ferrous minerals is:

$$\bar{S}_{\mathrm{min}} = \frac{\sum v_{\mathrm{oxides}} \bar{S}_{\mathrm{oxides}} (\sum v_{\mathrm{oxides}} \bar{V}_{\mathrm{oxides}} + \bar{V}_{\mathrm{min}})}{2 \times \bar{V}_{\mathrm{min}}} \qquad (16.21)$$

The basic techniques above, or their refinements, are used in geochemical applications. Unfortunately, the techniques for the prediction of enthalpies of formation or Gibbs energies of formation are less reliable. Most of the techniques involve relating the energy of formation of a mineral to the sum of the energies of formation for the oxides or hydroxides. When tested against measured values, there is usually a residual error.

16.2.2   General trends for $\bar{C}_P$, $\bar{V}$ and α

Since the temperature dependence of C_P and the pressure dependence of V play a central role in developing retrieval equations (section 16.2.3), we will consider these phenomena in some detail.

Heat capacity at 1 bar pressure   The heat capacities of minerals increase smoothly with increasing temperature up to a point where a phase transition is encountered. As the temperature approaches absolute zero, the heat capacity approaches zero. Typically, the value of $\bar{C}_P$ for any one mineral will change by between 50 % and 150 % over a temperature range of 1000 °C.

No two minerals show identical properties, although there are similarities between minerals of similar structure and composition. Figure 16.3 shows the temperature dependence of the heat capacities of some common minerals at 1 bar pressure.

Phase transitions   The order of phase transitions for minerals is usually difficult to categorise unambiguously due to the lack of precise data. Observed phase transitions range from abrupt changes for some minerals through to gradual changes occurring over hundreds of degrees Celsius.

The first-order transition is a limiting case characterised by step changes in heat capacity, entropy, enthalpy and volume. There is an abrupt change in the slope of the Gibbs function with respect to temperature at the transition point. Although step changes occur, the slopes of $\bar{C}_P$, $\bar{H}$, $\bar{V}$ and $\bar{S}$ on each side of the transition temperature are similar. In principle the heat capacity jumps to infinity at the transition temperature.

The standard molar heat capacities of first-order polymorphic transitions (ΔC_P) are in the order of 0.5 J mol^{-1}. Consequently, values of ΔC_P are imprecise since their values are similar to the uncertainties in calorimetric data.

At the transition point the change in the Gibbs energy for any phase transition is zero ($\Delta G_{\mathrm{tr}} = 0$). If the enthalpy for the transition (ΔH_{tr}) is measured, the entropy change can be calculated from:

$$0 = \Delta H_{\mathrm{tr}} - T \Delta S_{\mathrm{tr}} \qquad (16.22)$$

If the Clapeyron slope ($\mathrm{d}P / \mathrm{d}T$) is known at the transition temperature, then the volume change can be calculated from

$$\frac{\mathrm{d}P}{\mathrm{d}T} = \frac{\Delta S_{\mathrm{tr}}}{\Delta V_{\mathrm{tr}}}$$

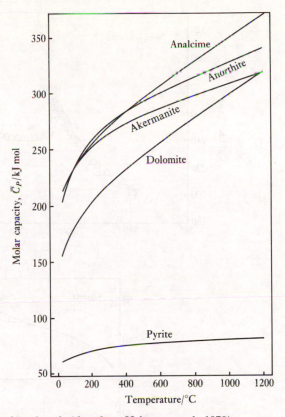

Figure 16.3   Heat capacities for minerals (data from Helgeson *et al.*, 1978).

The predicted volume change can be compared with the measured volume change as a check on internal consistency of the measurements.

For most minerals the value of the Clapeyron slope is independent of pressure, although experimental data are generally not sufficiently comprehensive to permit its precise evaluation. This makes predicting the effect of pressure on the temperature of phase transitions very uncertain.

A less common phase transition for mineral polymorphs is the second–order transition, characterised by a continuous gradient in the Gibbs function with respect to temperature at the transition point. This leads to abrupt changes in the slopes of entropy, volume and enthalpy at the transition temperature but no step change. The heat capacity shows a step change at the transition temperature.

A form of second–order phase transition is the λ transition characterised by a progressive increase in the gradient of C_P with temperature before the transition point. After the transition point, the heat capacity shows a more linear dependence on temperature. The gradual change in slope before the transition point is attributed to a progressive re-ordering of the atoms in the solid with increasing temperature.

Values of ΔC_P for λ transitions are larger than for first-order transitions, as indicated by data in figure 16.4 for quartz and nesquehonite. The data for quartz at elevated pressure make the point that ΔC_P is relatively independent of pressure. However, the temperature at which the transition occurs is dependent on pressure, indicating that the Clapeyron slope is non-zero. These two observations for quartz

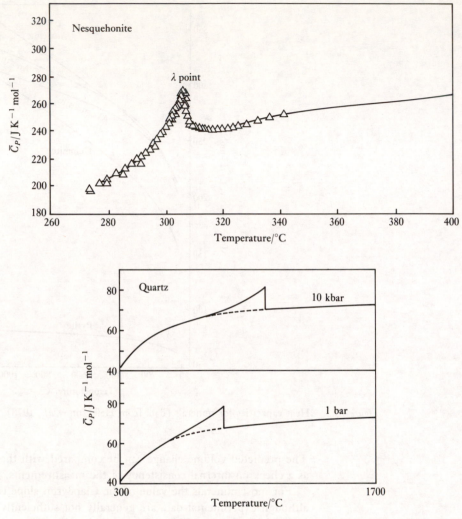

Figure 16.4   λ transitions for nesquehonite and quartz [data for nesquehonite ($\triangle$) from Robie and Hemingway, 1973, and theoretical fit (—) from Helgeson *et al.*, 1978; data for quartz after Berman, 1988].

seem to be true for most minerals although the reservation about the uncertainties in Clapeyron slopes still holds.

Volume   For most minerals, a pressure change of 40 000 bar decreases the molar volumes by 3–4%. The effect of temperature over the range 0–1000 °C increases the volume by a similar magnitude. Therefore, the effects of temperature and pressure are small, and, when T and P increase together, self-compensating. Consequently, it is common to regard the standard molar volume of minerals as independent of temperature and pressure, i.e. $\bar{V}^{\ominus}_{T_2, P_2} = \bar{V}^{\ominus}_{298.15\,K,\,1\,bar}$. This is a reasonable approximation for many geological applications, although its assumption may lead to errors in the prediction of the effect of pressure on equilibrium constants for solid–solid reactions where the molar volume changes for reactions are themselves small. A mineral for which the approximation is poor is quartz.

Using the assumption of constant molar volume, the effect of pressure on the apparent Gibbs energy of formation or the apparent enthalpy of formation is to increase these functions by approximately $\bar{V}^{\ominus}_{298.15\,K,\,1\,bar} \times (P - P_r)$. Since molar volumes for minerals are of the order of tens to hundreds of cubic centimetres per mole depending on the choice of formula unit for the mineral, we may expect the pressure contribution to be $< 100\,kJ\,mol^{-1}$ at pressures as high as $10\,000$ bar. Typically, this represents a change of around 5–$10\,\%$ in the respective energy of formation.

16.2.3   Suitable retrieval equations

Molar volume   One expression for the temperature and pressure dependence of molar volume below 50 kbar is:

$$\bar{V}^{\ominus}_{T,P} = \bar{V}^{\ominus}_{T_r,P_r}[1 + \alpha_V(T - 298)](1 - \beta_V P) \qquad (16.23)$$

where α_V and β_V are empirical fitting coefficients.

For the most precise calculations, Berman (1988) has proposed the following equation for the effect of temperature and pressure on the molar volume of minerals:

$$\bar{V}^{\ominus}_{(T,P)} = \bar{V}^{\ominus}_{(T_r,P_r)}[1 + v_1(P - P_r) + v_2(P - P_r)^2 + v_3(T - T_r) + v_4(T - T_r)^2] \qquad (16.24)$$

For example, the value of $\bar{V}_{(298.15\,K,\,1\,bar)}$ for anorthite ($CaAl_2Si_2O_8$) is 10.075 J bar^{-1} and the temperature/pressure fitting coefficients are $v_1 = -1.272 \times 10^{-6}$, $v_2 = 3.176 \times 10^{-2}$, $v_3 = 10.918 \times 10^6$ and $v_4 = 41.985 \times 10^{10}$. The fitting coefficients for other minerals are given by Berman (1988).

Molar volumes are usually quoted in $cm^3\,mol^{-1}$ and must be converted to the appropriate energy units when used to calculate the effect of pressure on the energies of formation.

Heat capacity   Equations used to describe the temperature dependence of the heat capacity, **at 1 bar pressure**, are usually polynomials in temperature. Some expressions for minerals which do not show λ transitions are as follows.

- The Maier–Kelly equation:

$$\bar{C}_{P(T,P_r)} = a + bT - cT^{-2} \qquad (16.25)$$

where a, b and c are temperature-independent coefficients for each mineral.
- An equation used by Berman and Brown (1985), of the form:

$$\bar{C}_{P(T,P_r)} = k_0 + k_1 T^{-0.5} + k_2 T^{-2} + k_3 T^{-3} \qquad (16.26)$$

where the empirical coefficients k_1 and k_2 are constrained to be less than or equal to zero.
- An equation used by Robie *et al.* (1979) which has some of the features of the above two equations:

$$\bar{C}_{P(T,P_r)} = \beta_0 + \beta_1 T + \beta_3 T^2 + \beta_4 T^{-0.5} + \beta_5 T^{-2} \qquad (16.27)$$

where the βs are empirical coefficients.

In general, simple expressions like the Maier–Kelly equation fit well, but not as well as equations with more coefficients. For the higher-order polynomials, the sacrifice is that they have more points of inflexion or turning points. This means that they are less well constrained outside the region of fit and are less well suited to extrapolation. The above three equations offer a reasonable compromise.

If a first-order phase transition occurs, it is usual to ignore the contribution to the heat capacity and to fit the selected equation over the entire temperature range. Only the values of ΔS_{tr} and ΔH_{tr} need be documented and these are usually assumed to be independent of temperature and pressure.

If a second-order phase transition or λ transition occurs, the marked difference in the temperature dependence of $\bar{C}_P$ before and after the transition makes quantitative treatment of λ transitions complicated. In such cases the temperature dependence of heat capacity is usually described by one of the above equations plus an extra component $\bar{C}_{P,\lambda}$. This term is itself temperature-dependent. Berman and Brown (1985) recommended a term of the form:

$$\bar{C}_{P,\lambda(1 \text{ bar})} = T(l_1 + l_2 T)^2 \tag{16.28}$$

where T is temperature in kelvin within the range $T_r < T < T_\lambda$ and T_λ is the temperature of the maximum in the plot of $\bar{C}_P$ versus T. It is common to find that the value of T_λ increases linearly with pressure and a suitable expression is:

$$T_{\lambda(P)} = T_{\lambda(1 \text{ bar})} k(P - 1) \tag{16.29}$$

where k has a specific value for each mineral. By defining $T_d = T_{\lambda(P)} - T_\lambda$ (1 bar) and $T' = T + T_d$, the expression for the temperature and pressure dependence of $\bar{C}_{P,\lambda(P)}$ is:

$$\bar{C}_{P,\lambda(P)} = T'(l_1 + l_2 T')^2 \tag{16.30}$$

16.2.4   Using retrieval equations

Equations for the temperature dependence of heat capacity and the pressure dependence of molar volume can be used to predict values for $[\bar{G}^{\ominus}_{(T_2,P_2)} - \bar{G}^{\ominus}_{(T_r,P_r)}]$, $[H^{\ominus}_{(T_2,P_2)} - H^{\ominus}_{(T_r,P_r)}]$ and $[S^{\ominus}_{(T_2,P_2)} - S^{\ominus}_{(T_r,P_r)}]$ for any desired temperature and pressure. The requirements are data at the reference pressure and the necessary fitting coefficients for the empirical expressions.

For maximum precision it may be necessary to use the most flexible equations for heat capacity. However, as an example we will use the Maier–Kelly equation to evaluate the above functions. From the relationship $\bar{C}^{\ominus}_{(T,1 \text{ bar})} = a + bT - c/T^2$, we obtain:

$$\int_{T_r}^{T_2} \bar{C}^{\ominus}_{P(T,1 \text{ bar})} \, dT = a(T_2 - T_r) + \frac{b(T_2^2 - T_r^2)}{2} + c\left(\frac{1}{T_2} - \frac{1}{T_r}\right) \tag{16.31}$$

and

$$\int_{T_r}^{T_2} \bar{C}^{\ominus}_{P(T,1 \text{ bar})} \, d \ln T = a \ln(T_2/T_r) + b(T_2 - T_r) + \frac{c}{2}\left(\frac{1}{T_2^2} - \frac{1}{T_r^2}\right) \tag{16.32}$$

If the molar volume of any mineral is independent of temperature, $(\partial \bar{V}/\partial T)_P = 0$.

This leads to

$$\int_{P_r}^{P} \bar{V}^{\ominus}_{(T_2,P)}\, dP = \bar{V}^{\ominus}_{(T_r,P_r)}(P_2 - P_r) \tag{16.33}$$

For greater precision we may choose to express the pressure dependence of molar volume with equation 16.24, from which:

$$\int_{P_r}^{P} \bar{V}^{\ominus}_{(T_2,P)}\, dP = \bar{V}^{\ominus}_{(T_r,P_r)}\{(v_1/2 - v_2)(P_2^2 - P_r^2) + v_2/3(P_2^3 - P_r^3)$$
$$+ [1 - v_1 + v_2 + v_3(T_2 - T_r) + v_4(T_2 - T_r)^2](P_2 - P_r)\} \tag{16.34}$$

The above three equations can be combined with the fundamental equations 16.9–16.11 to give:

$$[\bar{G}^{\ominus}_{(T_2,P_2)} - \bar{G}^{\ominus}_{(T_r,P_r)}] = \bar{S}^{\ominus}_{(T_r,P_r)}(T - T_r) + a[T - T_r - T\ln(T/T_r)]$$
$$+ \frac{(c - bTT_r^2)(T - T_r)^2}{2TT_r^2} + \bar{V}^{\ominus}_{(T_r,P_r)}(P - P_r) \tag{16.35}$$

$$[\bar{H}^{\ominus}_{(T_2,P_2)} - \bar{H}^{\ominus}_{(T_r,P_r)}] = a(T - T_r) + \frac{b}{2}(T^2 - T_r^2) + c\left(\frac{1}{T} - \frac{1}{T_r}\right) + \bar{V}^{\ominus}_{(T_r,P_r)}(P - P_r) \tag{16.36}$$

and

$$[\bar{S}^{\ominus}_{(T_2,P_2)} - \bar{S}^{\ominus}_{(T_r,P_r)}] = a\ln(T/T_r) + b(T - T_r) + \frac{c}{2}\left(\frac{1}{T^2} - \frac{1}{T_r^2}\right) \tag{16.37}$$

The above equations are useful for minerals that do not exhibit λ transitions. To extend the above equations to describe λ transitions it is necessary to add on the contributions from the additional heat capacity equation (16.28) to $[\bar{H}^{\ominus}_{(T_2,P_2)} - \bar{H}^{\ominus}_{(T_r,P_r)}]$ and $[\bar{S}^{\ominus}_{(T_2,P_2)} - \bar{S}^{\ominus}_{(T_r,P_r)}]$. These contributions are:

$$\Delta H_\lambda = \int_{T_r}^{T} \bar{C}_{P,\lambda}\, dT \quad \text{and} \quad \Delta S_\lambda = \int_{T_r}^{T} \bar{C}_{P,\lambda}\, d\ln T \tag{16.38}$$

Substituting equation 16.30 into the above gives:

$$\Delta H_\lambda = x_1(T - t_r) + \frac{x_2}{2}(T^2 - t_r^2) + \frac{x_3}{3}(T^3 - t_r^4) + \frac{x_4}{4}(T^4 - t_r^4) \tag{16.39}$$

and

$$\Delta S_\lambda = x_1 \ln(T/t_r) + x_2(T - t_r) + \frac{x_3}{2}(T^2 - t_r^2) + \frac{x_4}{3}(T^3 - t_r^3) \tag{16.40}$$

where

$$x_1 = l_1^2 T_d + 2l_1 l_2 T_d^2 + l_2 T_d^3 \tag{16.41}$$

$$x_2 = l_1^2 + 4l_1 l_2 T_d + 3l_2 T_d \tag{16.42}$$

$$x_3 = 2l_1 l_2 + 3l_2^2 T_d \tag{16.43}$$

$$x_4 = l_2^2 \tag{16.44}$$

$$t_r = T_r - T_d \tag{16.45}$$

The contribution to the Gibbs energy of a mineral due to a λ transition can be calculated at any temperature and pressure below the transition temperature from:

$$\Delta G_\lambda = \Delta H_\lambda - T\Delta S_\lambda \tag{16.46}$$

16.2.5   The effect of particle size on the Gibbs energy of minerals

For some minerals it has been observed that small particles are more soluble than large particles. The effect is only noticeable if particle radii are reduced to 10^{-6} m (1 μm) or less; many oxide and clay minerals have particle radii of this order.

Our knowledge of these effects is far from complete. However, we may rationalise the effects by noting that any factor which increases the molar Gibbs energy of a substance will increase its tendency to dissolve. Strictly, Gibbs energy of a particle should contain two contributions: first, a term for the bulk of the material which is equal to the value of the Gibbs energy per unit volume (G_V) multiplied by the volume of the particle. The second term is the product of the surface area of the particle and the interfacial energy between the particle and the solution (γ_S, in units of energy per unit area). Therefore, for a particle of radius r,

$$G = \tfrac{4}{3}\pi r^3 G_V + 4\pi r^2 \gamma_S \tag{16.47}$$

Values of γ_S are presumed positive.

If we neglect the surface area term, the value of G increases linearly with the cube of the radius (volume). We may divide both sides of the above equation by the mass ($4\pi r^3 \rho/3$) to obtain the Gibbs energy per unit mass. Assuming that the density ρ is independent of particle size, the value of G per unit mass remains constant. However, including the interfacial energy term increases the value of G by an amount proportional to the square of the radius, compared with the cube of the radius for the bulk term. At very small particle sizes this contribution becomes large compared with the bulk term but it decreases proportionally as the particle size increases. The precise contribution for any mineral will depend on the actual value of γ_S, which is unknown for most minerals. However, for some metal oxides and hydroxides the overall increase in the molar Gibbs energy has been observed to be as high as 20 kJ mol^{-1}.

16.2.6   Solid solution formation

Compositional considerations   Many minerals including silicates exhibit compositional variation. Indeed, some minerals with the same crystal structure can be clearly distinguished from each other on the basis of relatively minor compositional differences. We may think of compositional variation as the mixing of two or more different components. If mixing is ideal with respect to Raoult's law, the activity of any one component may be equal to its mole fraction. Ideal mixing is rarely found in solid solutions.

We note that solid solutions are not simple homogeneous mixtures as are gases or liquid mixtures. In solids the atoms reside at discrete crystallographic sites in the structure. Consequently, solid-solution formation is often replacement or mixing of atoms on sites, with the crystallographic structure usually remaining basically unchanged. For example, OH^- replaces F^- in amphiboles, micas and clays. In carbonates we observe Mg^{2+} replacing Ca^{2+}.

The way exchanging ions distribute themselves on sites has some bearing on the thermodynamic treatment. For example, in pyroxenes there are two different types of sites: a tetrahedral site usually containing Ca^{2+} or Na^+ and an octahedral site occupied by one or more of the ions Al^{3+}, Fe^{3+}, Mg^{2+} and Fe^{2+}. There are clearly many ways the ions can be distributed through the sites. If ions are replaced independently on each of the sites we have a different situation from that where substitution on one site is linked to substitution on another site. If, for example, the number of trivalent ions on the octahedral site is increased by replacement of a divalent ion, charge balance may be retained by a divalent ion on the tetrahedral site being simultaneously replaced by a monovalent ion. This type of replacement also occurs in feldspars, amphiboles, micas and clays. Such phenomena not only make the modelling of activity difficult but also make the choice of components unclear. In the pyroxene example the minerals diopside ($CaMgSiO_2O_6(s)$), hedenbergite ($CaFe^{II}Si_2O_6(s)$), acmite ($NaFe^{III}Si_2O_6(s)$), and jadeite ($NaAl^{III}Si_2O_6(s)$) could be components but there are many more appropriate formula units. It is not clear if any of these formula units are meaningful for the complex mixture and we should not expect the activities of the components to be equal to the mole fractions of the formula units.

In some geochemical applications it may be appropriate to estimate the activity or standard chemical potential of a specified component in a solid solution. In other applications, the standard chemical potential of the entire mixed solid may be required. The methods to calculate such properties have been discussed by Nordstrom and Munoz (1986) and Helgeson *et al.* (1978). We will summarise some of the essentials discussed in these reviews and place emphasis on the commonly used **mixing-on-sites (MOS) model**.

Calculating activities in solid solutions using multisite models

The MOS model assumes that all ions occupying a given crystallographic site mix independently on that site. For example, garnets have the general formula $X_3Y_2Si_3O_{12}$ where the symbols X and Y represent the formulae of ions occupying a cubic and an octahedral site respectively. Crystallographic measurements show that ions Mg^{2+}, Ca^{2+} and Fe^{2+} are found in the cubic site whereas Al^{3+} and Fe^{3+} mix on the octahedral site. A typical composition may be $(Mg_{0.12},Ca_{0.08},Fe^{II}_{0.8})_3(Al_{0.97}Fe^{III}_{0.03})_2Si_3O_{12}$. The choice of components in such a mixture is a matter of convenience. In the above system we may select the minerals pyrope ($Mg_3Al_2Si_3O_{12}(s)$), almandine ($Fe^{II}_3Al_2Si_3O_{12}(s)$), grossular ($Ca_3Al_2Si_3O_{12}(s)$) and andradite ($Ca_3Al_2Si_3O_{12}(s)$). There are other potential components arising from binary combinations of the ions in the two sites, all of which may be ascribed an activity.

The activity of any component in a mixture can be obtained from one form of the MOS model. The following equations are for the prediction of the activity of a component in a multisite solid assuming that all ions mix randomly on all sites, i.e. there is ideal mixing on sites. Essentially, **the exercise is to calculate the mole fraction of a specified component in terms of the mole fractions of ions on sites**. In accordance with ideal behaviour, the mole fraction of the component is then its activity.

For a system containing K sites we may define $X_{j,k}$ to be the mole fraction of the jth ion on the kth sites, where:

$$X_{j,k} = \frac{n_{j,k}}{\sum_j n_{j,k}} \qquad (16.48)$$

and $n_{j,k}$ is the number of moles of the jth ion on the kth type of site. If the number of moles of the jth ion on the kth type of site in the ith component is given the symbol $v_{i,j,k}$, the expression for the activity of the ith component is:

$$a_i = q_i \prod_k \prod_j X_{j,k}^{v_{i,j,k}} \qquad (16.49)$$

where q_i is a constant of proportionality given by:

$$q_i = \prod_k \prod_j \left(\frac{v_{i,j,k}}{v_{i,k}} \right)^{-v_{i,j,k}} \qquad (16.50)$$

where $v_{i,k}$ is defined as:

$$v_{i,k} = \sum_j v_{i,j,k} \qquad (16.51)$$

Considering the garnet example we will assign values of i, j and k to the components, ions and sites as follows.

i	j	k
Pyrope $= 1$	Mg $= 1$	$X = 1$
Almandine $= 2$	Ca $= 2$	$Y = 2$
Grossular $= 3$	$Fe^{II} = 3$	
Andradite $= 4$	Al $= 4$	
	$Fe^{III} = 5$	

In accordance with the general equation we must calculate the constants $v_{i,j,k}$ from $v_{1,1,1}$ (which represents the stoichiometric number of moles of Mg in pyrope on the cubic sites) through to $v_{4,5,2}$ (the stoichiometric number of moles of iron(III) in andradite on the octahedral sites). To do this we construct two $v_{i,j,k}$ matrices, one for $k = 1$ and the other for $k = 2$. Each matrix has five columns ($j = 1-5$) and four rows ($i = 1-4$). Most of the elements of these matrices will be zero because not all ions occupy all sites in the components. The only non-zero-valued coefficients are $v_{1,1,1}$, $v_{2,3,1}$, $v_{3,2,1}$ and $v_{4,2,1}$, which all equal 3, and $v_{1,4,2}$, $v_{2,4,2}$, $v_{3,4,2}$ and $v_{4,5,2}$, all of which equal 2. In general such simplifications are common and we note that the value of $v_{i,k} = v_{i,j,k}$, for any j, if only one type of ion occupies the kth type of site in the formula unit of the ith component. Under these conditions $q_i = 1$. In the garnet example $v_{i,1}$ always equals 3 whereas $v_{i,2}$ equals 2.

Similarly, there are 20 different mole fractions of the ions which can be calculated. These are $X_{1,1}$, representing Mg in the cubic site, through to $X_{5,2}$, representing Fe^{III} on the octahedral sites. Some of the mole fractions are zero since not all ion types reside on all sites.

To use equation 16.49 to evaluate the component activities we must recall that any number raised to the power zero is 1. Using the garnet composition

$(Mg_{0.12},Ca_{0.08},Fe_{0.8}^{II})_3(Al_{0.97}Fe_{0.03}^{III})_2Si_3O_{12}$, this gives:

$$\left.\begin{aligned}
a_1 &= X_1 = X_{1,1}^3 \times X_{4,2}^2 = 0.12^3 \times 0.97^2 = 1.626 \times 10^{-3} \\
a_2 &= X_2 = X_{3,1}^3 \times X_{4,2}^2 = 0.8^3 \times 0.97^2 = 4.817 \times 10^{-1} \\
a_3 &= X_3 = X_{4,1}^3 \times X_{4,2}^2 = 0.08^3 \times 0.97^3 = 4.817 \times 10^{-4} \\
a_4 &= X_4 = X_{4,1}^3 \times X_{5,2}^2 = 0.08^3 \times 0.03^2 = 4.608 \times 10^{-7}
\end{aligned}\right\} \quad (16.52)$$

In real garnets it is necessary to include the replacement of silicon in the tetrahedral sites (Si_3O_{12}) with aluminium. This extends the calculation to four components with three sites, seven ions and 28 mole fraction terms, most of which are zero.

The standard molar Gibbs energy of formation for the entire mixture can be calculated from the sum of the molar Gibbs energies of formation of the individual components plus the molar Gibbs energy of mixing (ΔG_{mix}). The Gibbs energy of mixing is given by:

$$\Delta G_{mix} = RT \sum_i \left(\ln q_i + \sum_k \sum_j v_{i,j,k} \ln X_{j,k} \right) \quad (16.53)$$

The entropy of mixing is given by:

$$\Delta S_{mix} = -R \sum_i X_i \left(\ln q_i \sum_k \sum_j v_{i,j,k} \ln X_{j,k} \right) \quad (16.54)$$

where X_i is the mole fraction of the ith component. (It is worth noting that the calculation of the entropy of mixing is the key to predicting the thermodynamic properties of solid solutions in both ideal and highly non–ideal systems.)

In using the above expressions the components must be selected carefully. A major requirement for selecting components is that the composition of the mixture can be evaluated from the sum of each component multiplied by the number of moles of the component. This is illustrated by a simple mixture containing 1 mol of NaCl and 2 mol of KNO_3. This can be described by these two components or alternatively by a mixture of 1 mol of $NaNO_3$, 1 mol of KCl and 1 mol of KNO_3. Here we see that the choice of components $NaNO_3$ and KCl dictates that we also include the component KNO_3 to maintain mass balance. This makes the point that the formulae of the components are chosen for convenience but the number of components and their individual amounts are constrained by mass balance.

The above expressions for ΔG_{mix} and ΔS_{mix} are for ideal site mixing. This is consistent with no heat of mixing ($\Delta H_{mix} = \Delta G_{mix} + T\Delta S_{mix} = 0$), which is often satisfied when the standard molar volumes of the components are within 5% of each other and when components have similar crystal structures. There is no provision in this model for exchange of atoms between energetically distinct sites or for the situation where q_i is temperature-dependent. This latter situation occurs when solid solutions also exhibit temperature-dependent order/disorder effects.

Excess Gibbs-energy models   At low temperatures miscibility gaps are often observed in solid solutions. This is attributed to positive values for the enthalpies of mixing which outweigh the $T\Delta S_{mix}$ terms. Positive enthalpies of mixing accompany several different phenomena. First, there is the mixing of ions of different size which results in considerable strain on the crystal lattice with subsequent modification of lattice energies. An extreme case is when the crystal structures of the pure components are significantly different from

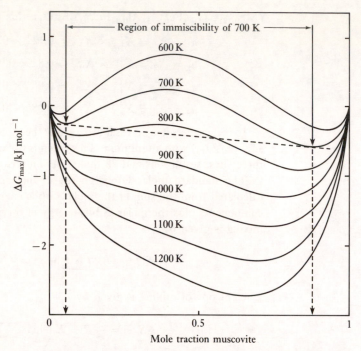

Figure 16.5   Gibbs energy of mixing for paragonite/muscovite solutions (after Nordstrom and Munoz, 1986; original data from Eugster *et al.*, 1972).

each other. This size factor may also be responsible for mixing between sites since it may be energetically favourable for ions to move between sites to accommodate replacements. A second important feature is the charge on the replacing ions. When ions of different charges replace each other there is usually a positive enthalpy of mixing; similarly when ions of the same magnitude mix the enthalpy of mixing is larger for highly charged ions. Finally, positive enthalpies of mixing accompany the mixing of ions of different bonding character.

Clearly, the thermodynamics of real solid solutions is complex, especially if temperature-dependent order/disorder effects are also acknowledged. The techniques to accommodate the complexities of such behaviour have been discussed by Helgeson *et al.* (1978) and Navrotsky(i) (1974, 1985, 1986). The most comprehensive and general treatments of non-ideal behaviour in solid solutions use equations for the excess Gibbs energy fitted to experimental variations in activity. The Margules equations given in section 10.5.2 serve the purpose well and have been used previously for an important number of minerals. Figure 16.5 shows calculated values of the excess Gibbs energy of mixing for binary solutions of paragonite and muscovite using a two-parameter expression incorporating temperature dependence as described by Nordstrom and Munoz (1986). The region of immiscibility at 700 K is indicated by vertical arrows. Immiscibility occurs at all temperatures below 1100 K.

Systems which show this simple behaviour are adequately described by excess energy models with only a limited number of parameters. This approach accommodates non-zero values of the enthalpy of mixing and its compositional variation. Unfortunately, order/disorder effects cannot be easily incorporated into such simple expressions.

Chapter 17

Electrochemical systems and redox

In the Earth's crust some elements may be found in several oxidation states; observed redox conditions vary from one natural environment to another, depending on the availability of oxygen. Redox reactions are well established as influencing the formation of rock-forming minerals and sedimentary processes. Consequently, this chapter is concerned with thermodynamic methods for characterising redox reactions.

Redox reactions are associated with the transfer of electrons between reactive species. They are accompanied by consumption or release of electrical energy and are therefore discussed within the context of electrochemical systems.

A simple electrochemical cell, consisting of two electrically conducting rods (electrodes) dipping into an electrolyte solution, is one such system. If the electrodes are connected by a high-impedance voltmeter, a stable potential difference between the electrodes will develop. In this chapter we show the reason why this potential arises and also the type of information that can be deduced from electrochemical potential measurements.

17.1   Some basic concepts

17.1.1   The Daniell cell

The best-known chemical cell is the Daniell cell, consisting of two solutions, one of zinc sulphate and the other of copper sulphate, separated by a small porous disc. A copper rod is partially immersed in the copper sulphate solution and a zinc rod is partially immersed in the zinc sulphate solution (figure 17.1). If a galvanometer is connected to both electrodes, completing the circuit, a current is observed. This current indicates that chemical reactions are taking place. If the galvanometer is replaced by a high-impedence voltmeter, the copper electrode is seen to be at a higher potential than the zinc.

It is observed that the zinc electrode dissolves, while copper is deposited on the copper electrode. The overall reaction is:

$$Zn(s) + Cu^{2+} \rightleftharpoons Zn^{2+} + Cu(s) \tag{17.1}$$

The individual electrode reactions are:

$$Zn(s) \rightleftharpoons Zn^{2+} + 2e^{-} \tag{17.2}$$

and

$$Cu^{2+} + 2e^{-} \rightleftharpoons Cu(s) \tag{17.3}$$

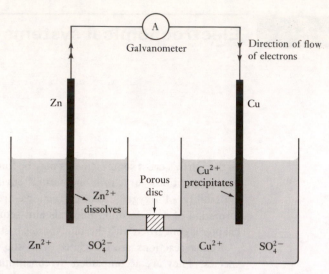

Figure 17.1   The Daniell cell.

Zinc ions enter the solution phase, leaving two electrons on the zinc electrode, whereas copper ions are depositing on the copper electrode from which each ion removes two electrons. A flow of electrons through the external circuit now occurs from the zinc to the copper electrode, in order to minimise the electron imbalance. This is the observed current. The rate of reaction is proportional to the current and if the electrical circuit is broken the reaction ceases.

Under most conditions where a current is flowing the measurable potential is unstable and irreproducible. If measured when no current is flowing, the potential rises to a stable reproducible maximum. This condition can be achieved if the external connections between the electrodes are made via a second cell and potentiometer which can be adjusted so that a current flows through the primary cell in the reverse direction to the spontaneous flow. It is possible to decrease the resistance of the potentiometer progressively so that the net current flowing through the system drops and reaches zero. Further reduction in the resistance of the potentiometer leads to a reversal of the current.

17.1.2   Some comments on the e.m.f. of cells

A condition for obtaining the maximum amount of usable energy from a system is that the system should operate reversibly (section 8.3). In the context of electrochemical cells, this condition is the point of zero current where the dissolution and precipitation processes are in equilibrium and capable of being reversed by an infinitesimal change in the external electrical potential. The value of the potential difference between the electrodes at the point of zero current is called the **reversible electromotive force** of the cell (e.m.f.) and is a measure of the maximum amount of electrical energy that can be extracted from the system.

It will be shown later that the e.m.f. of a cell is related to the Gibbs energy for the combined cell reactions (ΔG) by:

$$-\Delta G = nFE \tag{17.4}$$

where n is an integer equal to the number of moles of electrons exchanged between the chemical species in the reaction and F is the Faraday constant. Using equation 17.4, a fundamental electrochemical relationship can be deduced, i.e.

$$E = E^{\ominus} - \frac{RT}{nF} \ln \sum v_i \ln a_i \qquad (17.5)$$

where $E^{\ominus}$ is the **standard electrode potential** for the cell, v_i is the stoichiometric number of moles of the ith species in the complete cell reaction (positive for products and negative for reactants) and a_i its activity. This is the Nernst equation. We may also write for $E^{\ominus}$

$$E^{\ominus} = \frac{RT}{nF} \ln K \qquad (17.6)$$

where K is the equilibrium constant for the overall reaction.

Thus, **the e.m.f. of a cell is proportional to the chemical affinity of the cell reaction and is related to the activities of the relevant ions in solution.** When the e.m.f. of the electrode assembly is controlled by the activity of one ion, the electrode is said to be reversible with respect to that ion.

Using the example of the Daniell cell, and assuming the activities of the metal electrodes are unity, the fundamental relationship is:

$$E = E^{\ominus} - \frac{RT}{2F} \ln \frac{m_{Zn^{2+}} \gamma_{Zn^{2+}}}{m_{Cu^{2+}} \gamma_{Cu^{2+}}} \qquad (17.7)$$

where m and γ are the molalities and activity coefficients, respectively, of the subscripted species. We may formally separate equation 17.7 into reactions at the two individual electrodes, i.e.

$$E = \left[E^{\ominus}_{Cu^{2+}} + \frac{RT}{2F} \ln a_{Cu^{2+}} \right] - \left[E^{\ominus}_{Zn^{2+}} + \frac{RT}{2F} \ln a_{Zn^{2+}} \right] \qquad (17.8)$$

where $E^{\ominus}_{Cu^{2+}}$ and $E^{\ominus}_{Zn^{2+}}$ are the standard electrode potentials of copper and zinc respectively. In principle they cannot be determined individually; in practice methods exist for determining their relative values.

The above relationships give an interpretation of the e.m.f. of a cell in terms of the concentration of the reaction species. They also offer a way of measuring the equilibrium constant and therefore the thermodynamic properties of the species in the reaction.

There are two important points to be made. First, equation 17.5 is in terms of the ratios of the reactive ions. In principle, these ratios can be determined but the individual ion activities remain inaccessible. The second point is that in any real cell there are usually several reactions occurring at each electrode, and these usually prove notoriously difficult to define.

The sign convention for cell reactions

A sign convention exists which ensures that the e.m.f. of a complete cell has a positive value when the affinity of the reaction is positive and the reaction proceeds from left to right. That is:

1. The electrochemical cell and the cell reaction are written so that the uncharged metallic atoms appear on the same side of the equation as the corresponding metal electrode.

2. The sign of the e.m.f. is taken as the polarity of the right-hand electrode when the cell is an open circuit.

17.1.3   Types of electrodes

An electrode is a device at which an electron transfer is made to take place. There are several commonly used electrodes, some of which are more practical than others.

- **Metal/metal ion**: This is controlled by the reaction:

$$M(s) \rightleftharpoons M^{z+} + ze^- \tag{17.9}$$

The usefulness depends on the practicability of producing the metal in a well-defined state.

- **Gas/ion**: The commonest examples of this type are the hydrogen electrode, controlled by the half-cell reaction:

$$\tfrac{1}{2}H_2 \rightleftharpoons H^+ + e^- \tag{17.10}$$

and the chloride electrode, controlled by:

$$\tfrac{1}{2}Cl_2 + e^- \rightleftharpoons Cl^- \tag{17.11}$$

Physically, gas/ion electrodes are constructed so that the gas bubbles over a metal electrode into a solution containing the dissolved ion (in these examples, H^+ or Cl^-). The reaction takes place at the metal surface. The pure metal is to act as a reservoir for electrons and to catalyse the reaction while remaining inert to the ions, the gas and the solvent. Platinum or gold electrodes covered with a layer of platinum black are suitable for this purpose.

The **standard hydrogen electrode** is a gas/ion electrode at 298.15 K where the partial pressure of hydrogen is 1 bar and the activity of the hydrogen ion is unity. Figure 17.2 shows such an electrode connected to a second electrode to complete a circuit. The use of this standard hydrogen electrode is discussed later.

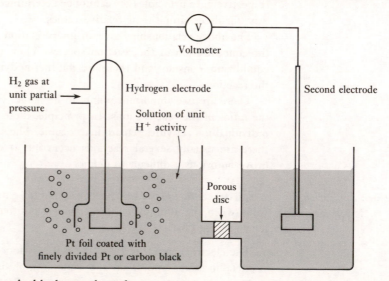

Figure 17.2   The standard hydrogen electrode.

- **Metal/insoluble salt**: This consists of a metal electrode with a coating of an insoluble salt of the metal. The metal-ion concentration is controlled by the solubility product of the insoluble salt. An example is silver wire coated with silver chloride, where the half-cell reaction can be regarded as two steps, i.e.

$$Ag(s) \rightleftharpoons Ag^+ + e^- \tag{17.12}$$

and

$$Ag^+ + Cl^- \rightleftharpoons AgCl(s) + e^- \tag{17.13}$$

By combining reactions 7.12 and 7.13, the half-cell reaction is:

$$Ag(s) + Cl^- \rightleftharpoons AgCl(s) + e^- \tag{17.14}$$

Since the chemical potentials of silver and silver chloride are independent of solution composition, the e.m.f. of the cell is controlled by the activity of the chloride ion in solution. The electrode is reversible to chloride. Other examples of this type of electrode are the calomel electrode:

$$2Hg(s) + 2Cl^- \rightleftharpoons Hg_2Cl_2(s) + 2e^-$$

and the silver oxide electrode:

$$Ag(s) + OH^- \rightleftharpoons \tfrac{1}{2}Ag_2O(s) + \tfrac{1}{2}H_2O + e^-$$

- **Ion/ion or redox**: All electrodes involve oxidation/reduction processes. However, the term 'redox electrode' is reserved for one where its interface is the site for a reaction involving an ion that can exist in two different oxidation states. Such a device is simply an inert metal electrode, possibly platinum, which acts as a sink for electrons from a redox reaction at the electrode/solution interface. A typical reaction is:

$$Fe^{2+} \rightleftharpoons Fe^{3+} + e^- \tag{17.15}$$

Box 17.1

Deriving the basic relationships

The relationship between the e.m.f. of a cell and the chemical affinity of the electrode reactions can be demonstrated by considering the Daniell cell. For the time being we will neglect the fact that the solution phases in the cell are separated by a porous membrane. If such a cell at equilibrium has an e.m.f. of E_{cell} and is balanced by an e.m.f. in the external circuit of E_{ext}, then no current will flow. If E_{ext} is lowered infinitesimally, the tendency of the cell reaction to proceed from left to right will overcome the opposition of the applied potential and electrons will flow from right to left. The amount of work done by the cell is:

$$dw = -E_{ext}\, dq = -E_{cell}\, dq \tag{17.16}$$

Since E_{ext} is infinitesimally different from E_{cell} we may drop the subscript on E. The flow of electrons is a measure of the progress of the reaction from left to right; therefore the extent of reaction $d\xi$ is positive. From Faraday's laws of electrolysis we may write:

$$dq = zF\, d\xi \tag{17.17}$$

continued

Box 17.1
continued

where $z = 2$ in this example. The electrical work done reversibly on the system is obtained by combining equations 17.16 and 17.17:

$$dw_{elect} = -zFE\, d\xi \tag{17.18}$$

In addition to electrical work there will also be work due to volume changes under constant pressure. In chapter 8 we saw that the change in the Gibbs energy, when a system at constant temperature interacts reversibly with its surroundings, is equal to the total work less that related to the change in the volume of the system. We may deduce that the electrical work is equal to the change in the Gibbs energy and write:

$$dG = -zFE\, d\xi \tag{17.19}$$

or

$$\frac{dG}{\xi} = -zFE \tag{17.20}$$

Recalling from section 11.2 that the affinity of any reaction is given by:

$$A\, d\xi = -dG \tag{17.21}$$

we may write:

$$A = zFE \tag{17.22}$$

The above equation provides the link between the e.m.f. of the cell and the thermodynamics of the cell reaction. The affinity of a reaction is expressed in terms of the chemical potential of the reactive species by:

$$A = -\sum_i v_i \mu_i \tag{17.23}$$

The next step is to express the chemical potential of any species by $\mu_i = \mu_i^{\ominus} + RT \ln a_i$, which leads to:

$$A = -\sum v_i \mu_i^{\ominus} - \sum v_i RT \ln a_i \tag{17.24}$$

The constant chemical potential terms can be collected together in the term $A^{\ominus}$ to give:

$$A = A^{\ominus} - \sum v_i RT \ln a_i \tag{17.25}$$

Substitution into equation 17.22 gives the fundamental relationship:

$$zFE = A^{\ominus} - \sum v_i RT \ln a_i \tag{17.26}$$

When the reactive species are in their standard states $\sum v_i RT \ln a_i$ is zero, and therefore the e.m.f. of the cell has the value $E^{\ominus}$ equal to $A^{\ominus}/zF$. We define $E^{\ominus}$ as the standard electrode potential of the cell and write:

$$E = E^{\ominus} - \frac{RT}{zF} \sum v_i \ln a_i \tag{17.27}$$

This is the Nernst equation.

A final point is that $A^{\ominus} = RT \ln K$ where K is the equilibrium constant of the cell. We may therefore write:

$$E^{\ominus} = \frac{RT}{zF} \ln K \tag{17.28}$$

Box 17.2   **Liquid–liquid junction potentials**

In box 7.1 the treatment is exact up to equation 17.22. However, from that point we have considered only the reactions at the electrode/solution interfaces and neglected the fact that there is another interface in the system. At the porous membrane in the Daniell cell there is an abrupt change in the composition of the solution phase such that the two solutions are different phases which are not in equilibrium. There will be a chemical potential gradient across this interface; therefore a potential difference will exist. This potential difference, called the **liquid–liquid junction potential**, E_j, will depend on the relative concentrations of the ions in both phases and cannot be eliminated or computed at concentrations other than infinite dilution. This means that the e.m.f. of a cell with a liquid–liquid junction is not given by equation 17.27 but by:

$$E = [E^{\ominus} + E_j] - \frac{RT}{zF} \sum v_i \ln a_i \qquad (17.29)$$

The equations for the Daniell cell are therefore:

$$E = [E^{\ominus} + E_j] - \frac{RT}{2F} \ln \frac{m_{Zn^{2+}} \gamma_{Zn^{2+}}}{m_{Cu^{2+}} \gamma_{Cu^{2+}}} \qquad (17.30)$$

or

$$E = E_j + \left[E^{\ominus}_{Cu^{2+}} + \frac{RT}{2F} \ln a_{Cu^{2+}} \right] - \left[E^{\ominus}_{Zn^{2+}} + \frac{RT}{2F} \ln a_{Zn^{2+}} \right] \qquad (17.31)$$

Equation 17.30 shows that it is not even possible to measure the ratio of the ion activities in concentrated solutions, even if the standard electrode potential of the cell is known, because there will always be the unknown quantity E_j present.

However, it is always possible to determine the value of $E^{\ominus}$ for any cell as follows. The e.m.f. for any cell reaction can be written:

$$E + \frac{RT}{zF} \sum v_i \ln m_i = E^{\ominus} + E_j - \frac{RT}{zF} \sum v_i \ln \gamma_i \qquad (17.32)$$

For the specific example this is:

$$E - \frac{RT}{2F} \sum \ln \left(\frac{m_{Cu^{2+}}}{m_{Zn^{2+}}} \right) = E^{\ominus} + E_j + \frac{RT}{2F} \sum \ln \left(\frac{\gamma_{Cu^{2+}}}{\gamma_{Zn^{2+}}} \right) \qquad (17.33)$$

A cell should be constructed with equimolar solutions of copper and zinc at molality m. If measurements are of E at progressively decreasing values of m, down to very dilute solution, a plot of the left-hand side of the above equation versus $m^{1/2}$ gives an intercept of $E^{\ominus}$ at zero concentration.

The principle of extrapolation to zero concentration can be used to evaluate $E^{\ominus}$ for any cell, not just the Daniell cell.

Box 17.3

Cells without liquid–liquid junctions

As examples we will consider two cells without liquid–liquid junctions. In these examples the liquid–liquid junction potential is eliminated by immersing both electrodes in the same solution.

The cell H_2–HCl–AgCl–Ag

This cell contains a single solution of hydrochloric acid (HCl) saturated with silver chloride.

The electrode on the left of the cell is the hydrogen electrode; the electrode on the right is silver. There is an equilibrium quickly established between the atoms of the silver electrode and the silver ions arising from the dissolved silver chloride. The overall cell reaction is:

$$\tfrac{1}{2}H_2 + AgCl \rightleftharpoons Ag(s) + H^+ + Cl^- \tag{17.34}$$

for which there is the transfer of one electron per mole of hydrogen ($z = 1$). It follows from equations 17.22 and 17.23 that:

$$zEF = -A = \tfrac{1}{2}\mu_{H_2} + \mu_{Ag^+} - \mu_{H^+} - \mu_{Ag(s)} \tag{17.35}$$

We recall that the chemical potential of solid silver is a constant at a fixed temperature and pressure and that the chemical potentials of the dissolved species and hydrogen are given by:

$$\mu_{H_2} = \mu_{H_2}^{\ominus} + RT \ln P_{H_2} \tag{17.36}$$

$$\mu_{H^+} = \mu_{H^+}^{\ominus} + RT \ln a_{H^+} \tag{17.37}$$

and

$$\mu_{Ag^+} = \mu_{Ag^+}^{\ominus} + RT \ln a_{Ag^+} \tag{17.38}$$

where P_{H_2} is the partial pressure of hydrogen. Substituting the above into equation 17.35 and collecting the constant chemical potential terms gives:

$$zEF = A^{\ominus} + RT \ln\left(\frac{P_{H_2}^{1/2} \times A_{Ag^+}}{a_{H^+}}\right) \tag{17.39}$$

At this stage we note two points. The first is that the activity of chloride ions is governed by the solubility product of silver chloride $K_{(S)}$ through the relationship:

$$a_{Ag^+} = \frac{K_S}{a_{Cl^-}} \tag{17.40}$$

The second point is that the mean molal activity of hydrochloric acid a_{HCl} is given by:

$$a_{H^+} \times a_{Cl^-} = a_{HCl}^2 \tag{17.41}$$

Substituting the above into equation 17.39 gives:

$$zEF = A^{\ominus} + RT \ln K_S + \tfrac{1}{2}RT \ln P_{H_2} - 2RT \ln a_{HCl} \tag{17.42}$$

If we define a term $\Delta G'$ as:

$$\Delta G' = (A^{\ominus} + RT \ln K_S + \tfrac{1}{2}RT \ln P_{H_2}) \tag{17.43}$$

then:

$$zEF = \Delta G' - 2RT \ln a_{HCl} \tag{17.44}$$

continued

Box 17.3
continued

The first two terms on the right-hand side of equation 17.43 are constants at fixed temperature and pressure. Therefore, if the partial pressure of hydrogen is fixed at exactly 1 bar, the term $\ln P_{H_2}$ is zero and $\Delta G'$ is a constant. By adopting the definition $E^\ominus = -\Delta G'/zF$ then:

$$E = E^\ominus - \frac{2RT}{zF} \ln a_{HCl} \tag{17.45}$$

The term $E^\ominus$ is dependent on temperature and pressure and is equal to the e.m.f. of the cell under conditions when the partial pressure of hydrogen is 1 bar and the mean ionic activity of the hydrochloric acid is unity. If the e.m.f. of the cell is measured at any other partial pressure of hydrogen P_{H_2} and has the value E', then the e.m.f. corrected to 1 bar is:

$$E = E'' - \frac{1}{2F} \ln p_{H_2} \tag{17.46}$$

To measure the value of $E^\ominus$ we first express the activity coefficient of HCl in terms of its molality and the mean stoichiometric activity coefficient of HCl, i.e.

$$E = E^\ominus - \frac{2RT}{zF} \ln (m_{HCl} \times \gamma_{\pm HCl}) \tag{17.47}$$

The above equation rearranges to give:

$$E + \frac{2RT}{zF} \ln m_{HCl} = E^\ominus - \frac{2RT}{zF} \ln \gamma_{\pm HCl} \tag{17.48}$$

As concentrations tend to zero, activity coefficients tend to unity and the left-hand side of equation 17.48 tends to $E^\ominus$. Therefore, a plot of $E + (2RT/zF) \ln m_{HCl}$ versus m_{HCl} gives an intercept of $E^\ominus$ at zero concentration.

The cell
$Ag-AgCl-KCl-Cl_2-Pt$

This cell is constructed from a solution of potassium chloride saturated with silver chloride. The left electrode is a silver electrode and the right electrode a platinum wire over which chlorine is bubbled. The half-cell reactions are:

$$Ag(s) + Cl^- \rightleftharpoons AgCl(s) + e^- \tag{17.49}$$

and

$$\tfrac{1}{2}Cl_2 + e^- \rightleftharpoons Cl_2 \tag{17.50}$$

from which the full-cell reaction, corresponding to the transfer of one electron is:

$$Ag(s) + \tfrac{1}{2}Cl_2 \rightleftharpoons AgCl(s) \tag{17.51}$$

The e.m.f. of the cell is:

$$E = E^\ominus + \frac{RT}{zF} \ln P_{Cl_2} \tag{17.52}$$

It is seen that the e.m.f. of the cell is dependent only on the partial pressure of chlorine.

17.2   The standard e.m.f. of half-cells

There are many different types of cells, each involving different reactive species and each having a unique standard electrode potential. For the purpose of compact tabulation it is convenient to select one electrode as a standard reference electrode and measure the e.m.f.s of other electrodes relative to this. The commonly chosen reference electrode is the hydrogen electrode.

17.2.1   Typical half-cells

The following is an example we will use to define and clarify the convention adopted for the standard hydrogen electrode.

The cell, written as $H_2-H^+:Zn^{2+}-Zn(s)$, is a hydrogen electrode on the left-hand side and a zinc electrode on the right-hand side, both immersed in an acidic solution. The complete cell reaction is:

$$H_2 + Zn^{2+} \rightleftharpoons 2H^+ + Zn(s) \qquad (17.53)$$

and is consistent with the transfer of two electrons ($z = 2$). The e.m.f. is derived from the relationship $zEF = A$ where:

$$A = \mu_{H_2} + \mu_{Zn^{2+}} - 2\mu_{H^+} - \mu_{Zn(s)} \qquad (17.54)$$

This yields:

$$zEF = \mu_{H_2}^{\ominus} + RT \ln P_{H_2} + \mu_{Zn^{2+}}^{\ominus} + RT \ln a_{Zn^{2+}} - 2\mu_{H^+}^{\ominus} - 2RT \ln a_{H^+} - \mu_{Zn(s)} \qquad (17.55)$$

Collecting the standard chemical potential terms in the term $A^{\ominus}$ gives:

$$zEF = A^{\ominus} + RT \ln\left(\frac{P_{H_2} a_{Zn^{2+}}}{a_{H^+}^2}\right). \qquad (17.56)$$

Defining $E^{\ominus}$ as $A^{\ominus}/zF$, we now have:

$$E = E^{\ominus} + \frac{RT}{zF} \ln P_{F_2} + \frac{RT}{zF} \ln\left(\frac{a_{Zn^{2+}}}{a_{H^+}^2}\right) \qquad (17.57)$$

The above relationship shows that the measured e.m.f. at a fixed temperature and pressure is related to the partial pressure of hydrogen and the activity ratios for Zn^{2+} and H^+ ions. By defining the standard hydrogen electrode as having a partial pressure of hydrogen of 1 bar and unit proton activity, the above equation reduces to:

$$E = E^{\ominus} + \frac{RT}{zF} \ln a_{Zn^{2+}} \qquad (17.58)$$

When the activity of the zinc ions is unity, then $E = E^{\ominus}$.

From equation 17.58 it appears that we can measure the activity of the zinc ions, which contradicts the view that ion activities are inaccessible. In reality **we cannot measure single ion activities** because it is not actually possible to construct a standard hydrogen electrode. The standard hydrogen electrode is hypothetical since we have no way of ensuring that the hydrogen ion activity is unity. Even if we know

$E^\ominus$, the best we can do is measure the Zn^{2+}/H^+ activity ratio. The above restriction does not stop us from calculating $E^\ominus$ using the extrapolation technique discussed previously.

The point to note is that the value of $E^\ominus$ has no absolute significance and represents the standard electrode potential of zinc **relative to the standard hydrogen electrode**. We emphasise this point by noting that E can be expressed in terms of individual electrode potentials from:

$$E = [E^\ominus_{Zn^{2+}} - E^\ominus_{Zn(s)}] - [2E^\ominus_{H^+} - E^\ominus_{H_2}] \frac{RT}{zF} \ln P_{H_2} + \frac{RT}{zF} \ln\left(\frac{a_{Zn^{2+}}}{a^2_{H^+}}\right) \quad (17.59)$$

The term $[E^\ominus_{Zn^{2+}} - E^\ominus_{Zn(s)}]$ is the standard potential for the zinc ion/zinc electrode, whereas the term $[2E^\ominus_{H^+} - E^\ominus_{H_2}]$ is the potential for the standard hydrogen electrode.

By convention, we define the potential for the standard hydrogen electrode as zero. In this case E becomes the standard e.m.f. of the zinc ion/zinc half-cell; more commonly it is called the **standard electrode potential** of zinc.

17.2.2   The sign convention for half-cells

By adopting the standard hydrogen electrode as a reference electrode we need to define a convention for writing cell reactions, since the e.m.f. of the cell may be positive or negative depending on how we construct the cell and the cell reaction. The IUPAC convention is as follows.

- Molecular hydrogen is always written on the left-hand side of the cell reaction and the hydrogen electrode on the left-hand side of the cell.
- The standard e.m.f. of this cell is defined as:

$$E^\ominus_{cell} = E^\ominus_{right} - E^\ominus_{left} \quad (17.60)$$

with $E^\ominus_{left} = 0$ for a standard hydrogen electrode.

In the zinc ion/zinc electrode example, the measured value of the e.m.f. is 0.763 V. Therefore the standard electrode potential of zinc is:

$$E^\ominus_{Zn(s)/Zn^{2+}} = -0.763/V \quad (17.61)$$

Once the above convention is adopted, the standard electrode potential of many electrodes can be calculated. For example, in the Daniell cell it is found that:

$$E_{cell} = E^\ominus_{Cu(s)|Cu^{2+}} - E^\ominus_{Zn(s)|Zn^{2+}} = 1.1/V \quad (17.62)$$

It follows that:

$$E^\ominus_{Cu(s)|Cu^{2+}} = E_{cell} + E^\ominus_{Zn(s)|Zn^{2+}} = 1.1 - 0.763 = 0.337/V \quad (17.63)$$

Similarly, from the cell:

$$Cu(s)|Cu^{2+}||Ag^+|Ag(s)$$

$$E_{cell} = E^\ominus_{Ag(s)|Ag^+} - E^\ominus_{Cu(s)|Cu^{2+}} = 0.462/V \quad (17.64)$$

from which:

$$E^\ominus_{Ag(s)|Ag^+} = E^\ominus_{cell} + E^\ominus_{Cu(s)|Cu^{2+}} = 0.462 + 0.337 = 0.799/V \quad (17.65)$$

Table 17.1   **Standard electrode potentials**

Half-Cell Reaction	$E^{\ominus}/V$
$Li^+ + e^- \rightleftharpoons Li(s)$	-3.05
$K^+ + e^- \rightleftharpoons K(s)$	-2.93
$Ca^{2+} + 2e^- \rightleftharpoons Ca(s)$	-2.87
$Na^+ + e^- \rightleftharpoons Na(s)$	-2.71
$Mg^{2+} + 2e^- \rightleftharpoons Mg(s)$	-2.37
$Al^{3+} + 3e^- \rightleftharpoons Al(s)$	-1.66
$Mn^{2+} + 2e^- \rightleftharpoons Mn(s)$	-1.18
$2H_2O + 2e^- \rightleftharpoons H_2(g)(s) + 2OH^-$	-0.83
$Zn^{2+} + 2e^- \rightleftharpoons Zn(s)$	-0.76
$Fe^{2+} + 2e^- \rightleftharpoons Fe(s)$	-0.44
$PbSO_4(s) + 2e^- \rightleftharpoons Pb(s) + SO_4^{2-}$	-0.31
$Pb^{2+} + 2e^- \rightleftharpoons Pb(s)$	-0.13
$2H^+ + 2e^- \rightleftharpoons H_2(g)$	**0**
$Cu^{2+} + e^- \rightleftharpoons Cu^+$	$+0.15$
$SO_4^{2-} + 4H^+ + 2e^- \rightleftharpoons SO_2(g) + 2H_2O$	$+0.2$
$AgCl(s) + e^- \rightleftharpoons Ag(s) + Cl^-$	$+0.22$
$Cu^{2+} + 2e^- \rightleftharpoons Cu(s)$	$+0.34$
$O_2(g) + 2H_2O + 4e^- \rightleftharpoons 4OH^-$	$+0.4$
$MnO_4^- + 2H_2O + 3e^- \rightleftharpoons MnO_2(s) + 4OH^-$	$+0.59$
$Fe^{3+} + e^- \rightleftharpoons Fe^{2+}$	$+0.77$
$Ag^+ + e^- \rightleftharpoons Ag(s)$	$+0.85$
$O_2(g) + 4H^+ + 4e^- \rightleftharpoons 2H_2O$	$+1.23$
$MnO_2(s) + 4H^+ + 2e^- \rightleftharpoons Mn^{2+} + 2H_2O$	$+1.23$
$MnO_4^- + 8H^+ + 5e^- \rightleftharpoons Mn^{2+} + 4H_2O$	$+1.51$
$PbO_2(s) + 4H^+ + SO_4^{2-} + 2e^- \rightleftharpoons PbSO_4(s) + 2H_2O$	$+1.70$

(Left margin, vertical: Increasing electronegativity)

Table 17.1 shows values of $E^{\ominus}$ for selected electrode reactions and some ion–ion reactions. The measured values of $E^{\ominus}$ in table 17.1 are for aqueous solutions and are not valid in other solvents. Also, the influence of liquid–liquid junction potentials may introduce uncertainties into the measurements.

The order in which the ion–element reactions fall in table 17.1 is called the **electrochemical scale**. If an element has a positive standard electrode potential the cell reaction tends to go to the right, i.e. the metal ion can be reduced by hydrogen to the metal. If an element has a negative standard electrode potential, the metal displaces hydrogen gas to form protons and ions. For redox half-cell reactions the value of $E^{\ominus}$ is a measure of the power of the oxidised form to act as an oxidising agent. In general, the more positive the value of $E^{\ominus}$ the stronger is the oxidising agent relative to the oxidising power of the hydrogen ion. The so-called **electropositive** elements have negative values for $E^{\ominus}$, whereas **electronegative** elements have positive $E^{\ominus}$ values.

17.3   Consolidating the thermodynamics of electrochemical cells

We may use standard electrode potentials of half-cell reactions to estimate the direction and extent of change in any mineral or solution reaction involving redox.

It has been shown that electrochemical energy can be related to the Gibbs energy of a cell reaction by the relationships:

$$\Delta G = -nEF \qquad (17.66)$$

and

$$\Delta G^{\ominus} = -nE^{\ominus}F \qquad (17.67)$$

We know that for a reaction to proceed spontaneously the value of ΔG should be negative, and that the value of $\Delta G^{\ominus}$ tells us the extent to which a reaction will occur under standard state conditions. The interrelationship between these two variables is:

$$\Delta G = \Delta G^{\ominus} + RT \sum_i v_i \ln a_i \qquad (17.68)$$

This means that a mixture of reactive species will spontaneously readjust their concentrations (activities), by reaction, proceeding towards a state where the value of ΔG is zero.

We can conclude from equation 17.67 that the values of E and $E^{\ominus}$ give the same information as ΔG and $\Delta G^{\ominus}$. That is, **if a complete reaction is constructed from combinations of half-cell reactions, the overall reaction will proceed spontaneously if E has a positive value.** The e.m.f. of a cell which has reached full equilibrium is zero. Also, the more positive the value of $E^{\ominus}$ the greater is the extent to which the reaction can proceed. This observation applies to half-cell reactions as well as complete–cell reactions.

17.3.1 The Nernst equation and E_H

The equation which is the analogue of equation 17.68 is the Nernst equation (derived in box 7.1):

$$E = E^{\ominus} - \frac{RT}{zF} \sum_i v_i \ln a_i \qquad (17.69)$$

It is conventional to use decadic logarithms rather than natural logarithms, in which case the value of $2.30258RT/zF$ is $0.0592/z$ at $298.15\,\text{K}$. Using the Fe^{3+}/Fe^{2+} couple from the reaction $Fe^{3+} + e^- \rightleftharpoons Fe^{2+}$ as an example we have:

$$E = 0.770 - 0.0592 \log \frac{a_{Fe^{2+}}}{a_{Fe^{3+}}} \qquad (17.70)$$

In any solution **we expect the ion activity ratios to adjust to achieve a zero value for E.** Alternatively, measuring the value of E for a system is a measure of the ratio of oxidised to reduced species in a cell. In geochemistry the symbol E_H is employed instead of E as a measure of the tendency of a reaction to proceed. The two symbols have identical meaning and are interchangeable. The subscript H is used to simply emphasize the fact that the standard e.m.f. is referenced to the standard hydrogen electrode.

The most common use of $E^{\ominus}$ is to rank half-cell reactions in terms of their reducing power. Under standard-state conditions any species on the left of a given half-cell reaction will react spontaneously with any species on the right of any half-cell reaction that is **above** it in table 17.1.

17.3.2   The electron activity

The variable E_H defines the redox state of any aqueous solution. A second variable which is frequently used for the same purpose is the electron activity. This variable defines the redox state in terms of the hypothetical activity of electrons in the aqueous solution (a_e). By analogy with pH we define pε as equal to:

$$p\varepsilon = -\log a_e \qquad (17.71)$$

To understand the significance of this definition we consider the full-cell reaction equation and the equilibrium constant for the reduction of Fe^{3+}:

$$Fe^{3+} + \tfrac{1}{2}H_2O(g) \rightleftharpoons Fe^{2+} + H^+ \quad \text{(equilibrium constant } K) \qquad (17.72)$$

This is composed of the two half-cell reactions:

$$Fe^{3+} + e^- \rightleftharpoons Fe^{2+} \quad \text{(equilibrium constant } K_2) \qquad (17.73)$$

and

$$\tfrac{1}{2}H_2(g) \rightleftharpoons H^+ + e^- \quad \text{(equilibrium constant } K_3) \qquad (17.74)$$

By convention **the change in the Gibbs energy for the oxidation of hydrogen is zero at any temperature and pressure**; therefore $K_3 = 1$. From this we can write:

$$K = K_2/K_3 = K_2 = \frac{a_{Fe^{2+}}}{a_{Fe^{3+}} \times a_e} \quad \text{or} \quad \frac{1}{e^-} = K \frac{a_{Fe^{3+}}}{a_{Fe^{2+}}} \qquad (17.75)$$

From this we may write:

$$p\varepsilon = \log K - \log \frac{a_{Fe^{2+}}}{a_{Fe^{3+}}} \qquad (17.76)$$

The above equation confirms that for the half-cell reaction the **ratio of the activities of the oxidised and reduced ions is defined by the value of pε.**

When the reacting ions are in their standard states and their activities equal unity, the value of pε is equal to the decadic logarithm of the equilibrium constant, i.e. for this example

$$p\varepsilon^{\ominus} = \log K \qquad (17.77)$$

We may therefore write:

$$p\varepsilon = p\varepsilon^{\ominus} - \log \frac{a_{Fe^{2+}}}{a_{Fe^{3+}}} \qquad (17.78)$$

For any half-cell reaction defined by the general equation:

$$\sum_i v_i A_i + ze^- = 0 \qquad (17.79)$$

the quantities pε and p$\varepsilon^{\ominus}$ are given by:

$$p\varepsilon = p\varepsilon^{\ominus} - \frac{1}{z} \log \prod_i a_i^{v_i} \quad \text{or} \quad p\varepsilon^{\ominus} - \frac{1}{z} \log \sum v_i a_i \qquad (17.80)$$

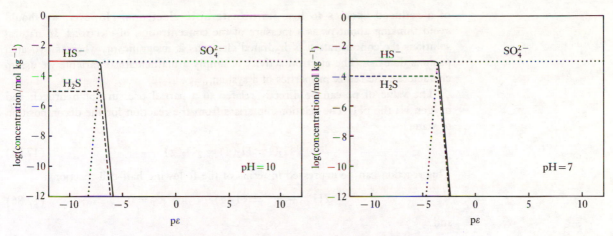

Figure 17.3 Sulphate species equilibria in basic and neutral media.

where in general:

$$p\varepsilon^{\ominus} = \frac{1}{z} \log K \tag{17.81}$$

As an example, the value of $p\varepsilon$ for the reduction reaction:

$$SO_4^{2-} + 9H^+ + 8e^- \rightleftharpoons HS^- + 4H_2O \qquad \log K = 34.0 \tag{17.82}$$

is given by:

$$p\varepsilon = \frac{1}{8} \log K - \frac{1}{8} \left(\frac{a_{HS^-}}{a_{SO_4^{2-}} \times a_{H^+}^9} \right) \tag{17.83}$$

Finally, the relationship between $p\varepsilon$ and E_H is deduced by comparing equations 17.69 and 17.80:

$$p\varepsilon = \frac{E_H F}{2.303 RT} \quad \text{and} \quad p\varepsilon^{\ominus} = \frac{E_H^{\ominus} F}{2.303 RT} \tag{17.84}$$

One use of the electron activity is as **master variable** in graphical representations of speciation. Figure 17.3 shows $p\varepsilon$ used as a master variable dictating the proportions of various sulphur-containing species derived from the sulphate ion. Total sulphur concentration is 10^{-3} mol dm^{-3} and the pH values considered are 7 and 10. We note that strong dependence of the species concentrations on pH and $p\varepsilon$ is well emphasised in such plots.

17.3.3 Some comments on the electron activity

In aqueous solutions the pH is a measure of the Brønsted acidity and is related to the chemical potential of the proton. The greater the pH, the smaller is the proton's chemical potential and the greater is the tendency of any Brønsted acid to lose its transferable protons. Similarly, the electron activity is a measure of the chemical potential of an electron. **The greater the value of $p\varepsilon$ the greater is the tendency** of a reduced species to lose its transferable electrons. However, we should avoid thinking about $p\varepsilon$ as a measure of the concentration of electrons. In natural solutions the concentration of hydrated electrons is insignificant, whereas hydrated protons do exist. The electron activity is simply a mathematical variable we use to characterise the redox properties of a system.

The value of $p\varepsilon$ can be directly related to a partial pressure of hydrogen and oxygen via the pH. The relationship arises from the reaction for the decomposition of water:

$$H_2O \rightleftharpoons H_2(g) = \tfrac{1}{2}O_2(g) \tag{17.85}$$

This reaction can be expressed in terms of the following half-cell reactions:

$$2H^+ + 2e^- \rightleftharpoons H_2(g) \qquad \log K_1 = 0 \tag{17.86}$$

and

$$O_2(g) + 4H^+ + 4e^- \rightleftharpoons 2H_2O \qquad \log K_2 \tag{17.87}$$

The above two relationships lead to:

$$\log P_{H_2(g)} = 0 - 2\,pH - 2p\varepsilon \tag{17.88}$$

and

$$\log P_{O_2(g)} = \log K_2 + 4\,pH + 4p\varepsilon \tag{17.89}$$

The implication is that $p\varepsilon$ and pH may be used interchangeably with $P_{O_2(g)}$ or $P_{H_2(g)}$ to define the redox conditions in natural waters. We may think of $p\varepsilon$ (or E_H) as a **measurable** parameter which defines the ratio of oxidised to reduced ionic species, and therefore as an indicator of the stabilities of minerals which are dependent on the solution concentrations of these species.

There are three problems with this approach. The first is that natural systems generally contain several redox couples, each defining a value of $p\varepsilon$. Only when all reactions are at full equilibrium will the values of $p\varepsilon$ for all redox couples be the same. In general the slow kinetics of redox reactions prevent this and we cannot truly consider the $p\varepsilon$ of a solution even though some individual couples may be in equilibrium. We must always think of the $p\varepsilon$ of individual couples.

The second problem is that $p\varepsilon$ or E_H can rarely be measured, even for a single redox couple. In principle it is possible to measure the E_H for a single redox couple if an electrode can be found which responds quickly and reversibly to the redox pair and to no other redox pairs in the solution. In practice few electrodes are perfectly selective and the measured E_H is a response to several reactions. For this reason the use of measured $p\varepsilon$ as a master variable is sometimes unjustified.

The final problem is that the concentrations of oxidized or reduced species are usually too low to give a stable electrode response. A good electrode, which responds to many metal-ion redox reactions but does not respond to SO_4^{2-}/H_2S°, N_2/NH_4^+, CO_2/CH_4 or N_2/NO_3^-, is the platinum electrode. This electrode has found applications in strongly reducing environments where minor species are sufficiently abundant to give a measurable electrode response.

In view of the difficulties in measuring meaningful values of $p\varepsilon$ or E_H, a common use of these parameters involves measuring the concentrations of the individual species and calculating $p\varepsilon$ or E_H. The parameters are used as simple thermodynamic variables which define the state of the redox couple. Unfortunately, the stoichiometries of many redox couples are difficult to evaluate, especially in high-temperature fluids or those containing decomposing organic matter which may be a source of electrons. In these cases the oxygen fugacity may be a better variable to measure or calculation even though it may be small. Redox reactions in mineral-fluid systems at high temperatures and pressures are often expressed in terms of oxygen or hydrogen fugacities even though there may be other sources of electrons.

Chapter 18

Graphical applications of chemical thermodynamics

In geological or other complex chemical systems we observe multiple reactions and phase transitions. Consequently, mineral stabilities and solution compositions may change with temperature, pressure or the overall composition of the system. To establish phase or compositional relationships, or to understand how a system responds to perturbations, we may use one or more convenient graphical techniques. These graphical techniques are the central theme of this chapter. It must always be remembered that such techniques refer to equilibrium relationships, and departures from equilibrium are common in natural systems.

18.1   General predominance diagrams

For any reaction the equilibrium constant defines the relationship between the activities of reacting species. This feature can be used to define simple relationships between the activities of any two species in a reaction if the activities of all other species are kept constant. This allows the construction of diagrams containing areas whose boundaries are defined by the chosen activity relationships. The boundaries are usually linear if plotted on logarithmic axes. Boundaries denoting solid/solution equilibria define conditions beyond which a mineral will dissolve or precipitate. Boundaries between solids define conditions at which phase changes occur. Within each bounded region, called a **predominance area**, a given substance dominates. However, even in the predominance areas for solids there will be aqueous species present, since all substances are soluble to some extent.

Such plots are termed **predominance diagrams**: there are several types. They are used to determine such matters as which mineral is stable under a specified set of conditions, or which mineral controls the activity of a species. We may also deduce the pH if a given mineral is in contact with a solution of known ion activity. This kind of information gives clues to the conditions under which rocks and minerals were formed.

Commonly, the term 'predominance diagram' is applied only to cases where one axis is pH and the other the logarithm of the activity of a second species. In this text we will not restrict the term to pH plots and will consider other suitable activities or activity ratios.

18.1.1   Congruent solubility diagrams

Simple activity diagrams

Consider a system containing water, carbonate ions and divalent copper ions at 298.15 K. Either tenorite ($CuO(s)$) or copper carbonate may be formed, depending on the pH and the relative proportions of the ions. The reactions and their equilibrium constants are:

$$Cu^{2+} + H_2O \rightleftharpoons CuO(s) + 2H^+ \tag{18.1}$$

$$\log K_1 = \log a_{CuO(s)} - \log a_{Cu^{2+}} + 2 \log a_{H^+} - \log a_{H_2O}$$

$$= -7.66 \tag{18.2}$$

$$Cu^{2+} + CO_3^{2-} \rightleftharpoons CuCO_3(s) \tag{18.3}$$

$$\log K_2 = \log a_{CuCO_3(s)} - \log a_{Cu^{2+}} - \log a_{CO_3^{2-}} = 9.63 \tag{18.4}$$

We may construct equations that define the relationship between two chosen variables (such as pH and $\log a_{Cu^{2+}}$) at each mineral solution boundary. The tenorite equilibrium in equation 18.2 gives one equation directly if we assume that the activities of all solids and water are unity, i.e.

$$\log a_{Cu^{2+}} = 7.66 - 2pH \tag{18.5}$$

The proton activity is absent from the equation for copper carbonate equilibrium (equation 18.2). It must be included by introducing the reaction for the formation of bicarbonate.

$$\left. \begin{array}{l} CO_3^{2-} + H^+ \rightleftharpoons HCO_3^- \\ \log K_3 = \log a_{HCO_3^-} - \log a_{CO_3^{2-}} - \log a_{H^+} = 10.33 \end{array} \right\} \tag{18.6}$$

To simplify the calculations we fix the activity of bicarbonate at 1×10^{-2}. Combining the equilibrium constants for equations 18.4 and 18.6 now gives:

$$\log a_{Cu^{2+}} = 2.7 - pH \tag{18.7}$$

Figure 18.1 shows the equilibrium lines for the two minerals plotted on the same axes. The position of the lines would differ if an alternative bicarbonate activity had been selected. Neglecting the malachite line for a moment, we see that a horizontal line, at any copper activity, can be drawn across the solution region. Its point of intersection with a mineral equilibrium line gives the pH at which the mineral will become stable. The line at $\log a_{Cu^{2+}} = -1$ shows that the solution phase is stable below pH 3.7, with copper carbonate precipitating above this. The line at $\log a_{Cu^{2+}} = -5$ shows that at pH 6.3 tenorite becomes the stable phase.

A final boundary we may consider is the solid/solid boundary between tenorite and copper carbonate. The reaction equation is derived by combining equations 18.1 and 18.2 to eliminate the Cu^{2+}. The resulting expression can be combined with equation 18.4 to eliminate the carbonate ion. Assuming all solids and water are in their standard states, this gives:

$$\left. \begin{array}{l} CuO(s) + H^+ + HCO_3^- \rightleftharpoons CuCO_3(s) + H_2O \\ \log K_4 = pH - \log a_{HCO_3^-} = 6.96 \end{array} \right\} \tag{18.8}$$

For the specified bicarbonate concentration this yields the vertical line, $pH = 6.96 + (-2) = 4.96$. A point to note is that we need not actually calculate

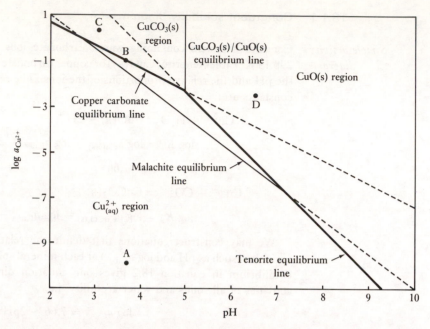

Figure 18.1   Solubility diagram for copper carbonate/tenorite equilibria.

a value for $\log K_4$ in order to define the $CuO(s)/CuCO_3(s)$ boundary. Inspecting the form of the equilibrium constant K_4, and recognising the fact that reaction 18.8 is derived from equations 18.1 and 18.3, shows us that the boundary must pass through the point of intersection of the $Cu^{2+}/CuO(s)$ equilibrium line and the $Cu^{2+}/CuCO_3(s)$ line.

Points within the stability regions can be interpreted. For example, point A is within a region where the solution is undersaturated with respect to both minerals, in which Cu^{2+} predominates. Point B represents a solution at the saturation level for copper carbonate whereas point C is within a region oversaturated with respect to copper carbonate but undersaturated with respect to tenorite. Point D represents a solution oversaturated with respect to both minerals but where the oxide is stable with respect to the carbonate. If the system at point C progressed to equilibrium, copper carbonate would precipitate. The mineral to precipitate is the one which gives the lowest value for the activity of copper.

There are three limitations to the approach. First, the assumption of fixed activity for some components may be unreasonable. For instance, it is difficult to vary the pH of a real system without changing the bicarbonate activity. Therefore, the plots do not represent actual chemical processes. Secondly, the activities of solids and liquid water in real systems may not be unity. It is usual to offset the diagrams using estimates for water activity or (more rarely) mineral activities. Thirdly, **these plots cannot yield any information about other dissolved species or stable mineral phases which may be present.** We must always know the minerals present before constructing such figures. This is a general point which applies to any phase diagram or predominance diagram. For example, in the copper system we may also consider the precipitation of malachite via the reaction:

$$2Cu^{2+} + CO_3^{2-} + 2H_2O \rightleftharpoons 2H^+ + Cu_2(OH)_2CO_3(s) \Big\}$$
$$\log K_4 = 4.4 \qquad\qquad\qquad\qquad\qquad\qquad\qquad\qquad (18.9)$$

By coupling the above reaction with the carbonate/bicarbonate equilibrium we produce the equation:

$$2 \log a_{Cu^{2+}} = 3.97 - 1.5 pH \tag{18.10}$$

This generates a line on figure 18.1 which intersects both the carbonate and the tenorite stability lines and will change any proposed interpretation. The dashed section of this line shows the region where malachite is unstable with respect to either tenorite or copper carbonate. Similarly, in the solution region it is necessary to define a domain at high pH in which the hydrolysed species $Cu(OH)^+$ is the predominant form.

More complicated solubility diagrams It is sometimes convenient to plot solubility relationships using ratios of activities rather than single-species activities. Consider an aqueous system containing magnesium and dissolved silica. Four magnesium-containing minerals plus quartz control the composition of the aqueous phase. These are:

$$Mg^{2+} + 2H_2O \rightleftharpoons Mg(OH)_2(s) + 2H^+ \; (\text{brucite}): \log K_1 = 16.7 \tag{18.11}$$

$$3Mg^{2+} + 4H_4SiO_4 \rightleftharpoons 6H^+ + 4H_2O + MgSi_4O_{10}(OH)_2(s)(\text{talc}): \log K_2 = 21.2 \tag{18.12}$$

$$3Mg^{2+} + 2H_4SiO_4 + H_2O \rightleftharpoons 6H^+ + Mg_3Si_2O_5(OH)_4(s) \; (\text{serpentine}): \\ \log K_3 = 34.0 \tag{18.13}$$

$$4Mg^{2+} + 6H_4SiO_4 \rightleftharpoons 8H^+ + H_2O + Mg_4Si_6O_{15}(OH)_2 \cdot 6H_2O(s) \; (\text{sepiolite}): \\ \log K_4 = 31.25 \tag{18.14}$$

$$H_4SiO_4 \rightleftharpoons 2H_2O + SiO_2(s) \; (\text{quartz}): \log K_5 = 4.0 \tag{18.15}$$

Assuming the activities of solids and water are unity, the complementary equilibrium constants can be obtained as:

$$-\log K_1 = \log \frac{a_{Mg^{2+}}}{a_{H^+}^2} \tag{18.16}$$

$$-\log K_2 = 3 \log \frac{a_{Mg^{2+}}}{a_{H^+}^2} + 4 \log a_{H_4SiO_4} \tag{18.17}$$

$$-\log K_3 = 3 \log \frac{a_{Mg^{2+}}}{a_{H^+}^2} + 2 \log a_{H_4SiO_4} \tag{18.18}$$

$$-\log K_4 = 4 \log \frac{a_{Mg^{2+}}}{a_{H^+}^2} + 6 \log a_{H_4SiO_4} \tag{18.19}$$

$$-\log K_5 = \log a_{H_4SiO_4} \tag{18.20}$$

These equations have been arranged to yield linear plots of $\log a_{Mg^{2+}}/a_{H^+}^2$ versus $\log a_{H_4SiO_4}$. Figure 18.2 shows the solubility relationships for this system. It is interpreted in the same way as the copper carbonate system in figure 18.1. We note that sepiolite stability line is always in the field of stability for another solid, indicating that sepiolite is always metastable at this temperature. Indeed, we may also wish to impose the solubility line for amorphous silica on figure 18.2 which would be a vertical line at $\log a_{H_4SiO_4} \sim 2.7$, well within the region of stability for quartz.

In some systems an additional consideration arises because some dissolved species

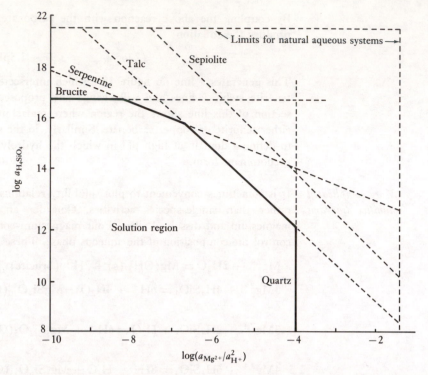

Figure 18.2   Solubility diagram for a magnesium silicate system at 298.15 K and 1 bar (after Drever, 1982).

form complexes in proportions which vary with conditions, especially pH. For
example, dissolved aluminium is distributed through several solution species of which
the most important are Al^{3+}, $Al(OH)_4^+$ and $Al(OH)_2^-$. The solubility of gibbsite,
$Al(OH)_3(s)$, is therefore influenced by the relative proportions of these species. A
useful coordinate for the equilibrium solubility plot is the log of the activity of the
total amount of dissolved aluminium, rather than the activity of any one aluminium
species. This can be computed as a function of pH by considering the following
reactions.

The common reaction equation for gibbsite is:

$$Al^{3+} + 3H_2O \rightleftharpoons 3H^+ + Al(OH)_3(s): K_1 = \frac{a_{H^+}^3}{a_{Al^{3+}}} = -8.75 \quad (18.21)$$

However, the formation reaction may be written in terms of any one of the three
aqueous aluminium species.

$$Al(OH)_2^+ + H_2O \rightleftharpoons H^+ + Al(OH)_3(s): K_2 = \frac{a_{H^+}}{a_{Al(OH)^{2+}}} = 1.41 \quad (18.22)$$

$$Al(OH)_4^- + H^+ \rightleftharpoons H_2O + Al(OH)_3(s): K_3 = \frac{1}{a_{H^+} \times a_{Al(OH)_4^-}} = 13.4 \quad (18.23)$$

We may define the activity of the total dissolved aluminium, $a_{Al(T)}$, as the sum of
the activities of the aluminium species, i.e.

$$a_{Al(T)} = a_{Al^{3+}} + a_{Al(OH)^{2+}} + a_{Al(OH)^{4-}} \quad (18.24)$$

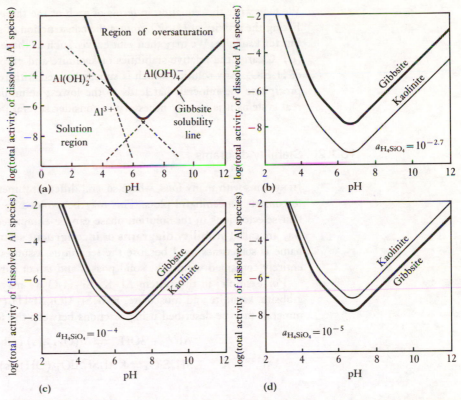

Figure 18.3   Solubilities of aluminium-containing species as a function of pH at 298.15 K and 1 bar (after Drever, 1982).

This can be calculated from the pH if the species concentrations are eliminated from equation 18.24 using equations 18.21–18.23, i.e.

$$a_{Al(T)} = \frac{a_{H^+}^3}{K_1} + \frac{a_{H^+}}{K_2} + \frac{1}{K_1 \times a_{H^+}} \tag{18.25}$$

Figure 18.3(a) shows the solubility diagram for gibbsite versus pH. The broken lines indicate the concentrations of the different aluminium species. At low pH, Al^{3+} is the major species and the solubility is dominated by the first term in equation 18.25. We predict an initial lowering of the solubility as pH increases because of the cubic dependence on a_{H^+}. As the pH is increased, the contributions from $Al(OH)_2^-$ and $Al(OH)_4^+$ become more important, until at high pH the third term in equation 18.25 dominates, and the solubility increases.

Drever (1982) describes the use of such a diagram to determine the mineral most likely to precipitate from a complex solution. Consider the gibbsite example in figure 18.3(a). If the reaction were to occur in a solution containing dissolved silica, the gibbsite stability line would be unaffected but a new line for the stability of kaolinite could be added. The reaction equation for kaolinite, $Al_2Si_2O_5(OH)_4(s)$, is:

$$\left. \begin{array}{l} 2Al^{3+} + 2H_4SiO_4 + H_2O \rightleftharpoons 6H^+ + Al_2Si_2O_5(OH)_4(s) \\[6pt] K = \dfrac{a_{Al^{3+}}^2 \, a_{H_4SiO_4}^2 \, a_{H_2O}}{a_{H^+}^6 \times a_{kaolinite}} = -7.45 \end{array} \right\} \tag{18.26}$$

We may write this equation in terms of each of the three known aluminium species. Fixing the activity of H_4SiO_4 allows the construction of an equation similar to 18.25 but for kaolinite. We may then plot curves such as those in figures 18.3(b), (c) and (d). Clearly, the relative stabilities of kaolinite and gibbsite depend on the activity of H_4SiO_4. Any solution which is supersaturated with respect to both minerals will precipitate the mineral that leads to the lowest aluminium activity in solution. In real systems it is usually necessary to consider several possible minerals.

18.1.2   Stability diagrams

In systems with many ions, where several different minerals may exist, it is difficult to construct meaningful plots. This may be avoided by considering reactions such that selected ions in the solution phase can be eliminated. Diagrams produced this way are called **stability diagrams** or **incongruent solution diagrams**. The latter name is sometimes used because the technique assumes that the eliminated ion is entirely contained within the solid phases and is not part of the solution phase.

For example, in the system Al–K–Si–H_2O we may consider the formation of gibbsite, kaolinite and muscovite, $KAl_2Si_3AlO_{10}(OH)_2(s)$. The formation of these minerals may be described using reactions between the following dissolved species:

$$Al^{3+} + 3OH^- \rightleftharpoons Al(OH)_3(s) \text{ (gibbsite)} \tag{18.27}$$

$$3Al^{3+} + K^+ + 3H_4SiO_4 \rightleftharpoons KAl_2Si_3AlO_{10}(OH)_2(s) + 10H^+ \text{ (muscovite)} \tag{18.28}$$

and

$$2Al^{3+} + 2H_4SiO_2 + H_2O \rightleftharpoons 6H^+ + Al_2Si_2O_5(OH)_4(s) \text{ (kaolinite)} \tag{18.29}$$

The above equations can be combined to eliminate the aluminium ion and give three reaction equations each containing two mineral phases. These are:

$$Al_2Si_2O_5(OH)_4(s) + 5H_2O \rightleftharpoons 2Al(OH)_3(s) + 2H_4SiO_4 \tag{18.30}$$

$$2KAl_2(Si_3AlO_{10})(OH)_2(s) + 3H_2O + 2H^+ \rightleftharpoons 2K^+ + 3Al_2Si_2O_5(OH)_4(s) \tag{18.31}$$

and

$$KAl_2(Si_3AlO_{10})(OH)_2(s) + 9H_2O + H^+ \rightleftharpoons 3H_4SiO_4 + K^+ + 3Al(OH)_3(s) \tag{18.32}$$

Assuming that water and minerals are in their standard states, the equilibrium constants can be arranged as:

$$\log K_{\text{gib/kaol}} = 2 \log a_{H_4SiO_4} - 5 \log a_{H_2O} = -10.39 \tag{18.33}$$

$$\log K_{\text{kaol/mus}} = 2 \log a_{K^+} - 2 \log a_{H^+} - 3 \log a_{H_2O} = 8.81 \tag{18.34}$$

and

$$\log K_{\text{gib/mus}} = 3 \log a_{H_4SiO_4} + \log a_{K^+} - \log a_{H^+} - 9 \log a_{H_2O} = -11.17 \tag{18.35}$$

The three equilibrium constants can be used to define the **stability boundaries between any two mineral phases**. At this stage we have several options. We may fix the activity of one ion and then plot the stability lines as a function of the other two. A second option is to plot the stability diagram as a function of $\log(a_{K^+}/a_{H^+})$ versus $\log a_{H_4SiO_4}$. We will choose the first option and fix potassium activity at 1×10^{-3}. This allows equations 18.33–35 to be re-arranged to define equilibrium relationships between pH and $\log a_{H_4SiO_4}$, i.e.

$$\log a_{H_4SiO_4} = -5.2 + 2.5 \log a_{H_2O} \qquad (18.36)$$

$$pH = 7.4 + 1.5 \log a_{H_2O} \qquad (18.37)$$

$$\log a_{H_4SiO_4} = -2.7 - \tfrac{1}{3}pH + 3 \log a_{H_2O} \qquad (18.38)$$

Constructing a stability diagram involves drawing one line at a time and systematically determining the regions of activity space where minerals are either oversaturated or undersaturated. A mineral will not exist in any region of undersaturation. We can define the regions where mineral phases can exist by interference. To construct such a figure it is convenient to start with any phase boundary which is a horizontal or vertical. Figure 18.4(a) shows a plot of the gibbsite/kaolinite boundary assuming a water activity of 1. Inspection of the reaction equation 18.30 shows that increasing the activity of dissolved silica promotes kaolinite whereas decreasing the activity promotes gibbsite. Therefore, any point above the line denotes a solution oversaturated with kaolinite and undersaturated with gibbsite. Figure 18.4(b) shows the addition of a stability line for the muscovite/gibbsite equilibria. Above this line muscovite is stable (gibbsite unstable), but below the line gibbsite is stable. To the left, this line extends into the region where gibbsite is unstable with respect to kaolinite. In this region gibbsite cannot exist and the gibbsite/muscovite equilibrium must be metastable. To denote this we use a broken line. Similarly, the right-hand portion of the kaolinite/gibbsite line is now in a region

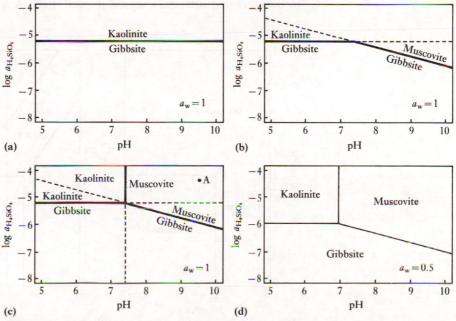

Figure 18.4   Predominance diagram for the Al–K–Si–H_2O system at 298.15 K and 1 bar (after Sposito, 1981).

where gibbsite is unstable with respect to muscovite. This part of the kaolinite/gibbsite equilibrium must also be metastable. The final equilibrium line, added to figure 18.4(c), is the muscovite/kaolinite boundary. We note that to the right of the line muscovite is stable (kaolinite unstable) but to the left muscovite is unstable (kaolinite stable). The broken portion of this line denotes metastability since neither muscovite or kaolinite is stable with respect to gibbsite. The heavy lines in figure 18.4(c) define the stability regions where one mineral is oversaturated and the rest undersaturated. For example, point A represents a solution oversaturated with respect to muscovite and undersaturated with respect to both gibbsite and kaolinite. Therefore, gibbsite or kaolinite in contact with a solution at composition A will dissolve to form muscovite.

In figure 18.4(c) the slopes of the lines are determined by the stoichiometries of the reactions, whereas their positions are controlled by the values of the equilibrium constants. Figure 18.4(d) is plotted for a water activity of 0.5 and shows that the positions of the lines will be different if water activities are not unity. The positions of the lines will also change with potassium activity and mineral activities. The most dramatic changes to the diagrams are brought about by the effect of temperature and pressure on the equilibrium constants. Plots at different temperatures and pressures show the conditions under which metastable minerals become properly stable.

Similar logic may be used to construct a stability diagram for the case where we chose to plot $\log(a_{K^+}/a_{H^+})$ versus $\log a_{H_4SiO_4}$. Figure 18.5(a) depicts the calculated stability fields.

A more realistic diagram would include lines for a range of other minerals as in figure 18.5(b). The quartz line that cuts across several stability fields is not a mineral/mineral equilibrium line, but a mineral/solution boundary derived from the quartz dissolution reaction. Any point to the right of the quartz line is oversaturated with respect to quartz and a second mineral. Point A is oversaturated with respect

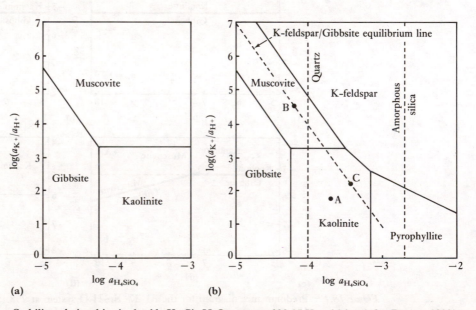

Figure 18.5   Stability relationships in the Al−K−Si−H$_2$O system at 298.15 K and 1 bar (after Drever, 1982).

to quartz and kaolinite, for example. The quartz line should define the upper limit for the activity of dissolved silica and points to the right of this line should be metastable. In practice quartz dissolves and precipitates very slowly and the line for amorphous silica is a more realistic limit for the activity of dissolved silica.

Other metastable equilibrium lines can also be depicted on such plots. Consider the reaction between gibbsite and K-feldspar:

$$K^+ + 3H_4SiO_4 + Al(OH)_3(s) \rightleftharpoons H^+ + 7H_2O + KAlSi_3O_8(s) \quad (18.39)$$

The stability line for this reaction is depicted in figure 18.5(b). We see that the line plots within the stability field of other minerals. At point B on the broken line both gibbsite and K-feldspar are unstable with respect to muscovite. At point C both gibbsite and K-feldspar are unstable with respect to kaolinite. We deduce that the gibbsite/K-feldspar equilibrium is unstable at all compositions at 298.15 K.

As we increase the number of components, graphical representations become more complex. We may add sodium to the $K-Al-Si-H_2O$ system by drawing a three-dimensional plot in which the third axis is $\log(a_{Na^+}/a_{H^+})$, as in figure 18.6. These are difficult to construct and interpret.

Another method involves fixing another species activity to reduce the number of coordinates further. A good candidate in the $K-Na-Al-Si-H_2O$ system is the activity of dissolved silica, which may be set at the limit imposed by the solubility of quartz or amorphous silica. Figure 18.7 shows a stability diagram for this system at 298.15 K and quartz saturation. The activity of dissolved silica is 10^{-4}. The system includes the reactions shown in table 18.1.

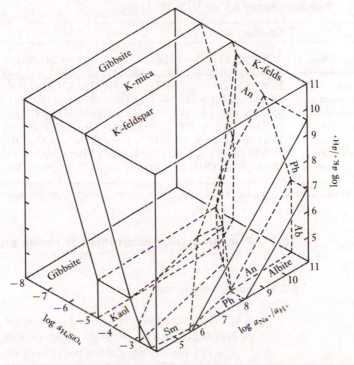

Figure 18.6   Three-dimensional representation of stability relationships in the $K-Na-Al-Si-H_2O$ system (from Drever, 1982). Ph phillipsite, Sm smectite, An analcime.

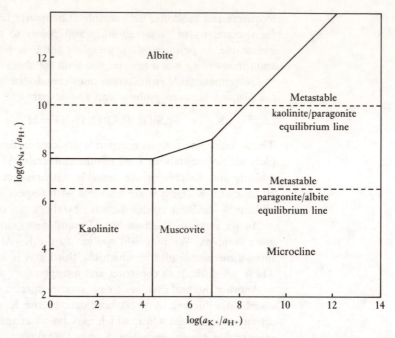

Figure 18.7 Stability relationships in the K–Na–Al–Si–H$_2$O system at a fixed activity of dissolved silica (10^{-4}) (from Drever, 1982).

Table 18.1 **Reactions in the K/Na/Al/Si/H$_2$O system**

Reaction	Equation	log K
Kaolinite/gibbsite	$Al_2Si_2O_5(OH)_4(s) + 5H_2O \rightleftharpoons 2Al(OH)_3(s) + 2H_4SiO_4$	−10.4
Pyrophilite/kaolinite	$Al_2Si_4O_{10}(OH)_2 + 5H_2O \rightleftharpoons Al_2Si_2O_5(OH)_4(s) + 2H_4SiO_4$	−6.35
Muscovite/kaolinite	$2KAl_3Si_3O_{10}(OH)_2(s) + 2H^+ + 3H_2O \rightleftharpoons 3Al_2Si_2O_5(OH)_4(s) + 2K^+$	8.81
Muscovite/microcline	$3KAl_3Si_3O_8(s) + 2H^+ + 3H_2O \rightleftharpoons KAl_3Si_3O_{10}(OH)_2(s) + 2K^+ + 6H_4SiO_4$	−10.3
Albite/paragonite	$3NaAlSi_3O_8(s) + 2H^+ + 12H_2O \rightleftharpoons NaAl_3Si_3O_{10}(OH)_2(s) + 2Na^+ + 6H_4SiO_4$	−10.7
Albite/kaolinite	$2NaAlSi_3O_8(s) + 2H^+ + 9H_2O \rightleftharpoons Al_2Si_2O_5(OH)_2(s) + 2Na^+ + 4H_4SiO_4$	−0.45
Muscovite/albite	$3NaAlSi_3O_8(s) + K^+ + 3H^+ + 12H_2O \rightleftharpoons KAl_3Si_3O_{10}(OH)_2(s) + 3Na^+ + H^+ + 6H_4SiO_4$	−5.08
Albite/microcline	$NaAlSi_3O_8(s) + K^+ + H^+ \rightleftharpoons KAl_3Si_3O_8(s) + Na^+ + H^+$	1.75
Paragonite/kaolinite	$2NaAl_3Si_3O_8(OH)_2(s) + 2H^+ + 3H_2O \rightleftharpoons 3Al_2Si_2O_5(OH)_4(s) + 2Na^+$	20

Note that an additional H$^+$ is added to each side of some of the equations to facilitate use of the ratios a_{K^+}/a_{H^+} and a_{Na^+}/a_{H^+}.

18.2   Predominance diagrams for redox equilibria

18.2.1   pε–pH diagrams

When redox reactions are involved in chemical processes, stability relationships can be depicted using a predominance diagram with the electron activity as one variable.

The electron activity defines the ratio of oxidised to reduced ionic species in the solution phase. The value of pε can therefore be used as an indicator of the stabilities of minerals which are dependent on the solution concentrations of these redox species.

In using the electron activity we must be aware of the limitations of the approach. The major limitation (section 17.3.3) is that natural systems generally contain several redox couples, each defining a separate value of pε. Only when all reactions are at full equilibrium will the values of pε for all redox couples be the same. In general, the slow kinetics of redox reactions prevent this and we cannot truly consider the pε of a solution even though some individual couples may be in equilibrium. We must always think of the pε of individual couples. Despite this reservation, pε–pH diagrams are instructive.

The stability of water   The regions in which water remains thermodynamically stable without decomposing to oxygen and hydrogen can be defined through the following reaction:

$$H_2O \rightleftharpoons H_2(g) + \tfrac{1}{2}O_2(g) \tag{18.40}$$

This reaction can be expressed in terms of the half-cell reactions

$$2H^+ + 2e^- \rightleftharpoons H_2(g): \log K_1 = \frac{f_{H_2(g)}}{a_{e-}^2 \times a_{H^+}^2} = 0 \tag{18.41}$$

and

$$O_2(g) + 4H^+ + 4e^- \rightleftharpoons 2H_2O: \log K_2 = \frac{a_{H_2O}}{a_{e-}^4 \times a_{H^+}^4 \times f_{O_2(g)}} = 83.12 \tag{18.42}$$

where f is the fugacity of the subscripted gas. The above two relationships lead to:

$$\log f_{H_2(g)} = 0 - 2pH - 2p\varepsilon \tag{18.43}$$

and

$$\log f_{O_2(g)} = \log K_2 + 4pH + 4p\varepsilon \tag{18.44}$$

If the fugacity of either oxygen or hydrogen is fixed at a selected value, and the equilibrium constant K_2 is known, then a linear relationship between pε and pH is defined. Figure 18.8 shows these linear relationships at selected fugacities for oxygen and hydrogen using the value of K_2 measured at 298.15 K and 1 bar total pressure. We recall that pressure is related to fugacity by $f = (P \times \chi)$ where χ is the fugacity coefficient. In view of this correction the actual partial pressures for the lines on figure 18.8 will not have identical values to the fugacities. The correction is minimal at these low pressures although at elevated pressures the correction should be made.

The upper limit for water stability is defined by the pε–pH line that has the same oxygen partial pressure as the total pressure. The lower limit for water stability is defined by the pε–pH line that has the same hydrogen partial pressure as the total pressure. If the partial pressure of either gas exceeds the total pressure, water will decompose. The equations defining the stability boundaries at 1 bar total pressure (derived from equations 18.43 and 18.44 assuming a unit fugacity coefficient), are:

$$p\varepsilon = pH \text{ (lower boundary)} \tag{18.45}$$

and

$$p\varepsilon = 20.78 - pH \text{ (upper boundary)} \tag{18.46}$$

The pH range 4–10 is the maximum range normally encountered in natural waters at 298.15 K and 1 bar. At pH 4 the pε may climb as high as 17 before water is

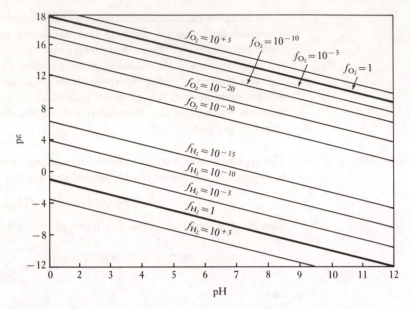

Figure 18.8   Contour lines of equal oxygen fugacity or hydrogen fugacity as a function of pε and pH at 298.15 K.

significantly oxidised to oxygen and at pH 10 the value of pε may be as low as -10 before water is reduced to produce hydrogen. At elevated partial pressures the stability range is increased whereas at elevated temperatures the range decreases (the equilibrium constant for reaction 18.38 increases with temperature). Figure 18.8 is the simplest example of a pε–pH diagram.

pε–pH diagrams involving solution species

To construct a pε–pH diagram we use the same procedures as for any other predominance diagram. We will illustrate this by constructing a pε–pH diagram for the species in the solution phase in the $Fe-H_2O$ system at 298.15 K and 1 bar.

The first step is to identify all the relevant iron species. In this case these are the hydrolysis products of Fe^{3+} and Fe^{2+} which have the formulae $Fe(OH)^{2+}$, $Fe(OH)_2^+$, $Fe(OH)_3^0$, $Fe(OH)_4^-$ and $Fe(OH)^+$. It is convenient to give the reaction equations for these in terms of stepwise formation reactions. Considering first the hydrolysis products for Fe^{3+}, we have:

$$Fe^{3+} + H_2O \rightleftharpoons Fe(OH)^{2+} + H^+ : \log K = -2.2 \qquad (18.47)$$

$$Fe(OH)^{2+} + H_2O \rightleftharpoons Fe(OH)_2^+ + H^+ : \log K = -3.5 \qquad (18.48)$$

$$Fe(OH)_2^+ + H_2O \rightleftharpoons Fe(OH)_3^0 + H^+ : \log K = -7.3 \qquad (18.49)$$

$$Fe(OH)_3^0 + H_2O \rightleftharpoons Fe(OH)_4^- + H^+ : \log K = -8.6 \qquad (18.50)$$

For the first reaction (18.4) we may write:

$$\log K = \log a_{Fe(OH)^{2+}} - pH - \log a_{Fe^{3+}} = -2.2 \qquad (18.51)$$

Under conditions where the activities of the two iron species are the same, this reduces to:

$$pH = 2.2 \qquad (18.52)$$

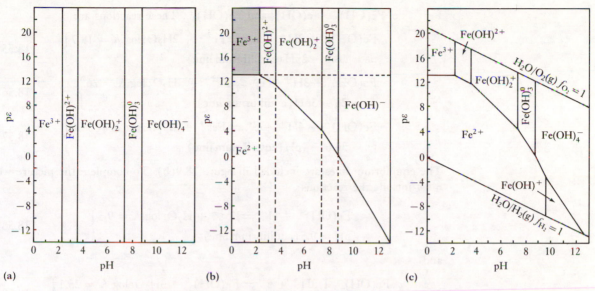

(b) (c)

Figure 18.9 $p\varepsilon$–pH predominance diagram for Fe^{2+}/Fe^{3+} equilibria at 298.15 K and 1 bar (after Nordstrom and Munoz, 1986).

In this equation there is no contribution from $p\varepsilon$: therefore a vertical straight line at pH 2.2 can be drawn on a pH–$p\varepsilon$ diagram. This line separates a region in which the iron is predominantly Fe^{3+} from one in which the iron is predominantly $Fe(OH)^{2+}$. Another predominance boundary is the one which separates $Fe(OH)^{2+}$ from $Fe(OH)_2^+$ and can be constructed using the equilibrium constant from equation 18.6. In addition to equation 18.52 there are three other vertical lines given by:

$$pH = 3.5 \ (Fe(OH)^{2+}/Fe(OH)_2^+ \ \text{equilibrium})$$

$$pH = 7.3 \ (Fe(OH)_2^{2+}/Fe(OH)_3^0 \ \text{equilibrium})$$

$$pH = 8.6 \ (Fe(OH)_3^0/Fe(OH)_4^- \ \text{equilibrium})$$

These are shown on figure 18.9(a).

We must now consider the reaction to form Fe^{2+} from the reduction of Fe^{3+}. The objective is to define boundaries on figure 18.9(a) which distinguish the regions in which Fe^{3+} and Fe^{2+} are predominant. The relevant redox reaction is:

$$Fe^{3+} + e^- \rightleftharpoons Fe^{2+}: \log K = 13.01 \qquad (18.53)$$

This gives:

$$p\varepsilon = 13.01 - \log a_{Fe^{3+}} - \log\left(\frac{a_{Fe^{2+}}}{a_{Fe^{3+}}}\right) \qquad (18.54)$$

When the activities of the two iron species are equal this gives:

$$p\varepsilon = 13.01$$

Thus, a horizontal line at $p\varepsilon = 13.01$ defines the predominance area for Fe^{3+}. As indicated in figure 18.9(b), the region of stability for Fe^{3+} is limited to the shaded area in the upper left-hand corner of the plot. The same exercise may be performed for the individual equilibria between Fe^{2+} and the other hydrolysis products of

Fe^{3+}, i.e. $Fe(OH)_2^+$, $Fe(OH)_3^0$ and $Fe(OH)_4^-$. These reactions are

$$Fe(OH)_2^+ + 2H^+ + e^- \rightleftharpoons Fe^{2+} + 2H_2O: \log K = 18.7$$
$$p\varepsilon = 18.7 - 2pH \, (\text{equilibrium line}) \qquad (18.55)$$

$$Fe(OH)_3^0 + 3H^+ + e^- \rightleftharpoons Fe^{2+} + 3H_2O: \log K = 26$$
$$p\varepsilon = 26 - 3pH \, (\text{equilibrium line}) \qquad (18.56)$$

$$Fe(OH)_4^- + 4H^+ + e^- \rightleftharpoons Fe^{2+} + 4H_2O: \log K = 34.6$$
$$p\varepsilon = 34.6 - 4pH \, (\text{equilibrium line}) \qquad (18.57)$$

The equilibrium lines are included in figure 18.9(b). To complete the picture we must include the reactions:

$$Fe(OH)^+ + H^+ \rightleftharpoons Fe^{2+} + H_2O: \log K = 9.5$$
$$pH = 9.5 \, (\text{equilibrium line}) \qquad (18.58)$$

and

$$Fe(OH)_4^- + 3H^+ + e^- \rightleftharpoons Fe(OH)_2^+ + 3H_2O: \log K = 25.1$$
$$p\varepsilon = 25.1 - 3pH \, (\text{equilibrium line}) \qquad (18.59)$$

Figure 18.9(c) shows the complete diagram plus the boundaries for the stability of water. In each predominance area a given species makes up over 50% of the total ion species. The predominance line defines the conditions where the relevant species activities are equal. This is different from solid/solution boundaries, which denote limits of solubility.

We should note that other boundaries such as the one between $Fe(OH)^+$ and $Fe(OH)_4^-$ could be included but they would not plot within the region of stability for water and are not considered here.

pε–pH diagrams for systems containing solid phases

The rules for constructing $p\varepsilon$–pH diagrams for systems with solid phases are no different from those with a single liquid phase. A system considered to be relevant to soil systems is that containing the aqueous species Fe^{2+} and Fe^{3+}, $Fe(OH)_2(s)$, $Fe(OH)_3(s)$ and $Fe_3(OH)_8(s)$. To construct a simple diagram we will neglect the aqueous hydrolysis products. The required reactions are those that define boundaries between the solution species and the solid phases, those that define boundaries within the solution phase, plus those that define boundaries between the individual solid phases. The following set will serve the purpose.

$$Fe^{3+} + e^- \rightleftharpoons Fe^{2+}: \log K_1 = 13.01 \qquad (18.60)$$

$$3H^+ + Fe(OH)_3(s) + e^- \rightleftharpoons Fe^{2+} + 3H_2O: \log K_2 = 15.87 \qquad (18.61)$$

$$H^+ + 3Fe(OH)_3(s) + e^- \rightleftharpoons Fe_3(OH)_8(s) + H_2O: \log K_3 = 5.27 \qquad (18.62)$$

$$8H^+ + Fe_3(OH)_8(s) + 2e^- \rightleftharpoons 3Fe^{2+} + 8H_2O: \log K_4 = 42.34 \qquad (18.63)$$

$$2H^+ + Fe_3(OH)_8(s) + 2e^- \rightleftharpoons 3Fe(OH)_2(s) + 2H_2O: \log K_5 = 3.79 \qquad (18.64)$$

$$Fe(OH)_2(s) + 2H^+ \rightleftharpoons Fe^{2+} + 2H_2O: \log K_6 = 12.85 \qquad (18.65)$$

Using the above equations, and assuming that water and all solids are in their standard states, we may define the following equilibrium lines:

$$p\varepsilon = \log K_1 + \log a_{Fe^{3+}} - \log a_{Fe^{2+}} \qquad (18.66)$$

$$p\varepsilon = \log K_2 - 3pH - \log a_{Fe^{2+}} \qquad (18.67)$$

$$p\varepsilon = \log K_3 - pH \qquad (18.68)$$

$$p\varepsilon = 0.5 \log K_4 - 4pH - 1.5 \log a_{Fe^{2+}} \qquad (18.69)$$

$$p\varepsilon = 0.5 \log K_5 - pH \qquad (18.70)$$

$$p\varepsilon = 0.5 \log K_6 - pH - 0.5 \log a_{Fe^{2+}} \qquad (18.71)$$

Strictly, we do not need to calculate the equilibrium constants for solid/solid equilibria. We require only the slopes of the equations since the lines must pass through the points of intersection of the complementary solution/solid lines.

To use these equations we may set the activity of Fe^{2+} to 10^{-5}. Figure 18.10 shows the resulting $p\varepsilon$–pH diagram and includes the limits of stability for water. Generally, this diagram shows that the stability fields are dominated by Fe^{2+} and $Fe(OH)_3(s)$ and changes in the $p\varepsilon$ or pH serve to dissolve or precipitate $Fe(OH)_3(s)$. Only under very reducing conditions, at high pH, will the solids dominate the system.

The high $p\varepsilon$/low pH region is of some interest. As expected, equation 18.60 defines a horizontal predominance line at $p\varepsilon = 13.01$ for the Fe^{3+}/Fe^{2+} species. Directly above this we see the $Fe(OH)_3(s)/Fe^{3+}$ boundary which seems to be an extrapolation of the $Fe(OH)_3(s)/Fe^{2+}$ boundary. We deduce this is true since the $Fe(OH)_3(s)/Fe^{3+}$ boundary is defined by the reaction

$$Fe(OH)_3(s) + 3H^+ \rightleftharpoons Fe^{3+} + 3H_2O \qquad (18.72)$$

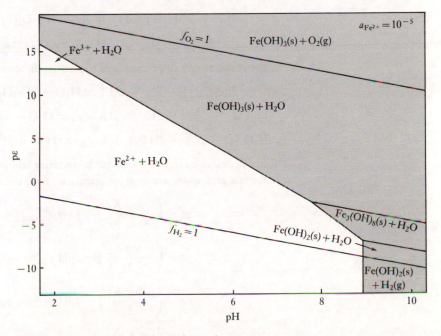

Figure 18.10 $p\varepsilon$–pH predominance diagram for Fe^{2+}/Fe^{3+} equilibria involving solids at 298.15 K and 1 bar (derived using equations 18.60–72).

This appears to have no terms in $p\varepsilon$. However, at fixed concentration of Fe^{2+} the concentration of Fe^{3+} must change with $p\varepsilon$ in accordance with equation 18.60. Therefore, we can eliminate all reference to Fe^{3+} by combining equations 18.60 and 18.72. This gives equation 18.61 which is just the previously defined $Fe(OH)_3(s)/Fe^{2+}$ reaction.

We note that the stability region for solids in figure 18.10 are actually two-phase regions. Between the upper and lower stability lines for water the two phases are solid plus water, whereas outside this region the solid will be stable but in equilibrium with the respective gas.

Additional complications in constructing $p\varepsilon$–pH diagrams

Figure 18.10 is one of the simplest $p\varepsilon$–pH diagrams encountered and its accurate construction is left as an exercise for the reader. However, as systems become progressively more complex the $p\varepsilon$–pH diagrams become difficult to construct. We will therefore consider some other systems in a little more detail.

Systems in contact with a soluble gas

Case 1

A complication arises if a species in the system is in equilibrium with a gas phase. An example of this is the system Fe–CO_2–H_2O where carbon dioxide can dissolve to form the carbonate ion which in turn precipitates as $FeCO_3(s)$ (siderite). Other possible mineral phases are magnetite $(Fe_3O_4)(s)$ and hematite $(Fe_3O_4)(s)$.

The boundaries to be defined are the boundaries between each solid and the aqueous ion Fe^{2+}, plus the three boundaries that separate the solid phases. The solid/solution boundaries are given by the reactions:

$$Fe^{2+} + CO_2(g) + H_2O \rightleftharpoons 2H^+ + FeCO_3(s): \log K_1 = -7.5$$

$$2Fe^{2+} + 3H_2O \rightleftharpoons 6H^+ + 2e^- + Fe_2O_3(s): \log K_2 = -23.79$$

$$3Fe^{2+} + 4H_2O \rightleftharpoons 8H^+ + Fe_4O_4(s): \log K_3 = -33.1$$

The above reactions can be combined to give the three solid–solid phase reactions:

$$2Fe_2O_3(s) + 2e^- + 2H^+ \rightleftharpoons H_2O + 2Fe_3O_4(s): \log K_4 = 5.17$$

$$Fe_2O_3(s) + 2e^- + 2H^+ + 2CO_2(g) \rightleftharpoons H_2O + 2FeCO_3(s): \log K_5 = 8.77$$

$$Fe_3O_4(s) + 2e^- + 2H^+ + 3CO_2(g) \rightleftharpoons H_2O + 3FeCO_3(s): \log K_6 = 10.53$$

The above reactions define the phase boundaries and predominance lines. Assuming all minerals and water are in their standard state, these are:

$$p\varepsilon = \left(\frac{-\log K_1}{2}\right) - \tfrac{1}{2}\log P_{CO_2(g)} - \tfrac{1}{2}\log a_{Fe^{2+}} \qquad (18.73)$$

$$p\varepsilon = \left(\frac{-\log K_2}{2}\right) - 3pH - \log a_{Fe^{2+}} \qquad (18.74)$$

$$p\varepsilon = \left(\frac{-\log K_3}{2}\right) - 4pH - \tfrac{3}{2}\log a_{Fe^{2+}} \qquad (18.75)$$

$$p\varepsilon = \left(\frac{\log K_4}{2}\right) - pH \qquad (18.76)$$

$$p\varepsilon = \left(\frac{\log K_5}{2}\right) - pH - \log P_{CO_2(g)} \qquad (18.77)$$

$$p\varepsilon = \left(\frac{\log K_6}{2}\right) - pH - \tfrac{3}{2}\log P_{CO_2(g)} \qquad (18.78)$$

where P_{CO_2} is the partial pressure of carbon dioxide gas in bars. An informative plot can be constructed by assuming values for the partial pressure of carbon dioxide and the activity of Fe^{2+}. Figure 18.11(a) shows all the boundaries for the case where $P_{CO_2(g)} = 1$ bar and $a_{Fe^{2+}} = 1 \times 10^{-6}$. Our problem is to decide which boundaries are metastable for the given conditions. We first inspect the solution/solid boundaries for $Fe^{2+}/Fe_3O_4(s)$ equilibria and for $Fe^{2+}/Fe_2O_3(s)$ equilibria. These lines intersect at the point P and define the upper bound to the stable predominance area for the ion Fe^{2+}. It is not clear whether one or other of these lines is metastable. However, inspection of the solid/solid boundary for $Fe_2O_3(s)/Fe_3O_4(s)$ equilibria shows that only $Fe_2O_3(s)$ is stable at points above the line AB and the $Fe^{2+}/Fe_3O_4(s)$ line must be metastable above the point P. A tip for testing whether a mineral is stable above or below a line is to look at the reaction equation and consider the effect of increasing pε with pH kept constant. For example, in the reaction $2Fe_2O_3(s) + 2e^- + 2H^+ \rightleftharpoons H_2O + 2Fe_3O_4(s)$, we see that increasing the electron activity (decreasing pε) pushes the reaction over to the right. Therefore, lower values for pε favour $Fe_3O_4(s)$ and this mineral must be stable below the equilibrium line. Using similar logic we see that the $FeCO_3(s)/Fe_3O_4(s)$ line CD must also be metastable since it plots above the line AB where $Fe_3O_4(s)$ is unstable. The

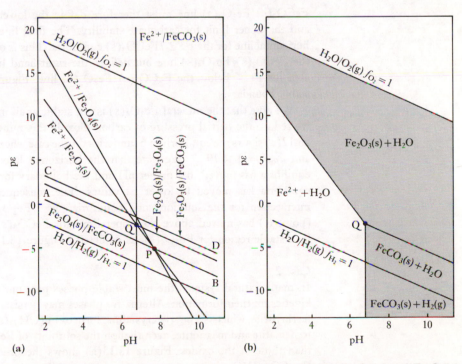

(a)   (b)

Figure 18.11   pε–pH predominance diagram for a system containing $FeCO_3(s)$, $Fe_3O_4(s)$ and $Fe_2O_3(s)$ at 298.15 K and 1 bar ($P_{CO_2} = 1$ bar, $a_{Fe^{2+}} \times 10^{-6}$; derived using equations 18.73–8).

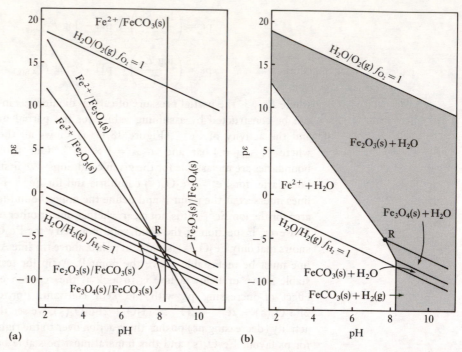

Figure 18.12   pε–pH predominance diagram for a system containing $FeCO_3(s)$, $Fe_3O_4(s)$ and $Fe_2O_3(s)$ at 298.15 K and 1 bar ($P_{CO_2} = 10^{-3}$ bar, $a_{Fe^{2+}} = 10^{-6}$; derived using equations 18.73–8).

$FeCO_3(s)/Fe_2O_3(s)$ line must therefore define the lower limit for oxide stability and the upper limit for carbonate stability. The final line we must consider is the horizontal line for the $Fe^{2+}/FeCO_3(s)$ equilibria. This is of course metastable above the $FeCO_3(s)/Fe_2O_3(s)$ line but defines the right-hand limit to the predominance area for Fe^{2+} below the $FeCO_3(s)/Fe_2O_3(s)$ line. Figure 18.11(b) shows all the stable boundaries.

We note that the mineral $Fe_3O_4(s)$ is not stable at all in this system. However, if we reduce the partial pressure of carbon dioxide this mineral is introduced. Figure 18.12(a) shows the equilibrium boundaries for the case where $P_{CO_2(g)} = 1 \times 10^{-3}$ bar and $a_{Fe^{2+}} = 1 \times 10^{-6}$. We note that the horizontal line for the $Fe^{2+}/FeCO_3(s)$ equilibria has moved to a higher pH and the boundary for the $FeCO_3(s)/Fe_2O_3(s)$ equilibria has moved to lower pε values. The consequence is that the point of intersection for the solution/solid equilibria $Fe^{2+}/Fe_2O_3(s)$ and $Fe^{2+}/Fe_3O_4(s)$ (point R) has moved above the upper limit for $FeCO_3(s)$. Thus, there must now be a stable region for $Fe_3O_4(s)$, as depicted in figure 18.12(b).

Case 2

In many natural systems the most stable phases may not exist because of the slow kinetics for their formation. Alternative phases may persist. For example, the mineral ferrihydrite, which is a poorly crystalline form of $Fe(OH)_3(s)$, may form in preference to hematite and magnetite, even though the solubility of $Fe(OH)_3(s)$ is much higher than those of the oxides. Figure 18.13(a) shows the predominance diagram for a $Fe–CO_2–H_2O$ system where $a_{Fe^{2+}}$ is 10^{-6} and P_{CO_2} is 1. To simplify the diagram we have assumed that $FeCO_3(s)$ and $Fe(OH)_3(s)$ are the only possible solid phases

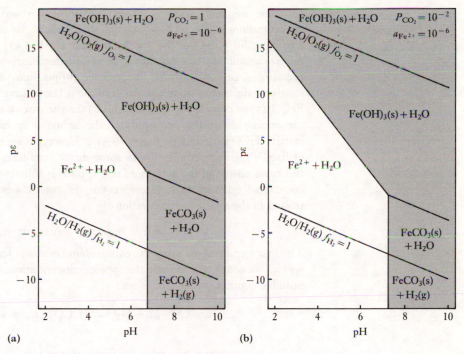

(a)   (b)

Figure 18.13 $p\varepsilon$–pH predominance diagram for the Fe–CO_2–H_2O system at 298.15 K and 1 bar (derived using equations 18.79–83).

and Fe^{3+} is not significant. The figure is constructed from the reactions:

$$2H^+ + FeCO_3(s) \rightleftharpoons Fe^{2+} + H_2O + CO_2(g): \log K_1 = 7.51 \quad (18.79)$$

$$3H^+ + Fe(OH)_3(s) + e^- \rightleftharpoons Fe^{2+} + 3H_2O: \log K_2 = 15.87 \quad (18.80)$$

These define boundaries between each solid and the aqueous ion Fe^{2+}, plus the boundary that separates the two solid phases. The solid/solution boundaries are given by:

$$0 = \log K_1 - 2pH - \log a_{Fe^{2+}} - \log P_{CO_2} \quad (18.81)$$

and

$$p\varepsilon = \log K_2 - 3pH - \log a_{Fe^{2+}} \quad (18.82)$$

The solid/solid boundary is derived from the combination of equations 18.79 and 18.80, yielding:

$$p\varepsilon = \log K_2 - \log K_1 + \log P_{CO_2} - pH \quad (18.83)$$

Figure 18.13(a) is directly comparable with figure 18.11(a) since the partial pressure of carbon dioxide and the activity of Fe^{2+} are the same in both cases. We note that replacing the stable oxide with the more soluble but kinetically favoured hydroxide leads to a larger predominance area for the aqueous ion Fe^{2+}.

In figure 18.13(b) the partial pressure of CO_2 is reduced to 10^{-2} bar and we see that the stability field of siderite has diminished.

Systems with pH-dependent soluble species

In some systems, such as those containing carbonate or sulphide ions, the concentrations of species have a complex dependence on other variables. In carbonate solutions, for example, the species present are $H_2CO_3^0$, HCO_3^- and CO_3^{2-}. The relative abundances are dependent on pH, the total carbonate content and the partial pressure of carbon dioxide. The relative proportions of the species influence the relative stabilities of minerals. An example is the relative stabilities of $Fe_3O_4(s)$ and $FeCO_3(s)$ as depicted in figure 18.11(b) for the case of constant partial pressure of carbon dioxide. In this first approach the carbonate species are not specified but the activity HCO_3^0 is fixed by fixing $P_{CO_2(g)}$. However, we may remove the constraint of fixed P_{CO_2} and replace it with some direct reference to the solution species. In this case, assuming the activity of one species is constant may not be realistic. The convenient quantity to hold constant is the total concentration of CO_2, which is related to the species concentrations by:

$$\sum CO_2 = m_{H_2CO_3^0} + m_{HCO_3^-} + m_{CO_3^{2-}} \tag{18.84}$$

The first step involves using the equilibrium constants for the reactions between the species, in order to eliminate the species concentrations from the above equations. Suitable equilibrium constants are:

$$CO_2(g) + H_2O \rightleftharpoons H_2CO_3^0 : K_1 = \frac{a_{H_2CO_3^0}}{P_{CO_2}}$$

$$CO_2(g) + H_2O \rightleftharpoons HCO_3^- + H^+ : K_2 = \frac{a_{HCO_3^-}}{P_{CO_2} \times a_{H^+}}$$

$$CO_2(g) + H_2O \rightleftharpoons CO_3^{2-} + 2H^+ : K_3 = \frac{a_{CO_3^{2-}}}{P_{CO_2} \times a_{H^+}^2}$$

Equation 18.84 can now be rearranged to give:

$$\sum CO_2 = P_{CO_2}\left(K_1 + \frac{K_2}{a_{H^+}} + \frac{K_3}{a_{H^+}^2} \right) \tag{18.85}$$

This shows that if the total carbonate is fixed, the partial pressure of $CO_2(g)$ must vary with pH. If activity coefficients are neglected, equation 18.85 may be rearranged and consolidated with the equilibrium constant for the reaction

$$Fe_3O_4(s) + 2e^- + 2H^+ + 3CO_2(g) \rightleftharpoons H_2O + 3FeCO_3(s): \log K_4 = 10.53 \tag{18.86}$$

giving

$$p\varepsilon = \tfrac{3}{2}\sum CO_2 + \tfrac{1}{2}\log K_4 - pH - \tfrac{3}{2}\log\left(K_1 + \frac{K_2}{a_{H^+}} + \frac{K_3}{a_{H^+}^2} \right) \tag{18.87}$$

The main point is that the $p\varepsilon$–pH stability boundary now has a non-linear dependence on pH, as shown in figure 18.14.

A general conclusion is that the oxide becomes stable relative to iron carbonate at high pH, whereas for the case of fixed P_{CO_2} (figure 18.12) the carbonate remains stable.

The redox states of sulphur

The sulphate/hydrogen sulphide boundary is perhaps the most important boundary in natural water systems. Many metal sulphides are very insoluble, providing there is enough hydrogen sulphide around, whereas many sulphates are soluble.

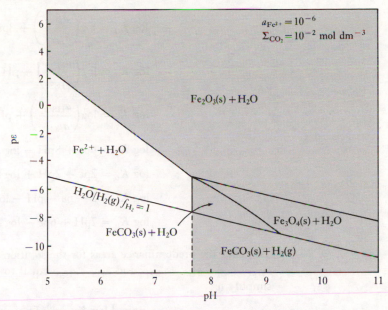

$$a_{Fe^{2+}} = 10^{-6}$$
$$\Sigma_{CO_2} = 10^{-2} \text{ mol dm}^{-3}$$

$Fe_2O_3(s) + H_2O$

$Fe^{2+} + H_2O$

$H_2O/H_2(g)\ f_{H_2} = 1$

$FeCO_3(s) + H_2O$

$Fe_3O_4(s) + H_2O$

$FeCO_3(s) + H_2(g)$

Figure 18.14   pε–pH predominance diagram for the Fe–CO$_2$–H$_2$O system at 298.15 K and 1 bar with total dissolved carbonate fixed at 10^{-2} mol dm^{-3} (derived using equations 18.84–7; after Drever, 1982).

Consequently, the geochemical fate of some metals may be controlled by the redox state of the sulphur species. Such reactions in turn have an influence on the stabilities of other minerals via the solution phase activities of these metal ions.

To describe the equilibrium solution chemistry of sulphur we require a set of reactions linking each solution species to each other, plus a set of reactions linking each solid to the solution species. The following set will serve the purpose well.

$$SO_4^{2-} + 8e^- + 10H^+ \rightleftharpoons H_2S^0 + 4H_2O: \quad \log K_1 = 41.04 \qquad (18.88)$$

$$SO_4^{2-} + 8e^- + 9H^+ \rightleftharpoons HS^- + 4H_2O: \quad \log K_2 = 34.08 \qquad (18.89)$$

$$H_2S^0 \rightleftharpoons H^+ + HS^-: \quad \log K_3 = -7.0 \qquad (18.90)$$

$$SO_4^{2-} + H^+ \rightleftharpoons HSO_4^-: \quad \log K_4 = 1.99 \qquad (18.91)$$

$$SO_4^{2-} + 6e^- + 8H^+ \rightleftharpoons S(s) + 4H_2O: \quad \log K_5 = 36.26 \qquad (18.92)$$

$$S(s) + 2e^- + 2H^+ \rightleftharpoons H_2S^0: \quad \log K_6 = 5.8 \qquad (18.93)$$

$$HS^- \rightleftharpoons S(s) + H^+ + 2e^-: \quad \log K_7 = 2.2 \qquad (18.94)$$

$$HSO_4^- + 6e^- + 7H^+ \rightleftharpoons S(s) + 4H_2O: \quad \log K_8 = 34.27 \qquad (18.95)$$

The first reaction is the fundamental reaction for sulphate/sulphide equilibria. The next three refer to sulphide species in solution. The final four refer to the precipitation of elemental sulphur and its link with dissolved species. There are no gas equilibria included here.

The pε–pH relationships are given by the equilibrium constants as follows:

$$\log K_1 = \log\left(\frac{a_{H_2S^0}}{a_{SO_4^{2-}}}\right) + 8p\varepsilon + 10pH \qquad (18.96)$$

$$\log K_2 = \log\left(\frac{a_{HS^-}}{a_{SO_4^{2-}}}\right) + 8p\varepsilon + 9pH \tag{18.97}$$

$$\log K_3 = \log\left(\frac{a_{HS^-}}{a_{H_2S^0}}\right) - pH \tag{18.98}$$

$$\log K_4 = \log\left(\frac{a_{HSO_4^-}}{a_{SO_4^{2-}}}\right) + pH \tag{18.99}$$

$$\log K_5 = 6p\varepsilon + 8pH - \log a_{SO_4^{2-}} \tag{18.100}$$

$$\log K_6 = 2p\varepsilon + 2pH + \log a_{H_2S^0} \tag{18.101}$$

$$\log K_7 = 2 - p\varepsilon - pH - \log a_{HS^-} \tag{18.102}$$

$$\log K_8 = 7pH + 6p\varepsilon - \log a_{HSO_4^-} \tag{18.103}$$

To define the predominance areas for the solution species we set the activity ratios $a_{H_2S^0}/a_{SO_4^{2+}}$, $a_{HS^-}/a_{SO_4^{2+}}$ and $a_{HS^-}/a_{H_2S^0}$ equal to 1. Equations 18.96–18.103 then simplify to:

$$p\varepsilon = \tfrac{1}{8}\log K_1 - \tfrac{10}{8}pH \tag{18.104}$$

$$p\varepsilon = \tfrac{1}{8}\log K_2 - \tfrac{9}{8}pH \tag{18.105}$$

$$pH = -\log K_3 \tag{18.106}$$

$$pH = \log K_4 \tag{18.107}$$

$$p\varepsilon = \tfrac{1}{6}\log K_5 - \tfrac{8}{6}pH + \tfrac{1}{6}\log a_{SO_4^{2-}} \tag{18.108}$$

$$p\varepsilon = \tfrac{1}{2}\log K_6 - pH - \tfrac{1}{2}\log a_{H_2S^0} \tag{18.109}$$

$$p\varepsilon = \tfrac{1}{2}\log K_7 - \tfrac{1}{2}pH - \tfrac{1}{2}\log a_{HS^-} \tag{18.110}$$

$$p\varepsilon = \tfrac{1}{6}\log K_8 - \tfrac{7}{6}pH - \tfrac{1}{6}\log a_{HSO_4^-} \tag{18.111}$$

Other assumptions are that above the SO_4^{2-}/H_2S^0 boundary all the sulphur is in the form of sulphate and that below the boundary and below pH 7 the sulphur is in the form of H_2S^0. These assumptions mean that the activity of the species can be set equal to the total sulphur concentration if activity coefficients are neglected.

Figure 18.15 shows the $p\varepsilon$–pH diagram for the case where the total sulphur is 0.001 mol kg^{-1}. The figure represents commonly encountered ranges of pH and $p\varepsilon$. The assumptions that above the SO_4^{2-}/H_2S^0 boundary all the sulphur is in the form of sulphate and that below the boundary and below pH 7 the sulphur is in the form of H_2S^0 are invalid close the actual boundaries. This makes the precise positions of the boundaries uncertain, but this is not a significant error when plotted on this scale. We see that sulphate is the major species on the diagram and only under strongly reducing conditions will H_2S^0 or HS^- be formed. The vertical line at pH 7 denotes the condition where H_2S^0 and HS^- have equal activities. Elemental sulphur has a limited field of stability at low pH.

A major problem with interpretation is that the reduction of sulphate to hydrogen sulphide is very slow unless catalysed by bacteria. This ensures that sulphate has a temporary existence in reducing media.

In order to extent the $p\varepsilon$–pH diagrams to include solid phases we simply construct

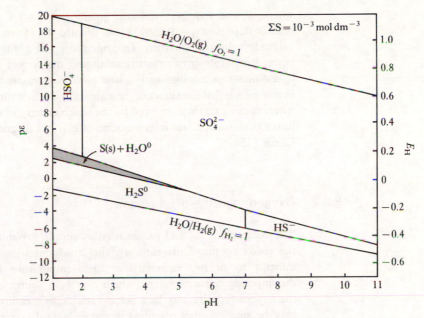

Figure 18.15
$p\varepsilon$–pH diagram for the sulphur/sulphide/sulphate at 298.15 K and 1 bar (derived using equations 18.104–11).

the phase boundary between each solution species and the respective solid, plus any solid/solid boundaries if they are within the $p\varepsilon$–pH region of interest.

To include the minerals $FeS_2(s)$ and $Fe_2O_3(s)$ which may be stable in the presence of SO_4^{2-} we need the equations:

$$2SO_4^{2-} + 4Fe^{2+} + 16H + 14e^- \rightleftharpoons FeS_2(s) + 8H_2O: \log K_1 = 85.74$$
$$(18.112)$$

and

$$Fe_2O_3(s) + 4SO_4^{2-} + 38H^+ + 30e^- \rightleftharpoons 2FeS_2(s) + 19H_2O: \log K_2 = 195.2$$
$$(18.113)$$

These give rise to the boundaries defined by:

$$p\varepsilon = \tfrac{1}{14} \log K_1 + \tfrac{2}{14} \log a_{SO_4^{2-}} + \tfrac{1}{14} \log a_{Fe^{2+}} - \tfrac{16}{14}pH \qquad (18.114)$$

and

$$p\varepsilon = \tfrac{1}{30} \log K_2 + \tfrac{4}{30} \log a_{SO_4^{2-}} - \tfrac{38}{30}pH \qquad (18.115)$$

The boundaries can be plotted by setting the sulphate and hydrogen sulphide activities equal to the total sulphur concentration (as before) and selecting a value for the activity of Fe^{2+} in the system.

E_H–pH diagrams
E_H–pH diagrams can provide the same type of information as $p\varepsilon$–pH diagrams. The diagrams differ only in the variable on the vertical axis, E_H, which is related to $p\varepsilon$ by:

$$E_H = \frac{2.30258RT}{F} p\varepsilon \qquad (18.116)$$

where F, the Faraday constant, equals $96484.6 \, J \, V^{-1} \, mol^{-1}$. At 198.15 K the relationship is $E_H = 0.059155 \, p\varepsilon$. Figure 18.15 contains an E_H axis for the sulphide/sulphate equilibria. In principle E_H is a directly measurable property of any redox couple or of an entire solution if the system is at equilibrium. In practice it is difficult to measure and is best used as a master variable. The only difficulty in using E_H is that the slopes of the stability boundaries are dependent on temperature, whereas using $p\varepsilon$ they are fixed by the stoichiometry of the reaction. Many examples in the construction and interpretation of E_H–pH diagrams are given by Garrels and Christ (1965).

18.2.2   Oxygen fugacity–pH diagrams

The parameters E_H and $p\varepsilon$ are used as simple variables which define the state of the redox couple. Unfortunately, the stoichiometries of many redox couples are difficult to define, especially in high-temperature fluids or those containing decomposing organic matter (which may be a source of electrons). In these cases the oxygen fugacity may be a better variable to measure or calculate even though it may be small. Redox reactions in mineral–fluid systems at high temperature and pressure are often described in terms of oxygen or hydrogen fugacities, even though there may be other sources of electrons.

An oxygen fugacity–pH diagram is another form of predominance diagram. Here, we plot the value of f_{O_2}, rather than $p\varepsilon$ or the activity of another species, against pH. At low pressures and high temperatures the partial pressure may be used in preference to fugacity.

No new principles are involved in constructing f_{O_2}–pH diagrams. The exercise involves manipulating reaction equations to eliminate unwanted variables and leave expressions in terms of those required. The simplest example is provided by the sulphate/hydrogen sulphide equilibria described previously in terms of $p\varepsilon$–pH using equations 18.88–18.95. The electrons in equation 18.88 can be eliminated by first replacing water using:

$$2H_2O \rightleftharpoons 2H_2(g) + O_2(g): \log K_1 = -82.1 \qquad (18.117)$$

and the eliminating both $H_2(g)$ and electrons using:

$$2H_2(g) \rightleftharpoons 2H^+ + 2e^-: \log K_2 = 0 \qquad (18.118)$$

This gives:

$$SO_4^{2-} + 2H^+ \rightleftharpoons H_2S^0 + 2O_2(g): \log K_3 = -125.16 \qquad (18.119)$$

The relationship between the partial pressure of oxygen and pH is given by:

$$\log P_{O_2(g)} = \frac{1}{2}\left[\log K_3 - \left(\frac{a_{H_2S^0}}{a_{SO_4^{2-}}}\right)\right] - pH \qquad (18.120)$$

This equation defines the $a_{H_2S^0}/a_{SO_4^{2-}}$-boundary. The boundaries between other species in equations 18.88–18.95 can be defined similarly.

Figure 18.16 shows the sulphate/hydrogen sulphide equilibria based on the same assumptions as in figure 18.15. Each boundary in figure 18.15 has an analogous boundary in figure 18.16 and the interpretation is identical. The lower line of stability

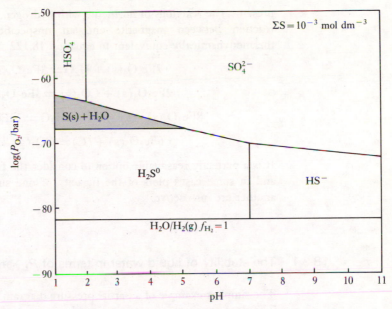

Figure 18.16 Partial pressure–pH diagram for sulphur/sulphide/sulphate equilibria at 298.15 K and 1 bar (derived using equations 18.117–20).

for water $[H_2O/H_2(g)$ equilibrium] has been plotted, but the upper line not considered since it would be far above the top of this plot.

To extend the plot to include the minerals $FeS_2(s)$ and $Fe_2O_3(s)$ would require little more than rearranging equations 18.112 and 18.113 to eliminate electrons.

18.3 Partial pressure diagrams

The concentration of free molecular oxygen is very low in most supercritical aqueous fluids and magmas, so the oxygen fugacity is virtually undetectable. We can see this by calculating the partial pressure of oxygen for the case when any two oxygen-containing iron minerals are in equilibrium. Consider the reaction:

$$2FeO(s) + \tfrac{1}{2}O_2(g) \rightleftharpoons Fe_3O_4(s): \log K = \log\left(\frac{a_{Fe_3O_4(s)}}{a_{FeO(s)}^2}\right) - \tfrac{1}{2}\log P_{O_2(g)}$$

(18.121)

The equilibrium constant for this reaction at 298.15 K and 1 bar is 49.25; therefore the value of $\log P_{O_2(g)}$ is -98.5, assuming the solids are in their standard states.

Despite the fact that oxygen fugacity is a controlling thermodynamic variable, its low value suggests that other species may be responsible for the transfer of electrons. For example, the oxidation of magnetite to hematite, written as:

$$4Fe_3O_4(g) + O_2(g) \rightleftharpoons 6Fe_2O_3(s)$$

(18.122)

can be re-written in terms of species other than oxygen if they are thought to be present. The reactions of magnetite with hydrogen, carbon dioxide, methane or the reaction between magnetite and an unspecified source of carbon are all thermodynamically equivalent to equation 18.122. That is,

$$2Fe_3O_4(s) + H_2O \rightleftharpoons 3Fe_2O_3(s) + H_2(g) \tag{18.123}$$

$$2Fe_3O_4(s) + CO_2(g) \rightleftharpoons 3Fe_2O_3(s) + CO(g) \tag{18.124}$$

$$8Fe_3O_4(s) + CO_2(g) + 2H_2O \rightleftharpoons 12Fe_2O_3(s) + CH_4(g) \tag{18.125}$$

$$6Fe_2O_3(s) + (C) \rightleftharpoons CO_2(g) + 4Fe_3O_4(s) \tag{18.126}$$

It is a perfectly reasonable option to consider the fugacities of any gaseous species, and in some cases plots of the fugacity of one substance against the fugacity of another are instructive.

18.3.1 The stability of liquid water in terms of P_{H_2} and P_{O_2}

The simplest example of a partial pressure diagram is that for the stability of water. The decomposition of water is given by:

$$2H_2O = 2H_2(g) + O_2(g) \tag{18.127}$$

Assuming water is in its standard state, the equilibrium constant has the form:

$$K = P_{H_2}^2 \times P_{O_2} \tag{18.128}$$

Thus, if the partial pressure of one gas is known, the other can be calculated.

If the partial pressure of either oxygen or hydrogen exceeds the external pressure, then liquid water will spontaneously decompose to its constituent gases. Thus, a plot of P_{H_2} versus P_{O_2} calculated from the equilibrium constant shows the stability region for water.

At 298.15 K and 1 bar, $\log K \simeq -83.1$, whereas on the saturation line at 573.15 K (300 °C and 85.8 bar), $\log K \simeq -35.6$. Figure 18.17 shows the P_{H_2}–P_{O_2} plot for the two situations: clearly the equilibrium lines differ in position and the stability range is less at the elevated temperature and pressure.

18.3.2 Some partial pressure diagrams for mineral systems

Two examples are given.

Fe/O₂(g)/CO₂(g)   Minerals in the iron/oxygen/carbon dioxide system are described by:

$$Fe_2O_3(s) + 2CO_2(g) \rightleftharpoons 2FeCO_3(s) + \tfrac{1}{2}O_2(g): \log K_1 = -31.8 \tag{18.129}$$

$$Fe_3O_4(s) + 3CO_2(g) \rightleftharpoons 3FeCO_3(s) + \tfrac{1}{2}O_2(g): \log K_2 = -30.9 \tag{18.130}$$

and

$$Fe(s) + CO_2(g) + \tfrac{1}{2}O_2 \rightleftharpoons FeCO_3(s): \log K_3 = 49.0 \tag{18.131}$$

The partial pressures of oxygen and carbon dioxide can then be linked by the

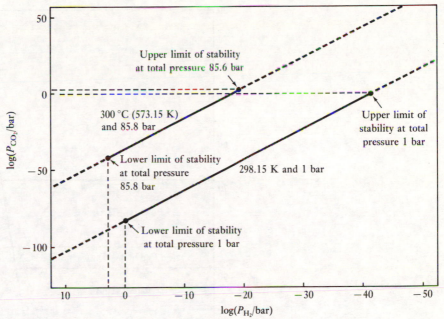

Figure 18.17   Regions of stability for liquid water at different total pressures (derived using equations 18.132–4).

complementary relationships:

$$\log P_{O_2(g)} = 2 \log K_1 + 4 \log P_{CO_2(g)} \qquad (18.132)$$

$$\log P_{O_2(g)} = 2 \log K_2 + 6 \log P_{CO_2(g)} \qquad (18.133)$$

and

$$\log P_{O_2(g)} = -2 \log K_3 - 2 \log P_{CO_2(g)} \qquad (18.134)$$

Figure 18.18 shows the stability diagrams for this system at 298.15 K and 1 bar with the lower stability line for water included. Note that this line for water is horizontal because there is no dependence on $P_{CO_2(g)}$.

Cu/O₂(g)/   This system is included because it introduces water as a reactive component.
CO₂(g)/H₂O   Minerals in the copper/oxygen/carbon dioxide system are described by:

$$2Cu(s) + \tfrac{1}{2}O_2(g) \rightleftharpoons Cu_2O(s): \log K_1 = 25.65 \qquad (18.135)$$

$$Cu_2O(s) + \tfrac{1}{2}O_2(g) \rightleftharpoons 2CuO(s): \log K_2 = 18.8 \qquad (18.136)$$

$$3Cu(s) + \tfrac{3}{2}O_2(g) + 2CO_2(g) + H_2O \rightleftharpoons Cu_2(OH)_2(CO_3)_2(s): \log K_3 = 73.8 \qquad (18.137)$$

$$3Cu_2O(s) + \tfrac{3}{2}O_2(g) + 4CO_2(g) + 2H_2O \rightleftharpoons 2Cu_3(OH)_2(CO_3)_2(s): \log K_4 = 70.4 \qquad (18.138)$$

$$Cu_2O(s) + \tfrac{1}{2}O_2(g) + CO_2(g) + H_2O \rightleftharpoons Cu_2(OH)_2CO_3(s): \log K_5 = 22.7 \qquad (18.139)$$

$$2CuO(s) + CO_2(g) + H_2O \rightleftharpoons Cu_2(OH)_2CO_3(s): \log K_6 = 3.8 \qquad (18.140)$$

$$3Cu_2(OH)_2CO_3(s) + CO_2(g) \rightleftharpoons 2Cu_3(OH)_2(CO_3)_2(s) + H_2O: \log K_7 = 2.5 \qquad (18.141)$$

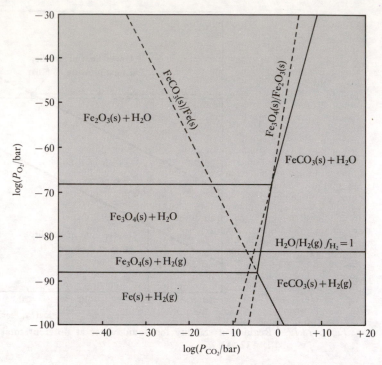

Figure 18.18   Partial pressure diagram for the $Fe/CO_2/H_2O$ system at 298.15 K and 1 bar (after Garrels and Christ, 1965).

The complementary partial pressure relationships, used to construct figure 18.19, are:

$$\log P_{O_2(g)} = -2 \log K_1 \tag{18.142}$$

$$\log P_{O_2(g)} = -2 \log K_2 \tag{18.143}$$

and

$$\log P_{O_2(g)} = -\tfrac{2}{3} \log K_3 - \tfrac{4}{3} \log P_{CO_2(g)} \tag{18.144}$$

$$\log P_{O_2(g)} = -\tfrac{2}{3} \log K_4 - \tfrac{8}{3} \log P_{CO_2(g)} \tag{18.145}$$

$$\log P_{O_2(g)} = -2 \log K_5 - 2 \log P_{CO_2(g)} \tag{18.146}$$

$$\log P_{CO_2(g)} = -\log K_6 \tag{18.147}$$

$$\log P_{CO_2(g)} = -\log K_7 \tag{18.148}$$

This resembles an ordinary plot of mineral stabilities. However, we expect the relative positions of the lines to change with water activity. For example, azurite is metastable with respect to malachite when the water activity is unity. Inspection of equation 18.140 shows that azurite $[Cu_2(OH)_2(CO_3)_2(s)]$ is stabilised by the reduction of water activity since water is produced in the reaction. In contrast, equation 18.141 shows that malachite $[Cu_2(OH)_2CO_3(s)]$ is destabilised by low water activities since water is consumed in the reaction. We expect the stability lines of malachite and azurite to switch position if the water activity is sufficiently low.

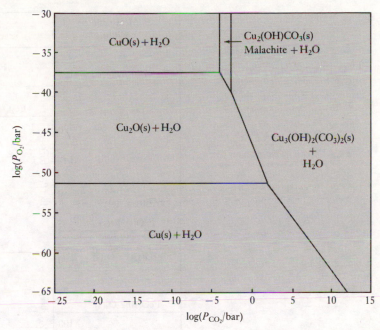

Figure 18.19   Partial pressure diagram for the $Cu/CO_2/H_2O$ system at 298.15 K and 1 bar (after Garrels and Christ, 1965).

18.3.3   Evaluating metastability using P_{O_2}

The fugacity of oxygen can be used as a 'yardstick' to distinguish metastable equilibria from stable equilibria following a procedure described by Garrels and Christ (1965, chapter 6).

In a mineral assembly containing iron–oxygen compounds, for example, the actual iron mineral present will depend on the oxygen fugacity. Elemental iron can be transformed through a sequence of oxide phases if the oxygen fugacity is increased. Table 18.2 shows the equilibrium partial pressures for the mineral couples at 298.15 K and 1 bar. The table contains all permutations of mineral/mineral and mineral/element couples.

The relationship between the reactions can be seen by a bar plot indicating the position of the equilibrium partial pressure for the individual reactions. Each value of P_{O_2} represents the point at which a transition occurs as partial pressure is progressively increased.

Table 18.2   **Equilibrium partial pressures for iron/oxygen minerals at 298.15 K and 1 bar (from Garrels and Christ, 1965)**

Reaction	log K
$Fe(s) + \frac{1}{2}O_2(g) \rightleftharpoons FeO(s)$	-85.6
$3Fe(s) + 2O_2(g) \rightleftharpoons Fe_3O_4(s)$	-88.9
$2Fe(s) + \frac{3}{2}O_2(g) \rightleftharpoons Fe_2O_3(s)$	-86.6
$3FeO(s) + \frac{1}{2}O_2(g) \rightleftharpoons Fe_3O_4(s)$	-98.5
$2FeO(s) + \frac{1}{2}O_2(g) \rightleftharpoons Fe_2O_3(s)$	-68.2
$3Fe_3O_4(s) + \frac{1}{2}O_2(g) \rightleftharpoons 3Fe_2O_3(s)$	-68.2

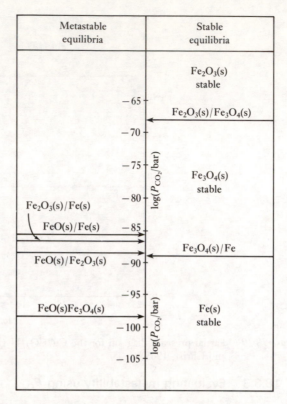

Figure 18.20   Bar graph showing P_{O_2} in equilibrium with various iron compounds at 298.15 K and 1 bar (after Garrels and Christ, 1966, chapter 6).

Figure 18.20 shows that some transitions are metastable. The first transition encountered, FeO(s) to Fe_3O_4(s), is metastable since FeO(s) cannot be oxidised before Fe(s) itself has been oxidised.

This type of analysis can be performed for other mineral reactions. Other measures of activity may be used instead of partial pressure.

18.3.4   General comments on partial pressure diagrams

In order to be more realistic, the simple partial-pressure diagrams may be extended to include more components. Indeed it may even be informative to introduce planar representations of three-dimensional plot using a further axis. For example, the system Fe–O_2(g)–CO_2(g)–S(g) could be represented three-dimensionally using the partial pressures of the three gases. However, such representations are difficult to construct and interpret.

Partial-pressure diagrams are useful for comparing the relative stabilities of minerals, estimating likely changes in mineral assemblies as partial pressures change, and attempting to predict the redox conditions responsible for observed mineral assemblies. However, they have limitations since data may not be available for real minerals which show compositional variation. Also, with the exception of carbon dioxide, predicted partial pressures are lower than the limits of detection and cannot be verified by experiment.

18.4   Phase diagrams

This section is concerned with interpreting the features of phase diagrams for solid/solid and solid/liquid systems.

The reader should be familiar with the principles defined in section 12.4 for the equilibrium relationships within liquid/liquid and liquid/vapour systems. These principles apply directly to solid/solid and solid/liquid systems. However, we will start by recalling some basic ideas. For a detailed review of the entire subject the reader is referred to Ehrlers (1972).

18.4.1   General principles

A phase diagram for a system is a representation of the assembly of phases that exist as a function of temperature, pressure and usually a composition variable. Common phase diagrams represent equilibrium systems, where the number of phases and their individual compositions are consistent with a minimum in the Gibbs energy for the system.

At equilibrium, the number of phases in a system is constrained by the Gibbs phase rule $P + F = C + 2$, where P is the number of phases, F the number of degrees of freedom and C the number of components.

In single-component systems the only equilibrium reactions involve boiling, melting, polymorphic phase transitions and sublimation. These are adequately represented by two-dimensional stability fields in terms of temperature and pressure, or temperature and volume. Occasionally, two-dimensional representations involving temperature, pressure and volume are used. These are difficult to construct but are quite revealing in terms of phase stability.

The information that can be extracted from unary phase diagrams has been discussed in chapter 12 and we will go directly to multicomponent systems.

For binary systems, where $C = 2$, the phase rule becomes $F = 2 + 2 - P$. When only one phase exists ($P = 1$), three intensive properties are necessary to define the system ($F = 3$). Consequently, temperature, pressure and the mole fraction of one component may be varied independently. If two phases are present, at the melting point of a mineral for example, we have only two degrees of freedom. If three phases exist then only one intensive property may be varied. The extension to multicomponent systems is logical.

Fixing of one or more intensive properties gives the **condensed phase rule**:

$$P + F' = C + 2 - R \tag{18.149}$$

where R is the number of variables made constant and F' is the number of remaining degrees of freedom. It is common to fix pressure and therefore lose one degree of freedom. This allows us to depict the phase behaviour of a binary system on a temperature–composition diagram. At the melting point of a binary mixture there are two phases and only one degree of freedom. Thus at any chosen temperature the composition must be fixed, and a line in temperature–composition space defines the melting point curve. If three phases exist, a single point in temperature–composition space defines the state of the system, and if any variable is changed a phase change must occur.

An important difference between liquid/vapour systems and solid/liquid systems is that the compressibilities of solids are low; therefore, their melting points change slowly with pressure. If the temperature is fixed at the melting point of one pure mineral, there is unlikely to be any accessible pressure at which another mineral can be made to melt. This makes pressure–composition diagrams difficult to construct since the composition boundaries at fixed temperature will be very steep. Consequently, only temperature–composition diagrams are used to depict solid/liquid equilibria.

18.4.2   Binary phase diagrams for immiscible solids

The most common phase diagrams are for liquid melts in equilibrium with immiscible solid phases.

General features   The temperature–composition diagram for a mixture of diopside ($CaMgSi_2O_6(s)$) and anorthite ($CaAl_2Si_2O_8(s)$) shows the features of this simplest type of solid/liquid system (figure 18.21). The abscissa is the mass fraction of the anorthite, although plots in terms of mole fractions are sometimes encountered.

We observe four distinct regions. These are a homogeneous mixed liquid phase bounded by the points AEB and unbounded in temperature, a region containing pure anorthite and a mixed liquid bounded by AEFA, a region containing pure diopside and a mixed liquid bounded by BEDB, and a region below the line DEF where the solid phases exist separately.

The lines AE and BE are the **solubility lines** for anorthite and diopside respectively. These lines are equivalent to the **depression of freezing point curves** and the two terms may be used interchangeably. The point E defines a temperature

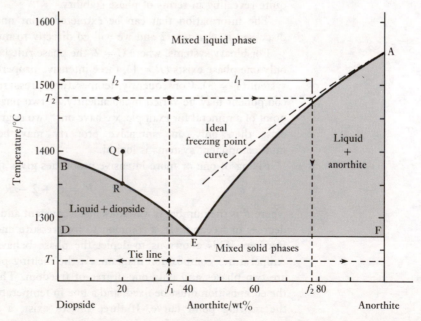

Figure 18.21   Temperature–composition diagram for diopside/anorthite mixtures (data from Bowen, 1928, page 26).

at which the maximum depression of the freezing point has occurred for both minerals. This is called the **eutectic point** and the composition corresponds to the eutectic mixture. No liquid mixture can exist below the eutectic point.

A horizontal tie line at a specified temperature can be used to determine the state of the system at any overall composition. We first select a composition and draw a vertical line to intersect with the tie line. If the point of intersection is in a single-phase region, the system is stable. If the point of intersection falls in a two-phase region, the compositions of each phase can be determined by extrapolating horizontally in both directions until a boundary is reached – either a phase boundary or the edge of the plot. Vertical lines at these points define the composition of the respective phases. For example, at the overall composition f_1 and temperature T_1 the system consists of two pure solid phases because the tie lines can be extrapolated to both edges of the plot without encountering a phase boundary. At the overall composition f_1 and the higher temperature T_2 the system exists as a liquid melt of composition f_2 in equilibrium with pure anorthite. The level rule may be applied to determine the relative amounts of each phase present in any two-phase region. For example, at the higher temperature T_2 on figure 18.21 the relative proportions of the mixed liquid phase and pure diopside can be determined from

$$\frac{n_{diopside}}{n_{liquid}} = \frac{l_1}{l_2} \simeq \frac{45}{60} = 0.75 \qquad (18.150)$$

where n is the mass of the subscripted phase.

Monotectic systems   If one component has a melting point much higher than the other, the eutectic point will be close to the vertical axis of the substance with the lowest melting point. Although there is a eutectic point it may be virtually undetectable. Such systems are called **monotectic**. An example is the silicon/tin system, where the melting points are $1420\,°C$ and $232\,°C$ respectively. The eutectic mixture is almost pure tin.

Polymorphic phase   Changes in temperature commonly initiate polymorphic phase transitions. For pure
transitions   components in binary systems these are depicted by horizontal phase boundaries at specified temperatures. The horizontal broken line around $1130\,°C$ in figure 18.28 (insert) defines the transition temperature for α- to β-wollastonite.

Cooling history   A simple phase diagram can give information about the course of crystallisation of a mixture, in addition to the equilibrium properties. Consider the point Q on figure 18.21. If the two-component mixture at this composition cools slowly, the temperature drops until the crystallisation of diopside occurs at the point R. With further cooling diopside crystallises out and the composition of the liquid progresses along the line RE towards the eutectic point. At the eutectic point anorthite begins to crystallise along with the remaining diopside in constant proportions. Since there are two solid phases and one liquid phase, and pressure is fixed, the phase rule dictates that temperature must also be fixed. Consequently, the temperature remains constant until crystallisation is complete. This plateau in temperature is called the **eutectic arrest**. The eutectic point is the analogue of the melting point for a pure substance. The cooling curve for such a system is shown in figure 18.22.

The melting of a mixture, of any composition, will occur by the reverse process. That is, the temperature can be increased until the eutectic temperature is reached, at which point both minerals start to melt simultaneously. The liquid phase will be

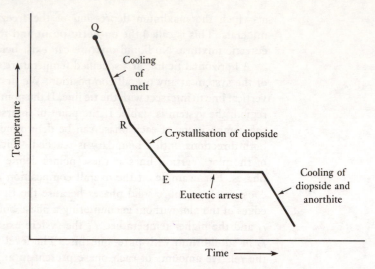

Figure 18.22   Cooling curve for a diopside/anorthite mixture (after Ehlers, 1972).

at the eutectic composition. More heating causes both phases to melt at a fixed ratio and constant temperature until one solid phase is eliminated. Upon further heating the temperature increases and the composition of the liquid phase progresses along the respective freezing point curve until the last crystal has melted.

Theoretical phase diagrams   A phase diagram may be determined experimentally or predicted from the known thermodynamic properties of the systems. For many chemical systems at atmospheric pressure the predicted phase diagrams are reliable. Unfortunately, most phase diagrams of geological interest described assemblages involving solids and liquid melts at high temperatures and pressures where thermodynamic properties for mixtures are less well characterised. For instance, we may predict the depression of freezing point curve for anorthite (line AE in figure 18.21) using equation 12.56, i.e.

$$\ln \frac{a_{an(l)}}{a_{an(s)}} = \frac{\Delta H_f}{R}\left(\frac{1}{T_f^\bullet} - \frac{1}{T_f}\right) \qquad (18.151)$$

where T_f is the freezing point of pure anorthite, ΔH_f the enthalpy of fusion of **pure** anorthite and $a_{an(l)}$ and $a_{an(s)}$ the activity of anorthite in the homogeneous liquid melt and solid phase respectively. By using this equation we are assuming that ΔH_f is independent of temperature, which is not unreasonable. Since the solid phase is immiscible we may assume that $a_{an(s)}$ is unity. The above equation can be rearranged to predict the temperature of melting, T_f, at any activity $a_{an(l)}$, i.e.

$$T_f = \frac{1}{(1/T_f^\bullet) - R \ln a_{an(l)}/\Delta H_f} \qquad (18.152)$$

In the absence of any information concerning the activity of anorthite in the melt we assume it is equal to the mole fraction (i.e. ideal mixing where $a_{an(l)} = x_{an(l)}$). The broken line shows the predicted values of T using the measured value of ΔH_f equal to 2200 J mol^{-1}. In passing we note that the predicted temperatures are plotted on figure 18.21 at specific percentages by weight, whereas the calculation of T_f requires mole fractions of anorthite to be used.

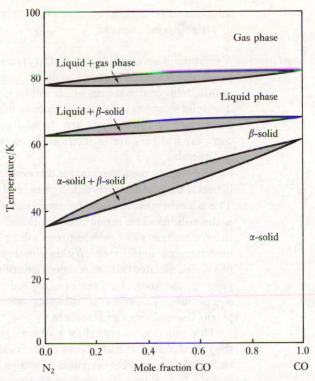

Figure 18.23   Phase equilibria for nitrogen carbon monoxide mixtures (after Parker, 1988, page 34).

The prediction is poor mainly because the assumption of ideality is incorrect. Characterising non-ideality in liquid melts is difficult because of the high temperatures and pressures involved. Significant advances have been made but the subject is less well developed than for systems under ambient conditions (Bottinga *et al.*, 1981; Noordstrom and Munoz, 1986). Consequently, the most useful phase diagrams are those constructed experimentally.

Multiphase diagrams   Binary phase diagrams need not be restricted to the presence of two phases. Figure 18.23 shows a phase diagram for mixtures of nitrogen and carbon monoxide at low temperatures. We observe liquid/vapour and solid/liquid equilibria and complete miscibility in all phases. An additional feature is a polymorphic phase transition in the mixed solid phase (α to β conversion). Note that the transition temperature of the mixed solid phase is dependent on composition; for a pure solid it would be constant.

18.4.3   Solid solutions in binary systems

Some mixtures of minerals, when melted and cooled slowly, give rise to a solid of a single crystallographic type rather than a mixture of the crystals of the pure minerals. A **solid solution** of the minerals is said to have been formed. A solid solution may be simply an ionic impurity in a mineral. In contrast, some solid solutions reflect complete mixing of two minerals to give an entire family of minerals of variable compositions.

Complete solid solutions   A mixture of anorthite ($CaAl_2Si_2O_8(s)$) and albite ($NaAlSi_3O_8(s)$), in almost any proportion, can be melted and cooled to form a solid solution. (There is some limited immiscibility in this system which will be neglected for the purpose of this exercise.) The solid solutions define the plagioclase feldspars ($NaAlSi_3O_8$–$CaAl_2Si_2O_8$) where the sodic and calcic forms are thought of as end-members of the family. Most common feldspars are of intermediate composition rather than being close to an end-member.

The melting point diagram for this system is shown in figure 18.24. The boundary between the homogeneous liquid phase and the liquid + solid phase is the liquidus. The boundary between the homogeneous solid phase and the liquid + solid phase is the **solidus**. The region between the solidus and the liquidus is a region where the system exists as a homogeneous solid phase (plagioclase) in equilibrium with a homogeneous liquid melt. Within this region, and on the phase boundaries, the phase rule dictates that the compositions of both phases are fixed if temperature and pressure are fixed. Tie lines can be used to link the composition of the phase in equilibrium. Examples of minerals which show complete solid solution are plagioclases, olivines and melilites.

This behaviour is typical for a system where there is only limited non-ideality in the mixed phases. It is analogous to the ideal and moderately non-ideal liquid/liquid and liquid/gas systems described in section 12.4. The procedures for using tie lines and the lever rule apply equally well. For example, at a temperature of 1200 °C and overall composition W the composition of the liquid phase is given by the intersection

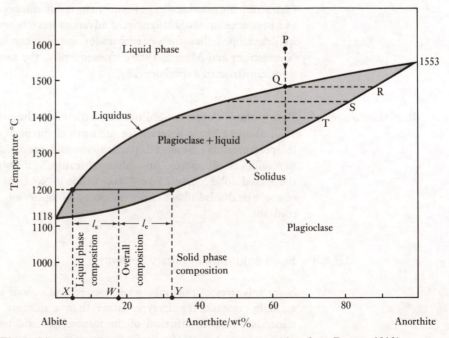

Figure 18.24   The melting point diagram for anorthite/albite mixtures (data from Bowen, 1913).

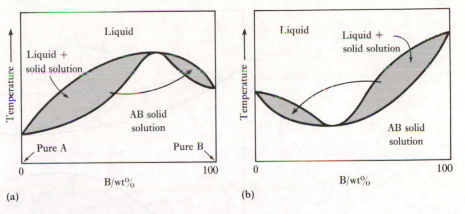

Figure 18.25   Temperature–composition diagrams for non-ideal systems.

of the tie line with the liquidus at point X, and the composition of the mixed solid at point Y on the solidus. The ratio of the mass of liquid to the mass of the plagioclase phase is given by the ratio of the lengths l_s and l_l.

At this point we note another difference between the properties of solids and liquids: mixing in the solid phase is very slow. Consider the liquid mixture in figure 18.24 at the point P, undergoing cooling. When the temperature has dropped to the point of intersection with the phase boundary, a solid will precipitate. This solid will be rich in anorthite and have the composition defined by the point R. The liquid will be enriched in the albite component. With further cooling, the precipitated mineral will become progressively richer in albite (points S, T, etc.). The precipitated mineral will not mix with the first precipitate to give a homogeneous solid solution. Rather, the growing solid phase will show compositional variation through the deposited layers. It is common to find compositionally zoned plagioclase crystals with the rims more sodic than the centre. The zones may be sharply defined or gradational, depending on the cooling history. Systems with eutectics are less likely to exhibit zoning since the composition of the solid phases will be constant once the eutectic has been reached.

If solid solutions exhibit significant non-ideality we may expect maxima or minima in the melting point diagrams, as in figure 18.25. Maxima and minima signify upper and lower critical solution temperatures.

A typical melting or crystallisation sequence will lead first to the mixture separating into the two phases at the appropriate phase boundary. When the composition of the liquid phase becomes equal to that at a minimum or maximum, the liquid and solid phases will remain constant until the phase change has been completed. The minimum in figure 18.25(b) is not a eutectic since two solid phases grow at a eutectic; here only one solid phase occurs.

Partial solid solution   Minerals which show total or partial immiscibility are more common than those that show complete solid solution formation. One example is the mixture of albite and potassium feldspar under high water vapour pressure (figure 18.26). This system is complicated by the formation of leucite in the dry melt. The water vapour is present to prevent this formation. This simplified figure shows the lower critical solution temperature at about 70% albite. However, at even lower temperatures in the solid region we see a phase boundary, the **solvus**, dividing the solid solution domain from

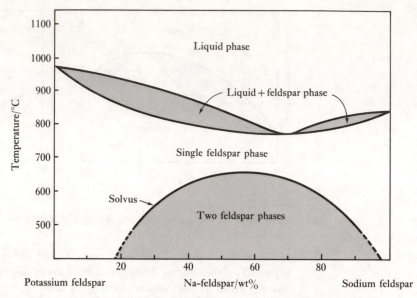

Figure 18.26   Temperature–composition diagram for an alkali-feldspar system (data from Bowen and Tuttle, 1950).

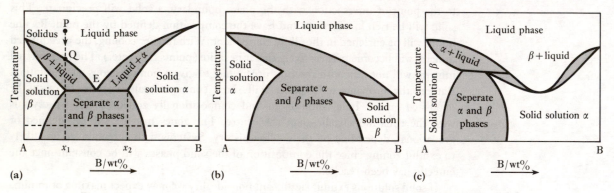

Figure 18.27   Temperature–composition diagrams for partially miscible solids.

a region in which two separate feldspars exist. In this lower region the separate feldspars approach the composition of the end-members as temperature is decreased. The slow kinetics of unmixing lead to intergrowths of crystals of different composition. Occasionally, the cooling rate may be slow enough to allow the formation of separate crystals.

If the solvus for any system occurs at a high enough temperature it may intersect with the liquid phase boundaries. Figure 18.27(a)–(c) shows some typical examples of this behaviour. Figure 18.27(a) depicts a system exhibiting a eutectic point where the progressive cooling of a melt at point P would give rise to the separation of a liquid phase and a solid solution β, the composition of which can be determined from the horizontal tie line at Q to the solidus. Further cooling leads to the composition of both phases changing until the eutectic point E is reached, at which the solid phase would separate into two solid phases of composition x_1 and x_2. As described in section 18.4.2, the temperature and composition will be maintained constant until all the liquid has crystallised. Further cooling leads to progressive changes in the compositions and relative proportions of the two separate phases.

18.4.4 Complex binary systems

The properties of more complex binary systems can usually be understood in terms of combinations of the simple effects. However, there are several other features which are commonly observed.

Reactivity If two minerals A and B are mixed, there may be a reaction to form a mineral product AB.

$$A + B \rightleftharpoons AB \tag{18.153}$$

The product can be described in terms of the two reactant minerals, so the system is still a two-component one. Figure 18.28 shows the hypothetical situation where there is complete immiscibility between the three solid phases.

The condensed phase rule phase dictates that there is no point in temperature–composition space at which all three minerals and a liquid phase can coexist. If two minerals and the mixed liquid phase exist, both temperature and composition must be fixed at a unique value which will be a eutectic. Figure 18.28 shows two eutectics arising from the coexistence of two minerals, either AB + A or AB + B. Such a system can be treated as two separate binary systems, each interpreted in the same way as the simple eutectic mixture in section 18.4.2. An example is a mixture of $CaSiO_3(s)$ (wollastonite) and $CaAl_2O_4(s)$ (calcium aluminate), which react to form $Ca_2Al_2SiO_7(s)$ (gehlenite) as depicted in the insert in figure 18.28. This figure shows the impossibility of the simultaneous crystallisation of wollastonite and calcium aluminate. An additional feature is that pure gehlenite, when heated to 1595 °C, melts to give a liquid of the same composition. This behaviour is termed **congruent melting** and is indicative of a mineral that does not decompose below its melting point. A further example of congruent melting with a double eutectic is the silica/nepheline system where albite [$NaAlSi_3O_8(s)$] is the AB product formed from the reaction $2SiO_2(s) + NaAlSiO_4(s) \rightleftharpoons NaAlSi_3O_8(s)$.

Incongruent melting Some minerals decompose upon melting. For example, in the silica/leucite system potassium feldspar can be thought of as a product of the reaction between silica and leucite ($KAlSi_2O_6(s)$). We consider this as a binary system and expect a double eutectic. However, at temperatures above 1150 °C, K-feldspar decomposes into a silica-rich liquid and leucite by the reaction $KAlSi_3O_8(s) \rightleftharpoons SiO_2(s) + KAlSi_2O_6(s)$. This is called **incongruent melting** since the liquid does not have the same composition as the original solid. The temperature at which incongruent melting occurs must be an invariant point since three phases (silica, K-feldspar and liquid) exist. The condensed phase rule dictates that the temperature and phase compositions must remain constant until the precipitation is complete. This occurs at a **peritectic** point, rather than a eutectic point.

Figure 18.29 is the phase diagram for such a system, where an AB product may form in some regions of temperature–composition space but decomposes before it can melt. The eutectic on the right-hand side is present, and normal eutectic behaviour

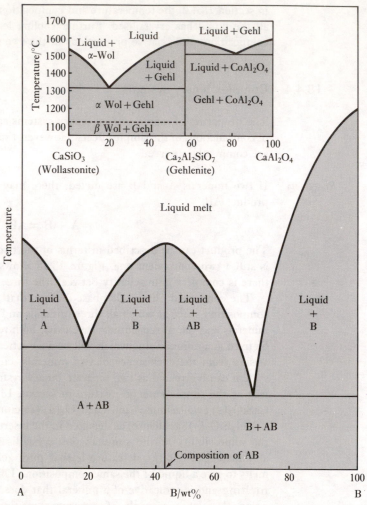

Figure 18.28   Systems with a double eutectic (data from Rankin and Wright, 1915; figure after Ehlers, 1972).

is expected. For example, a liquid melt at point H, if cooled, would progress to point I and precipitate component B. Upon further cooling more B precipitates and the composition of the fluid progresses along the line IE until the eutectic composition is reached at point E. At this point the AB product as well as B will precipitate and a eutectic arrest is observed until all the liquid is gone. The process of melting follows the reverse path.

Considering the left-hand side of the figure we see that the left-hand eutectic side is missing. Consequently, when a melt of composition G cools it initially precipitates solid A at point J. Further cooling leads to point P where the component A is no longer stable and formation of the product AB begins as result of a reaction between component A and the liquid. This is the peritectic point where three phases exist (A, AB and liquid). The consequence of a peritectic point is that liquid mixtures close in composition to the AB solid cannot produce the mineral directly by cooling and crystallisation. The growth of AB must be via precipitation of A and subsequent consumption of A by reaction with the remaining liquid. Since we started at the

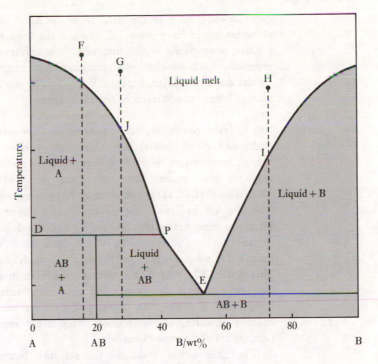

Figure 18.29   Incongruent melting.

point G, all the component A will be consumed, leaving a two–component mixture of AB and the excess liquid not consumed. At this point cooling will continue through a region of stable AB plus liquid until the system solidifies forming solids B and AB. If the original composition of the melt had been at point F, the peritectic point would have been reached via precipitation of A. The following consumption of A would leave an excess of solid A with all the liquid being consumed. Further cooling below the peritectic point is simply cooling of the solid mixture of A plus AB. Only when the composition of the melt is exactly that of the mineral AB will this mineral be the only product of cooling, although its formation will still be preceded by the formation of mineral A.

The precise sequence of mineral formation during cooling can always be determined by drawing vertical lines down from the starting point and noting the minerals encountered before the final temperature is reached. Such lines are called **isopleths**. The phase diagram shows that there are five possible mineral assemblies which may be encountered: pure A, pure B, pure AB, a mixture of A and AB, and finally a mixture of B and AB. Mixtures of A, B and AB cannot exist.

Mineral sequences predicted in this way may be incorrect if reactivity is slow. For example, if a mixture initially at point G cools it should precipitate mineral A, progress to the peritectic point P, redissolve A and never reach the eutectic point. However, if the dissolution of mineral A is slow, or cooling is rapid, there will be an apparent excess of B in the liquid and the system may continue to precipitate the AB mineral. The composition of the liquid may then progress along the line PE to the eutectic point, finally precipitating the AB + B eutectic mixture alongside the undissolved A. Since the liquid phase is now gone, mineral A cannot redissolve to attain the proper equilibrium composition. Two assemblages, of different compositions, have precipitated from the same melt. Thermodynamic arguments alone could not have predicted this and the lever rule may not be used to estimate relative proportions of the minerals. It is common to observe slow redissolution, especially when simultaneous precipitation of a mineral may form a protective layer on the dissolving layer, separating it from the reactive liquid. Quenching to form a glass is often encountered in these situations.

Salt hydrates   Salt hydrates commonly decompose below their melting points into water plus a less hydrated (or anhydrous) salt. Thus, salt hydrates are examples of substances that exhibit incongruent melting. Figure 18.30(a) shows the phase diagram for sodium sulphate plus water. We may think of solid sodium sulphate decahydrate ($Na_2SO_4 . 10H_2O(s)$) as a product of the reaction between water and sodium sulphate. The line AB depicts the depression of the freezing point curve for water. The solid phases existing below this line are ice and the decahydrate. The line BC shows the increase in solubility of the decahydrate with increasing temperature. At the point C a transition occurs and the decahydrate decomposes to form the anhydrous salt plus water. The line CD shows the decrease in solubility of the anhydrous salt as temperature is decreased. The area constrained by DCE and unbounded in temperature is a two-phase region where the system separates into a solid (anhydrous sodium sulphate) and a solution. This is the region where the reaction product (decahydrate) becomes unstable.

The broken line indicates the stability boundary for a metastable hydrate ($Na_2SO_4 . 7H_2O(s)$) which may form if a solution in the composition range BC is cooled too rapidly to allow the formation of the decahydrate.

Salt hydrates rarely exhibit congruent melting. An exception is the ferric chloride/water system depicted in figure 18.30(b). The different hydrates have the formulae $Fe_2Cl_6 . 12H_2O(s)$, $Fe_2Cl_6 . 7H_2O(s)$, $Fe_2Cl_6 . 5H_2O(s)$ and $Fe_2Cl_6 . 4H_2O(s)$. As can be seen from the phase diagram, these are thermodynamically stable, rather

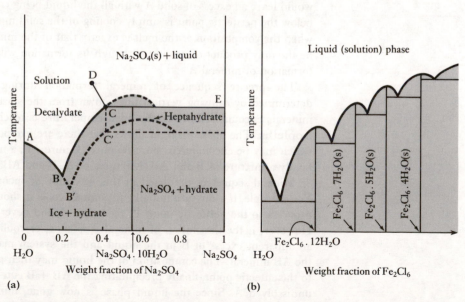

Figure 18.30   Phase diagrams for soluble salt hydrates and water (after Ferguson and Jones, 1973).

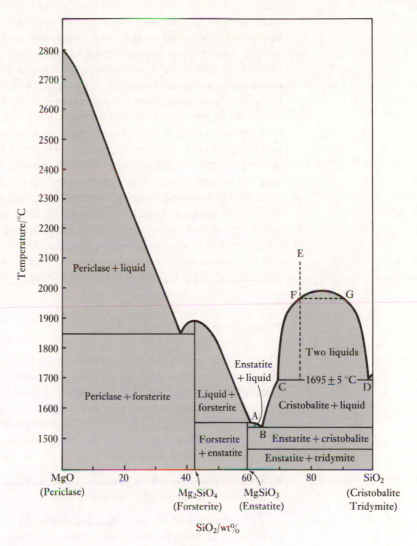

Figure 18.31   Phase diagram for the binary system MgO/SiO_2 showing liquid phase immiscibility (data from Bowen and Anderson, 1914; figure after Ehlers, 1972).

than metastable, phases and each hydrate has a congruent melting point. The peculiar feature of congruent melting is that a given hydrate can have two equilibrium solubilities at any given temperature.

Liquid immiscibility   A liquid melt may show immiscibility in certain temperature–composition ranges, adding complication to the phase diagram. For example, figure 18.31 shows a binary mixture of MgO and SiO_2.

Most features in this figure have been encountered previously. We note that two intermediate compounds, forsterite $(Mg_2SiO_4(s))$ and enstatite $(MgSiO_3(s))$, are formed. Forsterite melts congruently with a eutectic point between it and periclase $(MgO(s))$. Enstatite melts incongruently at 1557 °C to yield forsterite and a mixed liquid phase. Point A is the peritectic and point B the eutectic between enstatite and crystobalite. The additional feature on this diagram is a region of liquid immiscibility bounded by the dome connecting points C and D and the line CD.

The cooling curve through this two-phase region is interesting. If a liquid at point E is cooled, it separates into two fluids of composition F and G upon encountering the phase boundary. We may think of this as the formation of a second silica-rich liquid phase from the original mix. The liquid phases may have a density difference, in which case the two liquids will separate out. Upon further cooling the compositions of the liquids change until the silica-rich phase has the temperature and composition D and the magnesium-rich phase has reached the point C. Below the line CD the two liquid phases become unstable relative to a mixed liquid phase plus the mineral crystobalite. For the temperature to drop below the line CD, the silica-rich phase must therefore dissolve in favour of the new stable phases. The system will maintain a constant temperature until this transition is complete, after which crystobalite precipitates and the liquid composition progresses along the line CB. There will be a eutectic arrest at the point B, after which enstatite and crystobalite will precipitate in eutectic proportions. At about 1470 °C crystobalite undergoes a polymorphic phase transition to form tridymite.

18.4.5 Three-component solid systems

The study of ternary phase diagrams involves no principles other than those for binary mixtures. However, when more components are included in a system the types of behaviour become many and varied. This section therefore gives a description of some of the simplest behaviour and the reader is referred to more comprehensive texts for greater detail (Ehrlers, 1972; Krauskopf, 1967).

The types of behaviour can be separated into two classes, both of which include systems of geological interest.

- Systems with miscible or partially miscible solids.
- Systems with completely immiscible solids.

The properties of the first class are similar to those for liquid/liquid equilibria (see section 12.4 on the general properties of multicomponent equilibria). This class includes systems of great complexity.

Mixtures in the second class show simple binary and ternary eutectic points and will be discussed in some detail. We will consider ternary systems constrained by the condensed phase rule with pressure fixed. The maximum number of degrees of freedom in this system is four; this occurs when only one phase is present. These variables are usually temperature, pressure and the concentrations of two of the components. The concentration of a third component is fixed if the mass fractions or mole fractions of the other two are known.

We recall from section 12.4.2 that a complete representation for a system containing three components requires the intensive variables to be plotted on a three-dimensional graph with an equilateral triangle for its base. Each apex of the triangle denotes a pure component (i.e. A = 100%, B = 100% or C = 100%) and the sides represent the binary limits. Such plots are instructive but often difficult to interpret quantitatively. We will discuss ternary systems using two-dimensional plots of horizontal sections of the prism. Each plot represents the system at fixed temperature and pressure.

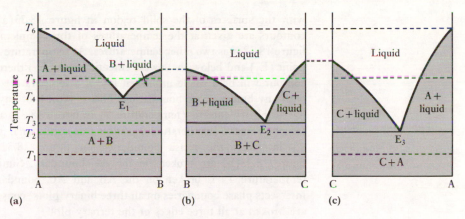

Figure 18.32   Three binary eutectics (after Ferguson and Jones, 1973).

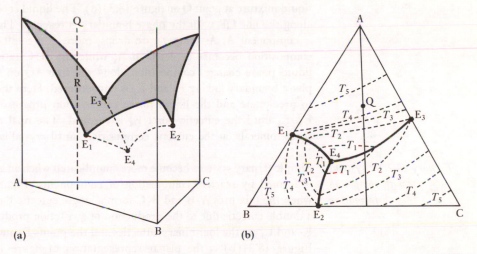

Figure 18.33   A ternary eutectic mixture (after Ferguson and Jones, 1973).

Ternary eutectic systems   The simplest ternary eutectic mixture can be understood in terms of three binary mixtures. Figure 18.32 shows properties of the binary mixtures A/B, A/C and B/C which help to visualise the properties of an A/B/C mixture. The three binary eutectic points E_1, E_2 and E_3 are clear.

The projection in figure 18.33(a) shows the three binary systems defining the limits of the ternary behaviour. The unique feature in figure 18.3 is the ternary eutectic point at the composition E_4. It is clear that the point E_4 represents a mixed liquid phase in contact with the three pure solid phases A, B and C. The condensed phase rule dictates that there are no degrees of freedom, and the point is invariant in temperature–composition space.

Figure 18.33(b) shows the two–dimensional representation of figure 18.33(a). The lines joining E_4 to each of the binary eutectic points are called **cotectic lines**. They are **not** isothermals but represent phase boundaries at which there is one degree of freedom, denoting simultaneous crystallisation of two components. The dotted lines are isothermal contours formed by the intersection of isothermal planes with the surfaces of the solid region in figure 18.33(a). In figure 18.33(b), the contours are given at the temperatures of the complementary binary eutectics in figure 18.32 plus two other temperatures. At temperatures above the highest eutectic point (E_1) and below the lowest eutectic point (E_4) there will be no contours. The contour will be a single unbroken line if a temperature tie line does not intersect with any binary phase boundaries. This is typical of any temperature between the lowest binary eutectic (temperature T_2 in figure 18.32) and the temperature of the ternary eutectic (temperature T_1 fulfils this condition). The tie line at temperature T_3 intersects the phase boundaries on figure 18.32(c) so the contours on figure 18.33(b) are broken on the AC boundary. Similarly, the contour line at temperature T_4 is broken on the AB and AC boundaries. The temperature T_5 intersects phase boundaries on all three binary plots; consequently, the contour plots are broken at all three edges of the ternary plot.

In order to predict the cooling curve for such a system it is helpful to consult both the three-dimensional and planar representations. Consider the cooling of a liquid mixture at point Q on figure 18.33(a). The liquid cools at constant composition along the line QR until the phase boundary is reached. The first solid to precipitate is component A. As temperature drops, more A precipitates and the liquid phase composition becomes depleted in A, while the B:C ratio remains constant. The liquid phase changes composition along the line AQ on figure 18.33(b) until the phase boundary linking E_3 and E_4 is encountered. Here the minerals A and B begin to precipitate and the liquid phase composition progresses along the cotectic line E_3–E_4 until the eutectic point E_4 is reached. The final melt then precipitates all three minerals, at the eutectic composition, until crystallisation is complete.

Ternary eutectic systems with compound formation

Simple ternary systems become more complicated when an additional solid compound is formed by reactive combination of components. Figure 18.34 depicts the case when the mixtures A/B and B/C form a single eutectic but the mixture A/C forms a double eutectic due to the production of a reaction product D. The points E_1, E_2, E_3 and E_4 are the four binary eutectics and the points E_5 and E_6 the ternary eutectics. Figure 18.34(b) is the planar representation of figure 18.34(a). Point U is the composition of compound U on the AC axis. The line BU represents a vertical plane

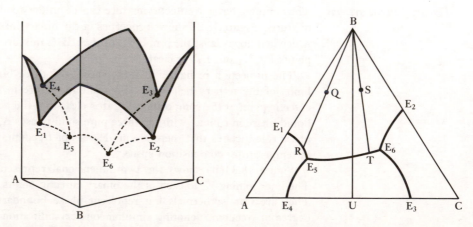

Figure 18.34   A ternary eutectic mixture with a reaction product (after Ferguson and Jones, 1973).

separating the planar diagram into two parts. The cooling curve for any composition is dependent on which side of this boundary the initial composition falls. A liquid at point Q will progress to the point R by precipitating mineral B, and then to the ternary eutectic E_5 by precipitating both B and A. In contrast a liquid at point S will progress to E_6, via point T, precipitating first B, then B and U, on the way.

A further complication arises when the reaction product exhibits incongruent melting, i.e. decomposes before its melting point (Ehrlers, 1972, chapter 3).

Chapter 19

The computer simulation of chemical equilibria

The development of high-speed digital computers has advanced the use of techniques to simulate the progress of chemical reactions. Computer simulations can help us determine which reactions may occur under defined conditions, the extent to which they occur and the expected changes a system will show if perturbed in some way. Computer techniques are especially useful when applied to multicomponent systems containing tens or hundreds of reactive species. For such systems the outcome of the reactions is difficult to estimate by other methods.

19.1 Different types of computer simulation

Commonly, geochemical simulations are from one of the following three categories.

19.1.1 Equilibrium simulations

These are used to predict the **final equilibrium composition of an assembly of reactive chemical species**. Such equilibrium calculations are commonly called **speciation** calculations. Figure 19.1 shows two of the simplest speciation calculations. The first (Figure 19.1(a)) is an acid/base titration curve, where the concentration of H^+ (and OH^-) in a 0.01 molal hydrochloric acid solution is calculated at progressively increasing concentrations of sodium hydroxide. Figure 19.1(b) shows the carbonate species which exist in a 0.025 molal solution of sodium carbonate at different pH values. Here, we see that the major species in acid conditions is carbonic acid (HCO_3^0). As pH is increased the bicarbonate ion concentration increases to a maximum at the expense of carbonic acid, and at very high pH the carbonate ion dominates. Such changes in speciation will affect the solubility of minerals which depends on the concentrations of these ions in solution.

Equilibrium calculation is the most common type of simulation used in geochemical studies. There are several computer codes designed for these purposes, many of which have been described by Nordstrom *et al.* (1979). They have found applications in the study of natural waters (Ohman, 1983; Liden, 1983), brines (Harvie and Wear, 1980; Wood, 1975, 1976), soil solutions (Mattigod and Sposito, 1979; Sposito and Mattigod, 1980) and to many hydrothermal systems. Equilibrium calculations are the main theme of this chapter.

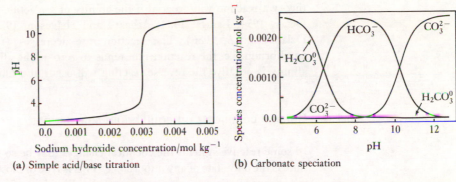

(a) Simple acid/base titration   (b) Carbonate speciation

Figure 19.1   Simple speciation calculations.

19.1.2   Path simulations

Path simulations are designed to predict the sequence of events during a specified reaction, assuming equilibrium is obtained instantaneously for all secondary reactions between the products of the primary reaction. For example, if a specified mineral is dissolving slowly in a multicomponent aqueous fluid we may simulate the equilibrium composition of the system as the extent of reaction for the primary mineral changes. The simulation would show how the secondary minerals and solution species (including pH) change with the extent of dissolution of the specified mineral (Helgeson *et al.*, 1966, 1969; Helgeson, 1968). A practical example of path simulations was given by Helgeson *et al.* (1969), who simulated the dissolution of K-feldspar. The predicted sequence of events, which involved an initial precipitation of gibbsite and kaolinite and a rise in pH followed by the dissolution of kaolinite and the precipitation of K-mica, was consistent with experimental data of Busenberg and Clemency (1976).

A path simulation is like an equilibrium simulation performed several times, with the overall composition of the system changing each time in accordance with the extent of the primary reaction. Several computer codes designed for this purpose are to be found in the literature and have found applications in the study of ore deposition, diagenetic processes, weathering processes, metasomatic processes, evaporation and many hydrothermal reactions (Helgeson *et al.*, 1970; Fritz, 1975, 1981; Parkhurst *et al.*, 1980; Plummer *et al.*, 1983; Reed, 1982). In particular, the EQ3/6 code (Wolery, 1979, 1983) has found extensive use in geochemistry.

19.1.3   Kinetic simulations

A full kinetic simulation predicts the final equilibrium composition of a reactive mixture and the **rate** at which species are consumed or produced on the way to equilibrium. A kinetic simulation is like a path code but with full coupling between the thermodynamic constraints and rate equations for slow reactions. Generalised kinetic codes will describe the dissolution and precipitation of many phases simultaneously (Lichtner, 1985; Lichtner *et al.*, 1983; Helgeson and Murphy, 1983). Usually, the coupling between the rate of reaction and the system thermodynamics is via the affinity of the reaction (section 11.2). In accordance with classical kinetic theory a reaction will proceed if the affinity of a reaction is positive (Prigogine and Defay, 1954; Lasaga, 1984; Aagaard and Helgeson, 1982; Helgeson *et al.*, 1984; Delaney *et al.*, 1986). The reaction rate decreases as the reaction approaches equilibrium, i.e. the reaction rate tends to zero as the affinity tends to zero. If the affinity is negative the reverse reaction will occur at a finite rate.

19.1.4   General comments on equilibrium calculations

In some reactive systems the chemical changes may be so rapid that the equilibrium state is the only state of any importance. This is true of most ionic speciation reactions in aqueous solutions, some reactions at solid/liquid interfaces and some phase changes at high temperature. In these cases it is possible to predict the precise composition and activities of the species present. Composition may then be used to calculate the volume, the density, the compressibility or the total heat capacity of the system. using an equilibrium simulation, the sensitivity of these variables to changes in temperature, pressure or concentration of a selected component may then be evaluated.

In general, reactions that involve the growth or consumption of solid phases are slow compared with reactions that occur within aqueous phases. In the former cases the value of an equilibrium calculation will diminish in favour of a path or kinetic calculation. However, even when the equilibrium state is not achieved rapidly, equilibrium calculations are useful since they define the overall direction of change and show whether a reaction is likely to occur. Equilibrium calculations are sometimes an invaluable aid in designing or interpreting kinetic experiments.

There are two approaches taken for the calculation of equilibrium compositions. The essential feature of these will be described in later sections. One approach involves solving a set of simultaneous non-linear equations which are simple mass balance equations coupled with equilibrium constants. The second approach is a direct minimisation of the Gibbs energy for a system subject to the constraints of conservation of mass, fixed temperature and fixed pressure. The two approaches are equivalent but use different computational schemes.

There are many variations on the two basic themes and the user would normally choose the approach best suited to the problem in question. Consequently, the procedures described here are neither comprehensive nor the best available. Rather, they give an insight into the methods.

19.2   Equilibrium simulation using the equilibrium constant/mass balance method

We may think of a chemical system as composed of a set of species existing in equilibrium. The species may be anything known to exist, such as minerals, gas or liquid molecules, simple ions or stable ligands such as SO_4^{2-} or CO_3^{2-}. Any element in the system can exist in one or more of the species composing the system. We can therefore construct a general mass balance equation for each element which equates the number of moles of the element in the system to the summation of all species containing the element. For a system containing N species this is:

$$T_k = \sum_{i=1}^{N} q_{i,k} n_i \qquad (19.1)$$

where T_k is the total amount of the kth element in the system, n_i is the number of moles of the ith species and $q_{i,k}$ is the number of moles of the kth element in the ith species.

The above equation forms the basis of the general technique which is usually modified for specific applications. The equation need not be applied solely to elemental mass balance. It is equally correct to replace an element with a combination of elements. For example, in a system containing sodium carbonate and liquid water, the species that exist could be liquid H_2O, $Na_2CO_3(s)$ in the solid phase, and the ions Na^+, CO_3^{2-} and $NaCO_3^-$ in the liquid phase. If we choose to neglect liquid water and not to consider hydrolysis species, there are mass balance equations for only Na^+ and CO_3^{2-}. Thus,

$$T_{Na^+} = n_{Na^+} + n_{NaCO_3^-} + 2n_{Na_2CO_3(s)} \qquad (19.2)$$

and

$$T_{CO_3^{2-}} = n_{CO_3^{2-}} + n_{NaCO_3^-} + n_{Na_2CO_3(s)} \qquad (19.3)$$

We will call the formula unit representing the sum of the relevant species a **primary component**. In any equilibrium calculation it is assumed that the values of T_k will be known.

The ultimate objective is to calculate the species concentrations in the mass balance equations. We are treating the species concentrations as unknowns in a set of simultaneous equations. There are generally more species than mass balance equations and it is necessary to introduce more equations to make the problem solvable. To demonstrate the general procedure we will consider the general use of an aqueous electrolyte solution that can dissolve minerals and gases in addition to forming complexes or ion-pairs in the aqueous phase.

19.2.1 Basis set species and mass balance

It is convenient to separate the species into two types. First we select a set of **primary species**, sometimes called **basis set species**. The remaining species, called **secondary species**, are those which arise as a result of reactive combination between the primary species. The distinction between the primary and secondary species is arbitrary and selected for convenience. The only requirement is that the reaction equations for the secondary species contain **only** primary species as reactants. in practice, species which exist at relatively high concentrations or are unlikely to break down into their elements are chosen as primary species. For example, with aqueous solutions containing sodium carbonate we may select Na^+ and CO_3^{2-} rather than Na^+, C and O as primary species. If necessary, the formation of carbon and oxygen could be expressed as a redox reaction of the carbonate ion. Alternatively, if the redox conditions were such that carbon and oxygen were present in large quantities compared with carbonate, it may be more convenient to select C and O as primary species.

If the free (uncomplexed) aqueous ions are chosen to be the primary species, all complexes and solid species can be expressed in terms of reactions between the free ions. For a system containing J free ions including H^+ and OH^-, and I complexes, a general reaction equation can be written for the formation of the ith complex. If we define $v_{i,j}$ as the number of moles of the free hydrated ion c_j in the reaction to form 1 mole of complex p_i, and n_i as the number of moles of water in the reaction, the general reaction equation is:

$$\sum_{j=1}^{J} v_{i,j} c_j + n_i H_2O \rightleftharpoons p_i \qquad (19.4)$$

The constants $v_{i,j}$ and n_i are positive if the species are consumed and negative if produced in the reaction. An equation of the same form is appropriate for the formation of the rth solid in an assembly of n solid phases. The solids present must be pre-determined using the technique described later (section 19.2.4).

By selecting the components to have the same elemental formulae as the basis set species there will be one equation for each basis set species. The general mass balance equation for the kth component has the form:

$$T_k = m_{ck} + \sum_{i=1}^{I} v_{i,k} \, m_{pi} + \sum_{r=1}^{N} v_{s(r,k)} S_r \qquad (19.5)$$

where m_{ck} is the molal concentration of the free hydrated ion, m_{pi} is the molal concentration of the ith complex, S_r is the molal concentration of the rth solid (mol kg^{-1}) and $v_{s(r,k)}$ the appropriate stoichiometry number for the species in the reaction to form the rth solid. Equation 19.4 is similar to equation 19.1 except that there is now a specific link to the reaction equations from which the secondary species arise via the terms $v_{i,j}$. This point will be exploited later.

Note that we have reversed the notation concerning the sign of the stoichiometry numbers from that adopted in section 11.2.1. This ensures that the mass balance equation is a summation of positive quantities. This exercise is purely cosmetic.

19.2.2   The proton condition

The general mass balance equation 19.5 applies to all components except protons or hydroxides. There is no mass balance equation for either H^+ or OH^-. Instead, we use the condition of electroneutrality to define the difference between the free ionic concentrations of these two species. If neither the proton nor hydroxide form complexes or solid phases by reaction with any other species, this difference is simply the total charge imbalance in the system. That is:

$$m_{OH^-} - m_{H^+} = \sum_{k=1(k \neq k(OH), k(H))}^{J} z_k T_k \qquad (19.6)$$

where z_k is the charge formally associated with the component. **The summation over all components $(\sum z_k T_k)$ is the charge imbalance in the entire system excluding protons or hydroxides.** For example, in a solution containing water plus dissolved hydrochloric acid, the summation would be equal to the chloride concentration.

We note that the concentrations of the proton and hydroxide are related by the autoprotolysis of water, i.e.

$$H^+ + OH^- \rightleftharpoons H_2O \qquad (19.7)$$

which has a mass action coefficient of the form

$$K_w^m = 1/(m_{H^+} \times m_{OH^-}) \qquad (19.8)$$

and an equilibrium constant,

$$K_w = a_w/(m_{H^+}\gamma_{H^+} \times m_{OH^-}\gamma_{OH^-}) \qquad (19.9)$$

The equilibrium constant is related to the mass action coefficient by:

$$K_w^m = \frac{K_w \gamma_{H^+} \gamma_{OH^-}}{a_w} \qquad (19.10)$$

It is therefore possible to eliminate either m_{H^+} or m_{OH^-} from equation 19.6 using equation 19.8. Eliminating the proton gives:

$$m_{OH^-} = \frac{1}{m_{OH^-} K_w^m} + \sum_{k=1(k \neq k(OH), k(H))}^{J} z_k T_k \qquad (19.11)$$

In a simple system, with no complex formation in the solution phase, the above equation could be solved for m_{OH^-} for any given charge imbalance. This is the basis of the calculation of a simple acid/base titration curve. In the specific case of electroneutrality ($\sum z_k T_k = 0$), the value of m_{OH^-} is controlled only by the value of K_w^m. At 298.15 K and 1 bar, and at infinite dilution, K_w^m has the value 10^{14}. This leads to value of m_{OH^-} equalling the expected value of 10^{-7} mol kg^{-1}. For neutral solutions at higher ionic strengths, activity coefficient effects change the value of K_w^m and move the equilibrium concentration of m_{OH^-} slightly.

The electroneutrality equation becomes more complicated when complexes or solids are present. If these reaction products consume protons in their formation the hydroxide concentration will increase. If these products consume hydroxide ions the proton concentration will increase. Therefore, on the right-hand side of the charge balance equation we must **subtract 1 mole of charge for every mole of hydroxide consumed by reaction** and **add 1 mole of charge for every mole of protons consumed by reaction**. This can easily be generalised if we note that any reaction of the proton can be written in terms of the hydroxide ion, and vice versa. For convenience we will denote the index number k of the hydroxide as 2 and express all relevant reactions in terms of the hydroxide ion. If there are I complexes in the solution phase and N solids, we may write:

$$m_{c2} = \frac{1}{m_{c2} K_w} + \sum_{k=3}^{J} z_k T_k - \sum_{i=1}^{I} v_{i,2} \, m_{pi} - \sum_{r=1}^{N} v_{s(r,2)} S_r \qquad (19.12)$$

We note that $v_{i,2}$ or $v_{s(r,2)}$ will be zero for any species that does not have the hydroxide ion in its reaction equation.

Finally, we may consolidate equations 19.11 and 19.12 to give:

$$\sum_{k=3}^{J} z_k T_k = m_{c2} - \frac{1}{m_{c2} K_w} + \sum_{i=1}^{I} v_{i,2} \, m_{pi} + \sum_{r=1}^{R} v_{s(r,2)} S_r \qquad (19.13)$$

This has a similar form to the general mass balance equation (equation 19.5).

An important point is that by electing to eliminate explicitly the proton from the expression, we must write all reactions in terms of the hydroxide ion. We are treating the proton as if it were a complex formed from the hydroxide ion.

19.2.3   Consolidating mass balance with the equilibrium constraints

At this point the number of equations from mass balance and the proton condition (equations 19.5 and 19.13) equals the number of primary species; but the number of unknowns equals the primary species plus all secondary species including solids. To determine all unknowns we must generate more equations or reduce the number of unknowns. With this example system we can do both.

Complexes and ion-pairs   The concentrations of the complexes may be eliminated from the mass balance equations using the general thermodynamic equilibrium constant for the reaction. This is formulated as:

$$K_i^a = \frac{m_{pi}\gamma_{pi}}{a_w^{n_i} \prod_{j=1}^{J} (m_{cj}\gamma_{cj})^{v_{i,j}}} \tag{19.14}$$

where m_{cj} and m_{pj} are the molal concentrations of the free ion and the complex respectively, γ is the respective single-ion activity coefficient, a_w is the activity of water and n_i is the number of moles of water in the reaction to form the ith complex.

The general equilibrium constant can be expressed as a mass action quotient in terms of concentrations only, i.e.

$$K_i^n = \frac{m_{pi}}{\sum_{j=1}^{J} m_{cj}^{v_{i,j}}} \tag{19.15}$$

So,

$$K_i^m = \frac{a_w^{n_i} K_i^a}{\gamma_{pi}} \prod_{j=1}^{J} \gamma_{cj}^{v_{i,j}} \tag{19.16}$$

The concentrations of the complexed species (m_{pi}) can now be eliminated from equation 19.5 using equation 19.15 to give **a set of equations whose unknowns are the concentrations of the free ions and the concentrations of the solids**:

$$T_k = m_{ck} + \sum_{i=1}^{I} v_{i,j} K_i^m \prod_{j=1}^{J} m_{cj}^{v_{i,j}} + \sum_{r=1}^{N} v_{s(r,k)} S_r \tag{19.17}$$

If the system contained no solid phases, the number of equations would now equal the number of unknowns and the problem would be soluble. However, for situations where up to N solids are present we must introduce more equations.

Solid phases   The concentrations of the free hydrated ions which react to form the rth solid are controlled by the solubility product for the solid $(K_s)_r$, i.e.

$$(K_s)_r = a_w^{n_r} \prod_{j=1}^{J} (m_{cj}\gamma_{cj})^{v_{s(r,j)}} \tag{19.18}$$

where n_r is the number of moles of water in the reaction to form 1 mole of the rth mineral and $v_{s(r,j)}$ the number of moles of the jth hydrated ion in this reaction. This constant has a complementary mass action quotient of the form:

$$(K_s^m)_r = \prod_{j=1}^{\mathcal{J}} (m_{cj})^{v_{s(r,j)}} \qquad (19.19)$$

The relationship between the mass action coefficient and the solubility product is:

$$(K_s^m)_r = \frac{(K_s)_r}{a_w^{n_r} \prod_{j=1}^{\mathcal{J}} \gamma_{cj}^{v_{s(r,j)}}} \qquad (19.20)$$

The inclusion of one mass action quotient for each solid phase matches the number of equations to the unknowns.

More comprehensive treatments require the consideration of non-ideal mixing in the solid phases using theories of solid solutions (Wolery, 1983).

The proton condition   The proton condition, with complexes eliminated, can be arranged into a form similar to equation 19.17, i.e.

$$\sum_{j=3}^{\mathcal{J}} z_j T_j = m_{c2} + \sum_{i=1}^{I} v_{i,2} K_i^m \prod_{j=2}^{\mathcal{J}} m_{cj}^{v_{i,j}} + \sum_{r=1}^{N} v_{s(r,2)} S_r \qquad (19.21)$$

Using this equation we are treating the proton as a complex formed between OH^- and water and including it in the summation over all complexes. It is equally permissible to denote the hydroxide ion as the complex ion and re-write the above equation in terms of proton concentrations.

The numerical scheme   Assuming ideal behaviour in all phases it is possible to solve the above equations using a technique described in section 19.2.6. However, for non-ideal electrolytes we must correct the mass action quotients using the relevant equatiions for activity coefficients and the activity of water. These corrections are made using equations which require **known** species concentrations. It is these very concentrations we are trying to calculate. A way to overcome this problem is to adopt the following procedure.

1. Estimate ionic strength, activity coefficients and water activity.
2. Calculate mass action quotients using equations 19.16, 19.10 and 19.20.
3. Select the mineral assembly (see section 19.2.6).
4. Solve equations 19.18, 19.17 and 19.21 to yield values of m_{c1} to m_{cj} and S_1 to S_N.
5. Calculate m_{pi} with equation 19.15 using predicted values of m_{c1} to m_{cj} leading to an improved estimate of ionic strength, activity coefficients and water activity.
6. Repeat processes 2–6 until changes in the calculated species concentrations are below a specified value.

This is a simple scheme and there are alternative ways of addressing the problem (Morel and Morgan, 1972; Sposito and Mattigod, 1980; Westall *et al.*, 1976; Wolery, 1983).

19.2.4   Selecting the mineral assembly

The general mass balance equation 19.17 requires the mineral assembly to be specified prior to the numerical solution. There is no universal method for choosing the minerals to be included. The common method of selection is based on equation

19.18, which specifies the relationship between the concentrations of the reacting species for any mineral at equilibrium.

A **saturation index** for the rth possible mineral (SI_r) is defined as the ionic activity product for the reaction to form 1 mole of mineral divided by the solubility product for the reaction, all raised to the power of the reciprocal of the sum of the number of moles of reactant in the reaction, i.e.

$$SI_r = \left[\frac{\prod_{j=1}^{J} (m_j \gamma_j)^{\nu_{s(r,j)}}}{(K_s)_r} \right]^{(1/\sum_{j=1}^{J} \nu_{s(r,j)})} \tag{19.22}$$

$(K_s)_r$ is the solubility product and $\nu_{s(r,j)}$ is the number of moles of component j in the mineral. The quotient is raised to the power $(1/\sum_{j=1}^{J} \nu_{s(r,j)})$ to make reactions of different stoichiometries directly comparable. Clearly, when $(SI)_r$ is unity the aqueous phase is in equilibrium with the mineral; when $(SI)_r$ is greater than unity the solution is oversaturated. This saturation index is a variant of that defined in section 16.1.3.

Prior to solving the set of simultaneous equations, the saturation indices for all possible minerals are estimated and those minerals which are oversaturated are included in the mineral assembly. For oversaturated minerals which contain a common ion, the mineral chosen is the one with the greatest value of $(SI)_r$. This prevents numerical difficulties arising due to unwieldy mineral assemblies. It also prevents violation of the Gibbs phase rule.

The simultaneous equations should be solved with the estimated mineral assembly and saturation factors subsequently re-calculated. If minerals are predicted to have negative concentrations they should be removed from the calculation and the equations re-solved. Similarly, if saturation factors for minerals not included in the assembly are exceeded, these minerals should be added. It is better to add one mineral at a time, starting with the most oversaturated.

19.2.5　Gas solubility

We will consider a simple extension of the above to a solution/mineral assembly which can dissolve a specified gas from an infinitely large reservoir containing the gas at a fixed partial pressure. This is the equivalent of opening the system to an atmosphere of fixed composition. The required prediction is the amount of gas dissolved and its effect on the solution speciation and the mineral assembly.

The chemical constraint on such a system is that for each gas present at equilibrium the concentration of one specific aqueous complex is fixed. For example, carbon dioxide controls the concentration of carbonic acid ($H_2CO_3^0$) via the reaction:

$$CO_2(g) + H_2O \rightleftharpoons H_2CO_3^0 \tag{19.23}$$

The equilibrium constant for this reaction is:

$$K_{CO_2} = \frac{m_{H_2CO_3^0} \gamma_{H_2CO_3^0}}{\hat{p}_{CO_2} a_w} \tag{19.24}$$

where $\hat{p}_{CO_2}$ is the partial pressure of carbon dioxide in the atmosphere. Thus, at constant ionic strength, a fixed value of $\hat{p}_{CO_2}$ fixes the concentration of $H_2CO_3^0$. This in turn influences the total concentration of the primary species from which the specific complex is derived. In this example the carbonate is the primary complex which is modified by the reaction:

$$CO_3^{2-} + 2H_2O \rightleftharpoons H_2CO_3^0 + 2OH^- \qquad (19.25)$$

Consequently, the total quantity of carbonate in the system is the amount which, after speciation, gives rise to the exact quantity of carbonic acid imposed by the gas solubility. This is the condition we are trying to impose.

In this case the total concentration of carbonate now becomes an unknown and we must introduce another equation into the general scheme. This equation is obtained by considering the equilibrium constant for the specified gas (CO_2) and the equilibrium constant linking the fixed complex ($H_2CO_3^0$) to the specified single ion (CO_3^{2-}). We note that in both equations the concentration of the fixed complex is common, a fact that may be used to produce the linking equation.

Generalising for the $\hat{j}$th gas in equilibrium with the $i(\hat{j})$th complex which modifies $T_{k(\hat{j})}$ [the total concentration of the $k(\hat{j})$th component]; the complete equilibrium constraint is:

$$K_{g,\hat{j}} \hat{p}_j a_w^{n\hat{j}} = K_{i(\hat{j})}^a a_w^{ni(\hat{j})} \sum_{j=1}^{J} (m_{cj} \gamma_{cj})^{\nu(i(\hat{j}),j)} \qquad (19.26)$$

where $K_{g,\hat{j}}$ is the equilibrium constant for the gas solubility reaction, $\hat{p}_j$ the partial pressure of the gas and $n_{i(\hat{j})}$ the number of moles of water in this reaction. Both the left- and right-hand sides of this equation equal the concentration of the complex $p_{i(\hat{j})}$. An analogous expression can be formulated in terms of mass action quotients corrected for water and ion activities.

19.2.6 Numerical techniques

The numerical technique involves finding roots of the simultaneous equations by first estimating the roots and then improving the estimations to a required accuracy. The technique most commonly used is Newton's method.

Some comments on Newton's method

Consider a simple function with one dependent variable and one independent variable given by $y = f(x)$, for example:

$$4 = x^3 - 6x \qquad (19.27)$$

The required solution is the value of x that satisfies the equation. A simple graphical solution can be obtained by plotting the function $y = x^3 - 6x$ and selecting the value of x that gives a value of y close to 4. This is equivalent to a point, on the plot of $f(0) = 4 - x^3 + 6x$, that has the value zero. Newton's technique is a way of estimating the roots of the expression $f(0) = 0$. Figure 19.2 shows a plot of $f(0)$ versus x for an arbitrary function. The value of $f(0)$ at an estimate of the solution x' is given by point P. If we calculate of the slope of $f(0)$ at this point we can extrapolate the tangent to the x-axis. The point of intersection is the next-best estimate of the solution to the equation. Repeating the process gives a sequence of improved estimates. We stop this iterative process when $f(0)$ differs from zero by an acceptably small number. In practical calculations the step size for each iteration and the convergence criteria must be considered carefully (Smith and Missen, 1982).

This iterative procedure can be defined algebraically as follows. Consider an estimate of the solution at the point x for which the value of $f(0)_{(x)}$ is not too far

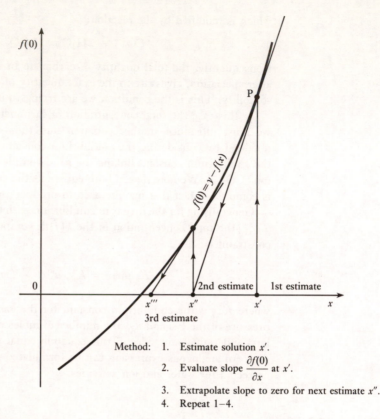

Method:
1. Estimate solution x'.
2. Evaluate slope $\dfrac{\partial f(0)}{\partial x}$ at x'.
3. Extrapolate slope to zero for next estimate x''.
4. Repeat 1–4.

Figure 19.2   Newton's method.

from zero. Let the exact solution, $f(0)_{(x=h)} = 0$, be at $(x + h)$ where h is a small value. From Taylor's expansion (section 8.1.1) we may write:

$$f(0)_{(x+h)} = 0 = f(0)_{(x)} + h\,\frac{\partial f(0)_{(x)}}{\partial x} + \frac{h^2}{2!}\frac{\partial^2 f(0)_{(x)}}{\partial x^2}\ldots \qquad (19.28)$$

Neglecting higher-order terms we have an expression for the value of h in terms of the first derivative of $f(0)$ at the estimated point x, i.e.

$$h \sim -\frac{f(0)_{(x)}}{\partial f(0)_{(x)}/\partial x} \qquad (19.29)$$

The value of h is the correction we must make to x in order to solve the equation. In practice, the estimate of x may not be close to the solution; then the value of $x + h$ is just a better estimate which requires improvement by further applications of Newton's formula.

The approach for one unknown in an expression can be generalised to the problem of N equations with N unknowns. The generalised technique can be used within the numerical scheme described in section 19.2.3 to predict the equilibrium concentrations of all species in a defined system.

Box 19.1

A summary of the simultaneous equations

The set of simultaneous equations for the system composed of a single electrolyte phase (no solids or gases) with J primary species (including the proton and the hydroxide species) are as follows. The proton condition is:

$$f(0)_2 = m_{c2} + \sum_{i=1}^{I} v_{i,k} K_i^m \prod_{j=2}^{J} m_{cj}^{v_{i,j}} + \sum_{r=1}^{N} v_{s(r,k)} S_n - \sum_{j=3}^{J} z_j T_j = 0 \quad (19.30)$$

There are a further $J - 2$ mass balance equations of the form:

$$f(0)_k = m_{ck} + \sum_{i=1}^{I} v_{i,k} K_i^m \prod_{j=2}^{J} m_{cj}^{v_{i,j}} + \sum_{r=1}^{N} v_{s(r,k)} S_n - T_k = 0 \quad (19.31)$$

The introduction of G soluble gases requires G further equations of the form:

$$f(0)_{J+\hat{j}} = K_{i(\hat{j})}^a a_w^{ni(\hat{j})} \prod_{j=2}^{J} (m_{cj} \gamma_{cj})^{v(i(\hat{j}),j)} - (K_g)_{\hat{j}} p_{\hat{j}} a_w^{n\hat{j}} = 0 \quad (19.32)$$

Similarly, the existence of R separate minerals introduces R equations of the form:

$$f(0)_{J+G+r} = \prod_{j=2}^{J} (m_{cj})^{v_{s(r,j)}} - (K_s^m)_r = 0 \quad (19.33)$$

The basis set

The unknowns in equation can be represented by the set x_2 to $x_{\hat{N}}$, where $\hat{N} = J + G + N$; x_2 is equal to the concentration of hydroxide, x_3 to x_J equal the concentrations of the free hydrated ions m_2 to m_J, x_{J+1} to x_{J+G} equal the concentrations of the primary components $T_{k(\hat{j})}$ which become unknowns upon introduction of the $\hat{j}$th gas, and x_{J+G+1} to x_{J+G+R} are the concentrations of the R solids.

The multidimensional Newton–Raphson technique

Equations 19.30–19.33 constitute a set of $\hat{N} - 1$ functions, $f(0)_k$ where $k = 2$ to $\hat{N}$, with $\hat{N} - 1$ unknowns. Taylor's theorem, truncated to the Jacobian (first derivative), gives the relationship between a function $f(0)_k$ evaluated at an estimate to the solution $(x_2 \ldots x_{\hat{N}})$ and the same function $f(0)_k^*$ evaluated at an adjacent point $(x_2^* \ldots x_{\hat{N}}^*)$, i.e.

$$f(0)_k^* = f(0)_k + \sum_{n=2}^{\hat{N}} (x_n^* - x_n) \frac{\partial f(0)_k}{\partial x_n} \quad (19.34)$$

where $\partial f(0)_k / \partial x_n$ is the derivative at the point $(x_2 \ldots x_{\hat{N}})$. We require a solution $(x_2^* \ldots x_{\hat{N}}^*)$ where $f(0)_k^* = 0$, $(k = 2 \ldots \hat{N})$. Therefore,

$$0 = f(0)_k + \sum_{n=2}^{\hat{N}} (x_n^* - x_n) \frac{\partial f(0)}{\partial x_n} \quad (19.35)$$

These can be expressed in matrix form, i.e.

$$\begin{vmatrix} -f(0)_2 \\ -f(0)_3 \\ \vdots \\ -f(0)_{\hat{N}} \end{vmatrix} = \begin{vmatrix} \dfrac{\partial f(0)_2}{\partial x_2} & \dfrac{\partial f(0)_2}{\partial x_3} & \cdots & \dfrac{\partial f(0)_2}{\partial x_{\hat{N}}} \\ \dfrac{\partial f(0)_3}{\partial x_2} & \dfrac{\partial f(0)_3}{\partial x_3} & \cdots & \dfrac{\partial f(0)_3}{\partial x_{\hat{N}}} \\ \vdots & \vdots & & \vdots \\ \dfrac{\partial f(0)_{\hat{N}}}{\partial x_2} & \dfrac{\partial f(0)_{\hat{N}}}{\partial x_3} & \cdots & \dfrac{\partial f(0)_{\hat{N}}}{\partial x_{\hat{N}}} \end{vmatrix} \begin{vmatrix} x_2^* - x_2 \\ x_3^* - x_3 \\ \vdots \\ x_{\hat{N}}^* - x_{\hat{N}} \end{vmatrix}$$

continued

Box 19.1
continued

> This matrix can be solved by a method of Gaussian elimination with back substitution to yield values for $(x_2^* - x_2)$ through to $(x_N^* - x_N)$. These are increments in the values of the basis set components which can be added to the existing values of x_k^* to yield the improved estimates of the roots. The process of evaluating the functions $f(0)$ and the Jacobians to yield the subsequent estimates of the roots is repeated until the incremental changes are less than a prescribed value. This approach gives problems if the diagonal elements of the Jacobian matrix tend to zero. Divisions by zero can be avoided if the Jacobian matrix is pivoted before the step of Gaussian elimination and unpivoted after this step. The technique is described elsewhere (Monro, 1982).

19.2.7   pH and pε as master variables

In many calculations it is convenient to consider the change in speciation as one **master variable** is changed systematically. Both pH and pε can be used as master variables.

The simplest way of using pH as a master variable is to fix the pH at each stage in the numerical scheme, rather than allowing the proton activity to be calculated using the proton condition. This requires only minor modification of the general scheme. The approach is reasonable but it has the drawback that the condition of charge balance, which should be imposed by the proton condition, is contravened. In order to fix pH properly in an equilibrium calculation we should vary the total amount of primary component until we obtain the pH that we desire. For example, in a high–pH solution we may change value of T_k for sodium in order to fine-tune the pH. To build this feature into the numerical scheme we simply make the total concentration of sodium a variable rather than a fixed quantity. This adds no more unknowns to the system since the proton (or hydroxide) concentration is no longer a variable.

It is possible to use the electron activity as a master variable and predict the extent to which a primary component can be reduced or oxidised under conditions of fixed redox potential E_H. The redox potential of a system is formally related to pε by:

$$p\varepsilon = -\log a_e = \frac{FE_H}{2.303RT} \tag{19.36}$$

where a_e is the electron activity, F is the Faraday constant, R the gas constant and T absolute temperature.

Since redox is a process of electron transfer, a general reaction equation can be formulated using the same conventions as for complex reactions (equation 19.4), i.e.

$$\sum_{j=1}^{J} v_{i,j}c_j + n_i H_2O + n_{ei} e^- \rightleftharpoons p_i \tag{19.37}$$

where $v_{i,j}$ is the number of moles of the free hydrated ion c_j in the reaction to form 1 mole of complex p_i, n_i is the number of moles of water in the reaction and n_{ei} is the number of moles of electrons. In this convention **a redox product is treated as if it were simply an aqueous complex**.

The general equilibrium constant is:

$$K_i^a = \frac{m_{pi}\gamma_{pi}}{a_w^{n_i} a_e^{n_{ei}} \prod_{j=1}^{J} (m_{cj}\gamma_{cj})^{v_{i,j}}}.$$
(19.38)

The electron activity is a measure of the tendency of the system to provide electrons rather than an analogue of a solution concentration. Consequently, the electron activity is treated as a master variable and not a basis set component. This results in the mass action quotient being identical to equation 19.15, i.e.

$$K_i^m = \frac{m_{pi}}{\prod_{j=1}^{J} m_{cj}^{v_{i,j}}}$$
(19.39)

We may use the above mass action quotient in the same way as for any other complex or ion-pair. However, the relationship between K^m and K^a becomes

$$K_i^m = \frac{a_w^{n_i} a_e^{n_{ei}} K_i^a}{\gamma_{pi}} \sum_{j=1}^{J} \gamma_{cj}^{v_{i,j}}$$
(19.40)

In simulations of redox reactions at fixed temperature and pressure the electron activity is fixed prior to any calculation. This fixes the term $a_e^{n_{ei}} K_i^a$. Therefore, in the section of the numerical scheme that corrects the mass action quotient for non-ideal behaviour, the term $a_e^{n_{ei}} K_i^a$ replaces the equilibrium constant.

Redox reactions modify pH. For example, the reduction of iron(III) to iron(II) in accordance with the reaction:

$$Fe^{3+} + e^- \rightleftharpoons Fe^{2+}$$
(19.41)

results in a change in the total charge balance in the true solution components since e^- does not physically exist. Consequently, for simulations with fixed pε, the numerical scheme must have the facility to modify continually the charge balance contribution to the proton condition (equation 19.21) at each iterative stage in the solution to the mass balance equations. This is simply another variation on the basic theme described in sections 19.2–19.2.6.

We must always be cautious about simulating redox processes with equilibrium codes since redox reactions are hardly ever in full equilibrium. Additionally, the redox state of geochemical environments can rarely be determined quantitatively.

19.3   Equilibrium simulations using direct minimisation of the Gibbs energy

An alternative method for predicting the distribution of species at equilibrium involves direct minimisation of the Gibbs energy. This method requires an expression for the Gibbs function for the entire system in terms of the composition. The equilibrium composition is the set molal compositions that minimises G at specified temperature and pressure.

19.3.1   The equations of constraint

For a system containing N species, the total Gibbs energy of the system is given by:

$$G = \sum_i n_i \mu_i \tag{19.42}$$

where n_i and μ_i are the numbers of moles and the chemical potential of the ith species. In this case a species may be any substance known to exist: ions, complexes, solids, etc. The approach differs from that in section 19.2 since we do **not** subdivide the species into basis set and secondary species, nor do we distinguish between solids and dissolved species. However, we still retain the idea of J primary components.

The chemical potential of any species is given by $\mu_i = \mu_i^\ominus + RT \ln a_i$. The above two equations give the basic equation for the Gibbs energy:

$$G = \sum_i^N n_i \mu_i^\ominus + \sum_i^N n_i RT \ln a_i \tag{19.43}$$

The second constraint on the system is mass balance, given by equation 19.1 for the kth primary component:

$$T_k = \sum_{i=1}^N q_{i,k} n_i \tag{19.44}$$

where T_k is the total amount of the kth primary component in the system and $q_{i,k}$ is the number of moles of one kth primary component in the ith species. We recall that a primary component may be an element or stable combination of elements such as CO_3^{2-}. In this approach there is no explicit provision for chemical reactions.

The objective is to calculate the n_i values such that G is a minimum, subject to the mass balance constraints. To do this we must first select the species that exist and then introduce an expression for the dependence of the activity of any species on the composition. The example we will consider will be a simple electrolyte solution composed of hydrated ions and complexes. There are no solids present and the activity coefficients of the species are unity. If we replace the amount of any solution species by its molal concentration, and replace the chemical potential by the apparent standard Gibbs energy of formation of the species ($\Delta G_i^\ominus$, section 11.1.4), equations 19.44 and 19.43 become:

$$G = \sum_i^N m_i \Delta G_i^\ominus + \sum_i^N m_i RT \ln m_i \tag{19.45}$$

and

$$T_k = \sum_{i=1}^N q_{i,k} m_i \tag{19.46}$$

The values of G and T_k are now per kilogram of solvent. This step has allowed us to reconcile the concentrations of the species, which are in molal units, with the basic Gibbs energy equation which was expressed in terms of the total number of moles. The replacement of $\mu_i^\ominus$ by $\Delta G_i^\ominus$ is exact, since any differences between them will cancel out in their summations. We recall from chapters 13, 14 and 16 that for water and minerals $\Delta G_i^\ominus$ is the apparent Gibbs energy of formation of the pure substance at any temperature and pressure, whereas for dissolved ions $\Delta G_i^\ominus$ refers to the property of the specified ion in infinitely dilute solution.

19.3.2   Lagrange's method of multipliers

The first stage is to apply Lagrange's method of multipliers. Essentially this is a way of consolidating the basic function and the equations of constraint into a single equation which can be solved.

A maximum or minimum point in a function such as equation 19.45 is called a **stationary point**. Lagrange's method of multipliers states that if a function $u = f(x, y, z)$ is to be stationary subject to a further constraint given by the function $\phi(x, y, z) = 0$, then there exists a number λ such that:

$$\frac{\partial f}{\partial x} + \lambda \frac{\partial \phi}{\partial x} = 0$$

$$\frac{\partial f}{\partial y} + \lambda \frac{\partial \phi}{\partial y} = 0$$

$$\frac{\partial f}{\partial z} + \lambda \frac{\partial \phi}{\partial z} = 0$$

The above equations are sufficient to determine the three unknowns (x, y, z) and λ.

With our simple example we see that the molal concentrations are equivalent to x, y, z, etc. Also, each mass balance equation (19.46) is similar to a function like $\phi(x, y, z) = 0$ and can be multiplied by a constant λ_k (the undetermined multiplier). The resulting equations can be summed over all the primary species in a system to give:

$$\sum_k \lambda_k \left(T_k - \sum_{i=1}^N q_{i,k} m_i \right) = 0 \qquad (19.47)$$

Since this quantity is zero it may be added to the right-hand side of the expression for the Gibbs energy to give:

$$G = \sum_i m_i \Delta G_i^\ominus + \sum_i m_i RT \ln m_i + \sum_k \lambda_k \left(T_k - \sum_{i=1}^N q_{i,k} m_i \right) \qquad (19.48)$$

We now have a single expression. When this function is a minimum the partial derivative of G with respect to any m_i is zero and the mass balance constraint is satisfied, which is the condition we are seeking.

At this point we have two options.

- **Option 1:** Differentiating equation 19.48 with respect to each m_i gives a set of N equations of the form:

$$0 = \Delta G_i^\ominus + RT \log m_i + RT - \sum_k q_{i,k} \lambda_k \qquad (19.49)$$

In addition we have the J mass balance equations given by equation 19.48. The problem is now reduced to a set of $N + J$ non-linear simultaneous equations. The object to select a set of λ and m values which are the roots of the equations. This is most easily performed with a modified Newton–Raphson algorithm, as in section 9.2.6. For some problems of this nature it may even be possible to solve for λ analytically using simple matrix algebra.

This method of Lagrange is the simplest and most elegant way of determining the set of m_i (or n_i), and will work for many problems.

● **Option 2**: A second method involves minimising the function directly using a numerical technique. The methods of function minimisation is an entire subject in itself and there are many techniques available. They are reviewed elsewhere (Press *et al.*, 1986; Smith and Missen, 1982). However, the basic principle involves a method of successive approximation whereby the unknowns (m and λ) are progressively and systematically modified. Each modification is made such that $(\partial G/\partial m_i)$ is reduced after the modification is made. The optimum set of unknowns is obtained when the partial derivative G with respect to any m_i approaches zero.

The details of the above two techniques and the conditions under which they may be applied have been reviewed by Smith and Missen (1982).

The specific problem we addressed made no reference to activity coefficients. These can be included in two ways. The first is to include them in the basic equation for the Gibbs function (equation 19.55). Here each value of m_i would be replaced first by $m_i\gamma_i$, after which the activity coefficient itself would be replaced by an equation expressing its dependence on the set of m_i values. This introduces an algebraic complexity but no new principles. The second method is to replace all m_i with $m_i\gamma_i$ using an estimate for the activity coefficients. The values of m_i may be determined and used to improve the next estimate of the activity coefficients. The process is repeated until the estimates of the activity coefficients are changing by an insignificant amount. This is a crude technique but effective.

If the general technique is to be extended to include mineral assemblies, solids should be included in accordance with the rules for selection in section 9.2.4. The only modifications to the methods in section 19.3 should be to ensure that the concentrations of the minerals, or their activities, are included in a way which maintains dimensional consistency with the mass balance equations.

19.4   Examples of equilibrium calculations

An equilibrium calculation can be used to predict the entire speciation of a mixture of defined elemental composition. For some purposes this may be adequate knowledge. However, a particularly useful application is to predict changes in speciation as an important property of the system is varied. The variable could be pH, $p\varepsilon$, temperature or the total concentration of a substance, perhaps. Figures 19.3–19.8 give illustrative examples of such calculations.

The following calculations were performed using the mass balance/equilibrium constant approach. The procedures involved calculating the concentrations aqueous complexes and, where appropriate, solid phases. The calculated activity coefficients are therefore single-ion activity coefficients rather than stoichiometric coefficients as would be predicted from the Pitzer equations (14.165–14.172).

19.4.1   Calcium sulphate solubility

Figure 19.3(a) depicts the solubility of gypsum in solutions containing different concentrations of sodium chloride at 298.16 K and 1 bar. For such a speciation we

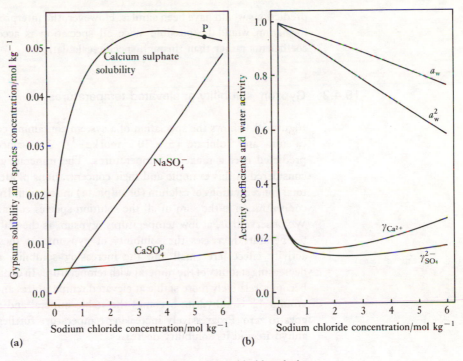

Figure 19.3   The solubility of calcium sulphate in sodium chloride solutions.

would define the total concentrations of calcium and sulphate to be equal and slightly higher than expected solubilities. The thermodynamic data for gypsum, anhydrite and the various solution species can then be used to predict speciation at different total concentrations of NaCl. The simulation is performed once for each total concentration of NaCl considered and the gypsum solubility is the total concentration of either calcium or sulphate in solution after mineral precipitation has been simulated. In figure 19.3(a) the solubility of gypsum rises dramatically with increasing concentrations of NaCl rising to a maximum. Figure 19.3(b) shows other thermodynamic data for the solution species which can be used to interpret the solubility behaviour. At low concentrations of NaCl, the activity coefficients for calcium and sulphate drop dramatically with increasing NaCl concentrations. As activity coefficients fall, the ion concentrations must increase to maintain equilibrium with the solid phase; therefore solubility increases. The increase in solubility is reinforced by existence of the species $NaSO_4^-$, which consumes sulphate in its formation leading to more gypsum being dissolved. Also, the solubility of any mineral consuming water in its precipitation reaction will increase as the water activity decreases. This effect is significant since gypsum has 2 moles of water in its formula and the contribution is from the square of the water activity. At the point P the water activity is so low that the stable mineral phase becomes anhydrite rather than gypsum. At higher NaCl concentrations the water activity has no further effect since anhydrite does not have water in its structure. The slight drop in solubility is explained in terms of the decrease in the activity coefficients of calcium and sulphate at increasing ionic strengths. We see that several effects must be considered to interpret the observed trends.

If the Pitzer equations have been used for such calculations, the solubility predictions would have been similar. However, the interpretation in terms of complex formation would be possible since all speciation is accommodated in the Pitzer coefficients rather than through specific calculation.

19.4.2   Gypsum solubility at elevated temperatures

Figure 19.4 shows the speciation of a system containing equal total concentration of calcium and sulphate ($5 \times 10^{-2}\,\mathrm{mol\,kg^{-1}}$). The concentrations of species are predicted over a range of temperatures. The minerals gypsum and anhydrite are considered in this example and their concentrations plotted along with those for the total concentrations of calcium (or sulphate) in solution. In this case the total solution concentration is the sum of all the solution species containing calcium or sulphate. We observe that, at low temperatures, gypsum is the stable mineral and increasing temperature increases the solubility of gypsum. At these low concentrations the activity effects are small and the increasing solubility of gypsum is due to the decreasing stability of the mineral with temperature. In contrast, the mineral anhydrite becomes relatively more stable at elevated temperatures and at a specific temperature around 42 °C anhydrite becomes the stable mineral and the gypsum concentration goes to zero. Progressively increasing temperature further increases the stability of anhydrite and its solubility decreases.

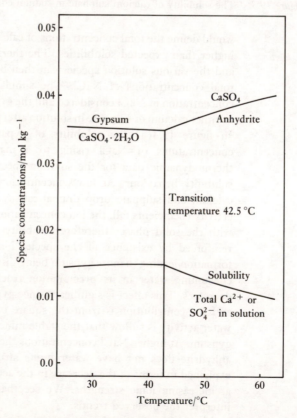

Figure 19.4   The solubility of gypsum at elevated temperatures.

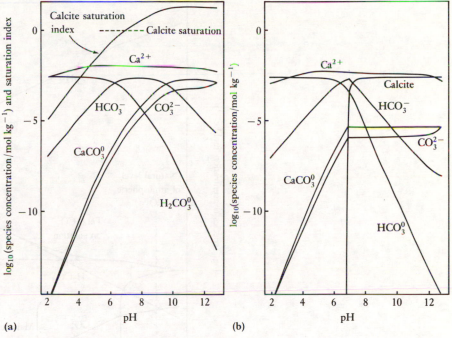

Figure 19.5   The solubility of calcite versus pH.

19.4.3   Calcite solubility as a function of pH

Figure 19.5 shows the pH used as a master variable influencing the solubility of calcite at 298.15 K and 1 bar. The speciation is of a system containing total calcium at 1×10^{-2} mol kg^{-1} and total carbonate at 2.5×10^{-3} mol kg^{-1}. The pH is adjusted in the range 2–12 by suitable additions of HCl or NaOH and the speciation predicted. Figure 19.5(a) shows the saturation factor for calcite versus pH along with the concentrations of various species. At low pH the solution is undersaturated with respect to calcite as a result of the low concentration of carbonate due to hydrolysis. As pH increases, the bicarbonate and carbonic acid concentrations change so that carbonate ion appears in increasing concentrations and the saturation factor for calcite increases. Above pH 6.7, the solution could precipitate calcite; however, in this speciation we have elected to prevent precipitation. Figure 19.5(b) shows a similar simulation where the precipitation of calcite is allowed. At low pH values, figures 19.5(a) and (b) are similar. Above pH 6.8, the differences in solution species reflect the fact that calcite is regulating the total solution concentrations in the second simulation.

19.4.4   Calcite and malachite solubility versus partial pressure of carbon dioxide

Figure 19.6 shows the use of the partial pressure of carbon dioxide being used as a master variable. The example is a simulation of species in a system containing copper(II) at 1×10^{-3} mol kg^{-1}, calcium at 1×10^{-4} mol kg^{-1} and carbonate at 1.1×10^{-3} mol kg^{-1}. This is an electrically neutral system. The minerals considered

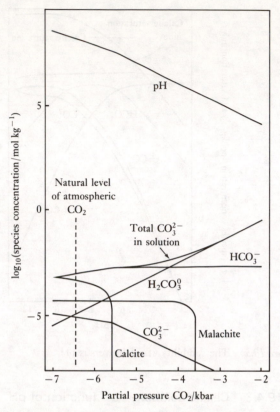

Figure 19.6   Calcite and malachite solubility versus partial pressures of carbon dioxide.

are calcite and malachite, and the speciation is performed with the system open to an infinitely large atmosphere containing carbon dioxide at specified partial pressures. At low partial pressures of CO_2 the system precipitates calcite and malachite due to the initially high ionic concentration of carbonate in solution. As the partial pressure of CO_2 is increased the effect is to drive carbonic acid into solution and reduce the pH by the dissociation of HCO_3^0 into H^+ and HCO_3^-. The consequence of reducing pH is that the carbonate ion is progressively consumed by hydrolysis and the concentrations of both calcite and malachite diminish with the system eventually becoming undersaturated with respect to both minerals. This example is almost counter-intuitive since we would expect to form more of the two carbonate minerals as carbon dioxide is added to the system.

19.4.5   The electron activity as a master variable

We may use the negative logarithm of electron activity ($p\varepsilon$) as a master variable in the same way as pH is used.

Figure 19.7 shows the aqueous concentrations of iron(II) and iron(III) in a system containing 1×10^{-2} mol kg^{-1} Fe^{II}Cl$_2$. As the electron activity is increased from -5 (highly reducing) to $+20$ (highly oxidising) the concentrations of iron(II) and iron(III) change dramatically. The value of $p\varepsilon$ at which the activities of iron(II)

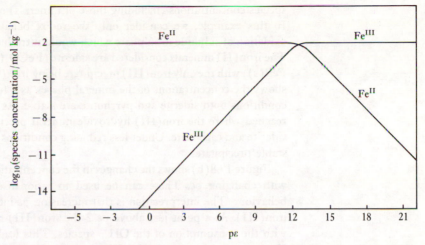

Figure 19.7   Iron(II)/iron(III) equilibria versus pε.

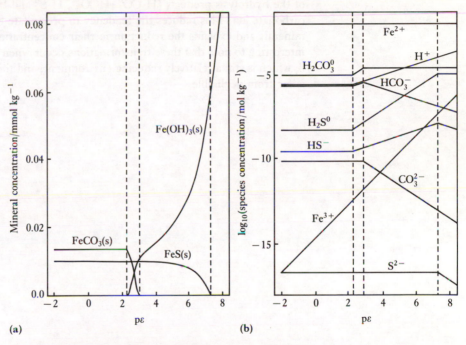

(a)

(b)

Figure 19.8   Mineral equilibria in an iron-containing system versus pε.

and iron(III) are equal is equivalent to the value of $p\varepsilon^{\ominus}$ for the iron(II)/iron(III) redox reaction.

It is clear that the relative stabilities of any minerals containing either form of iron will change with pε. This point is emphasised by figure 19.8, which depicts the speciation of a complicated mixture containing the sulphide ion, the iron species and the carbonate ion. The total concentrations are: sulphide 1×10^{-5} mol kg^{-1}, iron(II) 1×10^{-2} mol kg^{-1}, carbonate 2.5×10^{-5} mol kg^{-1} and sufficient chloride to ensure the system is electrically neutral. The simulation involves calculating the species concentrations (including those of minerals) over a range of fixed $p\varepsilon$ values. In this example, we consider only the redox behaviour of iron and neglect the oxidation of sulphide to sulphate which would introduce another level of complexity. The iron(II) minerals considered are siderite ($FeCO_3(s)$), pyrrhotite (approximately $FeS(s)$) with the only iron(III) precipitate being iron(III) hydroxide. Figure 19.8(a) shows the concentrations of the mineral phases as a function of $p\varepsilon$. Under reducing conditions both siderite and pyrrhotite are stable. As the $p\varepsilon$ is increased, a point is reached where the iron(III) hydroxide mineral precipitates at the expense of both siderite and carbonate. Under less reducing conditions iron(II) hydroxide is the only stable precipitate.

Figure 19.8(b) shows the changes in the concentrations of relevant solution species with changing $p\varepsilon$. These can be used to rationalise the changes in the mineral behaviour. The interpretation is that increasing $p\varepsilon$ increases in the concentration of iron(III). At a point just above $p\varepsilon$ 2 the iron(III) hydroxide starts to precipitate with the consumption of the OH^- species. This leads to a notable increase in the H^+ concentration as a result of the autoprotolysis of water. As the concentration of H^+ increases, the concentrations of both sulphide and carbonate decrease in favour of the hydrolysis products ($H_2CO_3^0$, HCO_3^-, H_2S^0 and HS^-) at the expense of the carbonate ion. This reduces the tendency to precipitate the carbonate and sulphide minerals and explains the reduction in their concentration with increasing $p\varepsilon$. It is interesting to note that these transformations occur when the value of $p\varepsilon$ is less than 8, which is still a relatively reducing environment, and the concentration of iron(II) is far from negligible.

Appendix 1 The effect of ion association on activity coefficients

Equation 14.153 (section 14.3.4) shows the change in the value of the stoichiometric activity coefficient for a salt composed of cation $A_j^{z_j+}$ and anion $A_k^{z_k-}$ due to complex formation between the ions. It is derived as follows.

Consider a solution derived from a single neutral salt where N^c complexes are formed from the reactions:

$$A_j + iA_k \rightleftharpoons A_i^c \qquad (A1.1)$$

where A_i^c is the ith complex. For example, aluminium forms complexes with chloride by the reactions:

$$Al^{3+} + Cl^- \rightleftharpoons AlCl^{2+} \qquad Al^{3+} + 2Cl^- \rightleftharpoons AlCl_2^+$$

$$Al^{3+} + 3Cl^- \rightleftharpoons AlCl_3^0 \qquad Al^{3+} + 4Cl^- \rightleftharpoons AlCl_4^-$$

In the general case, activities can be ascribed to all the solution species, i.e.

$$a_j = \bar{m}_j \bar{\gamma}_j, \ a_k = \bar{m}_k \bar{\gamma}_k \quad \text{and} \quad a_i^c = \bar{m}_i^c \bar{\gamma}_i^c \qquad (A1.2)$$

where $\bar{\gamma}_j$, $\bar{\gamma}_k$ and $\bar{\gamma}_i$ are true single ion activity coefficients as distinct from stoichiometric single ion activity coefficients and $\bar{m}_j$, $\bar{m}_k$ and $\bar{m}_i^c$ denote the true concentrations of the subscripted ions and the complexes respectively. The concentration of the ith complex ($\bar{m}_i^c$) can be related to the total concentration of the salt by:

$$\bar{m}_i^c = v_j \alpha_i m_{jk} \qquad (A1.3)$$

where α_i is the fraction of the concentration of cation $A_j^{z_j+}$ converted to the ith complex. The true concentrations of ions $A_j^{z_j}$ and $A_k^{z_k}$ will be lower than the total concentrations (m_j and m_k) but related to the concentrations of the neutral salt by:

$$\bar{m}_j = v_j \left(1 - \sum_i \alpha_i\right) m_{jk} \quad \text{and} \quad \bar{m}_k = \left(v_k - v_j \sum_i i\alpha_i\right) m_{jk} \qquad (A1.4)$$

Recalling equation 14.13, we note that a similar equation exists which describes the chemical potential of the salt in terms of the true ion concentrations and chemical potentials of the ions **including** complexes, i.e.

$$m_{jk}\mu_{jk} = \bar{m}_j \mu_j + \bar{m}_k \mu_k + \sum_{i=1}^{N^c} \bar{m}_i^c \mu_i^c \qquad (A1.5)$$

Substituting equations A1.4 into A1.5 gives:

$$\mu_{jk} = v_j \left(1 - \sum_i \alpha_i\right) m_{jk} + \bar{m}_k = \left(v_k - v_j \sum_i i\alpha_i\right) m_{jk} + \sum_{i=1}^{N^c} \bar{m}_i^c \mu_i^c \qquad (A1.6)$$

The chemical potential terms for the complexes can be eliminated since at equilibrium $\mu_i = \mu_j + i\mu_k$. Consequently, by introducing general expressions for chemical potential (e.g. $\mu_i = \mu_i^\ominus + RT (\ln \bar{m}_i \bar{\gamma}_i)$ etc. and following the same approach as in the derivation of equations

14.13–14.19 an activity coefficient product can be defined, i.e.

$$\bar{\gamma}_{\pm jk} = (\bar{\gamma}_j^{\nu_j} \bar{\gamma}_k^{\nu_k})^{1/\nu_{jk}} \tag{A1.7}$$

This function is similar to the mean stoichiometric activity coefficient but it refers to the activities of the true ionic species. It is related to the chemical potential for the salt by:

$$\mu_{jk} = \mu_{jk}^{\ominus} + \nu_{jk} RT \ln Q \tag{A1.8}$$

$$Q = m_{jk} \left\{ \left[\nu_j \left(1 - \sum_i \alpha_i\right) \right]^{\nu_j} \left(\nu_k - \nu_j \sum_i i\alpha_i \right)^{\nu_k} \right\}^{1/\nu_{jk}} \gamma_{\pm jk}. \tag{A1.9}$$

The fundamental equation for the chemical potential without explicit reference to ion association is (equation 14.16):

$$\mu_{jk} = \mu_{jk}^{\ominus} + \nu_{jk} RT \ln[m_{jk}(\nu_j^{\nu_j} \nu_k^{\nu_k})^{1/\nu_{jk}} \gamma_{\pm jk}] \tag{A1.10}$$

Since **the chemical potential of the salt must be independent of the way the components are formulated**, the relationship between the mean stoichiometric activity coefficient and the true ionic activity coefficients can be determined from equations A1.8–A1.10, i.e.

$$\log \gamma_{\pm jk} = \log \bar{\gamma}_{\pm jk} + \frac{1}{\nu_{jk}} \log\left(\frac{\bar{m}_j^{\nu_j} \bar{m}_k^{\nu_k}}{m_j^{\nu_j} m_k^{\nu_k}} \right) \tag{A1.11}$$

Appendix 2 Fitting coefficients for activity coefficient expressions

Tables A2.1–A2.12 include a selection of fitting coefficients for the activity coefficient expressions of Pitzer (1973) and Helgeson *et al.* (1981) (sections 14.3.6 and 14.3.5).

Table A2.1   **Crystallographic radii, electrostatic radii and Born interaction parameters for equations 14.183–14.193**

Ion	$r_i^c \times 10^8$	$r_i^{el} \times 10^8$	$\omega_i^{abs} \times 10^{-5}$	Ion	$r_i^c \times 10^8$	$r_i^{el} \times 10^8$	$\omega_i^{abs} \times 10^{-5}$
H^+	2.14	3.08	0.5387	Fe^{3+}	0.64	3.46	4.3186
Li^+	0.68	1.64	1.0249	Al^{3+}	0.51	3.33	4.4872
Na^+	0.97	1.91	0.8692	Au^{3+}	0.90	3.72	4.0169
K^+	1.33	2.27	0.7314	La^{3+}	1.14	3.96	3.7733
Rb^+	1.47	2.41	0.6889	Gd^{3+}	0.97	3.79	3.9426
Cs^+	1.67	2.61	0.6361	In^{3+}	0.81	3.63	4.1164
NH_4^+	1.37	2.31	0.7178	Ga^{3+}	0.62	3.44	4.3437
Tl^+	1.47	2.41	0.6889	Tl^{3+}	0.95	3.77	3.9635
Ag^+	1.26	2.20	0.7547	F^-	1.33	1.33	1.2483
Au^+	1.37	2.31	0.7187	Cl^-	1.81	1.81	0.9173
Cu^+	0.96	1.90	0.8738	Br^-	1.96	1.96	0.8471
Mg^{2+}	0.66	2.54	2.6146	I^-	2.20	2.20	0.7547

Table A2.1 **Continued**

Ion	$r_i^c \times 10^8$	$r_i^{el} \times 10^8$	$\omega_i^{abs} \times 10^{-5}$	Ion	$r_i^c \times 10^8$	$r_i^{el} \times 10^8$	$\omega_i^{abs} \times 10^{-5}$
Sr^{2+}	1.12	3.00	2.2137	OH^-	1.40	1.40	1.1859
Ca^{2+}	0.99	2.87	2.3140	HS^-	1.84	1.84	0.9023
Ba^{2+}	1.34	3.22	2.0625	NO_3^-	2.81	2.81	0.5908
Pb^{2+}	1.20	3.08	2.1562	HCO_3^-	2.10	2.10	0.7906
Zn^{2+}	0.74	2.62	2.5348	HSO_4^-	2.37	2.37	0.7005
Cu^{2+}	0.72	2.60	2.5543	ClO_4^-	3.59	3.59	0.4625
Cd^{2+}	0.97	2.85	2.3302	ReO_4^-	4.23	4.23	0.3925
Hg^{2+}	1.10	2.58	2.2286	SO_4^{2-}	3.15	3.15	2.1083
Fe^{2+}	0.74	2.62	2.5348	CO_3^{2-}	2.81	2.81	2.3634
Mn^{2+}	0.80	2.68	2.4780				

r_i^{el} = relative electrostatic radius/cm; r_i^c = crystallographic radius/cm; ω_i^{abs} = Born interaction parameter/cal mol^{-1}.

Table A2.2 **Short-range interaction parameters β_{ij} for equations 14.183–14.193 at 298.15 K and 1 bar**

Cation	Anion				Cation	Anion			
	F^-	Cl^-	Br^-	I^-		F^-	Cl^-	Br^-	I^-
H^+	−0.083	0.019	0.056		Ca^{2+}	−0.493	−0.257	−0.183	
Li^+	−0.182	−0.049	−0.024		Sr^{2+}	−0.503	−0.266	−0.204	
Na^+	−0.155	−0.096	−0.075		Ba^{2+}	−0.529	−0.292	−0.227	
K^+	−0.132	−0.112	−0.101		Pb^{2+}	−0.554	−0.324	−0.238	
NH_4^+	−0.131	−0.114	−0.107		Zn^{2+}	−0.462	−0.259	−0.186	
Cu^+	−0.153	−0.076	−0.047		Cu^{2+}	−0.485	−0.277	−0.202	
Mg^{2+}	−0.451	−0.256	−0.167		Cd^{2+}	−0.463	−0.256	−0.180	
					Fe^{2+}	−0.414	−0.282	−0.157	
					Mn^{2+}	−0.499	−0.298	−0.210	
					Al^{3+}		−0.481		

Table A2.3 **Solute–solvent interaction coefficients for equations 14.183–14.193**

Solute	$B_n \times 10^6$	Solute	$B_n \times 10^6$	Solute	$B_n \times 10^6$	Solute	$B_n \times 10^6$
HCl	1.4560	$FeCl_2$	1.4565	$LaCl_3$	1.0875	HNO_3	1.1295
NaCl	1.7865	$MnCl_2$	1.4375	HF	1.7870	$NaNO_3$	1.4600
LiCl	1.9421	$PbCl_2$	1.3303	NaF	2.1175	$LiNO_3$	1.6157
KCl	1.6487	$ZnCl_2$	1.4565	KF	1.9797	KNO_3	1.3222
NH_4Cl	1.6351	$CuCl_2$	1.4630	HBr	1.3858	$NaHCO_3$	1.6598
CuCl	1.7911	$CdCl_2$	1.3883	HI	1.2934	H_2SO_4	1.0619
$MgCl_2$	1.4831	$HgCl_2$	1.3544	NaI	1.6239	Na_2SO_4	1.2822
$CaCl_2$	1.3829	$AlCl_3$	1.2065	KI	1.4861	K_2SO_4	1.1904
$SrCl_2$	1.3494	$FeCl_3$	1.1792	KOH	1.9173	$MgSO_4$	1.1807
$BaCl_2$	1.2990	$GdCl_3$	1.1158	NaOH	2.0551	$ZnSO_4$	1.1608

B_n = solute–solvent interaction coefficient for the nth solute/kg mol^{-1}.

Table A2.4   **Extended term parameters for equations 14.194–14.197 at 298.15 K and 1 bar (low-concentration range)**

Solute	β_k	Max. $I/mol\,kg^{-1}$	Maximum Error in $\gamma_\pm$	Solute	β_k	Max. $I/mol\,kg^{-1}$	Maximum Error in $\gamma_\pm$
HCl	0.125	2.0	0.006	FeCl$_2$	0.087	6.0	0.012
NH$_4$Cl	0.020	7.4	0.009	MnCl$_2$	0.074	6.0	0.015
KCl	0.024	5.0	0.012	AlCl$_3$	0.098	6.0	0.032
NaCl	0.064	5.0	0.011	HBr	0.152	1.8	0.005
MgCl$_2$	0.106	6.0	0.015	HI	0.187	2.5	0.003
SrCl$_2$	0.064	3.6	0.006	KOH	0.120	4.5	0.010
CaCl$_2$	0.077	3.0	0.008	NaOH	0.085	4.5	0.004
BaCl$_2$	0.039	3.0	0.003	HNO$_3$	0.062	6.0	0.011

$\beta_k = $ *Extended term parameter* $/kg\,mol^{-1}$ *for kth solute, equivalent to* B_γ *in the original (Helgeson et al., 1981).*

Table A2.5   **Extended term parameters for equations 14.194–14.197 (high-concentration range)**

Solute	β_k	Max. $I/mol\,kg^{-1}$	Maximum Error in $\gamma_\pm$	Solute	β_k	Max. $I/mol\,kg^{-1}$	Maximum Error in $\gamma_\pm$
HCl	0.139	12.0	0.030	AlCl$_3$	0.104	10.8	0.066
LiCl	0.146	12.0	0.036	HBr	0.187	11.0	0.066
MgCl$_2$	0.117	12.0	0.058	HI	0.202	8.0	0.041
SrCl$_2$	0.076	10.5	0.039	KOH	0.123	12.0	0.026
CaCl$_2$	0.093	12.0	0.058	NaOH	0.098	11.0	0.054
BaCl$_2$	0.041	5.4	0.008				

$\beta_k = $ *Extended term parameter* $/kg\,mol^{-1}$ *for kth solute, equivalent to* B_γ *in the original (Helgeson et al., 1981).*

Table A2.6   **Pitzer coefficients for 1:1 interactions**

Species	$\beta(0)_{i,j}$	$\beta(1)_{i,j}$	$C^\phi_{i,j}$	Max. $mol\,kg^{-1}$
HBr	0.1960	0.3564	0.00827	3
HCl	0.1775	0.2945	0.00080	6
HClO$_4$	0.1747	0.2931	0.00819	5.5
HI	0.2362	0.392	0.0011	3
HNO$_3$	0.1119	0.3206	0.0010	3
KBr	0.0569	0.2212	−0.00180	5.5
KCl	0.04835	0.2122	−0.00084	4.8
KClO$_3$	−0.0960	0.2481	—	0.7
KF	0.08089	0.2021	0.00093	2
KOH	0.1298	0.320	0.0041	5.5
NH$_4$Cl	0.0522	0.1918	−0.00301	6
NH$_4$ClO$_4$	−0.0103	−0.0194	—	2
NH$_4$NO$_3$	−0.0154	0.1120	−0.0003	6
NaNO$_3$	0.0068	0.1783	−0.00072	6
NaOH	0.0864	0.253	0.0044	6

From Bradley and Pitzer (1979)

Table A2.7   **Pitzer coefficients for 2:1 interactions (equations 14.165 and 14.166)**

Species	$\frac{4}{3}\beta(0)_{i,j}$	$\frac{4}{3}\beta(1)_{i,j}$	$\frac{2^{5/2}}{5}C^{\phi}_{i,j}$	Max. mol kg^{-1}
$BaCl_2$	0.3504	1.995	−0.03654	1.8
$CaCl_2$	0.4212	2.152	−0.00064	2.5
$Ca(ClO_4)_2$	0.6015	2.342	−0.00943	2
$Ca(NO_3)_2$	0.2811	1.879	−0.03798	2
$CoCl_2$	0.4857	1.967	−0.02869	3
$CuCl_2$	0.3955	1.855	−0.06792	2
$FeCl_2$	0.4479	2.043	−0.01623	2
K_2SO_4	0.0666	1.039	—	0.7
$Mg(NO_3)_2$	0.4895	2.113	−0.03889	2
$MnCl_2$	0.4363	2.067	−0.03865	2.5
$(NH_4)_2SO_4$	0.0545	0.878	−0.00219	5.5
Na_2CO_3	0.2530	1.128	−0.09057	1.5
Na_2SO_4	0.0261	1.484	0.01075	4
$NiCl_2$	0.4639	2.108	−0.00702	2.5
$Pb(NO_3)_2$	−0.0482	0.380	0.01005	2
$ZnCl_2$	0.3469	2.190	−0.1659	1.2
$Zn(NO_3)_2$	0.4641	2.255	−0.02955	2

From Bradley and Pitzer (1979)

Table A2.8   **Pitzer coefficients for 3:1 interactions**

Species	$\frac{3}{2}\beta(0)_{i,j}$	$\frac{3}{2}\beta(1)_{i,j}$	$\frac{3^{5/2}}{2}C^{\phi}_{i,j}$	Max. mol kg^{-1}
$AlCl_3$	1.0490	8.767	0.0071	1.6
$CrCl_3$	1.1046	7.883	−0.1172	1.2
$LaCl_2$	0.8834	8.40	−0.0619	3.9
$La(ClO_4)_3$	1.158	9.80	0.0016	2.0

From Bradley and Pitzer (1979)

Table A2.9   **Pitzer coefficients for 2:2 interactions ($\alpha_1 = 1.4$, $\alpha_2 = 12.0$)**

Species	$\beta(0)_{i,j}$	$\beta(1)_{i,j}$	$\beta(2)_{i,j}$	$C^{\phi}_{i,j}$	Range mol kg^{-1}
$CaSO_4$	0.20	2.65	−55.7	—	0.004−0.011
$CdSO_4$	0.2053	2.617	−48.07	0.0114	0.005−3.5
$CuSO_4$	0.2340	2.527	−48.33	0.0044	0.005−1.4
$MgSO_4$	0.2210	3.343	−37.23	0.0250	0.006−3.0
$MnSO_4$	0.201	2.980	?	0.0182	0.1−4.0
$NiSO_4$	0.1702	2.907	−40.06	0.0366	0.005−2.5
$ZnSO_4$	0.1949	2.883	−32.81	0.0290	0.005−3.5

From Bradley and Pitzer (1979)

Table A2.10   **Temperature derivatives of coefficients for 1:1 salts**

Species	$\left(\dfrac{\partial \beta(0)_{i,j}}{\partial T}\right)_P \times 10^4$	$\left(\dfrac{\partial \beta(1)_{i,j}}{\partial T}\right)_P \times 10^4$	$\left(\dfrac{\partial C^\phi_{i,j}}{\partial T}\right)_P \times 10^5$	Max. mol kg^{-1}
CsCl	8.28	15.0	-12.25	3.0
KF	2.14	5.44	-5.95	5.9
KI	9.914	11.86	-9.44	7.0
HCl	-3.081	1.419	6.213	4.5
HClO$_4$	4.905	19.31	-11.77	6.0
HI	-0.230	8.86	-7.32	6.0
KCl	5.794	10.71	-5.095	4.5

Temperature coefficients based on linear dependence, e.g.

$$\beta(0)_{i,j(T)} = \beta(0)_{i,j(25°C)} + (T - 25)\left(\frac{\partial \beta(0)_{i,j}}{\partial T}\right)_P$$

(Bradley and Pitzer, 1979)

Table A2.11   **Temperature derivatives of coefficients for 2:1 and 1:2 salts**

Species	$\dfrac{3}{2}\left(\dfrac{\partial \beta(0)_{i,j}}{\partial T}\right)_P \times 10^3$	$\dfrac{4}{3}\left(\dfrac{\partial \beta(1)_{i,j}}{\partial T}\right)_P \times 10^3$	$\dfrac{2}{3}\left(\dfrac{\partial C^\phi_{i,j}}{\partial T}\right)_P \times 10^4$	Max. mol kg^{-1}
BaCl$_2$	0.854	4.31	-2.9	1.8
CaCl$_2$	-0.230	5.20	—	0.1
Ca(ClO$_4$)$_2$	1.106	6.77	-5.83	4.0
Ca(NO$_3$)$_2$	0.706	12.25	—	0.1
CuCl$_2$	-3.62	11.3	—	0.6
K$_2$SO$_4$	1.92	8.93	—	0.1
MgCl$_2$	-0.572	4.87	—	0.1
Mg(NO$_3$)$_2$	0.687	5.99	—	0.1
Na$_2$SO$_4$	3.156	7.51	-9.20	3.0

Temperature coefficients based on linear dependence, e.g.

$$\beta(0)_{i,j(T)} = \beta(0)_{i,j(25°C)} + (T - 25)\left(\frac{\partial \beta(0)_{i,j}}{\partial T}\right)_P$$

(Bradley and Pitzer, 1979)

Table A2.12   **Temperature derivatives of coefficients for 3:1 and 2:2 salts**

Species	$\left(\dfrac{\partial \beta(0)_{i,j}}{\partial T}\right)_P \times 10^3$	$\left(\dfrac{\partial \beta(1)_{i,j}}{\partial T}\right)_P \times 10^2$	$\left(\dfrac{\partial \beta(2)_{i,j}}{\partial T}\right)_P \times 10$	$\left(\dfrac{\partial C^\phi_{i,j}}{\partial T}\right)_P \times 10^3$	Max. mol kg^{-1}
CaSO$_4$	—	5.46	-5.15	—	0.02
CdSO$_4$	-2.79	1.71	-5.22	2.61	1.0
CuSO$_4$	-4.4	2.38	-4.73	4.80	1.0
LaCl$_3$	0.253	0.798	—	-0.371	3.6
MgSO$_4$	-0.69	1.53	-2.53	0.523	2.0
ZnSO$_4$	-3.66	2.33	-3.33	3.97	1.0

From Bradley and Pitzer (1979)

The coefficients for the Helgeson equations 14.183 and 14.193, valid at 298.15 K and 1 bar, are:

$$A_\gamma = 0.5091 \text{ kg}^{1/2} \text{ mol}^{-1}$$

$$B_\gamma = 0.3282 \times 10^8 \text{ kg}^{1/2} \text{ mol}^{-1} \text{ cm}^{-1}$$

Appendix 3 Thermodynamic data

It can be argued that the most useful thermodynamic parameters are equilibrium constants for complementary reactions. These can be used to perform numerical simulations of reactivity or to construct graphical representations of equilibria. Some thermodynamic data sources tabulate equilibrium constants; others tabulate the standard energies of formation from which the equilibrium constant can be computed. The chemical and geochemical literature is a rich source of such data for gases, liquids, minerals and aqueous species. There are many sources which may appear to yield conflicting values for some data and the reader is left with the difficult task of selecting the best data for the application. This appendix deals with some of the issues of concern when selecting suitable data.

A3.1   General comments on data sources

To select, data the reader is recommended to use the most recent compilations of critically reviewed data or data produced by well-established retrieval equations. These usually contain reliable data and are themselves extensive literature sources. Useful and recommended compilations are as follows. Particularly useful references produced most recently are printed in **bold type**. Full details are given in Appendix 4.

A3.1.1   Minerals

Robinson *et al.* **(1982)**; Robie *et al.* (1979); Helgeson *et al.* (1978); **Berman, 1988; Stull and Prophet (1971 + supplement); Haas** *et al.* **(1981)**; Johnson *et al.* (1992); CODATA (1978); **Benson and Teague (1982)**; Berman *et al.* (1985); **Berman and Brown, 1985.**

When selecting data for minerals it must be remembered that many data are for pure and highly crystalline phases. In real geological systems there are few perfect mineral phases. many phases will contain impurities and crystal defects or exhibit non-stoichiometry, solid solution formation, and order/disorder phenomena. A notable example is dolomite: the heat capacity for ordered dolomite is less than for disordered dolomite at elevated temperatures. The data collections and techniques given by Helgeson *et al.* (1978), Berman (1988) and Johnson *et al.* (1992) attempt to address such problems.

Kinetic factors should also be considered when selecting data. For example, the solution chemistry of Fe^{3+} is such that the precipitation of the stable phase, $Fe_2O_3(s)$, is slow compared with the metastable phase, ferihydrite. This mineral is a disordered form of iron hydroxide, $Fe(OH)_3(s)$, which is often precipitated preferentially at low temperatures. If only precipitation reactions are being considered, we would not use data for the stable phase $Fe_2O_3(s)$.

A3.1.2   Elements and aqueous species

Johnson *et al.* **(1992)**; CODATA (1978); **Benson and Teague (1982)**; **Phillips (1982)**; **Hogfeldt (1982)**; **Perrin (1979)**; Smith and Martell (1974–1976); Helgeson

(1969);   Shock and Helgeson (1988, 1990); Shock *et al.* (1989, 1991);   Robie *et al.* (1979); Hultgren *et al.* (1973);   Sverjenski *et al.* (1991);   Oelkers and Helgeson (1988).

A3.1.3   Water

Burnham *et al.* (1969);   Pitzer (1983);   Busey and Mesmer (1978);   Fine and Millero (1973);   Fisher and Barnes (1972);   Gregorio and Merlini (1969); Guildner *et al.* (1976);   Helgeson and Kirkham (1974a);   Holloway *et al.* (1971); Keenan *et al.* (1969);   Marshall and Franck (1981);   Olofsson and Hepler (1975); Olofsson and Olofsson (1977, 1981);   Rivkin *et al.* (1978);   Uematsu and Franck (1980);   Johnson *et al.* (1992);   Levelt Sengers *et al.* (1983);   Haar *et al.* (1980);   Johnson and Norton (1991);   Oshry (1949);   Owen *et al.* (1961);   Heger *et al.* (1980);   Gildseth *et al.* (1972).

A3.1.4   Ion and salt activity coefficients

Robinson and Stokes (1959);   Pitzer (1979);   Zematis *et al.* (1980);   Helgeson *et al.* (1981);   Goldberg (1981);   Hamer and Wu (1972);   Horvath (1985).

A3.2   Consistency of thermodynamic data

An important constraint on thermodynamic data is that of **consistency**. Consistent simply means in agreement, and for thermodynamic data we must ensure that the data are in agreement with the following constraints (*CODATA, 1987*).

- Consistency with the laws of thermodynamics.
- Consistency of units.
- Consistency of definitions.
- Consistency of standard states.
- Consistency of common scales and fundamental constants.
- Consistency between different experimental measurements.

A3.2.1   Consistency with laws of thermodynamics

The primary consideration is concerned with consistency between the basic thermodynamic relationships. For example, the standard Gibbs energy change for any reaction at a specified temperature or pressure is related to the standard entropy change and the standard enthalpy for the reaction via the relationship $\Delta G^{\ominus}_{T,P} = \Delta H^{\ominus}_{T,P} - T\Delta S^{\ominus}_{T,P}$. The equilibrium constant for the reaction must be related to the standard Gibbs energy change by $\Delta G^{\ominus}_{T,P} = -RT \ln K$. For example, the reaction to precipitate siderite from solution has the form:

$$Fe^{2+} + CO_3^{2-} \rightleftharpoons FeCO_3(s) \tag{A3.1}$$

At 298.15 K and 1 bar the standard thermodynamic properties are given by:

$$\Delta G^{\ominus} = \Delta H^{\ominus} - T\Delta S^{\ominus} \tag{A3.2}$$

where the standard enthalpy change for the reaction is related to the standard enthalpies of formation by:

$$\Delta H^{\ominus} = \Delta H^{\ominus}_{f,FeCO_3(s)} - \Delta H^{\ominus}_{f,Fe^{2+}} - \Delta H^{\ominus}_{f,CO_3^{2-}} \tag{A3.3}$$

The standard entropy change for the reaction is related to the standard third-law entropies by:

$$\Delta S^{\ominus} = \bar{S}^{\ominus}_{f,FeCO_3} - \bar{S}^{\ominus}_{f,Fe^{2+}} - \bar{S}^{\ominus}_{f,CO_3^{2-}} \tag{A3.4}$$

and the standard Gibbs energy change for the reaction is related to the standard Gibbs energies of formation by:

$$\Delta G^{\ominus} = \Delta G^{\ominus}_{f,FeCO_3} - \Delta G^{\ominus}_{f,Fe^{2+}} - \Delta G^{\ominus}_{f,CO_3^{2-}} \tag{A3.5}$$

Tabulated data for the properties of the reaction, including the equilibrium constant, **must** be consistent with tabulated data for the energies of formation and the third-law entropies. However, this is not enough, since values may be high or low by approximately compensating amounts and still be consistent. Further constraints arise from the partial derivatives of the thermodynamic potentials with respect to temperature and pressure. For example, equations 9.81 and 9.89 show that additional constraints on entropy and enthalpy are imposed by the heat capacity through the relationships:

$$\left(\frac{\partial S}{\partial T}\right)_P = \frac{C_P}{T} \quad \text{or} \quad dS = \frac{C_P}{T} dT \tag{A3.6}$$

$$\left(\frac{\partial G}{\partial T}\right)_P = C_P \quad \text{or} \quad dH = C_P dT \tag{A3.7}$$

Therefore, if we have measured values of C_P at various temperatures, the temperature dependence of the standard enthalpy of formation and the standard entropy are constrained. To impose the above constraints it is common to use integrated equations of state for the temperature dependence of C_P, and the pressure dependence of volume, as described in section 9.2. Datasets based on this principle will be more consistent than datasets which quote values for thermodynamic data which are average values taken from a range of literature sources. However, the integrations still require thermodynamic properties under the reference conditions to be defined, and the use of an equation of state does not itself ensure complete consistency.

The issue is further resolved if we note that thermodynamic data must be consistent **between** reactions as well as **within** specified reactions. For example, the reaction to precipitate calcite is:

$$Ca^{2+} + CO_3^{2-} \rightleftharpoons CaCO_3(s) \tag{A3.8}$$

A consistent dataset would include the standard entropy, the standard heat capacity and the standard energies of formation for all species including the mineral over the specified range of temperature and pressure. However, carbonate may also react with Fe(II) to form siderite in accordance with equation A3.1. Consequently, the thermodynamic properties of carbonate must be consistent with both reactions, and isolated consideration of only one reaction would not ensure this.

Thermodynamic data which have been selected or calculated in a way which ensures thermodynamic consistency between many species are the most reliable; they are then termed **internally consistent**. There are two approaches to the construction of such data. The first involves characterising one set of data precisely, assuming these data are correct, and using the values to estimate data from other reactions. For example, we may calculate a value for the standard Gibbs energy of formation for carbonate which is consistent with the precipitation reaction for $CaCO_3(s)$. From solubility data for siderite and phase equilibrium data we may evaluate a standard Gibbs energy change for the siderite reaction and a standard Gibbs energy of formation for siderite itself. By adopting the previously determined value for the standard Gibbs energy of formation for carbonate it is possible to calculate a value for the standard Gibbs energy of formation for Fe^{2+} which ensures that all data are consistent with both reactions.

When this logic is applied to extensive data collections the major drawback is that compounded experimental errors are transmitted through the calculations. This 'knock-on' effect may generate erroneous results for some species. Modern mathematical programming techniques have been developed to evaluate thermodynamic data **simultaneously** from experimental measurements. The procedures employ error minimisation techniques which average out error over all species being considered (Berman *et al.*, 1986). This approach has the feature that very accurate measurements may be undervalued by the averaging of errors with less accurate measurements. In practice care is taken to place constraints on calculated data which are consistent with the expected experimental uncertainty. The merits of different mathematical techniques for data extraction and compilation are considered in several articles (Robinson *et al.*, 1982; Halbach and Chatergee, 1984; Holland and Powell, 1985; Haas and Fisher, 1976; Berman *et al.*, 1986).

When judging consistency we must always remember that all data have experimental uncertainty. The rules for defining experimental error are given by Topping (1972) and Chatfield (1983) and are discussed in a geochemical context by Nordstrom and Munoz (1986).

A final point is that internal consistency within one data-set does not ensure that data between different data-sets are consistent.

A3.2.2   Consistency of units

In 1960 the international authority on units (the General Conference of Weights and Measures) proposed a rationalised system of metric units called the **International System of Units**, which was abbreviated to **SI** from the French 'Systeme International d'Unites'. In this system there are seven basic units from which all others can be derived. Many authors argue that all scientists in all disciplines should adhere to the SI system since it allows the standardisation of data collection and retrieval and avoids potential conflict. This is a sound principle. However, in geochemistry an extensive amount of useful data was collected prior to the establishment of the system. It is essential therefore that workers are familiar with both the **SI** system and the necessary numerical factors for unit conversions. In addition to maintaining consistency of units between systems, it is essential to maintain consistency within a unit system. For example, we may measure volume in units of cm^3, dm^3 or m^3. However, when using equations of state all volumes should have units of energy per unit pressure. This type of computational error can creep in even when one adheres to a specific system of units.

Tables A3.1–A3.7 will be useful.

Table A3.1   **SI base units**

Base quantity	Name of unit	Abbreviation
Length	Metre	m
Mass	Kilogram	kg
Time	Second	s
Electrical current	Ampere	A
Temperature	Kelvin	K
Amount of substance	Mole	mol
Luminous intensity	Candela	cd

Table A3.2   **Derived SI units and quantities**

Physical quantity	Name of unit	Abbreviation	Relation to fundamental unit and other derived units
Force	Newton	N	$N = kg\ m\ s^{-2}$
Energy	Joule	J	$J = kg\ m^2\ s^{-2} = V\ C$
Pressure	Pascal	Pa	$Pa = kg\ m^{-1}\ s^{-2} = N\ m^{-2}$
Power	Watt	W	$W = kg\ m^2\ s^{-3} = J\ s^{-1} = V\ A$
Amount of electricity	Coulomb	C	$C = A\ s$
Electromotive force	Volt	V	$V = kg\ m^s\ s^{-3}\ A^{-1} = W\ A^{-1} = J\ C^{-1}$
Electrical resistance	Ohm	Ω	$\Omega = V\ A^{-1}$
Volume	Litre or cubic decimetre	l, L or dm^{-3}	$l = 10^{-3}\ m^3$

Table A3.3   **SI prefixes**

Name	Abbreviation	Multiple
Tera-	T	$\times 10^{12}$
Giga-	G	$\times 10^{9}$
Mega-	M	$\times 10^{6}$
Kilo-	k	$\times 10^{3}$
Hecto-	h	$\times 10^{2}$
Deca-	da	$\times 10$
Deci-	d	$\times 10^{-1}$
Centi-	c	$\times 10^{-2}$
Milli-	m	$\times 10^{-3}$
Micro-	μ	$\times 10^{-6}$
Nano-	n	$\times 10^{-9}$
Pico-	p	$\times 10^{-12}$
Femto-	f	$\times 10^{-15}$
Atto-	a	$\times 10^{-18}$

A3.2.3   Consistency of definitions

An example of a potential inconsistency arising from definitions concerns the use of apparent and true energies of formation (section 11.1.4). Data for minerals and ions tabulated by Helgeson *et al.* (1978, 1981) are apparent energies of formation for minerals and ions, whereas Robie *et al.* (1979) tabulate true energy of formation. With the exception of experimental variation these are identical at 298.15 K and 1 bar. As described in section 11.1.4, the temperature and pressure dependence of the true energies of formation includes the terms for changes in the properties of the elements. These are not included in apparent energies of formation since they cancel out in reactions. For example, the temperature and pressure dependence of the true Gibbs energy of formation is given by:

$$\Delta G^{\ominus}_{f,(T,P)} = \Delta G^{\ominus}_{f,(T_r,P_r)} + [\bar{G}_{(T,P)} - \bar{G}_{(T_r,P_r)}] + \sum_i v_i [\bar{G}_{i,(T,P)} - \bar{G}_{i,(T_r,P_r)}] \qquad (A3.9)$$

where $\Delta G_{f,(T,P)}$ is the Gibbs energy of formation at the defined temperature and pressure,

Table A3.4   **Some useful conversion factors**

Energy	Length
$1\,J = 1\,V\,C = 1\,W\,s$	$1\,\text{Ångstrom (Å)} = 10^{-10}\,m$
$\quad = 10\,cm^3\,bar$	$1\,\text{micrometre }(\mu m) = 10^{-6}\,m$
$\quad = 0.2390\,cal\ (calorie)$	$\quad\text{or micron }(\mu)$
$\quad = 10^7\,ergs$	$1\,\text{millimicron }(m\mu) = 10^{-9}\,m$
$\quad = 9.869 \times 10^{-3}\,L\,atm$	$1\,\text{nanometre (nm)} = 10^{-9}\,m$
$\quad = 9.484 \times 10^{-4}\,BTU\ (British\ Thermal\ Units)$	$\quad = 10\,Å$
$\quad = 6.25 \times 10^{18}\,eV\ (electronvolts)$	
$1\,cal = 4.4184\,J$	
$\quad = 41.2929\,cm^3\,atm$	
$1\,eV = 1.6 \times 10^{-19}\,J$	

Pressure	Volume
$1\,Pa = 1\,N\,m^{-2}$	$1\,cm^3 = 0.1\,J\,bar^{-1}$
$1\,bar = 10^5\,Pa$	$\quad = 0.02391\,cal\,bar^{-1}$
$1\,atm = 101325\,Pa$	$1\,m^3 = 1 \times 10^2\,kJ\,bar^{-1}$
$\quad = 1.01325\,bar$	$\quad = 23.91\,cal\,bar^{-1}$

Table A3.5   **Conversion factors and values of the gas constant (R) in various units**

	$cm^3\,g^{-1}$	$cm^3\,mol^{-1}$	$J\,g^{-1}\,bar^{-1}$	$J\,mol^{-1}\,bar^{-1}$	$therm\,cal\,g^{-1}\,bar^{-1}$	$therm\,cal\,mol^{-1}\,bar^{-1}$
$1\,cm^3\,g^{-1} =$	1	18.0153	0.1	1.80153	0.23901	0.430584
$1\,cm^3\,mol^{-1} =$	0.05551	1	0.005551	0.1	0.001327	0.023901
$1\,J\,g^{-1}\,bar^{-1} =$	10	180.1477				
$1\,J\,mol^{-1}\,bar^{-1} =$	0.55508	10				
$1\,therm\,cal\,g^{-1}\,bar^{-1} =$	41.8393	735.579				
$1\,therm\,cal\,mol^{-1}\,bar^{-1} =$	2.32243	41.8393				

	$J\,g^{-1}$	$J\,mol^{-1}$	$therm\,cal\,g^{-1}$	$therm\,cal\,mol^{-1}$
$1\,J\,g^{-1} =$	1	18.0153	0.23901	4.30584
$1\,J\,mol^{-1}$	0.05551	1	0.01327	0.23901
$1\,therm\,cal\,g^{-1} =$	4.1840	75.3760	1	18.0153
$1\,therm\,cal\,mol^{-1} =$	0.23225	4.1840	0.05551	1

R	Units	R	Units	R	Units
0.46151	$J\,g^{-1}\,K^{-1}$	0.110306	$therm\,cal\,g^{-1}\,K^{-1}$	4.6151	$cm^3\,bar\,g^{-1}\,K^{-1}$
8.31424	$J\,mol^{-1}\,K^{-1}$	1.98719	$therm\,cal\,mol^{-1}\,K^{-1}$	83.14241	$cm^3\,bar\,mol^{-1}\,K^{-1}$

The molal units shown in the table are referred to a molecular mass of H_2O equal to $18.0153\,g\,mol^{-1}$, which is consistent with the 1961 table of relative atomic weights based on $^{12}C = 12$ exactly.

Table A3.6   **Values of some important constants**

Constant	Symbol	Value
Molar gas constant	R	$8.314511 \text{ J K}^{-1}\text{ mol}^{-1}$
Avogadro's constant	L	$6.02214 \times 10^{23}\text{ mol}^{-1}$
Boltzmann constant	$k = R/L$	$1.380658 \times 10^{-23}\text{ J K}^{-1}$
Faraday constant	$F = Le$	$9.648531 \times 10^{4}\text{ s A mol}^{-1}$
Charge on an electron (proton)	e	$1.602177 \times 10^{-19}\text{ s A}$
Permittivity of a vacuum	$\varepsilon_0 = 1/c^2\mu_0$	$8.851419 \times 10^{-12}\text{ J}^{-1}\text{ C}^2\text{ m}^{-1} \simeq$ $(1/36\pi) \times 10^{-9}\text{ J}^{-1}\text{ C}^2\text{ m}^{-1}$
Gravitational acceleration	g	9.80665 m s^{-2}
Speed of light	c	$2.99792458 \times 10^{8}\text{ m s}^{-1}$
Permeability of a vacuum	μ_0	$4\pi \times 10^{-7}\text{ J s}^2\text{ C}^{-2}\text{ m}^{-1}$
Pi	π	3.14159265359
Base of natural logarithms	e	2.71828182846
Natural logarithm of x	$\ln x$	$2.302585 \times \log_{10} x$
Ionic product for water at 298.15 K and 1 bar	K_w	$1.008 \times 10^{-14}\text{ mol}^2\text{ kg}^{-2}$
Critical point for water		$647.286 \text{ K and } 22088.0 \text{ kN m}^{-2} \text{ (22.88 bar)}$
Temperature of triple point of water		$273.16 \text{ K and } 611.3 \text{ N m}^{-2} \text{ (0.006113 bar)}$

and $\bar{G}_{i,(T,P)}$ is the Gibbs energy for the ith element. The variable v_i is the number of moles of the ith element in the reaction following the reaction notation in section 11.2.1. The terms $\sum_i v_i[\bar{G}_{i,(T,P)} - \bar{G}_{i,(T_r,P_r)}]$ are those explicitly excluded from the apparent Gibbs energy of formation. Consequently, it is incorrect to use data for ions from Helgeson *et al.* (1981) with data for minerals from Robie *et al.* (1979) at elevated temperatures and pressures. In a similar way the reader should be aware of the differences in the thermodynamic data for water tabulated by Burnham *et al.* (1969) and those data by Helgeson and Kirkham (1974a). This is discussed in chapter 13.

A3.2.4   Consistency of standard states

A potential source of inconsistency arises from the difference between a standard state at 1 bar (1×10^5 Pa) and that 1 atmosphere (1.0135×10^5 Pa or 1.0135 bar). The effect of changing between these standard states has an almost insignificant effect on the thermodynamic properties of condensed phases. For gas phases the corrections are small but greater than for condensed phases. For an ideal gas behaving in accordance with the relationship $\bar{V} = RT/P$ we may deduce that:

$$\left(\frac{\partial \bar{V}}{\partial T}\right)_P = R/P \quad \text{and} \quad \left(\frac{\partial^2 \bar{V}}{\partial T^2}\right)_P = 0 \tag{A3.10}$$

From the standard thermodynamic relationships in section 9.1.10 we see that:

$$\left(\frac{\partial \bar{C}_P}{\partial P}\right)_T = T\left(\frac{\partial^2 \bar{V}}{\partial T^2}\right)_P = 0 \tag{A3.11}$$

$$\left(\frac{\partial \bar{S}}{\partial P}\right)_T = -\left(\frac{\partial \bar{V}}{\partial T}\right)_P = -R/P \tag{A3.12}$$

$$\left(\frac{\partial \bar{H}}{\partial P}\right)_T = \bar{V} - T\left(\frac{\partial \bar{V}}{\partial T}\right)_P = 0 \tag{A3.13}$$

$$\left(\frac{\partial \bar{G}}{\partial P}\right)_T = \bar{V} \tag{A3.14}$$

Table A3.7   **Standard atomic weights of the elements, based on $^{12}C = 12.000$ (definition)**

Atomic number	Name of element	Symbol	Relative atomic mass	Atomic number	Name of element	Symbol	Relative atomic mass
1	Hydrogen	H	1.008_0	53	Iodine	I	126.904 5
2	Helium	He	4.002 60	54	Xenon	Xe	131.30
3	Lithium	Li	6.94_1	55	Caesium	Cs	132.905 5
4	Beryllium	Be	9.012 18	56	Barium	Ba	137.3_4
5	Boron	B	10.81	57	Lanthanum	La	138.905_5
6	Carbon	C	12.011	58	Cerium	Ce	140.12
7	Nitrogen	N	14.006 7	59	Praseodymium	Pr	140.907 7
8	Oxygen	O	15.000_4	60	Neodymium	Nd	144.2_4
9	Fluorine	F	18.998 4	61	Promethium	Pm	—
10	Neon	Ne	20.17_9	62	Samarium	Sm	150.4
11	Sodium	Na	22.9898	63	Europium	Eu	151.96
12	Magnesium	Mg	24.305	64	Gadolinium	Gd	157.2_5
13	Aluminium	Al	6.9815	65	Terbium	Tb	158.925 4
14	Silicon	Si	28.08_6	66	Dysprosium	Dy	162.5_0
15	Phosphorus	P	30.973 8	67	Holmium	Ho	164.930 3
16	Sulphur	S	32.06	68	Erbium	Er	167.2_6
17	Chlorine	Cl	35.453	69	Thulium	Tm	168.934 2
18	Argon	Ar	9.94_8	70	Ytterbium	Yb	173.0_4
19	Potassium	K	39.102	71	Lutetium	Lu	174.97
20	Calcium	Ca	40.08	72	Hafnium	Hf	178.4_9
21	Scandium	Sc	44.955 9	73	Tantalum	Ta	180.947_9
22	Titanium	Ti	47.9_0	74	Tungsten	W	183.8_5
23	Vanadium	V	50.941_4	75	Rhenium	Re	186.2
24	Chromium	Cr	51.996	76	Osmium	Os	190.2
25	Manganese	Mn	54.938 0	77	Iridium	Ir	192.2_2
26	Iron	Fe	55.84_7	78	Platinum	Pt	195.0_9
27	Cobalt	Co	58.933 2	79	Gold	Au	196.966 5
28	Nickel	Ni	58.7_1	80	Mercury	Hg	200.5_9
29	Copper	Cu	63.54_6	81	Thallium	Tl	204.3_7
30	Zinc	Zn	65.3_7	82	Lead	Pb	207.2
31	Gallium	Ga	69.72	83	Bismuth	Bi	208.980 6
32	Germanium	Ge	2.5_9	84	Polonium	Po	(210)
33	Arsenic	As	4.921 6	85	Astatine	At	(210)
34	Selenium	Se	78.9_6	86	Radon	Rn	(222)
35	Bromine	Br	79.904	87	Francium	Fr	(223)
36	Krypton	Kr	83.80	88	Radium	Ra	(226)
37	Rubidium	Rb	85.467_8	89	Actinium	Ac	(227)
38	Strontium	Sr	87.62	90	Thorium	Th	(232)
39	Yttrium	Y	8.905_9	91	Protactinium	Pa	(231)
40	Zirconium	Zr	91.22	92	Uranium	U	(238)
41	Niobium	Nb	92.906 4	93	Neptunium	Np	(237)
42	Molybdenum	Mo	95.9_4	94	Plutonium	Pu	(242)
43	Technetium	Tc	98.906 2	95	Americium	Am	(243)
44	Ruthenium	Ru	101.0_7	96	Curium	Cm	(247)
45	Rhodium	Rh	102.905 5	97	Berkelium	Bk	(247)
46	Palladium	Pd	106.4	98	Californium	Cf	(249)
47	Silver	Ag	107.868	99	Einsteinium	Es	(254)
48	Cadmium	Cd	112.40	100	Fermium	Fm	(257)
49	Indium	In	114.82	101	Mendelevium	Md	(258)
50	Tin	Sn	118.6_9	102	Nobelium	No	(259)
51	Antimony	Sb	121.7_5	103	Lawrencium	Lr	(260)
52	Tellurium	Te	127.6_0				

Number reflects natural isotopic composition of carbon. From Cotton and Wilkinson (1972)

Clearly, for any ideal gas the pressure dependence of $C_P^\ominus$ and $\Delta H_f^\ominus$ is zero and will be insignificant for real gases. The pressure dependence of $\Delta G_f^\ominus$ and $S^\ominus$ can be calculated for an ideal gas using:

$$\Delta(\Delta G_f^\ominus) = \int_{1.0135}^{1.000} \bar{V}\,\mathrm{d}P = RT \int_{1.0135}^{1.000} \frac{1}{P}\,\mathrm{d}P = -37.8\ \mathrm{J\,mol^{-1}} \qquad (A3.15)$$

and

$$\Delta(\bar{S}^\ominus) = -R \int_{1.0135}^{1.000} \frac{1}{P}\,\mathrm{d}P = 0.11\ \mathrm{J\,mol^{-1}\,K^{-1}} \qquad (A3.16)$$

Both the above corrections are small, but not negligible, with the major correction being that for $\Delta G_f^\ominus$. The corrections for non-ideal gases will be of similar magnitudes. For condensed phases the corrections are important only if one of the reference elements is a gas under standard state conditions, oxygen being the most common example.

A more serious example of conflict between standard states is given by the properties of water. The usual standard state for water in electrolyte solutions is the same as the reference state for dissolved ions. This is the infinitely dilute solution at any temperature and pressure. This standard state ensures that the activity of **pure water** is unity under all conditions. An alternative standard state is the **hypothetical ideal gas** at 1 bar pressure and any temperature. In this standard state the gas has unit fugacity ($f^\ominus = 1$); therefore the activity of water and the fugacity are equal at all temperatures and pressures. A third standard state is the pure liquid at 1 bar pressure and any temperature. With this convention $a = f/f^\ominus$, where $f^\ominus$ is equal to the fugacity of water at 1 bar and the specified temperature. The activity is clearly unity in this standard state and no longer equals the fugacity, which has a non-unit value. With this standard state the activity of liquid water is dependent on pressure but independent of temperature. Water activities can be calculated at pressures other than 1 bar using:

$$\ln a = \frac{\Delta G - \Delta G^\ominus}{RT} \qquad (A3.17)$$

where $\Delta G^\ominus$ equals the apparent Gibbs energy of formation at 1 bar and the specified temperature. The point to note is that **we chose the standard state to suit our purposes and change it when appropriate**.

When calculating standard Gibbs energies for chemical reactions it is essential to ensure consistency between the standard states of the reacting species.

For example, if $\Delta G^\ominus$ for a reaction is calculated from the summation of the apparent Gibbs energies of formation for species, which are themselves dependent on T and P, the value of $\Delta G^\ominus$ will change with T and P. This is the normal convention for chemical reactions. Consequently, the activity coefficients for substances must also be consistent with this convention i.e. the standard states for substances must be unconstrained in temperature and pressure in which case the activity coefficients have no temperature and pressure dependence other than through the temperature and pressure dependence of molecular or ionic interactions. For example, Debye–Hückel type expressions for the activity coefficients of ions and the osmotic coefficients of water are based on the standard states being unconstrained in T and P. As stated above, this means that the activity of pure water is unity at all temperatures and pressures, as are the activity coefficients of ions in the hypothetical molar solution. This is consistent with the temperature and pressure dependency of reaction being correctly incorporated in the value of $\Delta G^\ominus$. If, however, we elected to define the standard state for the reacting species to be at 298.15 K and 1 bar then $\Delta G^\ominus$ and the equilibrium constant would have no dependence on T and P. In this case, the temperature and pressure dependence of reaction would be ascribed to the activity coefficients and the usual Debye–Hückel expressions may not be used without modification.

A3.2.5   Some comments on thermodynamic measurements and model dependence

Available thermodynamic data for minerals come from a variety of experimental techniques. The most common source, although not necessarily the most accurate, is direct calorimetric measurement. Calorimetry allows us to measure heat capacities of phases or the enthalpies of appropriate chemical reactions. From these it is possible to determine third-law entropies, Gibbs energies of formation and standard enthalpies of formation at appropriate temperatures and pressures. Such data are usually supported with direct measurements of molar volumes, compressibilities and expansibilities. Some details on such measurements are discussed by Nordstrom and Munoz (1986).

If data for Gibbs energy of formation are available for a mineral and its complementary aqueous ions it is possible to calculate a solubility product for the solubility/dissolution reaction. In principle, the value of the predicted solubility product can be validated by the independent measurement of solubility. Any measurement which is validated in this way has an additional degree of consistency to those previously discussed. Solubility measurements are more precise than calorimetric measurements. However, solubilities must be measured in solutions of finite concentration and the determination of a solubility product requires a knowledge of activity coefficients. For example, the solubility of gypsum is in accordance with the reaction:

$$Ca^{2+} + SO_4^{2-} + 2H_2O \rightleftharpoons CaSO_4 \cdot 2H_2O(s) \tag{A3.18}$$

The equilibrium constant is:

$$K = \frac{1}{a_{Ca^{2+}} \times a_{SO_4^{2-}} \times a_{H_2O}^2} \tag{A3.19}$$

Adopting the convention for stoichiometric activity coefficients $\gamma_\pm$ (section 14.1.3), the equilibrium constant and solubility product K_s are:

$$\frac{1}{K_s} = K = \frac{1}{(m_{Ca^{2+}}\gamma_{Ca^{2+}})(m_{SO_4^{2-}}\gamma_{SO_4^{2-}})a_{H_2O}^2} = \frac{1}{(m_{Ca^{2+}}m_{SO_4^{2-}})\gamma_{\pm CaSO_4}^2 a_{H_2O}^2} \tag{A3.20}$$

The simplest method for calculating the equilibrium constant involves making solubility measurements in the presence of a soluble salt, such as KCl, over a range of ionic strengths in a solution where the only source of Ca^{2+} and SO_4^{2-} is gypsum solubility. In this case the term $(m_{Ca^{2+}}m_{SO_4^{2-}})$ in equation A3.20 can be set equal to the square of the solubility S and equation A3.20 rearranges to:

$$2\log S = \log K_s - 2\gamma_{\pm CaSO_4}. \tag{A3.21}$$

Neglecting water activity, we can replace the activity coefficient term with the Debye–Hückel law ($\log \gamma_{\pm CaSO_4} = -A_\gamma I^{0.5}$) to give:

$$\log S = \tfrac{1}{2}\log K_s + A_\gamma I^{0.5} \tag{A3.22}$$

Thus a plot of $\log s$ versus ionic strength $I^{0.5}$ should give a straight line at low concentrations, with the value of $\log S$ at zero ionic strength equal to $\tfrac{1}{2}\log K_s$. This makes the calculated value K_s dependent on the use of the Debye–Hückel model as well any uncertainty in the extrapolation of solubility measurements to zero ionic strength.

A second model-dependent method involves using a precise expression for activity coefficients. If the solubility of gypsum is measured at a single concentration we may use the Pitzer equation (section 14.3.5) to estimate directly the value of $\gamma_{\pm CaSO_4}$ at the measured electrolyte concentration. In this case the estimated value of K_s may be free of extrapolation errors but will reflect any uncertainty in the activity coefficient model. The solubility of gypsum in pure water is around 1.5×10^{-2} mol kg^{-1} at 298.15 K and at such low concentrations the Pitzer equation will be adequate. If the solubility is measured in a concentrated brine there may be greater uncertainty in the predicted activity coefficient and subsequent error in the calculated equilibrium constant. The situation becomes worse at elevated temperatures and pressures where activity coefficient data are generally less well characterised and activity coefficient models less valid.

A third approach which may yield a different value for k_s involves predicting the true single-ion concentrations and activities at the measured electrolyte concentrations with a chemical speciation model (chapter 19). Equation A3.20 then becomes:

$$\frac{1}{K_s} = K = \frac{1}{(\bar{m}_{Ca^{2+}} \bar{\gamma}_{Ca^{2+}})(\bar{m}_{SO_4^{2-}} \bar{\gamma}_{SO_4^{2-}}) a_{H_2O}^2} \qquad (A3.23)$$

where $\bar{m}_{Ca^{2+}}$ and $\bar{m}_{SO_4^{2-}}$ are the true concentrations of the specified ions. They are, in this case, lower than the total ion concentrations due to formation of the ion-pair $CaSO_4^0$. The true single-ion activity coefficients $\bar{\gamma}_{Ca^{2+}}$ and $\bar{\gamma}_{SO_4^{2-}}$ may be predicted using an expression such as that by Davies (1962) (equation 14.182) and a suitably chosen value for the equilibrium constant for formation of the ion-pair. The mineral solubility equilibrium constant predicted this way will only be consistent with the calculations using the Pitzer equation or the simple extrapolation technique if the expressions for predicting ion and salt activities are themselves thermodynamically consistent. This situation is rarely observed, especially at elevated temperature and pressure.

In the third approach an expression for ion activity coefficient was used along with an assumed constant for ion-pair formation to evaluate a solubility product. The inverse calculation is also possible; i.e. if the solubility product is known, and an expression for single-ion activity coefficients adopted, the value of the equilibrium constant for the ion-pair may be calculated. Such a constant will be consistent with the selected expression for activity coefficients but not necessarily consistent with other expressions which may be found in the literature. In more complex chemical systems where there may be several ion-pairs in solution, the procedure for calculation of all the necessary ion-pair constants is algebraically quite complex but it is entirely possible. Equilibrium constants for ion-pairs can be calculated by a range of other techniques including the analysis of stoichiometric activity coefficients or conductivity measurements. All such techniques have some degree of model dependency.

Appendix 4 References, data sources and further reading

A4.1 References and data sources

Data sources are indicated by an asterisk (*).

Aagaard P and Helgeson H C, 1982, Thermodynamic and kinetic constraints on reaction rates among minerals and aqueous solutions, I. Theoretical considerations. *Am. J. Sci.* **282**: 273

Abbott M M and Van Ness T K, 1989, *Theory and Problems of Thermodynamics*, McGraw-Hill

* **Adams L H**, 1931, Equilibrium in binary systems under pressure: I. An experimental and thermodynamic investigation of the system $NaCl–H_2O$ at $25\,°C$. *J. Am. Chem. Soc.* **53**: 3769–813

Akogi M, Ito E and Navrotski J, 1989, *J. Geophys. Res.* **96**, 2157

Argersinger W J, Davidson A W and Bonner O D, 1950, Thermodynamics and ion exchange phenomena. *Trans. Kansas Acad. Sci.* **53**: 404–10

Barrer R M and Klinowski J, 1974, Ion exchange selectivity and electrolyte concentration, *J. Chem. Soc., Faraday Trans. 1*, **70**: 2080–91

* **Benson L V and Teague L S**, 1982, Lawrence Berkeley Laboratory, University of Berkeley, Berkeley, CA, LBLK-11448

* **Berman R G**, 1988, Internally-consistent thermodynamic data for minerals in the system $Na_2O–K_2O–CaO–MgO–FeO–Fe_2O_3–Al_2O_3–SiO_2–CO_2$, representation, estimation, and high temperature extrapolation. *Contr. Miner. Petrol.* **89**: 168–283

* **Berman R G and Brown T H**, 1985. The heat capacity of minerals in the system $K_2O–Na_2O–CaO–MgO–FeO–F_2O_3–Al_2O_3–SiO_2–H_2O–CO_2$. *J. Petrol.* **29**: 445–522

* **Berman R G, Brown T H and Greenwood H J**, 1985, An internally consistent database for the minerals in the system $Na_2O–K_2O–CaO–MgO–FeO–Fe_2O_3–Al_2O_3–SiO_2–H_2O–CO_2$. Atomic Energy Canada Ltd, Tech. Rep. 377

Berman R G, Brown T H and Perkins E H, 1987, GEO-CALC software for calculation and display of pressure–temperature–composition phase diagrams. *Am. Miner.* **72**: 861–2

Berman R G, Perkins E H, Engi M, Greenwood H J and Brown T H, 1986, Derivation of internally consistent thermodynamic data by the technique of mathematical programming, a review with applications to the system $MgO–SiO_2–H_2O$. *J. Petrol.* **27**: 1331–64

Bernal J D and Fowler R H, 1933, A theory of water and ionic solutions, with particular reference to hydrogen and hydroxyl ions. *J. Chem. Phys.* **1**: 515–48

Bockris J M and Reddy A K N, 1970, *Modern Electrochemistry*, Plenum Press, New York

Bjerrum N 1926, *Kgl. Danske Vid. Selsk., Mat.-Fys. Medd.* **9**: 1–48

Blum A and Lasaga A C, 1988, *Nature* **331**: 413

* **Bottinga Y, Weill D F and Richet P**, 1981, The thermodynamic modelling of silicate melts. In *Thermodynamics in Minerals and Melts*, Newton R C, Navrotski A and Wood B J (eds), Springer-Verlag, Berlin, pp 207–66

Bowen N L, 1913, The melting phenomena of the plagioclase feldspars. *Am. J. Sci.* **35**: 577–99

Bowen N L, 1928, *The Evolution of the Igneous Rocks,* Princeton University Press, Princeton, NJ

Bowen N L and Anderson O 1914, The binary system $MgO-SiO_2$. *Am. J. Sci.* **37**: 487–500

Bowen N L and Tuttle O F, 1950, The system $NaAlSi_3O_8-KAlSi_3O_8-H_2O$. *J. Geol.* **58**: 589–11

*   **Bradley D J and Pitzer K**, 1979, Thermodynamic properties of water and Debye–Huckel parameters to 350 °C and 1 kbar. *J. Phys. Chem.* **83**: 1599–603

*   **Burnham C W, Holloway J R and Davis N F**, 1969, Thermodynamic properties of water to 1000 °C and 10 000 bars. *Geol. Soc. Am. Spec. Paper* **132**: 96

Busenberg E and Clemency C U, 1976, The dissolution kinetics of feldspars 25 °C and 1 atm CO_2 partial pressure. *Geochim. Cosmochim. Acta,* **40**: 41–9

*   **Busey, R H and Mesmer R E**, 1978, Thermodynamic quantities for the ionisation of water in sodium chloride media to 300 °C. *J. Chem. Eng. Data* **23**: 175–6

*   **CODATA**, 1978, CODATA recommended key values for thermodynamics 1977, *CODATA Bulletin* **28**

CODATA, 1982, A systematic approach to the preparation of thermodynamic tables. *CODATA Bulletin* **47**

Chu S-Y and Sposito G, 1981, The thermodynamics of ternary cation exchange and the subregular model. *J. Soil. Sci. Soc. Am.* **45**: 1084–98

Cotton F A and Wilkinson G, 1972, *Advanced Inorganic Chemistry – A Comprehensive Test,* 3rd edn, John Wiley and Sons, New York

Davies C W, 1962, *Ion Association,* Butterworths, London

Debye P and Hückel E, 1923, The theory of electrolytes I. Lowering of freezing point and related phenomena. *Zeitschr. Physik.* **24**: 185–206

Delaney J M, Puigdomenech I and Wolery T J, 1986, Precipitation kinetics option for the EQ6 geochemical reaction path code. Lawrence Livermore National Laboratory, Livermore, CA, UCRL-53642

Ehlers E G, 1972, *The Interpretation of Geological Phase Diagrams,* Dover Publications, New York

Eisenberg D and Kauzmann W, 1969, *The Structure and Properties of Water,* Clarendon Press, Oxford

Elprince A M and Babcock K L, 1975. Prediction of ion-exchange equilibria in aqueous systems with more than two counter-ions. *Soil Sci.* **120**: 332–8

Eugster H P, Albee A L, Bence A E, Thompson J B Jr and Waldbaum D R, 1972, The two phase region and excess mixing properties of paragonite–muscovite crystalline solutions. *J. Petrol.* **13**: 147–79

Everett D H, 1959, *An Introduction to The Study of Chemical Thermodynamics,* Longman, Harlow

Fei Y, Saxena S K and Navrotski J, 1989, *J. Geophys. Res.* **96**: 6913

*   **Fine R A and Millero F J**, 1973, Compressibility of water as a function of temperature and pressure. *J. Chem. Phys.* **59**: 5529–36

*   **Fisher J R and Barnes H L**, 1972, The ion product of water to 350 °C, *J. Phys. Chem.* **76**: 90–9

Fletcher P and Sposito G, 1989, The chemical modelling of clay/electrolyte interactions for montmorillonite. *Clay Miner.* **24**: 375–91

Fletcher P and Townsend R P, 1980, Exchange of hydrated and amminated silver(I) ions in synthetic zeolites X,Y and mordenite. *J. Chromatography* **201**: 93–105

Fletcher P and Townsend R P, 1985, Ion exchange in zeolites. The exchange of cadmium and calcium in sodium X using different anionic backgrounds. *J. Chem. Soc., Farad. Trans.* **81**: 1731–44

Fletcher P, Franklin K R and Townsend R P, 1984. The thermodynamics of binary and ternary ion exchange in zeolites: the exchange of sodium, ammonium and potassium ions in mordenite. *Phil. Trans. Roy. Soc. A* **312**: 141–78

*Friedman H L and Krishnan C V, 1973, *Water: A Comprehensive Treatise*, Vol. 3, chapter 1, Franks E (ed), Plenum Press, New York. Detailed, with extensive data on properties of ions and electrolytes

Fritz B, 1975, Thermodynamic study and simulation of the reactions between minerals and solutions. Application to the geochemistry of alteration and of continental waters. EngD. thesis, University of Strasbourg. *Sci. Geol. Mem.* **41**

Fritz B, 1981, Thermodynamic study and simulation of hydrothermal and diagenetic reactions. PhD thesis, University of Strasbourg. *Sci. Geol. Mem.* **65**

Gaines G L and Thomas H C, 1951, Adsorption studies on clay minerals. II. A formulation of the thermodynamics of exchange adsorption. *J. Chem. Phys.* **21**: 714–18

Garrels R M and Christ C L, 1966, *Solutions, Minerals and Equilibria* Freeman Cooper, San Francisco

Gast R G, 1969, Standard free energies of exchange for alkali metal cations on Wyoming bentonite. *Soil Sci. Soc. Am. Proc.* **33**: 37

*Gildseth W, Habenschuss A, Spedding F H, 1972, Precision measurements of densities and thermal dilation of water between 5 °C and 80 °C. *J. Chem. Eng. Data* **17**: 402–9

*Goldberg R N, 1981, Evaluated activity and osmotic coefficients for aqueous solutions: thirty-six uni–bivalent electrolytes. *J. Phys. Chem. Ref. Data.* **10**

*Gouq-Jen Su, 1946, *Ind. Eng. Chem.* **38**: P 803. Extensive data on compressibilities of gases

*Grant S and Fletcher P, 1993, Chemical thermodynamics of cation exchange: theoretical and practical considerations. In *Ion Exchange and Solvent Extraction*, vol 11, Marinski J A and Marcus Y (eds), Marcel Decker, York p. 1–108

*Gregorio P and Merlini C, 1969, Thermodynamic properties of water in the critical region, *Thermotechnica*, **23**: 41–54

Guggenheim E A, 1935, The specific thermodynamics properties of aqueous solutions of strong electrolytes. *Phil. Mag.* **19**: 588–643

Guggenheim E A and Stokes R H, 1969, *Equilibrium Properties of Aqueous Strong Electrolytes.* Pergamon Press, New York

*Guildner L A, Johnson D P and Jones F E, 1976, Vapour pressure of water at its triple point: highly accurate value. *Science* **191**: 1261

*Haar D L, Gallagher J S and Kell G S, 1980, Thermodynamic properties for fluid water. In *Water and Steam: Their Properties and Current Industrial Applications*, Straub J and Scheffler K (eds), Pergamon, New York, pp. 69–82

Haas J L Jr and Fisher J R, 1976, Simultaneous evaluation and correlation of thermodynamic data, *Am. J. Sci.* **276**: 525–45

*Haas J L Jr, Robinson G R Jr and Hemingway B S, 1981, Thermodynamic tabulations for selected phases in the system $CaO–Al_2O_3–SiO_2–H_2O$ at 101325 kPa (1 atm) between 273.15 and 1800 K. *J. Phys. Chem. Ref. Data*, **10**: 575–669

Halbach H and Chatergee N D, 1984, An internally consistent set of thermodynamic data for twenty-one $CaO–Al_2O_3–SiO_2–H_2O$ phases by linear parametric programming. *Contr. Miner. Petrol.* **88**: 14–23

*Hammer W J and Wu Y-C, 1972, Osmotic coefficients and mean activity coefficients of uni–univalent electrolytes in water at 25 °C. *J. Phys. Chem. Ref. Data.* **1**: 1047–99

Harvie C E and Weare J H, 1980, The prediction of mineral solubilities in natural waters: the $Na–K–Mg–Ca–Cl–SO_4–H_2O$ system from zero to high concentration at 25 °C. *Geochim. Cosmochim. Acta*, **44**: 981–97

Harvie C, Eugster H P and Wear P, *Geochim. Cosmochim. Acta*, **46**: 1603

*Heger K, Uematsu M and Franck E U, 1980, The static dielectric constant of water at high pressures and temperatures up to 500 MPa and 550 °C. *Ber. Bunsenges. Phys. Chem.* **84**: 758–62

* **Helgeson H C**, 1969, Thermodynamics of hydrothermal systems at elevated temperatures and pressures. *Am. J. Sci.* **267**: 729–804

Helgeson H C, 1985, Some thermodynamic aspects of geochemistry. *Pure Appl. Chem.* **57**: 31–44

Helgeson H C, 1968, Evaluation of irreversible reactions in geochemical processes involving minerals and aqueous solutions, I. Thermodynamic relations. *Geochim. Cosmochim. Acta* **33**: 853

* **Helgeson H C and Kirkham D H**, 1974a, Theoretical prediction of the thermodynamic behaviour of aqueous electrolytes at high pressures and temperatures: I. Summary of the thermodynamic/electrostatic properties of the solvent. *Am. J. Sci.* **274**: 1089–198

* **Helgeson H C and Kirkham D H**, 1974b, Theoretical prediction of the thermodynamic behaviour of aqueous electrolytes at high pressures and temperatures: II. Debye–Huckel parameters for activity coefficients and relative partial molal properties. *Am. J. Sci.* **274**: 1199–261

* **Helgeson H C and Kirkham D H**, 1976, Theoretical prediction for the thermodynamic behaviour of aqueous electrolytes at high pressures and temperatures: III. Equation of state for aqueous species at infinite dilution. *Am. J. Sci.* **276**: 97–240

Helgeson H C and Murphy W M, 1983, Calculation of mass transfer among minerals and aqueous solutions as a function of time and surface area in geochemical processes, I. Computational approach. *Math. Geol.* **15**: 109

Helgeson H C, Garrels R M and MacKenzie F T, 1966, Evaluation of irreversible reactions in geochemical processes involving minerals and aqueous solutions. *1966 Annual Meeting Geol. Soc. Am.* P 92

Helgeson H C, Garrels R M and MacKenzie F T, 1969, Evaluation of irreversible reactions in geochemical processes involving minerals and aqueous solutions, II. Applications. *Geochim. Cosmochim. Acta* **34**: 455–81

Helgeson H C, Brown T H, Nigrini A and Jones T A, 1970, Calculation of mass transfer in geochemical processes involving aqueous solutions. *Geochim. Cosmochim. Acta* **34**: 569–92

* **Helgeson H C, Delany J M, Nesbitt H W and Bird D K**, 1978, Summary and critique of the thermodynamic properties of rock-forming minerals. *Am. J. Sci.* **278A**: 1–299

* **Helgeson H C, Kirkham D H and Flowers G C**, 1981, Theoretical prediction of the thermodynamic behaviour of aqueous electrolytes at high pressures and temperatures: IV. Calculation of activity coefficients, osmotic coefficients, and apparent molal and standard and relative molal partial molal properties to 600 °C and 5 kbar. *Am. J. Sci.* **281**: 1249–516

Helgeson H C, Murphy W M and Aagard P, 1984, Thermodynamic and kinetic constraints on reaction rates among minerals and aqueous solutions, II. Rate constants, effective surface area and the hydrolysis of feldspar. *Geochim. Cosmochim. Acta* **48**: 2405

Hess P C, 1966, Phase equilibria of some minerals in the $K_2O–Na_2O–Al_2O_3–Si_2O–H_2O$ system at 25 °C and 1 atmosphere. *Am. J. Sci.* **264**: 289–309

* **Hogfeldt E**, 1982, **Stability Constants of Metal-Ion Complexes**, Part A, Inorganic ligands, IUPAC Chemical Data Series, No. 21, Pergamon Press, Oxford

Holland T J B and Powell R, 1985, An internally consistent thermodynamic dataset with uncertainties and correlations: 2 Data and results. *J. Metamorph. Geol.* **3**: 343–70

* **Holloway J R, Eggler D H and Davis N F**, 1971, Analytical expression for calculating the fugacity and free energy of H_2O to 10 000 bars and 1300 °C. *Geol. Soc. Am. Bull.* **82**: 2639–42

Horvath A L, 1985, *Handbook of Aqueous Electrolyte Solutions*, Ellis Horwood, New York

Hückel E, 1925, Zur theorie konzentrierterer wässeriger lösunger stacker Elektrolyte. *Phys. Zeitschr.* **26**: 93–147

* **Hultgren R, Desai P D, Hawkins, D T, Gleiser M, Kelly K K and Wagman D D**, 1973, *Selected Values of the Thermodynamic Properties of the Elements*, Am. Soc. for Metals, 1435 pp

* **Johnson J W and Norton D**, 1991, Critical phenomena in hydrothermal systems: State, thermodynamic, electrostatic and transport properties of H_2O in the critical region. *Am. J. Sci.* **291**: 541–648

* **Johnson J W, Oelkers E H and Helgeson H C**, 1992, SUPCRT 92: A software package for calculating the standard molal thermodynamic properties of minerals, gases, aqueous species and reactions from 1 to 5000 bars and 0 to 1000 °C. *Computers and Geosciences* **18**: 899–947

* **Keenan J H, Keyes F G, Hill P G and Moore J G**, 1969, *Steam Tables*, John Wiley and Sons, New York

Kirkwood J G, 1939, The dielectric polarization of polar liquids. *J. Chem. Phys.* **7**: 911–19

Krauskopf, K B, 1967, *Introduction to Geochemistry*, International Series in the Earth and Planetary Sciences, McGraw-Hill, New York

Kubaschewski O and Alcock C B, 1979, *Metallurgical Thermochemistry*, 5th edn, Pergamon Press, Oxford

Lasaga A C, 1984, Chemical kinetics of water–rock interactions. *J. Geophys. Res.* **89**: 4009

Latimer W M, Pitzer K S and Slansky C M, 1939, The free energy of hydration of gaseous ions, and the absolute potential of the normal calomel electrode. *J. Chem. Phys.* **7**: 108–11

* **Levelt Sengers J M H, Kamagar-Parsi B, Balfour F W and Sengers J V**, 1983, Thermodynamic properties of steam in the critical region. *J. Phys. Chem. Ref. Data* **12**: 1–28

Lichtner P, 1985, Continuum model for simultaneous chemical reaction and mass transport in hydrothermal systems. *Geochim. Cosmochim. Acta* **49**: 779

Lichtner P, Helgeson H C and Pruess K, 1983, Numerical modelling of fluid flow with simultaneous chemical reactions in hydrothermal systems. *Geol. Soc. Am. Abstr. with Programs*, **15**: 627

Liden J, 1983, Equilibrium approaches to natural water systems: a study of anoxic- and ground-waters based on *in situ* data acquisition. PhD thesis, University of Umea, Umea

Litzke M H and Stoughton R W, 1962, The calculation of activity coefficients from osmotic coefficient data. *J. Phys. Chem.* **66**: 508–9

Mahan B, 1964, *Elementary Chemical Thermodynamics*, Benjamin, New York

* **Marshall W L and Franck E U**, 1981, Ion product of water substance, 0–1000 °C, 0–10 000 bars: New international formulation and its background. *J. Phys. Chem. Ref. Data* **10**: 295–304

Mattigod S V and Sposito G, 1979, Chemical modelling of trace metal equilibria in contaminated soil solutions using the computer program GEOCHEM. In *Chemical Modelling in Aqueous Systems*, Jenne E A (ed), ACS Symp. Ser. No. 93, pp. 837–56

Monro D M, 1982, *Fortran 77*, Edward Arnold, London, chapter 14

Moore W J, 1972, *Physical Chemistry*, Longman, Harlow

Morel F and Morgan J, 1972, A numerical method for computing equilibria in aqueous chemical systems. *Environ. Sci. Technol.* **6**: 58

Navrotsky A, 1974, Thermodynamics of binary and ternary transition metal oxides in the solid state. *MTP International Reviews of Science, Inorganic Chemistry*, Series 2, Vol. 5, Sharp D W A (ed), Butterworths–University Park Press, Baltimore, MD, pp 29–70

Navrotsky A, 1985, Crystal chemical constraints on thermochemistry of minerals. In *Microscopic to Macroscopic. Reviews in Mineralogy*, Kieffer S W and Navrotski A (eds), **13**: 223–76

Navrotsky A, 1986, Cation distribution energetics and heat of mixing in $MgFe_2O_4$–$MgAl_2O_4$, $ZnFe_2O_4$–$ZnAl_2O_4$ and $NiAl_2O_4$–$ZnAl_2O_4$ spinels: Study by high temperature calorimetry. *Am. Miner.* **71**: 1160–9

Nordstrom D K and Munoz J, 1986, *Geochemical Thermodynamics*, Blackwell Scientific, Palo Alto

Nordstrom D K et al, 1979, A comparison of computerized chemical models for equilibrium calculations in aqueous systems. In *Chemical Modelling in Aqueous Systems, Speciation,*

Sorption, Solubility and Kinetics, Jenne E A (ed), ACS Symp. Ser. No. 93, Am. Chem. Soc., Washington DC, pp. 857–92

Oelkers E H and Helgeson H C, 1988, Calculation of the thermodynamic and transport properties of aqueous species at high pressures and temperatures: Dissociation constants for supercritical alkali metal halides at temperatures from 400 °C to 800 °C and pressures from 500 to 4000 bars. *J. Phys. Chem.* **92**: 1631–9

Oelkers E H and Helgeson H C, 1989, Calculation of the thermodynamic and transport properties of aqueous species at high pressures and temperatures. Standard partial molal properties of inorganic neutral species. *Geochimica et Cosomochimica Acta* **53**: 12157–83

Öhman L-O, 1983, Equilibrium studies of ternary aluminium(III) complexes. PhD thesis, University of Umea, Umea

*Olofsson G and Hepler L G**, 1975, Thermodynamics of ionization of water over wide ranges of temperature and pressure, *J. Solution Chem.* **4**: 127–43

*Olofsson G and Olofsson I**, 1977, The enthalpy of ionization of water between 273 and 323 K, *J. Chem. Therm.* **9**: 65–9

*Olofsson G and Olofsson I**, 1981, Empirical equations for some thermodynamic quantities for the ionization of water as a function of temperature. *J. Chem. Therm.* **13**: 437–40

*Oshry H I**, 1949, The dielectric constant of saturated water from the boiling point to the critical point. PhD Dissertation, University of Pittsburg, Pittsburg, PA, p. 13

*Owen B B, Miller R C, Milner C E and Cogan H L**, 1961, The dielectric constant of water as a function of temperature and pressure. *J. Phys. Chem.* **65**: 2065–70

Parker S P, 1987, *Physical Chemistry Source Book*, McGraw-Hill, New York

Parkhurst D L, Plummer L N and Thorstenson D C, 1980, PHREEQE – a computer program for calculating mass transfer for geochemical reactions in groundwater *US Geol. Survey* **4**: 193–204

*Perrin D D**, 1979, *Stability Constants of Metal-Ion Complexes*, Part B, Organic Ligands, IUPAC Chemical Data Series, No. 22, Pergamon Press, Oxford

*Phillips S L**, 1982, *Hydrolysis and Formation Constants at 25 °C*, Lawrence Berkeley Laboratory, University of Berkeley, Berkeley CA, LBL-14313

Pitzer K S, 1973, Thermodynamics of electrolytes, I. Theoretical basis and general equations. *J. Phys. Chem.* **77**: 286

Pitzer K S, 1979, *Activity Coefficients in Electrolyte Solutions*, Pytkowicz R (ed), CRC Press, Boca Raton, FL

*Pitzer K S**, 1983, Dielectric constant of water at very high temperature and pressure. *Proc. Nat. Acad. Sci. USA*, **80**: 4575–6

Plummer L N, Parkhurst D L and Thorstenson D C, 1983, Development of reaction models for groundwater systems. *Geochim. Cosmochim. Acta*, **47**: 665–85

Prigogine I and Defay R, 1954, *Chemical Thermodynamics*, translated by Everett D H, Longmans Green, London

Rankin G A and Wright F E, 1915, The ternary system $CaO–Al_2O_3–SiO_2$ with optical study. *Am. J. Sci.*, 4th Ser. **39**: 1–79

Reed M H, 1982, Calculation of multicomponent chemical equilibria and reaction processes involving minerals, gases and an aqueous phase. *Geochim. Cosmochim. Acta*, **46**: 513–28

*Rivkin S L, Aleksandrov A A and Kremenvskaya E A**, 1978, *Thermodynamic derivatives of Water and Steam*, transl. by Kastin J, Halstead Press, London

*Robie R A, Hemingway B S and Fisher J**, 1979, Thermodynamic properties of minerals and related substances at 298.15 K and 1 bar pressure and at higher temperatures. US *Geological Survey Bulletin* 1452, US Government Printing Office, Washington

*Robinson G R, Haas J L Jr, Schafer C M and Hazelton H T Jr**, 1982, Thermodynamic and thermophysical properties of selected phases in the $MgO–SiO_2–H_2O–CO_2$, $CaO–Al_2O_3–H_2O–CO_2$ and $Fe–FeO–Fe_2O_3–H_2O–SiO_2$ chemical systems with special emphasis on the properties of basalts and their mineral components. US Geol. Surv. Open File Rept 83–79

*Robinson R A and Stokes R H, 1959, *Electrolyte Solutions*, Butterworths. A book on the thermodynamic properties of electrolytes at low temperatures and pressures with good introductory sections and tables of activity coefficient data

Sewart T M, 1981, Metal complex formation in aqueous solutions at elevated temperature and pressures. *Phys. Chem. Earth* **13/14**: 113–29

*Shock E and Helgeson H C, 1988, Calculation of the thermodynamic and transport properties of aqueous species at high pressures and temperatures: Correlation algorithms for ionic species and equation of state predictions to 5 kbar and 1000 °C. *Geochim. Cosmochim. Acta* **52**: 2009–36

*Shock E and Helgeson H C, 1990, Calculation of the thermodynamic and transport properties of aqueous species at high pressures and temperatures: Standard partial molal properties of organic species. *Geochim. Cosmochim. Acta* **54**: 915–45

*Shock E, Helgeson H C and Sverjenski D A, 1989, Calculation of the thermodynamic and transport properties of aqueous species at high pressures and temperatures: Standard partial molal properties of inorganic neutral species. *Geochim. Cosmochim. Acta* **53**: 2157–83

Shock E, Helgeson H C and Sverjenski D A, 1991, Calculation of activity coefficients and degrees of freedom of neutral ion pairs in supercritical electrolyte solutions. *Geochimica et Cosomochimica Acta* **55**: 1235–51

*Shock E, Oelkers E H, Johnson J W, Sverjenski D A and Helgeson H C, 1992, Calculation of the thermodynamic and transport properties of aqueous species at high pressures and temperatures: Effective electrostatic radii, dissociation constants and standard partial molal properties to 1000 °C and 5 kbar. *J. Chem. Soc. Faraday Trans.* **88**: 803–26

*Smith R M and Martell A E, 1974–1976. *Critical Stability Constants*, Vols 1–4, Plenum Press, New York (Vol. 4, 1974, Inorganic complexes)

Smith W R and Missen R W, 1982, *Chemical Reaction Equilibrium Analysis*, Wiley Interscience, New York

*Sohnel O and Novotny P, 1985, *Densities of Aqueous Solutions of Electrolytes*, Elsevier, Amsterdam

Sposito G, 1981, Cation exchange in soils: An historical and theoretical perspective. In *Chemistry in the Soil Environment*, Stelly M (ed), Am. Soc. Agron. Spec. Publ. 40, pp. 13–30

Sposito G and Mattigod S V, 1980, GEOCHEM: A computer program for the calculation of chemical equilibria in soil solutions and other natural water systems. Kerney Foundation of Soil Sci., University of California, Riverside, CA

Stokes R H and Robinson R A, 1949, Ionic hydration and activity in electrolyte solutions. *Am. Chem. J.* **70**: 1870–8

*Stull D R and Prophet H, 1971, *JANAF Thermochemical Tables*. NSRDS–NBS 37, US Natl Bur. Standards; 1974 supplement by Chase M W, Curnutt J L, Hu A T, Prophet H and Walker L C *J. Phys. Chem. Ref. Data* **3**: 311–480; 1975 supplement by Chase M W, Curnutt J L, Prophet H, McDonald R A and Syverud A N *J. Phys. Chem. Ref. Data* **4**: 1–175; 1978 supplement by Chase M W, Curnutt J L, Prophet H, McDonald R A and Syverud A N *J. Phys. Chem. Ref. Data* **7**: 793–940

*Tanger J C and Helgeson H C, 1988, Calculation of the thermodynamic and transport properties of aqueous species at high pressures and temperatures: Revised equation of state for partial molal properties of ions and electrolytes. *Am. J. Sci.*, **288**: 19–98

Thompson D W and Tahir N M, 1991, The influence of a smectite clay on the hydrolysis of iron(III). *Colloids and Surfaces* **603**: 69–398

*Uematsu M and Franck E U, 1980, Static dielectric constant of water and steam. *J. Phys. Chem. Ref. Data* **9**: 1291–304

* **Walther J V and Helgeson H C**, 1977, Calculation of the thermodynamic properties of aqueous silica and the solubility of quartz and its polymorphs at high pressures and temperatures. *Am. J. Sci.* **277**: 1315–51

Westall J C, Zachary J L and Morel F M M, 1976, MINEQL: A computer program for the calculation of chemical equilibrium composition of electrolyte solutions. Tech. Note 18, Ralph M. Parsons Lab. MIT, Cambridge, MA

Wolery T J, 1979, Calculation of chemical equilibria between aqueous solutions and minerals: The EQ3/6 package. Lawrence Livermore National Laboratory, Livermore, CA, UCRL-52658

Wolery T J, 1983, EQ3NR, A computer program for geochemical aqueous speciation–solubility calculations: User's guide and documentation. Lawrence Livermore National Laboratory, Livermore, CA, UCRL-53414

Wood J R, 1975, Thermodynamics of brine–salt equilibria – I. The systems $NaCl$–KCl–$MgCl_2$–$CaCl_2$–H_2O and $NaCl$–$MgSO_4$–H_2O at 25 °C. *Geochim. Cosmochim. Acta*, **39**: 1147–63

Wood J R, 1976, Thermodynamics of brine–salt equilibria – II. The system $NaCl$–KCl–H_2O from 0 to 200 °C. *Geochim. Cosmochim. Acta* **40**: 1200–20

* **Zematis J F, Clark D M, Rafel M and Scrivener N C**, 1986, *Handbook of Aqueous Electrolyte Thermodynamics*, AIChE, New York

A4.2 Further reading

A4.2.1 *Recommended introductory texts to support chapters 1–8*

Adams D M, 1974, *Inorganic Solids*, John Wiley and Sons, London
Covers the physics of the solid state from the viewpoint of a chemist.

Atkins P W, 1978, *Physical Chemistry*, 4th edn, Oxford University Press, Oxford
A comprehensive introduction to physical chemistry.

Chang R, 1987, *Chemistry*, McGraw-Hill, New York
A simple introduction to chemical principles with good sections on redox and electrochemistry, less advanced than Moore or Atkins.

Chatfield C, 1983, *Statistics for Technology: A Course in Applied Statistics*, Chapman and Hall, London

Denbigh K, 1983, *The Principles of Chemical Equilibrium*, Cambridge University Press, Cambridge
Arguably the most informative introductory text on chemical thermodynamics available. Emphasis is placed on thermodynamic principles and mathematical formalisms. The book is detailed and comprehensive but very readable.

Drever J I, 1982, *The Geochemistry of Natural Waters*, Prentice-Hall, New Jersey
An excellent general overview on aqueous systems at low pressures including brief introductions to thermodynamics and kinetics. A great deal of emphasis is placed on geochemical processes as well as properties of geochemical systems. A little light on the thermodynamics.

Ferguson F D and Jones T K, 1973, *The Phase Rule*, Butterworths, London
A short and well-presented overview of the phase rule.

Francis P G, 1984, *Mathematics for Chemists*, Chapman and Hall, London
Contains clear descriptions of the mathematical principles necessary to understand thermodynamics.

Garrels R M and Christ C L, 1965, *Solutions, Minerals and Equilibria*, Freeman, Cooper and Co., San Francisco
An extensive overview of the application of thermodynamics to geochemical substances.

Mason B, 1958, *Principles of Geochemistry*, John Wiley and Sons, New York
A general book on geochemistry with good sections on physical properties of geochemical substances.

Moore W J, 1972, *Physical Chemistry*, 5th edn, Longman, London
A classic text on physical chemistry.

Murrel J N and Boucher E A, 1982, *Properties of Liquids and Solutions*, John Wiley and Sons, Chichester
Provides a broad coverage of the physical and molecular properties of the liquid state. Contains sections on the properties of water and aqueous electrolytes.

Parker S B (ed), 1988, *Physical Chemistry Source Book*, McGraw-Hill, New York
A compendium of physical chemistry: short and to the point. A good reference text.

Rossotti H, 1978, *The Study of Ionic Equilibria*, Longman, London
A short, simple review of reactions in aqueous solutions.

Topping J, 1972, *Errors of Observation and Their Treatment*, Science Paperbacks, Chapman and Hall, London
A short and easily readable introduction to statistics.

A4.2.2   *More advanced texts to support chapters 1–8*

Bockris J M and Reddy A K N, 1970, *Modern Electrochemistry*, Plenum Press, New York
A comprehensive review text on the electrochemical properties of dissolved electrolytes.

Fyfe W S, Price N J and Thompson A B, 1978, *Fluids in the Earth's Crust*, Elsevier, Amsterdam
An advanced text on metamorphic, tectonic and chemical transport processes at elevated temperatures and pressures. Subjects are described conceptually rather than with mathematical formulations.

Guggenheim E A, 1967, *Thermodynamics*, North-Holland, Amsterdam
An advanced treatment of thermodynamic principles presented using algebraic and mathematical arguments. Comprehensive but not an introductory text.

Nordstrom D K and Munoz J, 1986, *Geochemical Thermodynamics*, Blackwell Scientific, Palo Alto
A comprehensive and detailed review of chemical thermodynamics, with the emphasis placed firmly on geochemical applications. An advanced text written by experts on the subject. Contains an extensive collection of references to thermodynamic data in the literature.

Roberts J K and Miller A R, 1961, *Heat and Thermodynamics*, Blackie and Son, Glasgow
A detailed overview of thermodynamic principles. Not heavily mathematical but with the emphasis placed on the philosophy of classical thermodynamics.

Stumm W and Morgan J, 1981, *Aquatic Chemistry*, John Wiley, New York
A detailed introduction to the thermodynamics and kinetics of reactions in natural waters at low temperatures and pressures. This excellent book is comprehensive on ideas and the theories but has few data.

A4.2.3   *Introductory texts for chapters 9 onwards*

Brown N L, 1928, *The Evolution of the Igneous Rocks*, Princeton University Press, Princeton, NJ

An authoritative text on the application of the phase rule to geochemical systems. It includes a description of many experimental studies.

Ehrlers E G, 1972, *The Interpretation of Geological Phase Diagrams*, Dover Publications, New York

A book on the construction and interpretation of phase diagrams with a great many relevant examples from the geochemical literature.

Horvath A L, 1985, *Handbook of Aqueous Electrolyte Solutions*, Ellis Horwood, New York

A detailed overview of the thermodynamic and kinetic properties of electrolyte solutions with emphasis on the use of working expressions rather than details of theory. This is a comprehensive but uncritical guide to the literature and contains some useful thermodynamic data for salts at low temperatures and pressures.

Krauskopf K B, 1967, *Introduction to Geochemistry*, International Series in the Earth and Planetary Sciences, McGraw-Hill, New York

A comprehensive introduction to geochemistry.

Press W H, Flannery B P, Teukolsky S A and Vettering W T, 1986, *Numerical Recipes. The Art of Scientific Computing*, Cambridge University Press, Cambridge

A comprehensive review of general numerical techniques with details of working computer codes.

Robinson R A and Stokes R H, 1959, *Electrolyte Solutions*, Butterworths, London

A book on the thermodynamic properties of electrolytes at low temperatures and pressures with good introductory sections and tables of activity coefficient data.

Smith W R and Missen R W, 1982, *Chemical Reaction Equilibrium Analysis*, Wiley Interscience, New York

An overview of the mathematical techniques in computer modelling of chemical equilibria. Highly mathematical but with details of working computer codes.

Sposito G, 1981, *The Thermodynamics of Soil Solutions*, Clarendon Press, Oxford

A comprehensive overview of the application of thermodynamics to complicated natural solid/water systems at low temperatures and pressures. Much emphasis is placed on solubility, redox and ion exchange.

Zematis J F, Clark D M, Rafal M and Scrivener N C, 1986, *Handbook of Aqueous Electrolyte Thermodynamics*, AIChE, New York

A review of the alternative conventions for dealing with the thermodynamics of electrolytes at low temperatures and pressures. Contains some data on activity coefficients and mineral solubilities plus a comprehensive summary of the literature.

Index